OBSTETRICS

OBSTETRICS
BY TEN TEACHERS

SIXTEENTH EDITION

EDITED BY

GEOFFREY CHAMBERLAIN
MD FRCS FRCOG FRACOG(Hon)

A member of the Hodder Headline Group
LONDON • SYDNEY • AUCKLAND
Co-published in the USA by Oxford University Press, Inc., New York

First published in Great Britain 1917 as *Midwifery*
Eleventh edition published 1966 as *Obstetrics*
Sixteenth edition published 1995 by Edward Arnold,
Third impression 1997 by Arnold,
a member of the Hodder Headline Group
338 Euston Road, London NW1 3BH

Co-published in the United States of America by
Oxford University Press, Inc.,
198 Madison Avenue, New York, NY10016
Oxford is a registered trademark of Oxford University Press

Whilst the advice and information in this book is believed to be true and
accurate at the date of going to press, neither the author nor the publisher
can accept any legal responsibility or liability for any errors or omissions
that may be made. In particular (but without limiting the generality of the
preceding disclaimer) every effort has been made to check drug dosages;
however, it is still possible that errors have been missed. Furthermore,
dosage schedules are constantly being revised and new side effects
recognized. For these reasons the reader is strongly urged to consult the
drug companies' printed instructions before administering any of the
drugs recommended in this book.

British Library Cataloguing in Publication Data
A catalogue record for this book is available from the British Library

Library of Congress Cataloging-in-Publication Data
A catalog record for this book is available from the Library of Congress

ISBN 0 340 57313 9

Typeset in 10/11 Garamond by Wearset, Boldon, Tyne and Wear
Printed and bound in Great Britain by The Bath Press, Bath

CONTENTS

1 : OSCE

LIST OF CONTRIBUTORS

Mary Anderson FRCOG
Consultant Obstetrician and Gynaecologist
Lewisham Hospital, London

Geoffrey Chamberlain MD FRCS FRCOG
Formerly Professor of Obstetrics and
Gynaecology, St George's Hospital
Medical School, London

Tim Coltart PhD FRCS(Ed) FRCOG
Consultant Obstetrician and Gynaecologist
Guy's Hospital and Queen Charlotte's and
Chelsea Hospital, London

Gedis Grudzinskas MD FRCOG FRACOG
Professor, Academic Unit of Obstetrics and
Gynaecology, The Royal London Hospital and
St Bartholomew's Hospital, London

Frank Loeffler FRCS FRCOG
Consultant Obstetrician and Gynaecologist,
St Mary's Hospital and Samaritan Hospital
for Women, and Queen Charlotte's and
Chelsea Hospital, London

Malcolm Pearce MD FRCS FRCOG
Formerly Consultant Obstetrician
St George's Hospital, London

Charles Rodeck DSc FRCOG FRCPath
Professor of Obstetrics and Gynaecology
University College and Middlesex School of
Medicine, London

Marcus Setchell FRCS FRCOG
Consultant Obstetrician and Gynaecologist
St Bartholomew's Hospital and Medical Director,
Homerton Hospital, London

Nick Siddle MRCOG
Formerly Consultant Obstetrician and
Gynaecologist, University College and Middlesex
Hospitals, London

Philip Steer MD FRCOG
Professor of Obstetrics and Gynaecology
Chelsea and Westminster Hospital, London

SPECIAL CHAPTERS CONTRIBUTED BY

Elizabeth Hopper RN, RM Dip Couns
Sister in Charge, The Bereavement Unit
St George's Hospital, London

Tom Lissauer MBChir, FRCP
Consultant Paediatrician, St Mary's Hospital
London

Barbara Morgan FFA RCS
Consultant Anaesthetist, Queen Charlotte's and
Chelsea Hospital, London

Ruth Warwick MRCP FRCPath
Consultant Haematologist, North London Blood
Transfusion Centre, London

Ruth White MRCPsych MPhil
Consultant Psychiatrist, The Warstock Lane
Centre, Billesley, Birmingham

A HISTORY OF TEN TEACHERS IN OBSTETRICS

The book was first published under the title *Midwifery by Ten Teachers* by Edward Arnold (Publishers) Ltd in 1917. Comyns Berkeley, a consultant obstetrician and gynaecologist on the staffs of the Middlesex Hospital, the City of London Lying-in Hospital and the Chelsea Hospital for Women provided the inspiration; he was the first Editor and Director and continued as such until the fifth edition in 1935, remaining in an editorial capacity until his death in 1942.

The aim was to produce a textbook for students 'preparing for their final examination, and for others who have passed beyond the stage of examinations'. The original ten contributors, with experience as examiners, were all teachers who came from 'eight general hospitals with medical schools and three large lying-in hospitals', all in London. In order to achieve collective authorship, reflecting the views of all the teacher–contributors, the subjects were first portioned out among the ten writers and their chapters were then 'criticized, amended and partly rewritten' at numerous meetings of all ten. In 1948 the editors 'insisted on close collaboration at all stages of composition so as to express the unanimous opinion of the ten authors; only on occasion was it necessary to accept the views of a majority'. To each author representations and suggestions were made by the other nine 'to assist him in the preparation of his manuscript'. The pace of life today has meant that more responsibility for the contents falls on the editors, but meetings of all ten teachers are still held at which contentious and debatable issues are raised.

In a book which has gone through 16 editions spanning 76 years, the subjects dealt with reflect the changing practice of obstetrics during this time. For instance in the second edition in 1920, a chapter on the new subject of 'Ante-natal Hygiene' and a short account of recently discovered 'vitamines' were added; in the third edition in 1925 'several skiagrams were reproduced to emphasize the advances made in the application of photography to obstetric diagnosis'. In the fourth edition in 1931, 'the time-honoured position of the management of labour with pelvic disproportion as a final lesson was discarded and put in its proper place'; new matter on the medical induction of labour and on blood transfusion was included. In the fifth edition in 1935, endocrinology was introduced in relation to the physiology of menstruation and the ovarian cycle. Illustrations of the lower segment caesarean section operation appeared for the first time in the sixth edition in 1938 when metric measures were inserted after the English ones throughout. The classification of abnormal pelves and their effect on labour, tonic retraction of the uterus in obstructed labour and the use of sulphonamides in obstetrics were new topics introduced in the seventh edition in 1942.

With every edition there has been the rewriting and rearrangement of chapters, and the removal of redundant material and repetitions, although different points of view on subjects have been retained. Illustrations have been removed and others substituted. Modern work has been included. Certain subjects such as pre-eclamptic toxaemia (pregnancy induced hypertension) appear in every edition but, with more recent information available each time, alteration has been necessary reflecting modern therapy and practice. Due notice was taken in preparing new editions of letters from students and obstetricians and of criticism in the medical press. The three principles enunciated by Comyns Berkeley in the first edition in 1917 have, in the main, been adhered to. 'First, that the book should be written for medical students and young practitioners; secondly that it expressed the collective view of the ten contributors; and thirdly, that all ten were actively engaged in teaching obstetrics in the medical schools of hospitals in London' (ninth edition, 1955). However, in 1955 the editors decided that the section of the book devoted to the health and disease of the newborn infant should more properly be written by a paediatrician than by an obstetrician, and this has been the case in every edition since then.

In the eleventh edition in 1966 the title was changed to *Obstetrics*. The editors 'regretted the loss of an English word but thought that the new title was more in keeping with current usage'. They tried to describe modern obstetric practice and to keep abreast of recent advances in theoretical knowledge. They reduced the length of the chapters on major 'mechanical complications of labour' although they did not completely eliminate those sections because some readers practised in countries where adequate obstetric services were not generally available.

After more than 50 years of service to students and practitioners the plan of the contents was re-arranged. The twelfth edition in 1972 contained new chapters on Placental Insufficiency and Fetal Distress, Haemolytic Disease, Coagulation Disorders, Oxytocic Drugs and Therapeutic Abortion. New topics such as amniocentesis, fetal blood sampling, fetal heart monitoring, ultrasonics and prostaglandins were included.

By 1980 (thirteenth edition) it was considered that a third revolution in obstetric practice was taking place. The first was a general adoption of lower segment caesarean section permitting safe intervention during labour, often in the fetal interest; the second was the overcoming of puerperal sepsis. This third phase meant the introduction of many new methods of assessing the state (or well-being) of the fetus, prolonged labour was being prevented by the proper use of oxytocin and better relief of pain was being achieved with extradural anaesthesia. These advances were reflected by improvements in maternal and perinatal mortality.

While describing fully newer methods of obstetric management, the editors took care to retain all that was good in earlier practice, underlining the importance to students of a proper understanding of the principles on which critical decisions must depend.

In 1990 (fifteenth edition) emphasis was placed on modern imaging techniques and the assessment of fetal well-being. Because of the increasing litigation that was taking place in medicine generally, and in obstetrics in particular, a chapter on the medico-legal aspects of obstetrics was included. A haematologist wrote a chapter on coagulation disorders; a psychiatrist with a special interest in obstetrics wrote one on psychiatric disorders; and a paediatrician revised the paediatric section extensively.

The fourteenth edition (1985) was published for the first time in paperback, though maintaining the size and shape of the old hardback. The fifteenth edition went even further by being published in paperback in a new size and shape. It was published also in the Educational Low-Priced Books Scheme funded by the Overseas Development Administration as part of the British Government Overseas Aid Programme.

During the 76 years of its existence *Ten Teachers Midwifery/Obstetrics* has had 43 contributors, of whom 14 acted as editors. Six consultants wrote special chapters on their own subjects (paediatrics, psychiatric illness, blood coagulation and medico-legal problems). They have all played a part in achieving the undoubted success of the book, but perhaps three individual editors had more influence on it than the rest. Comyns Berkeley masterminded the concept of a book based on the collaboration of ten London teachers from the foundation in 1917 until his death in 1942; Frederick Roques turned awkward phrases from 1948 to 1961 into good English throughout the book; and Stanley Clayton made sure from 1955 to 1985 that no statement was made by a contributor unless it had good scientific support. Thus was achieved a well-balanced and up-to-date view of current obstetric practice written with the senior medical student and young doctor in mind.

T.L.T. Lewis   1995

Ten Teachers 1917–1994

OBSTETRICS

Name	1	2	3	4	5	6	7	8	9	10	11	12	13	14	15	16
*H. Russell Andrews	1	2	3	x												
J. D. Barris	1	2	3	4	5	6	x									
*Conyers Berkeley	1	2	3	4	5	6	7	x								
Victor Bonney	1	2	3	4	5	6	7	x								
Harold Chapple	1	2	3	x												
G. F. Darwall Smith	1	2	3	x												
Stanley Dodd	1	2	3	4	5	x										
*J. S. Fairbairn	1	2	3	4	5	x										
T. G. Stevens	1	2	3	4	5	x										
*Clifford White	1	2	3	4	5	6	7	8	x							
*Frank Cook				4	5	6	7	8	x							
Victor Lark				4	5	6	7	8	9	10	x					
Donald W. Roy				4	5	x										
*William Gilliatt						6	7	8	x							
Aubrey Goodwin						6	7	x								
Douglas MacLeod						6	7	8	9	x						
*A. J. Wrigley						6	7	8	9	10	x					
*John Beattie							7	8	9	10	x					
Humphrey Arthure								8	9	10	x					
Arthur C. H. Bell								8	x							
*Frederick W. Roques								8	9	10	x					
Stanley Clayton									9	10	11	12	13	14	x	
A. Briant Evans									9	10	11	x				
C. M. Gwillim									9	10	x					
*T. L. T. Lewis									9	10	11	12	13	14	15	
Robert Percival										10	11	12	x			
Donald Fraser											11	12	x			
Joseph M. Holmes											11	12	13	14	x	
Ian Jackson											11	12	x			
*George Pinker											11	12	13	14	15	
Philip Rhodes											11	12	x			
D. W. T. Roberts											11	12	13	14	x	
Charles P. Douglas												12	x			
J. M. Brudenell													13	14	15	
*Geoffrey Chamberlain													13	14	15	
Denys Fairweather													13	14	15	16

Name	9	10	11	12	13	14	15	16
John C. Hartgill					13	14	x	
Marcus Setchell					13	14	15	16
Ronald W. Taylor					13	14	15	16
Tim Coltart							15	16
Gedis Grudzinskas							15	16
Frank Loeffler							15	16
Philip Steer								16
Malcolm Pearce								16
Nick Siddle								16
Mary Anderson								16
Charles Rodeck							15	16

PAEDIATRICIANS

Name	9	10	11	12	13	14	15	16
Edward Hart	9	10	11	12	x			
R. J. K. Brown					13			
Dr R. Dinwiddie						14	x	16
Dr T. Lissauer							15	16

SPECIAL CHAPTERS

Name	9	10	11	12	13	14	15	16
Margaret Christie-Brown					13	14	x	
Ruth White								16
Mary Anderson							15	
G. Oppenheim							15	
Ruth Warwick								16
Liz Hopper								16
Barbara Morgan								16

CONTRIBUTORS: 43
SPECIAL CHAPTERS: 16
*EDITORS 14

PREFACE

This is the longest standing English textbook in obstetrics. The sixteenth edition has been changed and adapted to help undergraduates and those working for the Diploma of the Royal College of Obstetricians and Gynaecologists. We have brought in many new methods of obstetrical management and our contents have changed in accordance with alterations in the London medical scene. We thank all our contributors for their work in bringing this new edition to fruition.

We have taken the opportunity of asking Mr T.L.T. Lewis, the Editor Emeritus and senior contributor to this series of volumes, to write a short history of the Obstetrics by Ten Teachers. This we include for we think it would be of interest to students joining the subject who can feel they have some continuity with teaching back to the time of the First World War when this book was first published. We think that continuity in medicine is important and would like to pay tribute to the many previous editors and contributors of Obstetrics by Ten Teachers over its 80 years' existence.

We have lost some contributors by retirement and moving from London teaching scene. We welcome their replacements. We are most grateful to all the production staff at Edward Arnold for their courtesy and prompt help in publishing this volume.

Geoffrey Chamberlain
London 1995

INTRODUCTION

The woman expecting her baby in the 1990s is very different from the patients the student may have met in other branches of medicine. The pregnant woman is young and fit, while many others they have met are patients who are older and have disease processes. Most women in the obstetric scene have no disease and are going through a physiological process. This, however, can alter sharply to pathology and the borderline between the two must be watched carefully by obstetricians and midwives.

Modern women having babies are much better informed than they were years ago. They want to know more about what is happening and they and their partners have the right to be kept in the picture. This is not just a feminist fad but common sense, for the woman who knows what is likely to happen will be better prepared for actions in late pregnancy and labour. The mind is acutely attuned during pregnancy to listening to others. Among those who provide the woman with true information and sound ideas must be the professionals as well as the market-place, the magazines and broadcasting that often feed the extremes of thought. It is for everyone in obstetrics, from the professor to the medical student, to help to support women and their partners during the antenatal period, bringing them to the highest preparation in body and mind to the time of childbirth.

Most people outside obstetrics consider the essential part of having a baby is childbirth; this is a great event of a new life coming into the world. However, professionals know that life actually started some 38 weeks before this at conception. A most important part of obstetrics is the antenatal period and the care given then. If this goes well, it can remove many of the problems of childbirth and ensure an easier, happier and safer time at birth for both mother and baby. It is further probable that many postnatal influences are laid down in the fetus inside the uterus. This does not just apply to the genetic changes associated with chromosomal anomaly but to the functional structure and nutrition of the fetus as it grows. For example, the biggest single independent variable for raised blood pressure in the 50 to 60 year-old man is his birth weight and its ratio to placental weight, a measure of what happened inside the uterus between 14 and 20 weeks of gestation. This is more important than any of the subsequent factors that people pay so much attention to such as diet, high and low density lipoproteins and exercise. The nutrition of the mother and so of the fetus inside the uterus is the major feature affecting blood pressure (and therefore coronary artery disease and stroke), obstructive respiratory disease, clotting factors and the non-insulin dependent diabetes in older age. Intrauterine life is the most important time of existence and research is now being carried out intensively in this area.

All who would wish to study obstetrics must pay great attention to the antenatal period for it has been an underinvestigated and regimental area until the 1990s. If good antenatal care is carried out, then the woman will avoid some of the major problems that can make later pregnancy and labour hazardous. She will come to labour better prepared and with all the methods of modern analgesia available for her choosing, will be able to have a relatively painfree delivery of a healthy baby.

1

ANATOMY AND PHYSIOLOGY

THE EARLY DEVELOPMENT OF THE OVUM

An oocyte is released from the ovary during most normal menstrual cycles.

The cycle is described in two phases: the first half is known as the *follicular* or *proliferative phase*, during which the ovarian follicle enlarges and becomes distended with fluid. On about the 14th day of the cycle the follicle discharges an oocyte into the uterine (Fallopian) tube. The second half of the cycle is known as the *luteal* or *secretory phase*. In this phase the cells of the empty follicle become swollen with yellow lipid, and the follicle is now called the *corpus luteum*.

During the follicular phase the cells lining the follicle secrete oestrogens which cause proliferation of the glands and stroma of the endometrium. During the luteal phase the cells of the corpus luteum secrete both oestrogens and progesterone, and the combined action of these hormones causes further proliferation of the endometrium and also secretion of sugars, amino acids, mucus and enzymes by the endometrial glands. The endometrium reaches its maximum development in the late luteal phase, forming a decidua into which the oocyte, if it is fertilized, can embed.

The complex hormonal control and details of the anatomical changes in the ovary and endometrium during the menstrual cycle are described in *Gynaecology by Ten Teachers*.

OVULATION

In the 4th week of life germ cells migrate from the wall of the yolk sac to an area of mesenchyme on the posterior wall of the coelom. These form the primordial germ cells, which give rise to numerous primary oocytes, large round cells with relatively large, chromatin-rich nuclei. The primary oocytes become surrounded by a single layer of smaller flattened cells to form primordial follicles. At birth each ovary contains up to 2 million follicles, although many are lost before the menarche.

THE OVARIAN FOLLICLE

During maturation of the follicle the flat cells that surround the primordial ovum multiply, become rounded and arranged in several layers (the *granulosa cells*). Their growth is eccentric, so that the oocyte comes to lie at one side of the mass of granulosa cells, and eventually clear fluid appears among these cells, so that a follicle is formed with the oocyte placed to one side (*see* Fig. 1.1). The clump of granulosa cells that is directly related to the oocyte forms a hillock (the *cumulus*) that projects into the cavity of the follicle, and at this stage the oocyte is surrounded by a clear membrane, the *zona pellucida*, within which it can rotate. The granulosa cells that are immediately related to the zona pellucida become arranged in a

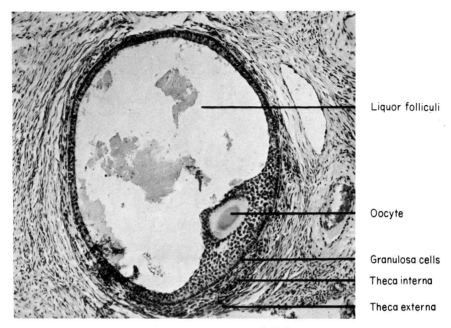

Liquor folliculi

Oocyte

Granulosa cells
Theca interna
Theca externa

Fig. 1.1 Ripe ovarian follicle

radial fashion and form the *corona radiata*. The oocyte itself enlarges slightly during maturation of the follicle, chiefly by increase in the volume of the cytoplasm, and reaches a diameter of about 0.15 mm.

The cells of the ovarian stroma which surround the granulosa cells also proliferate, becoming swollen by the accumulation of lipid. These cells form the *theca interna* and play an important part in the formation of the corpus luteum at a later stage. The ovarian stromal cells outside the theca interna become somewhat compressed by the growth of the follicle and form the *theca externa* (see Fig. 1.1).

As the follicle increases in size it approaches the surface of the ovary, where it is seen as a transparent vesicle up to 20 mm in diameter. The follicle eventually projects from the surface of the ovary, and at about the midpoint of the cycle the cell layers dehisce and the oocyte, still surrounded by the corona radiata, is discharged over 2–3 minutes.

CORPUS LUTEUM

After ovulation the walls of the follicle collapse and are thrown into folds; there may be a little haemorrhage into the now empty cavity. The granulosa cells become swollen by the accumulation of yellow lipid, and are now *granulosa–lutein cells*. Similar changes also occur in the theca interna, the cells of which are now termed *theca–lutein cells*. If the oocyte that was discharged from the follicle is not fertilized then degenerative changes (luteolysis) begin in the corpus luteum at about the 22nd day of the cycle. If the oocyte *is* fertilized some of its cells produce *human chorionic gonadotrophin* (*hCG*), a hormone which causes the corpus luteum to persist and enlarge further, maintaining its production of oestrogen and progesterone for about 12 weeks; around 10–13 weeks this function is taken over by the placenta.

FORMATION OF DECIDUA

The changes in the endometrium during the menstrual cycle occur in preparation for the possible reception of the fertilized ovum. If the ovum is fertilized and embeds in the endometrium all these changes are accentuated and the endometrium is described as the *decidua of pregnancy*. If fertilization and embedding do not occur the endometrium undergoes necrosis, except for the basal layer,

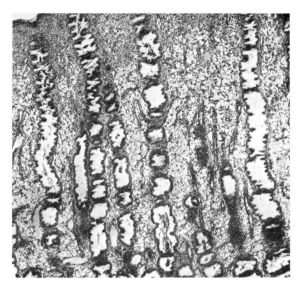

Fig. 1.2 Photomicrograph of endometrium at the end of the secretory or phase of the menstrual cycle

from which its regeneration in the next cycle occurs.

During the *follicular*, or *proliferative phase* of the menstrual cycle, under the influence of oestrogen produced by the follicle, the endometrium becomes more vascular, the cells of the glandular epithelium and stroma proliferate and the endometrium becomes thicker, with long straight glands.

In the *luteal*, or *secretory phase*, when the corpus luteum produces both oestrogen and progesterone, the gland cells become tall and columnar and pour out their secretions of glycogen and mucin into the lumina of the glands. The glands become convoluted (*see* Fig. 1.2) and the stromal cells swell. These changes do not affect the deepest parts of the endometrial glands; the stromal changes are most marked in the superficial layer and result in a great contrast between the basal layer of endometrium with straight glands, the spongy intermediate layer with distended convoluted glands and the superficial compact layer. The arterioles have a spiral arrangement, and there is a great increase in vascularity.

MATURATION OF THE OVUM

The nuclei of human body cells each contain 46 chromosomes; this number is maintained through-

out successive divisions of the cells. The chromosomes are responsible for the transmission of all inheritable qualities. The reproductive cells (the oocyte and the spermatozoon) differ from the ordinary body cells in that they each contain only 23 chromosomes. When fertilization occurs the oocyte and spermatozoon unite and the total of 46 chromosomes is re-established. The male and female cell each contribute half of the total, so that some inherited qualities are derived from the father and some from the mother. The number of chromosomes in the germ cells is reduced during their maturation by a special type of cell division called *meiosis*.

In ordinary division of body cells (*mitosis*) each chromosome replicates and the two halves of each chromosome separate and pass into two daughter cells.

In the meiotic division of germ cells the chromosomes first become arranged in pairs, then they replicate and one member of each pair passes to one of four daughter cells, so halving the number of chromosomes in the mature germ cells.

The ovum undergoes the first maturation division when it is still in the ovary. The primary oocyte divides by meiosis into two cells of unequal size – a large secondary oocyte and a small polar body; the latter comes to lie in the perivitelline space within the zona pellucida. After this meiotic (reducing) division the secondary oocyte contains only 23 chromosomes. The subsequent division of the secondary oocyte, which takes place in the uterine tube, is a mitotic division. At this division a second polar body is extruded, but the final mature ovum still contains only 23 chromosomes.

A similar meiotic division takes place in the spermatozoon, when the primary spermatocyte divides into two equal secondary spermatocytes, which each contain 23 chromosomes. By a further mitotic division each of these divides to form two spermatids, so that the original primary spermatocyte gives rise to four spermatozoa, each with 23 chromosomes.

The adult cells of a normal female contain two X chromosomes, and after meiotic division each oocyte will contain one X chromosome. The adult cells of a normal male contain one X and one Y chromosome; when meiotic division occurs these separate so that each secondary spermatocyte contains either an X or a Y chromosome, and eventually there will be two types of spermatozoa, those with an X chromosome (gynaecogenic) and those with a Y chromosome (androgenic). During fetilization, if conjugation of an X sperm and an X oocyte occurs, the final combination will be XX, and will give rise to female genetic structure; but if conjugation of a Y sperm and an X oocyte occurs the final combination will be XY, and will give rise to male genetic structure. There are recognizable sexual differences in the nuclei of adult cells, particularly seen as a small club-shaped projection of chromatin in the nuclei of polymorphonuclear leucocytes, and as a peripherally placed coiled mass of chromatin (Barr body) in other cells in the female. This Barr body is the inactive X chromosome of the cell.

TRANSIT AND FERTILIZATION OF THE OVUM

The mechanism by which the ovum reaches the lumen of the Fallopian tube has been much discussed. The ciliated cells of the fimbriae at the abdominal ostium of the tube set up a flow of fluid which can carry small particles from the rectovaginal pouch into the tube. The fimbriated end of the tube is brought into close contact with the ovary at the time of ovulation so that they embrace the ovarian surface around the ripe follicle; the mechanism of this is uncertain. Once the ovum has reached the cavity of the tube it is carried towards the uterine cavity by ciliary action. The tubes show peristaltic movements, but these are less important than the current produced in the oviduct fluid by the action of the cilia in moving the ovum down the tube.

Successful fertilization depends on intercourse occurring at the correct time in the cycle and semen of adequate quality being deposited in the region of the cervix.

Semen is a suspension of spermatozoa in seminal plasma, a combined secretion of the epididymis, seminal vesicle and prostate gland. It contains, among other substances, fructose (chiefly from the vesicle and an essential nutrient for spermatozoa) proteins, fibrinolytic and proteolytic enzymes (chiefly from the prostate), and prostaglandins. Immediately after ejaculation the semen coagulates, and then after about 15 minutes reliquefies; these changes are caused by the prostatic enzymes. Prostaglandins may cause tubal and uterine contractions.

Spermatozoa contain hyaluronidase, which is readily released into the seminal plasma. This assists the sperm to penetrate the cervical mucus and the corona radiata around the ovum.

Fertilization of the ovum normally takes place in the ampullary portion of the Fallopian tube, and spermatozoa reach the tube between 30 minutes and 3 hours after intercourse. The transit of the spermatozoa results partly from their own motility, and partly from uterine and tubal peristalsis, the direction of which is reversed at the time of ovulation. Some stimulation of action may be from seminal prostaglandins.

Sperm are attracted to the ovum by chemotaxis, and after intercourse many spermatozoa can be found in the tubes. Several may penetrate the zona pellucida, but as soon as one sperm makes its way into the ovarian cytoplasm the ovum separates from the zona pellucida and becomes impervious to further penetration. The head of the spermatozoon represents the nucleus of the male cell, and it fuses with the nucleus of the oocyte to form the segmentation nucleus, whose complement of 46 chromosomes is again complete. The fertilized ovum is carried down the tube by ciliary and peristaltic action, and reaches the uterine cavity 5–6 days after ovulation.

EARLY DEVELOPMENT AND EMBEDDING OF THE BLASTOCYST

After the formation of the segmentation nucleus the zygote starts to divide, and soon a solid clump of cells called the *morula* is formed. A cavity appears among the cells of the morula so that it becomes vesicular, and it is then termed the *blastocyst*. This stage of development is reached while the zygote is still in the Fallopian tube.

The outermost cells of the blastocyst form the *trophoblast*, which has the power of eroding and digesting the surface epithelium of the decidua. The fertilized ovum sinks into the thickness of the decidua, lying in a cavity in the stroma between adjacent endometrial glands. The glands are distorted and pushed aside by the enlargement of the growing zygote. The aperture through which the zygote entered the decidua is sealed over with a plug of fibrin. Maternal blood vessels are invaded and eroded by the trophoblast, and extravasation of maternal blood occurs around the zygote.

The structure of the blastocyst is indicated in Fig. 1.3(a). The inner cell mass, from which the embryo will develop, is seen projecting into the blastocyst, and two small cavities appear in the

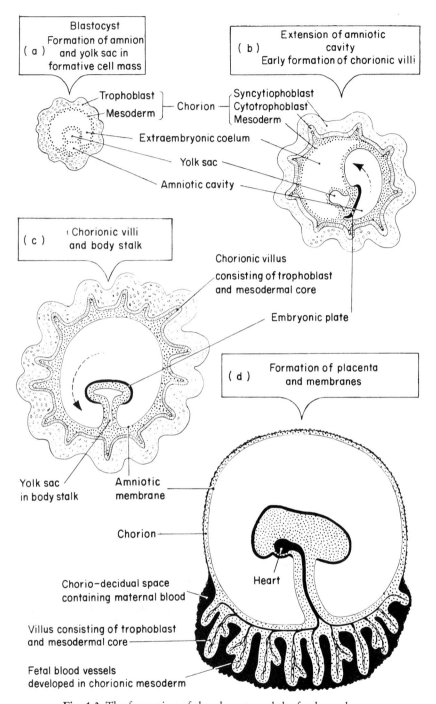

Fig. 1.3 The formation of the placenta and the fetal membranes

cells of the inner cell mass, from which the amniotic cavity and the yolk sac develop. The diagrams indicate the progressive extension of the amniotic cavity, which comes to surround and envelop the embryo, and ultimately the amniotic membrane, which covers the body stalk up to the point at which it becomes continuous with the embryonic ectoderm at the umbilicus.

Very soon, while eroding into the maternal decidua, the trophoblast becomes arranged in projecting masses, at first in a labyrinthine formation, but later arranged as villi which grow and branch. Maternal blood vessels are opened by the cytolytic action of the trophoblast, so that maternal blood lies in the intervillous spaces, and the embryo starts to secure its nutrition from this.

The trophoblast becomes differentiated into two layers. There is a thick outer layer of *syncytotrophoblast*, in which the nuclei are scattered in a mass of cytoplasm that has no evident division into separate cells (*see* Fig. 1.4). The inner layer of cytotrophoblast (*Langhans' layer*) is thinner, and consists of a single layer of rounded cells.

The blastocyst is lined with extraembryonic mesoderm, and it will be seen from the diagrams (*see* Fig. 1.4) that this is continuous with the mesoderm of the embryo itself. The extraembryonic mesoderm is continuous with the central tissue of the villi, so that each villus has an outer covering of trophoblast and a central mesodermal core. Lacunae appear in the mesoderm and gradually become joined to form a pattern of primitive blood vessels, extending through both the body of the embryo and the extraembryonic mesoderm. By this arrangement the placental circulation is ultimately formed; the fetal heart not only pumps blood through the tissues of the fetus itself, but also through the tissues of the placenta.

The combined layer of trophoblast and underlying mesoderm is termed the chorion. Within it is the amniotic membrane that bounds the amniotic cavity.

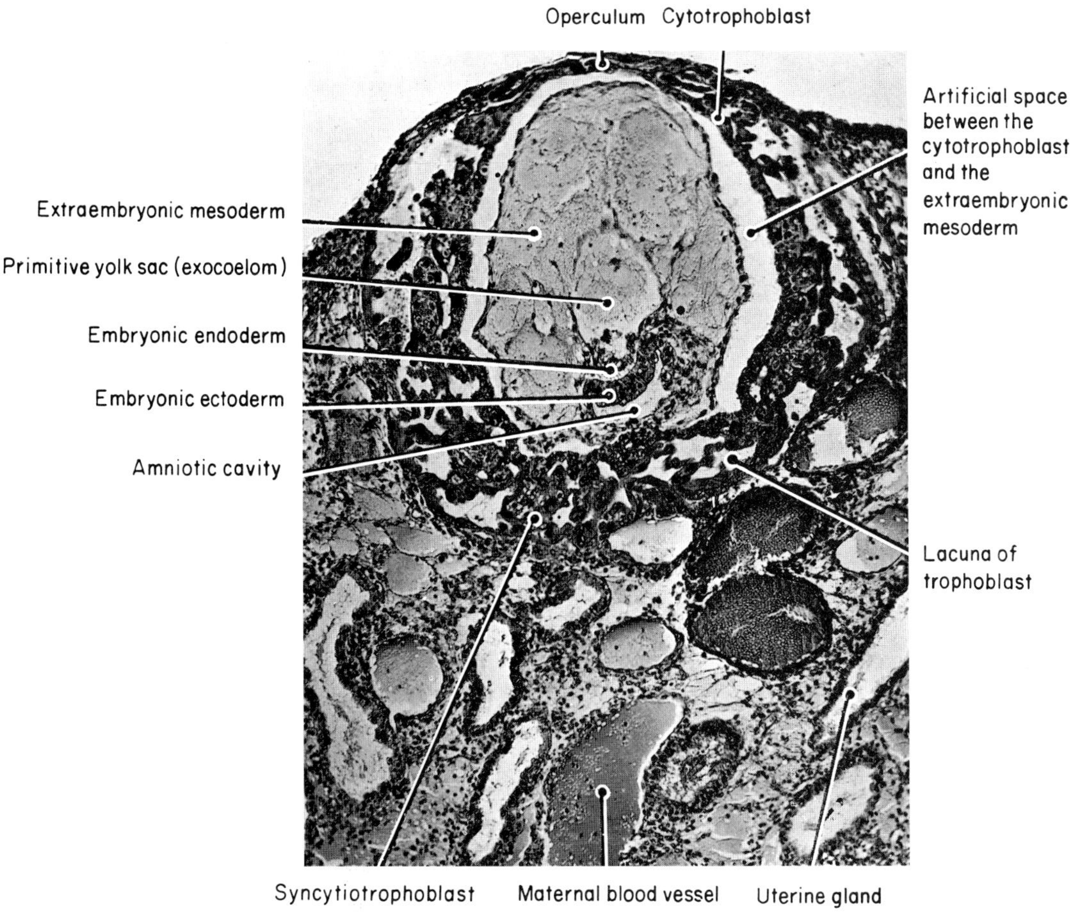

Fig. 1.4 Section of human embryo of 10–11 days' development

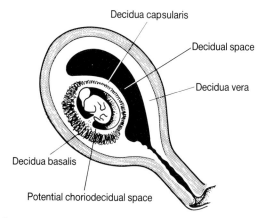

Fig. 1.5 Uterus containing an embryo of about 9 weeks

The zygote embeds in the stratum compactum of the decidua, the most frequent site of implantation being the upper and posterior part of the uterine cavity; it produces a slight projection into the uterine lumen. According to its relations to the embryo three parts of the decidua are distinguished (*see* Fig. 1.5):

Decidua basalis

This is the portion of the decidua which lies between the embryo and the muscular wall of the uterus. It bounds the deeper half of the implantation cavity and later on forms the site of attachment of the definitive placenta. A number of thin-walled sinuses pass through it, bringing blood to the intervillous spaces. It is the future placental bed.

Decidua capsularis

This component of the decidua intervenes between the embryo and the uterine cavity and bounds the superficial half of the implantation cavity. As the ovum grows it bulges into the cavity, and by the 12th week the growing embryo fills the cavity, so that the decidua capsularis becomes fused with the decidua vera.

Decidua vera

This decidua is not related to the site of implantation and lines the rest of the cavity of the uterus.

Two other terms need definition. The *decidual space* is the space bounded by the decidua vera and the decidua capsularis. It is obliterated by the 12th week of pregnancy. The *choriodecidual space* is the space between the chorionic villi and the decidua basalis. It contains maternal blood.

Ultrasound appearances

The early pregnancy can be well seen on ultrasound, especially if a vaginal probe is used. The gestation sac of 8–9 mm diameter is visible by 5 weeks menstrual age, and the fetus itself by 6 weeks. By 7 weeks the scan will detect fetal heart pulsations.

THE PLACENTA, CORD AND MEMBRANES

In the primitive mammalian placenta the fetal chorion is merely applied to the surface of the maternal decidua, so that exchange of nutrients and excretory products between maternal and fetal blood takes place through several layers (epitheliochorial placenta).

In the human placenta the trophoblast erodes into the decidua, so that the endothelium of the maternal blood vessels is destroyed and maternal blood is in direct contact with the chorion, without the intervention of any decidual tissue (haemochorial placenta).

At first the syncytiotrophoblast forms an open reticulated network, whose lacunae are filled with maternal blood, and so, in part, the trophoblast replaces the maternal endothelium. The trophoblast soon becomes arranged in trabeculae, which are covered by syncytiotrophoblast and have a core of cytotrophoblast. When the embryonic mesoderm appears it extends into each of these trabeculae and finally the vascularization of the mesoderm completes the formation of chorionic villi by about the 16th day after fertilization.

At some points the trophoblast comes into

Chorionic villi

Syncytiotrophoblast
covering villus

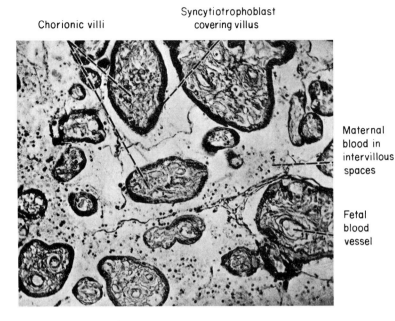

Maternal
blood in
intervillous
spaces

Fetal
blood
vessel

Fig. 1.6 Microscopical section of placenta at term ×170

direct contact with the decidua, thus anchoring the main villi to the maternal tissues at the junctional zone. Here, lying between the invading trophoblast and the decidua, a wavy layer of fibrin – the fibrinous layer of Nitabuch – can be seen. The number of anchoring villi is small but, by budding from both them and the chorion, true chorionic villi are formed. These differ from the anchoring villi in that their outer ends protrude free into the intervillous space. The capillaries in the terminal parts of the villi are very convoluted, but adjacent villi are not joined, i.e. the villi do not form a reticulum (*see* Fig. 1.6).

The trophoblast extends for a variable distance into the maternal spiral arterioles where they enter the intervillous space, partly replacing the maternal vascular endothelium. The vessels in the trophoblast are converted into funnel-shaped deltas, so that the peripheral resistance to maternal blood flow is reduced and the flow to the placental bed is increased. Conversely, if this extension of trophoblast does *not* occur, there is a reduction in placental bed blood flow. This may lead to intrauterine growth retardation, pre-eclampsia or hypoxia in labour.

Structure of chorionic villus

A villus has a core of mesoderm and a covering of trophoblast; there are numerous fetal blood vessels in the mesoderm core. In the main villous stems the arteries and veins have connective tissue walls, but in the terminal villi only capillaries are present. The arteries are branches of the umbilical arteries. They end as capillaries in the terminal villi, from where oxygenated blood is collected into venous radicals and passed back into the veins of the main villous stems, and from there to the umbilical vein of the cord (*see* Fig. 1.7).

From the time of their formation in the third week until the end of the first trimester the villi are covered by a single layer of cytotrophoblast and an outer layer of syncytiotrophoblast which is in immediate contact with the blood in the intervillous space. After the 20th week the cytotrophoblast begins to disappear until finally only a thin layer of syncytium remains.

Further development of the placenta

The chorionic villi are formed in immense numbers and constitute the bulk of the placental area. They form an arrangement over a great surface, so that thin-walled vessels carrying fetal blood are

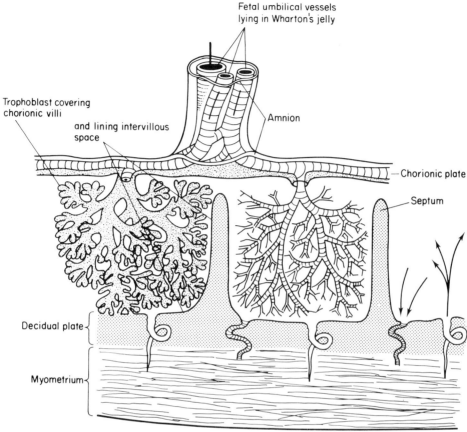

Fig. 1.7 The structure of the placenta. For clarity of only the fetal arterioles are shown in the mid section. The real villi are much more finely branched

only separated from the maternal blood by thin layers of villous connective tissue and syncitio-trophoblast.

At first villi are formed all over the surface of the gestation sac, so that an ovum of 4 weeks' growth appears as a translucent, thin-walled sac covered by a spherical halo of soft villi (*see* Fig. 1.8). As growth proceeds the decidua capsularis becomes thinner, and between the 12th and 16th weeks the villi on the capsular surface of the embryo degenerate rapidly, leaving this side of the chorion smooth (*chorion laeve*). In compensation, the villi on the surface opposed to the decidua basalis undergo a great deal of hypertrophy (*chorion frondosum*) and become matted into a solid disc, which is the fully developed placenta. This is formed by the 12th week, though proportionately it then comprises a much larger part of the ovular surface than it does at term.

Increase in size of the placenta

From the end of the 4th week, when the villi have penetrated towards the decidua basalis, there is no further invasion from the villi. Further growth in thickness of the placenta is now due to growth of the chorionic villi with an accompanying expansion of the intervillous sinuses. Until the end of the 16th week, the placenta grows both in thickness and circumference. Subsequently it continues to increase in size circumferentially until near term. The growth is proportionate with that of the fetus and of that part of the wall of the uterus to which the placenta is attached.

Fig. 1.8 Embryo showing chorionic villi

The placenta at term

The placenta at term is circular in shape, forming a spongy disc 20 cm in diameter, and about 3 cm thick. Its weight is usually about 500 g but there is a direct relationship with the fetal weight.

The placenta has a fetal and a maternal surface. The fetal surface is covered by smooth amnion underneath the chorion. The blood vessels are visible beneath this as they radiate from the insertion of the umbilical cord.

The maternal surface is rough and spongy, and presents a number of polygonal areas known as *cotyledons*, each being somewhat convex and separated from those adjoining it by a shallow groove. Each cotyledon is formed by, and corresponds to, a main villous stem and its system of branch villi. The number of cotyledons, between 15 and 20, depends on the number of end arteries into which the umbilical artery divides. A cotyledon is made up of between 10 and 20 *lobules*, each lobule corresponding to the opening of a maternal utero-placental vessel and is the functioning unit of the placenta. The colour is a dull red, with a thin, greyish, somewhat shaggy layer on the surface, which is the remnant of that part of the decidua basalis which has come away with the placenta. Numerous small grey spots are frequently seen on the maternal surface. These are due to the deposition of calcium in degenerate areas and are the result of normal senescence. They are more numerous in placentae after term, often amounting to 30 per cent of the surface. These deposits do not occur in the villi nor do they interfere with the maternal circulation in the intervillous spaces; they are of no apparent significance.

The chorion spreads away from the edge of the placenta to form the outer layer of the two membranes which enclose the fetus and amniotic fluid. Though the line of demarcation between the placental edge and the chorion is sharp they are essentially one structure, for the placenta is a specialized part of the chorion. The amnion, a thinner membrane which lines the fetal sac, can be separated from the chorion after birth; it can be stripped back to the root of the umbilical cord while the amnion stops at the edge of the placenta.

The umbilical cord usually reaches the fetal surface of the placenta at about the middle of its disc, although sometimes it is at its edge (*squash racket placenta*). It brings with it two umbilical arteries and one umbilical vein.

The area supplied by each umbilical artery varies. However, shortly after they have reached the placenta, there is always a communication between the two which serves to equalize the pressure in the two systems. Except for this communication, the main trunks into which the arteries divide are terminal, and each ends in a tuft of capillary vessels which drains into the corresponding tributary of the umbilical vein.

The substance of the placenta is made up almost entirely of a multitude of chorionic villi, most of which protrude in an arborescent manner into the intervillous blood spaces. Thus the placenta as a functioning organ is really a space containing maternal blood, bounded on the maternal side by a decidual plate, and on the fetal side by a chorionic plate from which the chorionic villi branch into the maternal blood (*see* Fig. 1.7).

The intervillous space is a lake of maternal blood which has left the maternal vessels to flow round in a space bounded by fetal trophoblast. To supply and drain this space arteries and venous sinuses perforate the decidual plate. It was formerly maintained that the blood flows to the edge of the placenta to be collected in a marginal sinus before leaving the intervillous space, but vessels serving the arterial inflow and venous drainage are now known to be scattered over the entire decidual plate.

The lobules of each coyledon consist of a group of villi which are based on one large fetal artery which branches and rebranches to supply the villi. The maternal blood enters the intervillous space via about 200 spiral arteries which perforate the decidual plate to spurt a jet of blood into the centre of a corresponding fetal lobule. The blood percolates through the branching villi of the lobule and then returns to the basal plate where it flows out through the decidual veins.

Placental bed

The placental bed is covered with decidua which the spiral arteries pierce to deliver blood to the maternal lake which bathes the villi. The ends of the arteries are narrow at first, but from about the 6th week after conception trophoblastic cells invade their walls and the arterial ends open out like river deltas. By 16 weeks, in the normal woman, this process of invasion is complete and each of the 200-odd spiral arteries ends in a funnel-like delta. This results in a much lower pressure at the distal end of the blood vessel, allowing greater flow. The trophoblast invasion may be absent in those who subsequently develop intrauterine growth retardation and pre-eclampsia.

FUNCTIONS OF THE PLACENTA

The placenta is the only means of transfer of anabolites and catabolites and, as such, is the main interface between the fetus and the outside world. The placenta:

- enables the fetus to take oxygen and nutrients from the maternal blood
- serves as the excretory organ where carbon dioxide and other waste products pass from the fetal to the maternal blood
- forms a barrier against the transfer of infection to the fetus, although some organisms, for example the rubella and HIV viruses are able to cross it
- secretes large amounts of hormones such as chorionic gonadotrophin and oestrogens. The maternal reaction to pregnancy is exemplified by the maintenance of the decidua and the growth of the uterus and breasts.

Placental transport

The transport of oxygen to the fetus is discussed later in this chapter.

In general the trophoblast and underlying endothelium of the fetal vessels behave as a semi-permeable membrane, allowing the free passage of water and soluble substances of relatively low molecular weight according to the laws of osmotic equilibrium, but there are also mechanisms of active transport, which allow more rapid diffusion of solutes, as well as the transfer of larger protein molecules.

Substances with a molecular weight of less than 1000 are, in general, able to pass to and from the placenta; most anaesthetic agents and drugs fall into this class. The concentrations of sodium chloride, magnesium, urea and uric acid are equal in maternal and fetal blood. A few substances, including some amino acids, nucleic acid, calcium and inorganic phosphorus, have a higher concentration in fetal than maternal blood; the trophoblast provides selective transfer of these substances, most of which are necessary for the building of fetal tissues. Glucose is found in higher concentration in maternal than in fetal blood, not because it does not cross the placenta freely, but probably because of its continual rapid utilization by the fetus. It enters the fetus by facilitated transfer at the placenta.

Much of the fat in fetal tissues is synthesized from carbohydrate, especially in late pregnancy. Synthesis of fatty acids, triglycerides, cholesterol and lipids mostly takes place in fetal tissue, although there is also a contribution by direct transfer of maternal lipid anabolites.

The serum iron concentration in the fetus is higher than that in the mother, but the mechanism of transfer is uncertain.

Oestrogens, androgens and thyroxine cross the placenta but insulin, parathyroid hormone and posterior pituitary hormones do not. The vagina of the newborn female child shows evidence of the oestrogenic action of maternal oestrogens *in utero*.

The syncytium has a high degree of selective activity, permitting passage of some IgG gamma-globulins, but not IgA or IgM globulins.

Despite the separation of the nucleated red cells in the vessels in the villi from the maternal red cells in the intervillous space, fetal red cells and fragments of villi may escape into the maternal circulation by intercellular passages in the syncytotrophoblast. The entry of red cells from a fetus whose blood group differs from that of its mother accounts for the development of haemolytic disease in certain circumstances (*see* Chapter 9).

Placental permeability increases as pregnancy progresses, probably because the trophoblast becomes thinner, and also because the villi become finer and more branched.

The functional efficiency of the placenta

Placental transfer will depend upon the maternal blood flow through the intervillous space, the effective surface area of the chorionic villi, and the fetal blood flow through them.

The maternal blood flow through the placenta is about 1000 ml per minute at term. The pressure in the intervillous space, which is of the order of 15 mmHg, will be affected by uterine contractions in late pregnancy, as well as in labour. As the uterus contracts the veins in the myometrium are occluded first, so that the venous outflow ceases. The inflow of blood will continue until the uterine pressure equals the pressure of the arterial inflow. These pressures are not usually present until labour. With a strong contraction the inflow also ceases, and the blood in the intervillous space is temporarily stagnant but the space does not empty because the veins are closed.

The capacity of the intervillous space is about 140 ml. The blood in it is exchanged about four times per minute. The area of the surface of the chorionic villi which is exposed to this 140 ml of blood has been estimated to be about 11 square metres (which is equivalent to 116 one foot square carpet tiles, laid on the floor). The blood therefore forms a very thin film over the surface of the villi. This large area, resulting from the fine branching of the villi, does not take into account the microvilli on the trophoblast, which can be seen with the electron microscope.

The volume of the fetal capillaries has been estimated to be about 60 ml and the fetal blood flow to be about 300 ml per minute, giving a total replacement of the fetal blood in the placenta of about five times per minute. The pressure in the fetal capillaries in the villi is of the order of 30 mmHg.

Placental hormones

Oestrogens, progesterone, chorionic gonadotrophin, lactogen, thyroid-stimulating hormone, a hormone resembling adrenocorticotrophin, and possibly relaxin (a hormone which causes relaxation of pelvic ligaments in some species) are secreted into the maternal blood by the placenta.

During pregnancy there is a progressive increase in the blood oestrogen and urinary oestrogen concentrations; these reach their peak just before the onset of labour. The oestrogen found during pregnancy is chiefly oestriol glucuronate. The placenta converts precursors, formed in the fetal adrenal gland and liver, to oestriol sulphate. This passes into the maternal blood before being conjugated with glucuronic acid in the maternal liver, and is finally excreted by the maternal kidney.

Until the end of the 12th week the corpus luteum continues to secrete progesterone. With the gradual cessation of function of the corpus luteum, the placenta becomes responsible for the secretion of progesterone which, like oestrogen, reaches a peak just before labour. The concentration of progesterone in maternal plasma is about 24 ng/ml at the 8th week, and 180 ng/ml at term.

The beta unit of chorionic gonadotrophin can be detected in the maternal plasma as early as 6 days after fertilization of the ovum, and is found in the urine soon after that. It reaches its peak concentration in the blood and urine at 10–11 weeks' gestation. The concentration then falls to a lower level at which it remains steady. The presence of chorionic gonadotrophin in the urine forms the basis of the routine tests for pregnancy. It eventually disappears from the urine after delivery if the fetus dies but may persist for a time if placental tissue is retained.

THE CHORION

After the definitive placenta is formed the rest of the chorion atrophies and persists only as a thin, friable membrane intervening between the amnion and the decidua. On its outer surface vestiges of decidual cells and of the trophoblastic layer that formerly covered it can be distinguished microscopically, but the bulk of it is a fragile connective tissue, loosely attached to the amnion.

THE AMNION

The first appearance of the amnion is a hollow space in the embryonic ectoderm. It is lined by cubical cells. These quickly become more columnar at the part which eventually forms the embryonic plate and the embryo. At first more or less spherical, the amniotic cavity soon becomes flattened down upon the embryo and closely applied to it. As the head and tail appear and the body walls fold round to enclose the embryonic coelom, the amnion attached at their margins is also carried round, so that embryo is pushed up and projects into the amniotic cavity. When the body cavity of the embryo is quite closed up, the amnion is attached all round the place at which the ventral stalk emerges (*see* Figs 1.5(b) and 1.5(c)).

At this period the embryo is relatively very small, and has the amnion closely applied to it, while the cavity of the blastocyst is relatively very large. Now the great change begins. The amniotic cavity enlarges out of proportion to the embryo, and becomes distended with fluid, while the embryo is gradually carried more and more into the amniotic cavity by elongation of the ventral stalk, which becomes the umbilical cord.

The enlargement of the amniotic cavity which brings about the complete investment of the umbilical cord also brings the amnion into close contact with the fetal surface of the placenta. This surface is, therefore, completely covered by the amnion. The amnion is loosely attached to the placenta and chorion, and can be separated up to the insertion of the cord.

The amnion is lined by a single layer of cubical or flattened epithelial cells which is attached to a layer of connective tissue. The epithelium contains granules, fat droplets and vacuoles. The connective tissue on the outer side of the amniotic membrane is closely applied to the similar connective tissue on the inner side of the chorion; the two merely stick together, and are not organically united. They can easily be separated from one another at all periods of pregnancy. That portion of the amnion which covers the umbilical cord, however, is very closely incorporated with the connective tissue of the cord and cannot be stripped off.

AMNIOTIC FLUID

The amniotic fluid is usually slightly turbid from the admixture of solid particles derived from the fetal skin and the amniotic epithelium. It may also be stained a green or brown colour if any meconium has been passed into it. The solid matter is composed of lanugo hairs, epithelial cells, sebaceous material from the fetal skin, and cast-off amniotic epithelial cells.

The volume of the amniotic fluid at term is about 800 ml, with a wide range from 400 to 1500 ml in normal cases. At 10 weeks the average volume is 30 ml, at 20 weeks 300 ml and at 30 weeks 600 ml. The rate of increase is therefore about 30 ml per week, but this falls off near term. This is evident on clinical examination. On palpation at 30 weeks there seems to be a lot of fluid relative to the size of the fetus; nearer term there is relatively less fluid, and when the expected date of delivery is passed the uterus seems to be full of baby.

At term the amniotic fluid has a specific gravity of 1010. It contains 99 per cent water and its osmolality is less than that of maternal or fetal plasma. The fluid has organic, inorganic and cellular constituents. Concentrations of some of the important contents near term are as follows:

sodium 130 mmol/l
urea 3–4 mmol/l
protein 3 g/l
lecithin 30–100 mg/l
alpha-fetoprotein 0.5–5 mg/l.

In addition traces of steroid and non-steroid hormones and of enzymes are present. It is mildly bacteriostatic.

The amniotic fluid is of both maternal and fetal origin, and the relative importance of the different mechanisms of production alters as pregnancy progresses. In very early pregnancy it is mostly secretion from the amnion.

Later, diffusion through the fetal skin accounts for much of the fluid, and there is increasing diffusion through the fetal membranes, including the part of the amnion that covers the fetal vessels in the cord and on the surface of the placenta. By the 20th week the skin loses its permeability, and from this time onwards the fetal kidneys play an increasing role in the production of amniotic fluid.

By term about 500 ml per day is secreted as fetal urine and tracheal fluid accounts for as much as 200 ml. Studies with radioisotopes have shown that near term 500 ml of water are exchanged hourly between the maternal plasma and the amniotic fluid. Disposal of the fluid is partly by absorption through the amnion into maternal plasma and partly by fetal swallowing and absorption in the intestine to enter the fetal plasma.

In cases of renal agenesis there is oligohydramnios (a deficiency of amniotic fluid) and in conditions of defective swallowing such as anencephaly and oesophageal atresia there is polyhydramnios (an excess of fluid).

Functions of the amniotic fluid

The fluid guards the fetus against mechanical shocks, equalizes the pressure exerted by uterine contractions and allows, at least in the early months, plenty of room for fetal movement. Since the temperature of the fluid is maintained by the mother, the fetus is not subjected to loss of heat. Fetal metabolism is devoted entirely to growth and differentiation, and not dissipated in making good heat loss. Some of the fluid is swallowed, but it can hardly be regarded as a source of nutrition for the fetus since the content of protein and salts is so small.

During labour, the amniotic fluid contained in the bag of forewaters forms a fluid wedge which, with the uterine contractions, dilates the internal os uteri and cervical canal. When the membranes rupture during labour fluid flushes the lower genital tract with fluid that is aseptic and bactericidal.

Investigation of the amniotic fluid

Samples of fluid can be obtained during pregnancy for various diagnostic purposes by abdominal amniocentesis. Not only can a variety of chemical estimations be made for normal and pathological constituents, but fetal amniotic cells can be obtained for tissue culture and chromosomal study.

THE UMBILICAL CORD

The umbilical cord forms the major connection between the fetus and the placenta. It is derived from the ventral stalk and receives a close covering of amniotic epithelium. The constituents of the umbilical cord are:

- *the covering epithelium*, which is a single layer of amnion
- *Wharton's jelly*, which is composed of cells with elongated anastomosing processes in a gelatinous fluid, and is part of the extraembryonic mesoderm
- *blood vessels*. At first these are four in number; two arteries and two veins. The embryonic umbilical veins fuse before the 3rd month leaving a single vessel. The two umbilical arteries are derived from the internal iliac (hypogastric) arteries of the fetus, and carry reduced blood from the fetus to the placenta. The umbilical vein carries oxygenated blood from the placenta to the fetus
- *the umbilical vesicle and its duct*, the shrivelled remnant of the yolk sac, may be found as a very small yellow body near the attachment of the cord to the placenta
- *the allantois*, which occasionally occurs as a blindly-ending tube just reaching into the cord, is continuous inside the fetus with the urachus and bladder.

The cord is approximately 50 cm in length, with a range of 7 to 180 cm. It is about 1 cm thick, but is not uniform, presenting nodes and swellings which are sometimes caused by dilatation of the umbilical vein, but more often by local increase in Wharton's jelly. At the earliest stage the cord is straight, but as early as the 12th week it shows a spiral twist.

In addition to the nodes mentioned above (sometimes called false knots) on rare occasions the cord has one or more true knots, due to the fetus passing through a loop in the cord. It is arguable if a true knot has ever caused significant impairment of the fetoplacental circulation. Rather, it might be that the impaired circulation allows a true knot to become tight. If at delivery of a healthy baby one attempts to tie tight a true knot in the cord, it will always slip loose when the tension is released. Only when the pressure in the vessels is reduced (e.g. by draining off the blood from the cord, or *in utero* by the fetal circulation failing) can the knot be tightened. This chicken and egg situation has long been a point for discussion among obstetricians and will likely remain so.

The cord is often coiled around the fetal body, limbs or neck, but seldom does it give rise to any serious trouble. The possible exception to this statement is when the cord may be coiled two or three times around the neck-nuchal cord; tension on the cord then occurs during the second stage of labour when the fetus descends in the birth canal. Such tension may obstruct the fetoplacental circulation and lead to fetal distress.

THE FETUS

SIZE OF THE FETUS

The maximum diameter of the embryonic sac at 5 weeks is 8 mm and it increases by about 1 mm per day up to 12 weeks. Obviously the size of the sac must be sufficient to enclose the flexed fetus.

The length rather than the weight of the aborted fetus is a more reliable index of its age. In the early weeks the measurement of the fetus is commonly taken from the vertex of the head to the coccyx (crown–rump length), but from the 20th week onwards the measurement is taken from the vertex to the heels. For antenatal care the biparietal diameter of the fetal head is a more useful measurement, and this can be determined accurately with ultrasound. Fig. 1.9 shows the approximate measurements:

A rough guide to the vertex–heel length of the fetus (in cm) after the 20th week is given by multiplying the number of lunar months by five.

Other characteristics of the growing fetus are useful in determining the age.

At 8 weeks the 2.3 cm fetus lies in the enlarging amniotic cavity, but the amnion is not yet in contact with the chorion and the extraembryonic coelom is not yet obliterated. The ventral stalk and yolk sac have united to form the umbilical cord which is invested by amnion, and the primitive

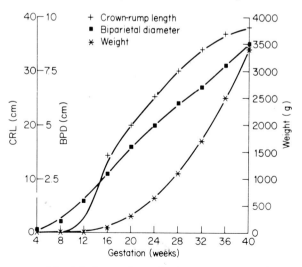

Fig. 1.9 Approximate fetal measurements

small intestine is contained in the dilated proximal part of the cord. The facial form has been completed by the formation of the nose and its separation from the mouth. The ears are completely formed externally, and the eyelids have appeared around the eyeball. The limbs are enlarging and show their jointed appearance, and the fingers and toes are formed. The flexion of the trunk has diminished, so that the vertex of the head rather than the back of the neck now forms the upper end of the embryo.

At 12 weeks the placenta is discoid. The amnion entirely fills the chorionic sac. The umbilical cord, still short and thick, shows a spiral twist. The primitive intestine is completely withdrawn into the body cavity. Nails have appeared on the fingers and toes, and the external sexual organs are differentiated. The crown–rump length is 6 cm and the weight about 14 g.

At 28 weeks the fetus weighs about 1100 g. The subcutaneous fat is becoming more evident, and the skin wrinkles begin to disappear. The testicles are in the inguinal canals. The eyelids have opened.

After this the weight of the fetus increases with comparative rapidity. The fetus becomes completely covered with *vernix caseosa*, a greasy substance composed of the secretion of sebaceous glands mixed with desquamated epithelial cells. The scalp hair increases in length. The short colourless hairs known as lanugo, which have previously appeared on the body and head, tend to disappear. The bright red colour of the skin changes to a pink flesh colour owing to the thickness of subcutaneous fat. Just before the 36th week one testicle has usually descended into the scrotum. The nails reach the ends of the fingers but not of the toes.

At 40 weeks the fetus measures 50 cm from vertex to heels and, as a rule, weighs between 2700 and 3600 g. The signs that the fetus has reached term are not always certain, but the length and weight are important. The nails usually project beyond the ends of the fingers and have reached the ends of the toes. The skin is pink and the lanugo has almost disappeared, except over the shoulders. The whole of the intestine contains

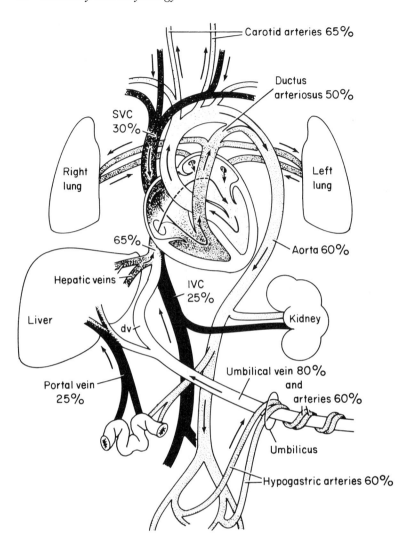

Carotid arteries 65%

Ductus arteriosus 50%

SVC 30%

Right lung

Left lung

65%

Aorta 60%

Hepatic veins

IVC 25%

Liver

Kidney

dv

Umbilical vein 80% and arteries 60%

Portal vein 25%

Umbilicus

Hypogastric arteries 60%

Fig. 1.10 Fetal circulation. SVC, superior vena cava; IVC, inferior vena cava; dv, ductus venous. The percentage figure represents the oxygen saturation of the blood

meconium. The umbilicus is almost at the centre of the body. Both testicles have descended into the scrotum. As a rule only one epiphysis has started to ossify, that at the lower end of the femur, but the centres of ossification of the upper epiphysis of the tibia and of the humerus may have appeared.

Babies weighing more than 4500 g at birth are rare. If a baby weighs less than 2500 g at term it is described as being of low birth weight (LBW) and on this account may have a diminished chance of survival. It is different with twins; both may weigh less than 2500 g, and they tend to be born before term, and yet both are likely to survive. The heaviest children are likely to be born when the mother's age is between 25 and 35 years. Very young mothers commonly have small babies. The weights of the children tend to increase in successive pregnancies, provided that the mother's age is under 35 years. On average male babies are heavier than females at birth.

FETAL CIRCULATION

The fetal circulation differs from that of the adult in several respects. Oxygen and nutrients are carried from the placenta to the fetus in the single large umbilical vein. Although there is free diffusion of oxygen across the thin barrier between the maternal and fetal circulations in the placenta, the oxygen saturation in the umbilical venous blood is already reduced from the 95 per cent level in maternal arterial blood to about 80 per cent because the active metabolism of the placenta has already used some of the oxygen. On entering the body of the fetus the vein communicates with the portal vein, so that some blood passes through the liver to the hepatic vein and inferior vena cava, but most of the oxygenated blood passes more directly through the ductus venosus to the vena cava (*see* Fig. 1.10). In the inferior vena cava it mixes with the stream of desaturated blood returning from the lower limbs and abdominal organs, so that the oxygen saturation is reduced to less than 70 per cent by the time the blood enters the right atrium. By a remarkable anatomical arrangement most of the blood entering the right atrium from the inferior vena cava is directed through the foramen ovale into the left atrium, and from thence it passes into the left ventricle and out into the ascending aorta. By this arrangement the coronary arteries and the brain receive the most highly oxygenated blood.

Blood returning from the head and upper limbs through the superior vena cava streams towards the tricuspid valve to enter the right ventricle, taking with it a portion of the blood which has entered through the inferior vena cava. The unexpanded fetal lungs offer such high resistance to blood flow that the greater part of the right ventricular output passes from the main pulmonary artery through the widely patent ductus arteriosus into the aorta, to be distributed to the lower part of the body of the fetus, and along the paired umbilical arteries to the placenta. The blood in the descending aorta has a saturation of about 60 per cent, which is lower than that in the ascending aorta because of the admixture of blood which has come from the ductus arteriosus.

In the fetus only 5 to 10 per cent of the total cardiac output goes directly to the lungs and returns to the left atrium via the pulmonary veins. Although the left ventricle is pumping the most highly oxygenated blood to the brain and coronary arteries, the right ventricle contributes an equal amount to the total cardiac output, and its wall is as thick as that of the left ventricle.

The greater part of the blood descending in the aorta leaves the body by the umbilical (hypogastric) arteries to be oxygenated in the placenta.

OXYGEN SUPPLY TO THE FETUS

As the fetus grows it requires ever increasing amounts of oxygen. This is achieved by:

- an increase in placental transfer of oxygen due to reduction in the diffusion distance
- an increase in the haemoglobin content of fetal blood
- fetal haemoglobin (HbF), which is made in the liver and spleen rather than the bone marrow, and has a different structure from adult haemoglobin (HbA).

Instead of two alpha- and two beta-globin chains as in HbA, HbF has two alpha and two gamma chains. The oxygen dissociation curve of HbF is shifted to the left so that it is more saturated than HbA at the low P_{O_2} levels present *in utero* (*see* Fig. 1.11).

The fetal red cell count rises progressively from 1.5 million per mm^3 at 10 weeks to 5.5 million at term. The cells are macrocytic, and the haemoglo-

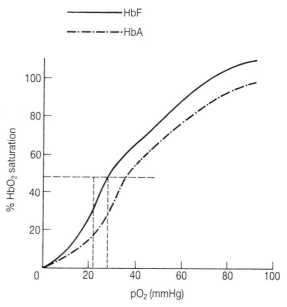

Fig. 1.11 The haemoglobin dissociation curves of adult haemoglobin (HbA) and fetal haemoglobin (HbF). The partial pressure of oxygen (pO_2) at which each haemoglobin is 50 per cent saturated is 21 mmHg for HbF and 27 mmHg for HbA

bin content at term is about 16 g per 100 ml. The maternal blood at term often contains less than 12 g per 100 ml.

The effect of these differences between maternal and fetal blood is that 100 ml of maternal blood can carry 16 ml of oxygen if it is fully saturated, but the same volume of fetal blood can carry 21 ml. In other words, the fetal blood can easily take up oxygen from the maternal blood. Maternal arterial blood is nearly fully saturated and carries about 15.7 ml of oxygen per 100 ml, whereas fetal blood in the descending aorta is only about 60 per cent saturated, but carries 13 ml of oxygen per 100 ml.

NUTRITION OF THE FETUS

Although the fetus is completely dependent on the maternal blood and the placenta for its nutritional requirements, it is a separate physiological entity, and it will take what it needs from the mother, even at the cost of reducing her reserves of some substances, for example calcium and iron. For nearly every substance the rate of fetal intake increases progressively, and is greatest in the late weeks of pregnancy.

ANATOMY OF THE NORMAL FEMALE PELVIS AND THE FETAL SKULL

Knowledge of the shape and dimensions of the normal female pelvis and of the fetal skull is essential for proper understanding of the mechanism of labour and its abnormalities.

THE PELVIS

Although some sexual differences may be recognizable at birth the female pelvic characteristics are chiefly developed between that time and puberty. Radiological surveys have shown that variations in the shape of the pelvis are very common. What is described as the normal female pelvis, the *gynaecoid pelvis*, occurs in only 40 per cent of white women. Other women have pelves with male characteristics (*android pelvis*), in some there is a slight increase in the anteroposterior diameter of the pelvis (*anthropoid pelvis*) while in the remainder the pelvis is slightly flattened (*platypelloid pelvis*). Only the typical gynaecoid pelvis will be described here.

The part of the pelvis above the brim is described as the false pelvis and that below the brim as the true pelvis. The obstetrician is only concerned with the latter. The true pelvis may be described in terms of the brim, the cavity and the outlet.

The pelvic brim

The pelvic brim (*see* Fig. 1.12) lies in one plane bounded in front by the symphysis pubis, on each side by the upper margin of the pubic bone, the illiopectineal line and the ala of the sacrum and posteriorly by the promontory of the sacrum. The brim is oval in shape, with the transverse diameter slightly greater than the anteroposterior (*see* Fig. 1.12).

The pelvic cavity

This is sometimes described in terms of an imaginary plane bounded in front by the middle of the symphysis pubis, on each side by the pubic bone, the obturator fascia and the inner aspect of the

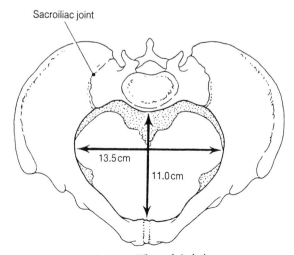

Sacroiliac joint

13.5 cm

11.0 cm

Fig. 1.12 The pelvic brim

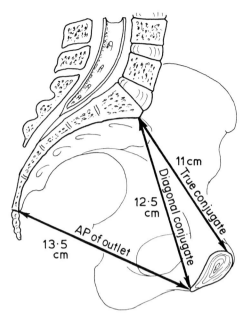

Fig. 1.13 Sagittal section of pelvis. The true and diagonal conjugate diameters are shown as it is the anteroposterior diameter of the outlet

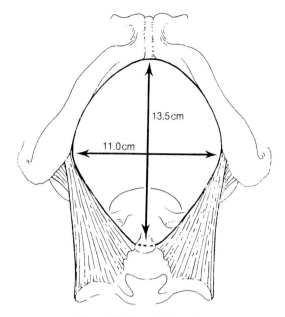

Fig. 1.14 The pelvic outlet

ischial bone and posteriorly by the junction of the second and third pieces of the sacrum. The ischial spines lie slightly below this plane. In the gynaecoid pelvis the cavity is circular and roomy because the sacrum is inclined backwards, is well curved and the sacrosciatic notches are wide.

The pelvic outlet

The pelvic outlet (*see* Fig. 1.14) is roughly diamond-shaped and is bounded in front by the lower margin of the symphysis pubis, on each side by the descending ramus of the pubic bone, the ischial tuberosity and the sacrotuberous ligament and posteriorly by the last piece of the sacrum (*not the coccyx, which is mobile*). Unlike the brim, the outlet does not have boundaries which lie in a single plane, but an imaginary plane of the outlet is described, passing from the lower margin of the symphysis pubis to the last piece of the sacrum. It will be noted that the ischial tuberosities lie well below this plane. In a gynaecoid pelvis the subpubic arch is wide and the tuberosities are far apart. The shape of the outlet is oval with the long axis in the anteroposterior diameter.

During pregnancy the ligaments of the sacroiliac

joints and the symphysis pubis become softened and there is slightly increased mobility at these joints. The sacrococcygeal joint allows the coccyx to move freely backwards during delivery.

The pelvic axis

The axis of the pelvis is an imaginary curved line which shows the path which the centre of the fetal head follows during its passage through the pelvis. It is obtained by taking several anteroposterior diameters of the pelvis and joining their centres (*see* Fig. 1.15).

Pelvic inclination

The pelvic inclination is the angle that any pelvic plane makes with the horizontal. In the erect position the brim is normally inclined at 60 degrees. In negroid women this angle may approach 90 degrees, referred to as steep inclination of the brim, and because of this the head may be slow to engage during labour.

The pelvic outlet is inclined at about 25 degrees to the horizontal.

The inclination of the sacrum is measured differently. It is the angle between the front of the first piece of the sacrum and the plane of the pelvic

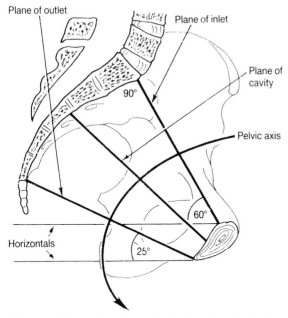

Fig. 1.15 The angles of the inlet (60°) and outlet (25°) to the horizontal in a normal standing woman are shown. The former is the angle of inclination (see text)

brim. It may affect the available space in the upper part of the pelvic cavity (*see* Fig. 1.15).

AVERAGE DIMENSIONS OF THE NORMAL PELVIS

It must be clearly understood that the pelvic diameters vary just as much as women's heights vary, and that the diameters will also be affected by the pelvic shape. What matters during delivery is not the absolute size of the pelvis but the size relative to that of the fetal head. However, in a woman of average height with a normal pelvis the following measurements are to be expected.

Diameters of the brim

The anteroposterior diameter of the brim (or true conjugate) is measured from the back of the upper part of the symphysis pubis to the promontory of the sacrum and is 11 cm (*see* Fig. 1.13). The transverse diameter of the brim measures 13.5 cm (*see* Fig. 1.12).

Diameters of the pelvic cavity

The plane of the cavity has already been defined.

The anteroposterior and transverse diameters both measure 12 cm, so in cross-section it is almost circular.

Diameters of the pelvic outlet

The outlet does not lie in a simple plane like the brim. The anteroposterior diameter of the outlet is measured from the lower part of the symphysis pubis to the lower end of the sacrum (*not* the coccyx) and is 13.5 cm (*see* Fig. 1.13). The transverse diameter is measured between the inner surfaces of the ischial tuberosities and is 11 cm (*see* Fig. 1.14).

The average normal measurements are summarized in Table 1.1.

The brim is a transverse oval, the cavity is round and the outlet is an anteroposterior oval.

Table 1.1 Average dimensions of the normal pelvis

	Anterioposterior (cm)	Transverse (cm)
Brim	13.5	11
Cavity	12	12
Outlet	11	13.5

CLINICAL EXAMINATION OF THE PELVIS

A pelvic examination used to be made early in pregnancy to confirm the diagnosis of pregnancy and its duration, to exclude abnormalities of the pelvic organs and to assess the capacity of the pelvis. If the woman has miscarried, or has threatened to miscarry previously, it is wise to postpone any internal examination until later. Further, since in the Western world most women have an ultra sound examination early in pregnancy, much of the need for an early clinical vaginal assessment is removed. A further pelvic examination may be performed at about the 36th week, when a better assessment of pelvic capacity can be made, since at this time the pelvic floor is more relaxed; the size of the fetal head can also be related to that of the pelvic brim at this time.

Vaginal assessment of the pelvis

The brim

The sacral promontory cannot be reached with an examining finger in a normal pelvis unless the woman is anaesthetized. If it can be felt it is likely that the true conjugate is considerably reduced, and in such a case it may be possible to estimate the diagonal conjugate with the exploring finger. This is the distance between the promontory and the lower margin of the symphysis pubis. The true conjugate may be derived by subtracting 1.5 cm from the diagonal conjugate (*see* Fig. 1.13).

The pelvic cavity

On vaginal examination a general idea of the capacity of the pelvic cavity can be gained. The anterior surface of the sacrum is palpated from above downwards, noting whether it is straight or concave. The position of the ischial spines may be assessed by palpation of the sacrospinous ligaments, which should be of a length that will accommodate three finger-breadths. The spines are sometimes unduly prominent, but it is the distance between them rather than their prominence that matters.

The pelvic outlet

The intertuberous diameter can be determined by external palpation, but vaginal examination gives the best assessment of the width of the subpubic arch and of the position of the sacrum.

The pelvic floor

Although this is not part of the bony pelvis it is mentioned here because it forms part of the birth canal and plays an important part in the mechanism of labour. The two levator ani muscles, with their fascia, form a musculofascial gutter during the second stage of labour, with the opening of the vagina looking forward between the sides of the gutter (*see* Fig. 1.16). The pelvic floor directs the most salient portion of the presenting part forwards under the subpubic arch.

THE FETAL SKULL

The fetal skull may be divided into the vault, face and base. By the time of birth the bones of the face and base are all firmly united, but the bones of the vault are not so well ossified, being joined only by unossified membranes at the sutures. During labour the bones of the vault can undergo *moulding*, by which the shape of the skull can be altered

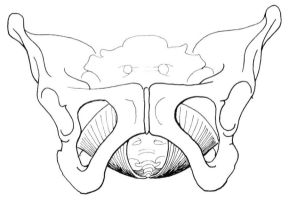

Fig. 1.16 The levator sling

by overriding the cranial bones with reduction of some of its diameters.

The bones which form the vault are the parietal bones, and parts of the occipital, frontal and temporal bones (*see* Fig. 1.17). At birth the frontal bone is divided into two parts. Three sutures are of obstetric importance:

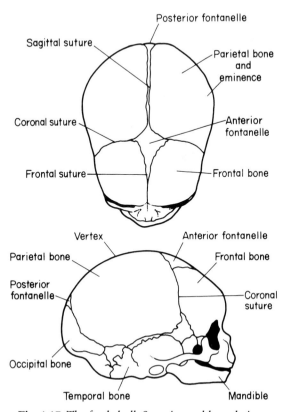

Fig. 1.17 The fetal skull. Superior and lateral views

- the sagittal suture lies between the superior borders of the parietal bones
- the frontal suture, which is a forward continuation of the sagittal suture, lies between the two parts of the frontal bone
- the coronal suture lies between the anterior borders of the parietal bones and the posterior borders of the frontal bones.

Fontanelles

The points of junction of the various sutures are termed *fontanelles*; the anterior and posterior fontanelles are important in obstetrics.

The anterior fontanelle lies where the sagittal, frontal and coronal sutures meet. The posterior fontanelle lies at the posterior end of the sagittal suture, between the two parietal bones and the occipital bone.

The position of these two fontanelles, when felt on vaginal examination, indicates in which direction the occiput is pointing and the degree of flexion or extension of the head.

The anterior fontanelle (or bregma) is much the larger of the two, is roughly kite-shaped, has four sutures running into it, is always patent at birth and takes about 20 months to close. The posterior fontanelle is triangular in shape, has three sutures running into it, in most cases cannot be felt as a space during labour and closes soon after birth.

The area of the fetal skull bounded by the two parietal eminences and the anterior and posterior fontanelles is termed the *vertex*. It is the part of the head which presents in normal labour.

DIAMETERS OF THE FETAL SKULL

The diameters of the fetal skull which are important in the mechanism of labour may be divided into vertical, longitudinal, and transverse.

Vertical and longitudinal diameters

The fetal head is ovoid in shape. In normal labour the head is well flexed so that the least diameters of the ovoid, namely the *suboccipitobregmatic* and *biparietal (transverse) diameters* are those which engage. The suboccipitobregmatic diameter is measured from the suboccipital region to the centre of the anterior fontanelle (bregma) and in the normal head at term is 9.5 cm (*see* Fig. 1.18).

If the head is less well flexed the *suboccipito-*

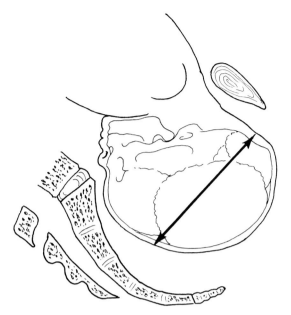

Fig. 1.18 The suboccipitobregmatic diameter of 9.5 cm, engaged in the pelvis when the head is fully flexed

frontal diameter is involved. This is taken from the suboccipital region to the prominence of the forehead and measures 10 cm. This is the diameter of the head which passes through the vulval orifice at the moment of delivery of the head (*see* Fig. 4.7).

With further extension of the head the *occipitofrontal diameter* engages. This is measured from the root of the nose (glabella) to the posterior fontanelle and is 11.5 cm. This is the diameter that meets the pelvis with a persistent occipitoposterior position (*see* Fig. 1.19).

The greatest longitudinal diameter is the *mentovertical*, which is taken from the chin to the furthest point of the vertex and measures 13 cm. This is the diameter which is thrown across the pelvis in a brow presentation (*see* Fig. 1.20) and is too large to pass through a normal pelvis.

Beyond this point further extension of the head so that the face presents results in a smaller vertical diameter, that is the other end of the ovoid presents. The *submentobregmatic diameter* is taken from below the chin to the anterior fontanelle and measures 9.5 cm (*see* Fig. 1.21).

Transverse diameter

The biparietal diameter, measured from one parietal eminence to the other, is 9.5 cm (*see* Fig. 1.22).

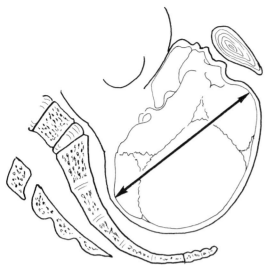

Fig. 1.19 The occipitofrontal diameter of 11.5 cm, engaged in the pelvis in a face to pubis position

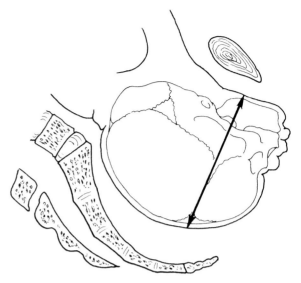

Fig. 1.21 The submentobregmatic diameter of 9.5 cm, engaged in the pelvis in a face presentation with the head completely extended

PROCESSES OF THE DURA MATER

The great folds of the dura mater, the falx cerebri and the tentorium cerebelli, act in some degree as internal ligaments, resisting too great deformation of the fetal head both in the longitudinal and the transverse directions. If moulding is excessive, or if the fetal head is subjected to severe and sudden stresses, these parts of the dura mater are liable to be torn; some of the great venous sinuses are then in danger of rupture. These are the inferior longi-tudinal sinus, running in the free edge of the falx cerebri and receiving the great cerebral veins of Galen from the brain, and the straight sinus running between the falx cerebri and the tentorium cerebelli.

DIAMETERS OF THE FETAL TRUNK

The *bisacromial diameter* is taken between the parts furthest apart on the shoulders and is 12 cm. The *bitrochanteric diameter* measures 10 cm.

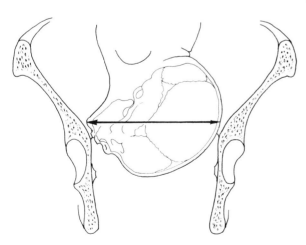

Fig. 1.20 The mentovertical diameter of 13 cm meeting the pelvis in a brow presentation

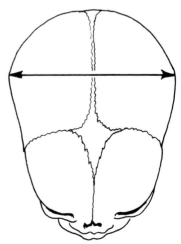

Fig. 1.22 Biparietal diameter, 9.5 cm

MATERNAL PHYSIOLOGY

Although the anatomical and physiological changes that occur during pregnancy chiefly involve the genital tract and the breasts, many other inter-related changes occur in other systems of the body.

Endocrine and paracrine changes

Placental proteins and hormones mostly increase in concentration in the blood through pregnancy. Some, however, like human chorionic gonadotrophin (hCG) are early pregnancy agents; after peaking at the time of their maximum need, their levels are reduced (*see* Fig. 1.23). Oestrogen and progesterone influence ovarian and pituitary function, maintain the decidua, initiate the growth of the myometrium, increase the vascularity of the whole genital tract and cause proliferation of the glandular tissue of the breasts. These substances have secondary effects, including the retention of water in the body, relaxation of smooth muscle, and possibly relaxation of pelvic ligaments.

Changes due to uterine enlargement

The increased blood flow through the uterus causes substantial changes in the maternal circulation, and the enlarged uterus alters the general posture and affects the mechanism of respiration.

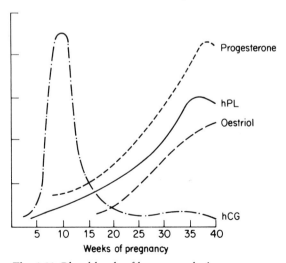

Fig. 1.23 Blood levels of hormones during pregnancy

Metabolic changes

The fetal requirements of oxygen and food substances, the growth of the uterus and preparation for lactation affect the mother's metabolism and dietary needs.

Details of some of these changes are as follows:

ENDOCRINE AND PARACRINE SYSTEM

Trophoblastic hormones and proteins

The trophoblast produces large amounts of hCG, particularly in the first trimester of pregnancy (*see* Fig. 1.23). High concentrations of this substance prolong the lifespan of the corpus luteum; this in turn continues to produce oestrogen and progesterone and so to maintain the uterine decidua until the output of oestrogen and progesterone from the placenta rises. The decidua (gestational endometrium) in turn increases the rate of synthesis of prostaglandins and secretory proteins.

The cardinal symptom of pregnancy, amenorrhoea, is dependent upon continuing production of progesterone and oestradiol by the corpus luteum to sustain the decidua (*see* Fig. 1.24). The luteotrophic action of rising concentrations of hCG extends the life of the corpus luteum and gonadal steroid production for at least 8 to 10 weeks, until the developing placenta is sufficiently mature to assume the synthesis of progesterone and oestriol. Thereafter, involution of the corpus luteum of pregnancy occurs.

It is interesting to note that the principal function of the ovary in early pregnancy is to produce these two hormones and little else, as evidenced by successful pregnancy following ovum donation in women with ovarian agenesis or premature menopause who seem only to require synthetic progesterone and oestrogen for the first trimester. In addition, the extensive structural changes in the endometrium as it becomes decidua are reflected by at least a quantitative alteration in the synthesis of secretory proteins by the various cell types of the endometrium. The most striking change is seen in the rate of production of the major secretory protein of the glandular epithelium, progestogen-dependent endometrial protein, and also of the

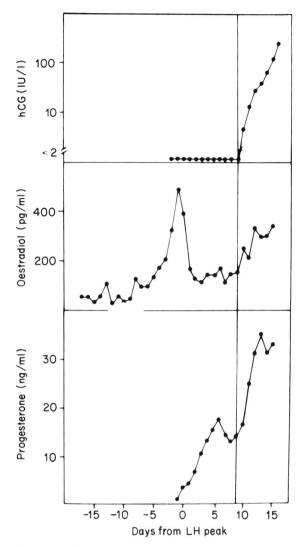

Fig. 1.24 Daily concentrations of oestradiol and progesterone throughout a spontaneous conception cycle indicating the changes in steroid hormones in relation to human chronic gonadotrophin (hCG)

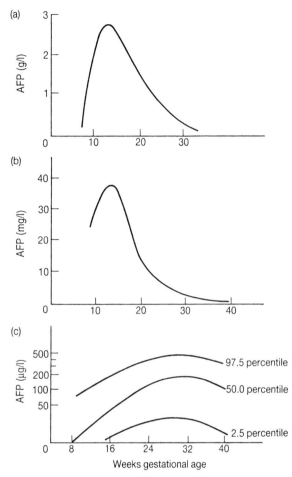

Fig. 1.25 The median concentrations of alpha fetoprotein (AFP) in (a) fetal serum, (b) amniotic fluid, and (c) maternal serum. Maternal serum is shown with 2.5 and 97.5 percentile limits

stromal cell types which produce insulin-like growth factor binding protein. Current research may reveal that serum levels of these proteins will provide the first non-invasive tests of endometrial and decidual function.

Embryonic development can also be reflected in the maternal bloodstream by changes in serum concentrations of alpha-fetoprotein (AFP) which in pregnancy is derived from endodermal tissues. AFP and other substances of embryonic origin are present in the highest concentrations in amniotic fluid, and in lesser amounts in maternal peripheral blood (*see* Fig. 1.25). These and other endocrine events are largely responsible for initiating the metabolic and physiological adaptations which prepare for pregnancy and parturition.

The common symptoms of pregnancy – morning sickness, breast changes and urinary frequency – which happen early in the first trimester are most likely to be caused by the hormonal changes, and as the second trimester is entered, abdominal swelling or distention due to the growing uterus is evident.

Large amounts of oestrogens and progesterone

are found in the blood and urine during pregnancy, at first secreted by the corpus luteum, then by the placenta after the 12th week. The oestrogen which appears in the largest quantity in maternal urine during pregnancy is oestriol. This is formed in the placenta from precursors which come from the fetal suprarenal cortex and liver.

The blood progesterone level rises during pregnancy, and therefore the urinary excretion of the metabolite pregnanediol increases. Even if the corpus luteum is removed in early pregnancy the hormone levels continue to rise as the placenta produces large quantities of these hormones. Oestrogen and progesterone maintain the decidua of pregnancy, cause the growth and hyperaemia of the uterus and lower genital tract, and the hyperaemia and development of the breasts. Oestrogen stimulates synthesis of prolactin via the pituitary gland, but together with progesterone it inhibits the lactogenic effect of human placental lactogen (hPL) and prolactin.

Other maternal endocrine changes

During pregnancy the anterior lobe of the pituitary gland undergoes hypertrophy, with an increase of both acidophil and basophil cells due to the influence of increased oestrogen. No anatomical changes are evident in the posterior lobe of the pituitary gland.

The thyroid gland is slightly enlarged during pregnancy, and the basal metabolic rate is increased.

There is an increased concentration of adrenal hormones in both the blood and the urine. The observation that rheumatoid arthritis often undergoes a remission during pregnancy led to research which identified and pioneered the use of glucocorticoids. There is also some increase in the excretion of mineralocorticoids (aldosterone).

THE UTERUS

During pregnancy the uterus is adapted to contain the growing fetus and placenta, and it also undergoes changes in preparation for its task of expelling the fetus during labour. The changes include development of the decidua, hypertrophy of the muscle coat, increased vascularity, formation of the lower segment and softening of the cervix.

At term the uterus is about 35 cm long and 23 cm in diameter. It weighs 1 kg, in contrast to the non-pregnant uterus which weighs 65 g. The pregnant uterus is usually slightly rotated on its long axis, so that the anterior surface faces a little to the right, and the fundus may also be inclined to one or other side, most often the right. The enlargement of the uterus is greatest at the fundus, so that the points of entry of the Fallopian tubes appear to lie well below the top of the uterus.

The enormous growth of the myometrium during pregnancy is caused by two factors – hormonal stimulation and distension. During early pregnancy the embryo does not fill the uterine cavity, and distension has no influence at this stage. An identical uterine enlargement occurs in cases of ectopic pregnancy when the embryo is outside the uterus. The growth is brought about by the action of oestrogen and progesterone. Later, as the fetus grows, the distension of the uterus owes much to the increasing mass of fetus, placenta and amniotic fluid.

In early pregnancy active mitosis can be seen in the connective tissue and muscle cells, and the enlargement of the uterus is largely due to an increase in the number of cells. Later in pregnancy the enlargement is due chiefly to hypertrophy of the individual cells, and cell division is less active. At term each muscle cell is about 10 times as long as it was before pregnancy. The blood supply to the uterus is greatly increased and, especially under the placental site, the veins become converted to large sinuses, thicker than a pencil.

At term the wall of the uterus appears thin in relation to the enormous enlargement of the cavity, but in fact it is still about 1 cm thick. In late pregnancy there are three layers in the muscular wall of the uterus (*see* Fig. 1.26):

- a thin outer layer of fibres that arches over the fundus and is continuous laterally with the muscle of the round ligaments
- a thick intermediate layer, consisting of a meshwork of interlacing fibres, which surrounds the blood vessels. The contraction of this layer will stop the blood flow in the vessels, and it is the strong contraction of this layer which prevents dangerous haemorrhage from the large placental sinuses in the third stage of labour
- a thin inner layer arranged in a circular fashion, especially around the internal os and around the tubal openings.

During pregnancy the uterus contracts from time to time (Braxton Hicks contractions), and contrac-

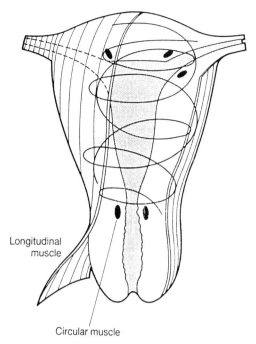

Longitudinal
muscle

Circular muscle

Fig. 1.26 Muscle layers in uterus

tions can be stimulated by abdominal palpation. These contractions are not so strong or regular as those of labour, and are painless.

The ovarian and uterine arteries and veins are greatly enlarged.

THE CERVIX

Although the cervix hypertrophies to some extent during pregnancy, this is not to the same extent as the body of the uterus. The cervix becomes softer because of increased vascularity and a great increase in the gland spaces. The glands are distended with mucus, and the pattern of the glands becomes far more complex, so that the cervix seems to contain a honeycomb filled with mucus; this is sometimes described as a mucus plug. On inspection the cervix has a purple tinge from venous congestion.

The cervix remains elongated until the end of pregnancy, although in multiparae the external os tends to be patulous. An effective internal os is of importance in retaining the embryo safely within the uterine cavity; with an incompetent cervix there is a considerable risk of abortion, or of premature rupture of the membranes.

Because of the great activity of the columnar epithelium of the cervix, and the increased secretion of mucus, it is common for the stratified epithelium on the vaginal surface of the cervix to be replaced by an outward extension of columnar epithelium – referred to as cervical ectropion. Such an appearance is not abnormal during pregnancy, and it will usually disappear in the puerperium.

The vagina and vulva

The increased vascularity already described in the cervix affects the vaginal walls a little later, and they eventually show the purple coloration right down to the vulva. The vaginal walls become softened and relaxed, and the same change occurs in the perineal body.

The watery transudation that normally occurs through the vaginal wall is increased during pregnancy. The hypertrophied cervical glands secrete more mucus, and this is added to the vaginal transudate and desquamated vaginal cells, so that the total discharge from the vagina is increased. The vaginal secretion is acid in reaction (pH 4.5 to 5) and is some protection against ascending infection.

Under the influence of placental hormones the vaginal epithelium is very active, and in a smear desquamated cells often have their lateral edges rolled over, so that they appear boat-shaped (navicular cells); these cells are usually clumped together. This normal epithelial activity should not be mistaken for carcinoma *in situ*.

As pregnancy advances the vulva shares in the increased vascularity and shows some swelling in consequence. Varicose veins may appear.

THE BREASTS

During pregnancy the secretion of oestrogen in large amounts causes thickening of the skin of the nipple and active growth and branching of the underlying ducts. The added action of progesterone causes proliferation of the glandular epithelium of the alveoli (*see* Fig. 1.27). Neither of these hormones causes the active secretion of milk, which only begins after delivery when the level of oestrogen falls and that of prolactin, from the anterior lobe of the pituitary gland, rises.

Slight changes in the breasts occur in the menstrual cycle, under the influence of oestrogen and progesterone, and the breasts may become tense

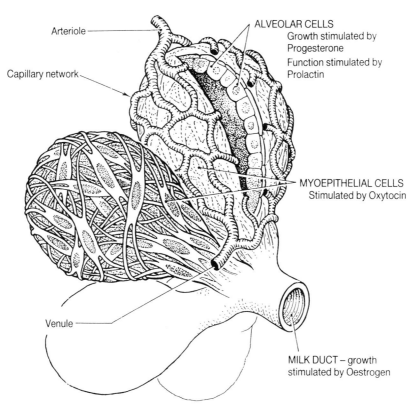

Arteriole

Capillary network

ALVEOLAR CELLS
Growth stimulated by
Progesterone
Function stimulated by
Prolactin

MYOEPITHELIAL CELLS
Stimulated by Oxytocin

Venule

MILK DUCT – growth
stimulated by Oestrogen

Fig. 1.27   Alveoli, the basic units of the mammary gland

and uncomfortable for a few days before the onset of the period. If pregnancy occurs these changes are more marked. The earliest change is a swelling of the breasts, especially at the periphery. The lobules of the gland can be felt easily and are harder than normal, these changes producing a knotty feeling in the breast. At the same time the breasts become a little tender, and the woman often describes a prickly sensation in them. The increased blood supply is shown by a very obvious network of veins under the skin. As a result of congestion, followed by actual growth of the glandular tissue, the breasts become more prominent. Plethysmographic studies have shown that each breast may increase in volume by about 200 ml, an increase of about one-third.

By about the 12th week of pregnancy the glands begin to secrete a clear fluid, which will appear in droplets if the breast is squeezed towards the nipple. Towards the end of pregnancy this secretion becomes more copious and is yellow in colour and creamy in consistence. It is then known as *colostrum*, and consists of water, fat, albumin, salts and colostrum corpuscles. The latter are cells shed whole from the gland acini, and filled with fat droplets. When the milk secretion is eventually established colostrum corpuscles are not found in it, because the fat is discharged from the secreting cells into the lumina of the acini, and the cells themselves are not detached.

Changes also occur in the nipple, which becomes larger and more readily erectile. The areolar skin is active and slightly raised above the surrounding skin. The areola often becomes pigmented, the change being most marked in women with darker hair and skin. Once the pigmentation has occurred it persists as a permanent change. The sebaceous glands on the areola are very active in pregnancy, and can be seen as a ring of about 12 to 20 small tubercles (Montgomery's tubercles).

At about the 20th week of pregnancy, in dark-skinned women, further pigmentation may occur on the skin beyond the margin of the areola. This is termed the secondary areola, and is not a uniform pigmentation, but takes the form of patchy streaks. The secondary areola disappears after pregnancy.

Sometimes outlying lobules of mammary tissue are found in the axillae; they enlarge during pregnancy and may form comparatively large swellings. Such an axillary tail has inadequate drainage to the main duct system, and so may become tense and painful when lactation begins.

THE ABDOMINAL WALL

The muscles of the abdominal wall become stretched to accommodate the enlarging uterus, and although, in perfect health, subsequent recovery is complete, in not a few multiparae some loss of tone of these muscles persists. In late pregnancy the umbilicus may be flattened out, or even protrude.

Stretching of the abdominal skin may cause the formation of *striae gravidarum*. These are due to rupture of the elastic fibres of the skin, and they appear as curved lines, roughly concentric with the umbilicus; they may also be seen on the loins or thighs, and sometimes on the breasts. At first the striae are pink or red, but after delivery they become silvery-white. Not all women develop striae; perhaps one-third do not. It is unusual to see striae in other conditions in which the abdominal wall is stretched, such as large ovarian cysts or ascites. The fact that they are frequently seen in Cushing's syndrome, when there is high level of glucocorticoids in the blood, as there is in pregnancy, has led to the suggestion that the striae are partly affected by the action of these hormones.

Pigmentation of the line from the pubes to the umbilicus (the *linea nigra*) may be seen, especially in dark-skinned women, and may persist in part after the pregnancy.

THE PELVIC JOINTS

The pelvic hyperaemia causes some softening and slight relaxation of the ligaments of the sacroiliac joints, and of the ligaments and fibrocartilage of the symphysis pubis, the mobility of which is slightly increased in pregnancy. In some animals a specific hormone, relaxin, derived from the ovary or in some species from the placenta, causes relaxation of the pelvic joints, but it is uncertain whether such a hormone plays any part in human physiology.

The changes so far described have mostly been those directly related to the genital tract and the breasts. Numerous other changes occur in the rest of the body, and nearly every system is involved in some change. Yet pregnancy is a physiological and not a pathological process, and many women both feel and appear to be in better general health during pregnancy than at other times. Given good previous nutrition and sound emotional adjustment, the physiological changes of pregnancy are not to be regarded as a strain on the mother's health, but merely as a temporary adaptation to a normal function.

MATERNAL METABOLISM DURING PREGNANCY

Weight gain

The body weight increases during pregnancy. The total gain varies between 7 and 17 kg in normal cases, with an average of 12.5 kg. After the 12th week the average normal gain is about 0.35–0.45 kg per week (*see* Figs 1.28 and 1.29).

A fetus weighing 3.4 kg, a placenta of 0.65 kg, amniotic fluid weighing 0.8 kg, a uterus of 1 kg, and an increase in the weight of the breasts of 0.8 kg would account for a total gain of 6.65 kg. The average additional gain of 6 kg represents the weight gained by the rest of the maternal tissues, due partly to fluid retention (1.5 kg), and partly due to increase in the body fat and protein.

During pregnancy, apart from the water contained in the fetus, placenta and amniotic fluid,

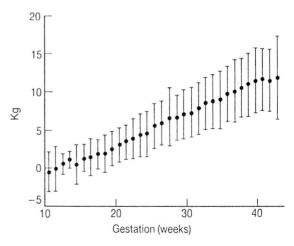

Fig. 1.28 Maternal weight gain in pregnancy (Mean ± 1SD)

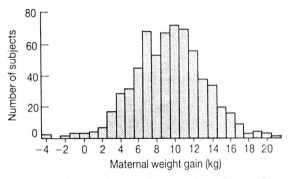

Fig. 1.29 The distribution of maternal weight gain from booking 36 weeks' gestation

there is a considerable retention of water in the maternal tissues, chiefly in the extracellular compartment, including the blood plasma. The following figures illustrate the typical increase in body water found in maternal tissues at term:

increase in intracellular water	550 ml
increase in extracellular water	
intravascular (plasma)	900 ml
extravascular	1850 ml
Total	3300 ml

Corresponding quantities of sodium are retained and it is believed that this retention of salt and water is due to the high concentrations of sex steroids during pregnancy, although the part played by mineralocorticoids has not yet been fully determined.

Metabolic changes

The basal metabolic rate during pregnancy is increased by between 10 and 25 per cent, but if allowance is made for the metabolism of the fetus and its supporting tissues, the basal metabolic rate of the maternal tissues is probably unaltered. Free thyroxine levels in the blood are normal or slightly reduced.

During pregnancy the total need for calories is increased by 80 000 kcal, to maintain the fetus and additional maternal tissues. Apart from fluid retention in cases of hypertension, abnormal weight gain during pregnancy and the puerperium is often due to simple overeating, chiefly by an excessive intake of carbohydrate.

During pregnancy the renal threshold for the excretion of sugar from the blood is often lowered, so that glucose may appear in the urine although the blood sugar level is normal. This condition is of

no importance but it must be distinguished from true or gestational diabetes, by a glucose tolerance test if necessary. Glycosuria is probably the result of increased glomerular filtration, which allows so much glucose to enter the tubules that they are unable to absorb it all. Lactose may appear in the urine during lactation, but is not found during pregnancy.

A high protein intake, amounting to a total extra requirement of 900 g, is needed to supply the growing fetus, placenta, uterus and breasts. In spite of an increased excretion of amino acids during pregnancy, sufficient nitrogen is retained for the maternal and fetal needs.

There are also changes in lipid metabolism during pregnancy; the total lipid requirement during pregnancy being 3.5 kg. Plasma levels of triglycerides, cholesterol and free fatty acids rise, and there is a greater tendency to ketosis.

The maternal diet must not only supply the protein, carbohydrate and fat required, but also the essential minerals and vitamins. An ordinary diet provides adequate amounts of most of these substances, but in the case of iron and calcium there is some risk of a deficit, particularly during the last trimester when the fetal uptake is greatest. The fetal body at term contains about 30 g of calcium, but even with this large demand, a mother taking a first class diet will maintain her calcium reserve during pregnancy. However, with a less favourable diet there may be a deficiency in late pregnancy which leads to decalcification of the maternal bones and dentine. Recalcification quickly occurs after lactation has been completed if the diet contains adequate calcium.

In the case of iron only about 1.2 mg is usually assimilated daily, an amount insufficient to meet the fetal needs, especially in late pregnancy. In addition the mother forms additional red cells during pregnancy. Her blood volume is increased, and although the amount of haemoglobin per millilitre may be reduced, the increase in blood volume outweighs this, so that the total amount of haemoglobin in the maternal body is increased. Even in health, and with a normal diet, the maternal iron reserves in the liver, spleen and marrow may therefore be reduced during pregnancy. Not all the iron so used is lost; apart from the iron in the fetal tissues and blood, and the maternal blood lost at delivery, the iron built up into additional maternal red cells during pregnancy is later returned to her reserves.

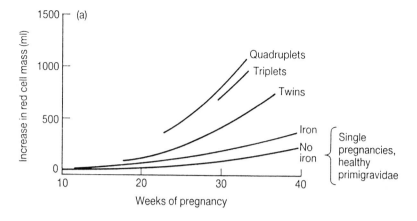

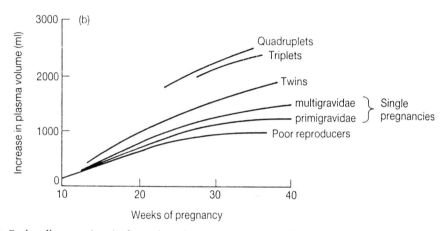

Fig. 1.30 (a) Red cell mass in single and multiple pregnancies, (b) plasma volume in single and multiple pregnancies

CHANGES IN THE BLOOD DURING PREGNANCY

The total blood volume is increased during pregnancy by about 30 per cent (*see* Fig. 1.30). The uterine wall and the maternal blood spaces in the placenta contain a large volume of blood, perhaps 800 ml. The increase in blood volume is an adaptation to supply the needs of the new vascular bed. Although the total number and volume of red cells increase by about 20 per cent, the plasma volume increases by about 50 per cent, with the result that the blood becomes more dilute and the red cell count and the haemoglobin concentration fall. A red cell count of 4 million per mm^3 and a haemoglobin concentration of 11 g per 100 ml are usually accepted as normal during pregnancy. Although such levels are commonly observed in completely healthy women in pregnancy, if supplemental iron is given in a form that is well absorbed then in many of these women the red cell count and haemoglobin concentration remain higher. The question which is still by no means decided is whether this gives any additional benefit. However, many women have poor iron reserves and inadequate diets, and for them iron supplements are essential.

A very high leucocytosis is often observed during labour and the early puerperium, but not during normal pregnancy, when the count does not exceed 11 000 per mm^3 (*see* Fig. 1.31). The platelet count is normal. The erythrocyte sedimentation rate is much increased during normal pregnancy. Readings of up to 100 mm per hour are not unusual without any detectable abnormality.

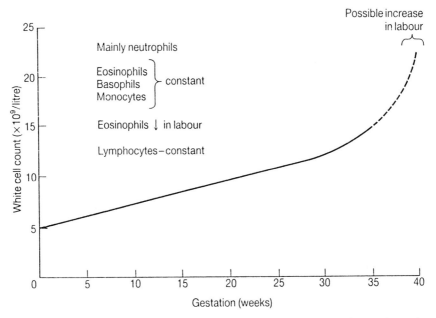

Fig. 1.31 There is an increased white cell count in pregnancy mostly made up of polymorphonuclear leucocytes

CHANGES IN THE CIRCULATION DURING PREGNANCY

The cardiac output rises from 4.5 l/min to 6 l/min during the first 10 weeks of pregnancy, remaining at the higher level until after delivery (*see* Fig. 1.32). The systolic blood pressure is unaltered during normal pregnancy but the diastolic pressure is reduced in the first and second trimester, return-ing to non-pregnant levels by term. The pulse rate rises by between 8 and 16 beats per minute so that the greatly increased cardiac output must be ach-ieved by the expulsion of 70–80 ml more blood from the heart at each beat. This extra cardiac work is well within the reserve of the normal heart. The heart is displaced upwards in late pregnancy and the apex is rotated outwards, so that the apex beat is displaced outwards and the electrical axis is

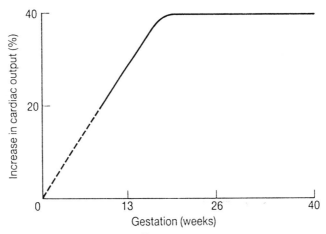

Fig. 1.32 Cardiac output pregnancy

altered. There may be slight left axis deviation in an electrocardiograph, but there is little evidence of muscular hypertrophy.

The large blood flow through the uterus may be regarded as an arteriovenous shunt across the main circulation, and the increase in cardiac output and in the blood volume may be related to this. In addition, there is increased renal blood flow and some dilatation of peripheral vessels. The hands and feet are often noticeably warm during pregnancy, and the skin capillaries are dilated.

The enlarged uterus interferes with the venous return from the legs, so that there is stasis in the large veins and slight oedema of the ankles may occur, even in normal pregnancy. Because of this interference with venous return, some women complain of faintness when lying on their backs (supine hypotension); haemorrhoids or varicose veins may appear for the first time or become worse during pregnancy.

Changes in the respiratory system during pregnancy

Pulmonary ventilation is increased by about 40 per cent, as a result of increased tidal volume. Oxygen requirements only increase by 20 per cent and the overbreathing leads to a fall in $P\text{CO}_2$. The low $P\text{CO}_2$ gives rise to a sensation of dyspnoea, which may be accentuated by elevation of the diaphragm. When the fetal head engages in the pelvis in late pregnancy this breathlessness diminishes (*see* Fig. 1.33).

Changes in the alimentary tract during pregnancy

The most striking change in the digestive function in pregnancy is nausea or morning sickness, which occurs to a greater or lesser degree in about a third of pregnancies. It begins at the 6th week and stops spontaneously before the 14th. Although it is generally limited to the early morning it may occur at other times of the day. Usually the symptoms are slight; excessive or prolonged vomiting is certainly pathological.

Appetite and thirst increase during pregnancy, but minor digestive upsets are common, probably due to the relaxant effect of progesterone on smooth muscle. Sometimes gastric or intestinal distension occurs, and especially in early pregnancy this causes a feeling of abdominal enlargement. Heartburn is a common complaint, and is caused by relaxation of the cardiac sphincter of the stomach. In a few of the more severe cases this symptom is caused by a hiatus hernia. The emptying time of the stomach is prolonged during pregnancy, and even more so during labour. This is of considerable importance in relation to the risk of vomiting during anaesthesia. The gastric acidity is often reduced, and peptic ulcers almost invariably become quiescent during pregnancy.

Constipation is not uncommon, and this, together with the pelvic hyperaemia and pressure of the enlarged uterus, may lead to the formation or increase in size of haemorrhoids.

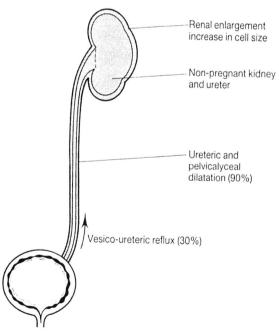

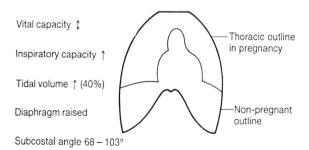

Vital capacity ↕

Inspiratory capacity ↑

Tidal volume ↑ (40%)

Diaphragm raised

Thoracic outline in pregnancy

Non-pregnant outline

Subcostal angle 68 – 103°

Fig. 1.33 Change in respiratory function in pregnancy

Fig. 1.34 Changes in urinary tract in pregnancy

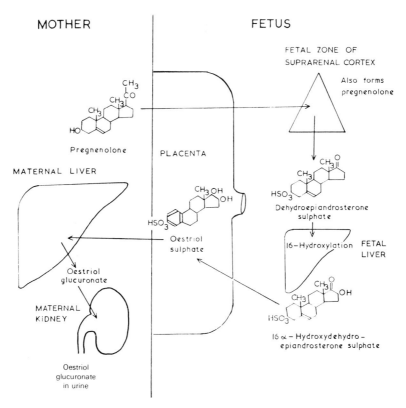

Fig. 1.35 Formation of oestriol in pregnancy

Changes in the urinary tract during pregnancy

There is progressive increase in the glomerular filtration rate and renal plasma flow, starting in early pregnancy and reaching 50 per cent at term. A progressive fall in plasma creatinine levels from 73 μmol/l to 47 μmol/l at term occurs, and there is a similar fall in plasma urea from 3.5 to 3.1 mmol/l. Values considered normal in a non-pregnant woman may therefore indicate impaired renal function during pregnancy. The cumulative water retention in pregnancy is 7.5 l, together with 950 mmol sodium. This is due largely to increased aldosterone, renin and angiotensin I and II.

Frequency of micturition occurs during the first 12 weeks, when the enlarging uterus is still in the pelvic cavity and presses on the bladder. It may also occur in the last month of pregnancy when the presenting part of the fetus is engaging in the pelvis.

From about the 16th week of pregnancy onwards there is considerable and progressive dila-tation of the renal pelvis and of the ureters down as far as the level of the pelvic brim. Direct measurement of the intraureteric pressure shows that it is lowered rather than raised, and the dilatation is chiefly caused by loss of muscle tone, although pressure from the uterus, displaced to the right by the descending colon, may explain the greater degree of change which is usually seen on the right side. The reduction in tone, which is thought to be caused by progesterone, is important in relation to pyelonephritis during pregnancy. The dilatation disappears during the puerperium, so long as there is no infection of the tract (*see* Fig. 1.34).

Changes in immune reactivity

The fetus contains antigens derived from the father and is therefore an allograft which is foreign to the mother. The reasons why the mother's immune system does not reject the fetus are not fully understood. One suggestion is that, as there is no expression of either classical class I or class II

MHC antigens by human chorionic villous trophoblast throughout gestation, cell mediated rejection is not activated. The immunological changes in pregnancy seem to be relatively minor. They include a 30 per cent increase in neutrophils, a decrease in helper T-cells, a slight reduction in IgG, an increase in IgD and a slight depression in cell-mediated immunity. These and other unknown factors may limit, rather than abolish, the immunological response and a restrained rejection process may be necessary to prevent excessive growth of the placenta. Many diseases which are thought to have an autoimmune basis tend to remit in pregnancy.

FURTHER READING

Austin C.R. and Stuart R.N. (1982) *Embryonic and Fetal Development*. Cambridge University Press, Cambridge.

Chamberlain G. and de Swick M. (1992) *Basic Science in Obstetrics and Gynaecology*. Churchill Livingstone, London.

Edward R. (1980) *Conception in the Human Fetus*. Academic Press, London.

Hylten F. and Chamberlain G. (1990) *Clinical Physiology in Obstetrics*. Blackwell Scientific Publications, Oxford.

Moore K. (1988) *The Developing Human*. W.B. Saunders Co., Philadelphia.

2

NORMAL PREGNANCY

THE DURATION OF PREGNANCY

In cases in which pregnancy has followed a single coitus the average duration of pregnancy from the date of intercourse is 266 days. If the calculation is made from the first day of the last menstrual period (LMP) the average duration is 280 days because ovulation most frequently occurs on the 14th day of a 28-day menstrual cycle. However, there is considerable variation in the duration of normal pregnancy, even in cases in which the menstrual cycles were previously regular and of normal length, and also in cases in which the date of a single coitus is known.

SYMPTOMS OF PREGNANCY

AMENORRHOEA

Amenorrhoea is the earliest symptom of pregnancy. If a healthy woman, whose menstrual periods were previously regular, has a sudden cessation of her periods the presumption must always be that she is pregnant unless some other cause of the amenorrhoea can be found. Amenorrhoea does not have the same significance in the case of a woman whose periods were previously irregular, nor may it have any significance in a woman of menopausal age. Pregnancy has been known to occur in a young girl before a menstrual period has been observed, and it can arise during a period of amenorrhoea, for example during lactation or following discontinuation of oral contraception.

Conversely, difficulty may arise if there is bleeding during early pregnancy. Such bleeding may come from the cavity of the uterus before the decidua capsularis fuses with the decidua vera, but it is not to be regarded as menstrual bleeding. Women with slight bleeding in the early weeks of pregnancy may prove to be completely normal, but such bleeding may be the first indication that the embryo is dead or dying.

BREAST SYMPTOMS

In the early weeks of pregnancy some tenderness and fullness of the breasts may be noticed.

FREQUENCY OF MICTURITION

During the first 12 weeks, when the uterus is still a pelvic organ, there is often some frequency of micturition, because the enlarging uterus presses on the bladder lightly, particularly when the woman is standing in the daytime.

ABDOMINAL ENLARGEMENT

Many women notice some abdominal fullness in early pregnancy at a time when the uterus is not much enlarged. This can only be the result of slight intestinal distension or a progesteronal effect on the muscles of the anterior abdominal wall. Later on the uterine enlargement becomes evident, and sometimes it is this that first brings the woman to the doctor, especially in a case in which the menstrual periods were previously irregular. On the other hand, a woman sometimes thinks that she is pregnant because of abdominal swelling from some other cause, such as fat or an ovarian cyst.

FETAL MOVEMENTS

A primigravida usually first feels fetal movements, called quickening, between the 16th and 20th weeks of pregnancy, but multiparae may recognize the movements about 2 or 3 weeks earlier. At first the movements are slight, and may be confused with wind. This very subjective symptom is not of much value in the diagnosis of pregnancy.

SIGNS OF PREGNANCY

It will be seen that it is not possible to make a certain diagnosis of pregnancy from any of the symptoms given above, although a combination of them may be highly suggestive. The clinical diagnosis of pregnancy, especially in the first third, depends on a combination of symptoms and signs, and in every case an examination to seek confirmatory physical signs is required. Only when the pregnancy has advanced far enough for the parts of the fetus to be recognized clearly, for fetal movements to be palpable, or the fetal heart sounds to be heard, can the physical signs be said to be absolute.

The ultrasound demonstration of the embryonic heart activity, which is now possible using a transvaginal probe, within 7 weeks of the LMP, or 4 weeks after implantation is an absolute sign of pregnancy. Subsequently, fetal structures can be easily visualized with ultrasound either by abdominal or vaginal routes.

SIGNS DUE TO CHANGES IN THE UTERUS

Enlargement of the body of the uterus

A slight enlargement of the body of the uterus is the earliest alteration which can be detected clinically, but it is difficult to be certain of this if the woman has had a previous pregnancy. On bimanual examination the body is felt to be globular, and as progressive enlargement occurs the diagnosis becomes more evident.

Softening of the uterus and cervix

Softening of the uterus due to its increased vascularity is a useful sign to the experienced obstetrician. Although there are many other causes for uterine enlargement they do not, as a rule, cause softening. Softening and blue discoloration of the cervix soon follow the softening of the uterus, and are usually complete by the 16th week. When they are marked they are reliable signs of pregnancy, but in a few cases the cervix remains firm throughout pregnancy.

Progressive enlargement of the uterus

By the 12th week the fundus of the uterus is usually palpable in the abdomen just above the symphysis pubis and progressive enlargement of the uterus follows (*see* Fig. 2.1). The fundus reaches the level of the umbilicus at about the 22nd week, and is just below the xiphisternum at the 36th week.

A tape measure can be used to measure uterine size. Once the fundus is distinguishable the fundal–symphysis length can be measured. A series of cross-sectional readings lead to Fig. 2.2.

If the presenting part of the fetus then sinks into the pelvis the fundus descends slightly, so that at term it is again at the level it occupied at the 34th week. The level of the fundus may be higher than expected from the duration of the pregnancy in cases of multiple pregnancy, polyhydramnios or fibroids, and at a lower level than expected with a transverse lie, an abnormal, growth-retarded or dead fetus. If the fundus is at an unexpected level an error in the dates is of course to be considered.

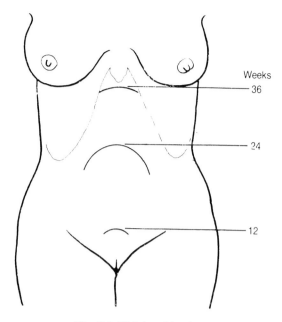

Fig. 2.1 Height of fundus

be felt even when the uterus is still in the pelvis. When the uterus rises up into the abdomen the contractions are more easily felt, and are reliable evidence that any swelling under examination is in fact a pregnant uterus.

SIGNS DUE TO THE PRESENCE OF THE FETUS

Fetal heart sounds

On auscultation of the abdomen with a Sonicaid the fetal heart sounds may be heard after the 12th week. The fetal heart rate varies between 120 and 160 beats per minute, and is therefore at roughly double the rate of the maternal pulse.

Palpation of fetal parts

Abdominal palpation of fetal parts is usually possible from the 24th week onwards, and at a later stage the definite recognition of the head, back and limbs of the fetus is an absolute sign of pregnancy.

Fetal movements

During palpation fetal movements may be felt, and

Painless contractions

The pregnant uterus varies in consistency on palpation because it has intermittent painless contractions. If the woman is easy to examine these may

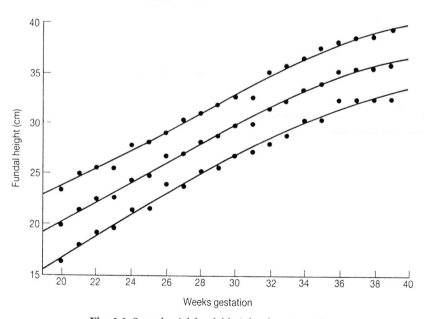

Fig. 2.2 Symphysial fundal height chart (±1 SD).

this is also an absolute sign. These movements can often be seen as well.

Funic souffle

The funic souffle is occasionally heard if the Sonicaid happens to lie directly over the umbilical cord. It is a soft, blowing murmur synchronous with the fetal heart sounds.

SIGNS DUE TO CHANGES IN THE BREASTS AND THE SKIN

These have already been described. The primary areola persists after the first pregnancy and this sign is therefore useless in the diagnosis of any subsequent pregnancy. Similarly, secretion can often be expressed from the breasts of any woman who has once been pregnant, even when she is not again pregnant.

HUMAN CHORIONIC GONADOTROPHIN LEVELS

Pregnancy tests depend on the detection of large quantities of the gonadotrophic hormone produced by the trophoblast. This is found in the maternal circulation and is excreted in maternal urine after implantation. Human chorionic gonadotrophin (hCG) is a glycoprotein. The alpha subunit is also found in other hormones (e.g. LH, FSH) but the beta subunit is specific for hCG. The test may be done on maternal blood (early) or maternal urine (after 4 weeks' gestation) (*see* Fig. 2.3).

Immunochemical tests (day 35 after LNMP)

If the woman is pregnant her urine will contain hCG soon after the time of implantation. When anti-hCG is added to the urine it combines with the hCG and is neutralized. The mixture is now tested for anti-hCG with a suitable indicator and if none is found the test is positive for pregnancy.

If the woman is not pregnant, the urine does not contain hCG and the anti-hCG remains unfixed and will react with the indicator. A change in the indicator shows that the woman is *not* pregnant.

Anti-hCG is obtained from rabbits or sheep which have been immunized against hCG. Various indicators have been used, including hCG-coated

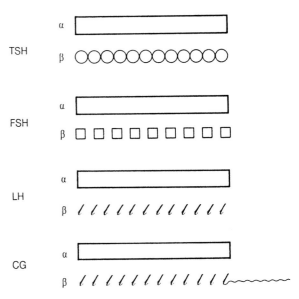

Fig. 2.3 Structural homology of human chronic gonadotrophin (hCG) and the three pituitary hormones, LH, FSH, and TSH. The α-chain of hCG is biochemically and immunologically similar to the α-chain of the three pituitary glycoproteins, hCG and LH are structurally similar to one another except for a structurally distinct tailpiece at the terminus of the chain of hCG

latex particles or hCG-coated red blood cells. If the woman is not pregnant the latex particles are precipitated or the red cells are agglutinated (*see* Fig. 2.4).

The urinary output of hCG, which parallels the maternal blood levels during pregnancy, rises rapidly to reach a peak at about 12 weeks and then falls to a much lower level (*see* Fig. 1.23). Tests for hCG may even give negative results after the 16th week, but by then the diagnosis is easily made clinically or by ultrasound.

TESTS PERFORMED ON MATERNAL SERUM OR PLASMA

Immunoassays, such as radioimmunoassays and enzyme-linked immunoassays can be used to detect hCG or its beta subunit at implantation, i.e. 7 to 10 days after conception, even before the period has been missed. These tests are not used for clinical purposes. Assays for the detection of the beta subunit are much less likely to give false-positive results due to cross-reaction with LH.

Radioimmunoassays are now being replaced by

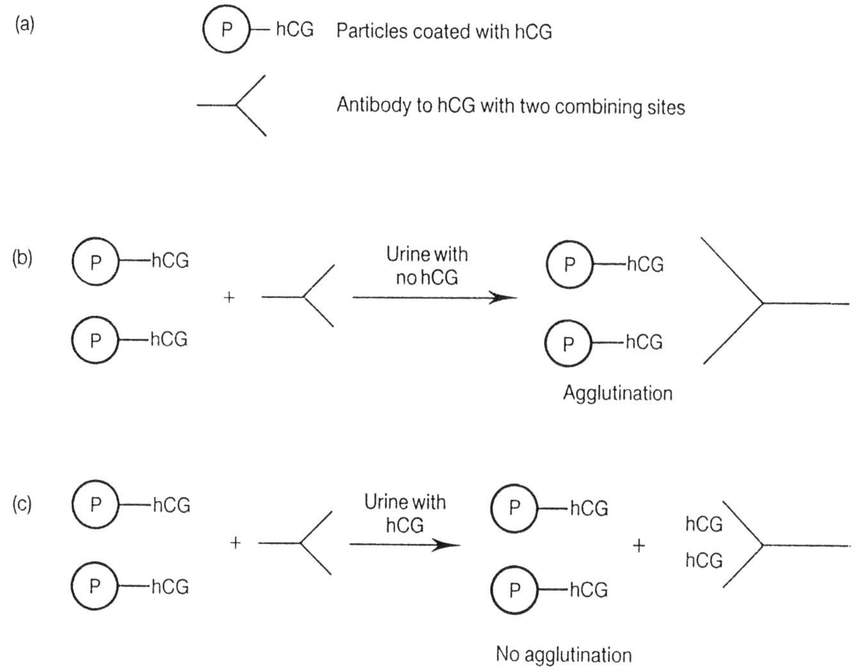

Fig. 2.4 The principle of agglutination inhibition assay: (a) the test reagents are particles (red cells or latex) coated with hCG, and an antibody to hCG; (b) when these are mixed with urine containing no hCG, the antibody binds adjacent particles thus causing agglutination; (c) when the reagents are mixed with urine containing hCG, the latter combines with the antibody and thus prevents it from binding and agglutinating the particles. The test can be carried out on slides or in tubes; in the latter a positive result is indicated by deposition of a ring at the bottom of the tube

enzyme-linked immunosorbent assays (ELISA) which involves a double reaction, with the hCG sandwiched between the solid-phase antibody and the enzyme-labelled antibody. When urine containing hCG is run into a tube which has been coated with antibody, binding occurs, and the enzyme-linked antibody becomes bound to the already captured hCG. The addition of an enzyme substrate, which is broken down by the bound enzyme, results in a blue colour – a positive end-point if pregnancy is present. In general ELISA tests are much more sensitive and more specific than the agglutination tests, they can detect pregnancy at lower levels of hCG and so can predict gestation at 8 to 10 days after fertilization (i.e. before the missed period) (*see* Fig. 2.5).

High concentrations of hCG are seen in multiple pregnancy, and in hydatidiform mole and choriocarcinoma when quantitative tests are routinely used.

Very sensitive hCG tests such as these may give positive results for many weeks after a pregnancy has been delivered, failed or terminated, as small fragments of trophoblastic tissue continue to metabolize hCG in minute, but still detectable, amounts.

False-positive results may occur rarely due to cross-reaction with LH when elevated levels are seen in the polycystic ovary syndrome, or together with elevated FSH in the climacteric or post-menopausally.

ULTRASOUND DIAGNOSIS OF PREGNANCY

With a real-time ultrasound scan the gestation sac can often be diagnosed as early as 5 weeks from the first day of the last menstrual period, and a week later echoes from the embryo within the sac can be obtained and cardiac pulsation may be recognized, particularly if a transvaginal transducer is used. Simple and easily portable machines (e.g. Sonicaid,

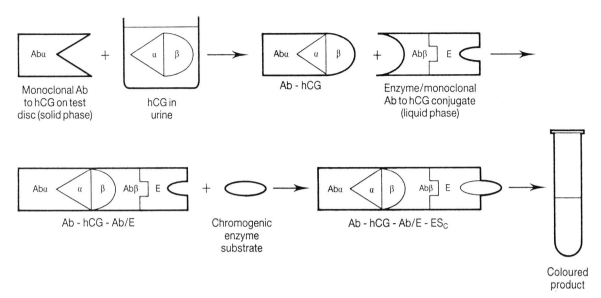

Fig. 2.5 Principle of enzyme-linked immunoassay (ELISA). In hCG ELISA, two antibodies are allowed to interact separately with the α and β subunits of the hCG molecule to form a sandwich around the entire hCG molecule. In ELISA, the first antibody, bound to the wall of the test tube, binds with α-subunit hCG. A second antibody binds with the β subunit. Patient serum containing hCG is then added to the polystyrene tube with the immobilized anti-hCG and allowed to incubate so that binding takes palce. The anti-β hCG, bound to an enzyme label, is added, and an additional incubation takes place. The sandwich, consisting of the polystyrene-bound anti-α-hCG, the hCG molecule, and the enzyme-labelled anti-β-hCG is formed. After each incubation, the tube is washed with removal of unbound enzyme. The labelled antibody forms a blue colour quantitatively in an enzyme-catalyzed colorimetric reaction. These results are determined visually by comparing the intensity of the blue color developed in a specimen to that of a positive reference hCG, human chorionic gonadotrophin α- and β-subunits; Abα, monoclonal antibody to the α-subunit of hCG; Abβ, monoclonal antibody to the β subunit of hCG. (Modified from Fletcher, with permission)

Doptone) are available which employ a different principle. Ultrasonic waves are emitted by the head of the machine, which is placed against the skin of the abdomen. If the waves strike a moving surface, such as flowing blood, the reflected waves have an altered wave length, according to the Doppler principle. The alteration is detected by the apparatus, and in the case of the fetal heart the rhythmic changes in wave length are converted into audible signals, which are surprisingly similar to the fetal heart sounds. Blood flowing in any large maternal vessel will give the same effect, but the characteristic rate of the fetal heart makes the source of the signals clear. With these machines the fetal heart beat can be recognized at the 10th week.

DIFFERENTIAL DIAGNOSIS OF PREGNANCY

The pregnant uterus has to be distinguished from other abdominopelvic swellings, of which the commonest are fibroids, ovarian cysts and a distended bladder. Pregnancy may of course be proceeding at the same time as any of these.

Uterine fibroids

When these tumours are deeply placed in the uterus they may enlarge it symmetrically, and they are sometimes soft enough to stimulate the consistency of the pregnant uterus. However they do not cause amenorrhoea and pregnancy tests are negative.

Ovarian cysts

In many cases, on bimanual palpation, the unenlarged uterus can be felt separately from the ovarian cyst. Ovarian neoplasms will not cause amenorrhoea except in the unlikely event of their being bilateral and totally destroying both ovaries. There is no clinical evidence of the presence of a fetus, and the pregnancy test is negative. Ultrasound examination will clearly show the cyst wall, an empty uterus and absence of fetal parts.

Distended bladder

The bladder may become distended and then be mistaken for the uterus if there is retention of urine from incarceration of a retroverted gravid uterus. The direction of the cervix and the passage of a catheter disclose the diagnosis.

Pregnancy associated with fibroids

It can sometimes be difficult to diagnose pregnancy in a uterus with multiple fibroids. Pregnancy tests and ultrasound examination will be of value in early pregnancy. Later on small tumours can easily be mistaken for fetal parts. A fibroid in the uterine wall is immobile, unlike a fetal limb which alters its position from time to time. When the uterus contracts fetal parts become more difficult to feel, whereas a fibroid may become more evident.

Pregnancy associated with an ovarian cyst

Two swellings are usually evident, the pregnant uterus and the cyst, but in some cases they are in such close juxtaposition that it is not easy to distinguish them. The combined swelling will then be larger than expected for the duration of pregnancy. Ultrasound scan will reveal the fetus within the uterus with the cyst as a separate cavity.

PSEUDOCYESIS

Pseudocyesis is a psychological disorder in which the woman has a false but fixed idea that she is pregnant. The term does not include the uncommon instances of wilful and conscious deception; the woman honestly believes that she is pregnant. Pregnancy fantasies may occur with other delusions in psychoses, but most women with pseudocyesis do not have serious mental illness.

It is frequently, but not always, seen near or after the menopause, and not invariably in women without children. There may be amenorrhoea, and the woman may declare that she has morning sickness and breast enlargement, and that she can feel fetal movements. The abdomen may appear distended, either by air collected in the stomach by aerophagy, by intestinal distension, by persistent contraction of the diaphragm with exaggerated lumbar lordosis or sometimes just by fat. The shape of the swelling is not that of the pregnant uterus, fetal parts cannot be felt and the fetal heart cannot be heard. A pregnancy test or ultrasound scan will be required but the difficulty is to convince the woman that she is not pregnant.

THE EXPECTED DATE OF DELIVERY

Since there is no exact means of knowing the time at which conception occurred, it is usual and convenient to calculate the date on which delivery is to be expected from the first day of the last menstrual period, with the assumption that ovulation took place about 14 days after that. A simple practical method is to count forward nine calendar months from the first day of the last period, and to add 7 days. This gives an average of 280 days, with a little variation as the lengths of the calendar months are not uniform. If the previous menstrual cycles were irregular or prolonged no reliance can be put on this method. Even if the previous cycles were regular, in as many as 40 per cent of cases labour begins more than 7 days before or after the calculated date (*see* Fig. 2.6).

An attempt has often to be made to estimate the probable date of delivery or the maturity of the gestation when the date of the last menstrual period is not known: if it has been forgotten, when conception occurred during a phase of amenorrhoea, (for example during lactation) or when conception took place soon after discontinuation of oral contraception. When decisions have to be made towards the end of pregnancy about the optimum moment for delivery an accurate ultrasound record made in early pregnancy may be found to be extremely useful; these early observations are more reliable than those made later.

Clinical observations

The date at which the woman first felt fetal movements gives a very rough indication of matu-

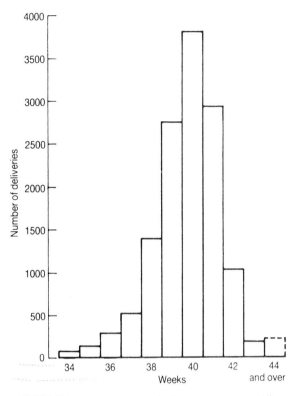

Fig. 2.6 Numbers of patients delivered each week in a sample of 13 634 single pregnancies (with certain dates of last menstrual period)

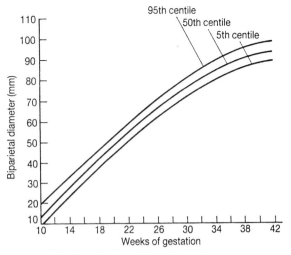

Fig. 2.7 The distribution of biparietal diameter measurements (5th, 50th and 98th centile) throughout pregnancy

the uterine fundus give only approximate estimates, with an error of perhaps plus or minus 4 weeks, and measurements of the abdominal girth are also an imprecise guide.

ULTRASOUND SCAN

Ultrasound measurements of the fetus provide a quick, safe and accurate method of dating a pregnancy, especially in the first 20 weeks. The crown–rump length can be measured in the first trimester, but the biparietal diameter, the abdominal circumference and the femur length are the most useful observations to record. The measurements are compared with standard curves showing normal increments of growth, and there is a straight-line relationship between the size of the fetal head and the duration of pregnancy before the 24th week (*see* Fig. 2.7).

rity. Primigravidae usually feel fetal movements between 18 and 20 weeks, and multiparae may recognize the movements a little sooner.

Bimanual examination by an experienced clinician performed between 6 and 12 weeks gives a fairly accurate assessment of uterine size. Subsequent repeated measurements of the height of

ANTENATAL CARE

Antenatal care and investigation is an important part of preventive medicine. Its object is to maintain the woman in health of body and mind, to anticipate difficulties and complications of labour, to ensure the birth of a healthy infant, and to help the mother rear the child.

If conducted properly it does more than detect and treat abnormalities. It should establish contact and promote understanding between the woman and those who will look after her in pregnancy and labour. Some hospital clinics have justly been criticized as impersonal and overcrowded. Long waiting times, uncongenial decor and furnishing, lack of privacy, poor facilities for children and inadequate time for the woman to talk to the doctor or midwife are faults which can often be rectified by better organization.

PRE-PREGNANCY COUNSELLING

It may be said that antenatal care should start before pregnancy. In a perfect world every woman might endeavour to secure that her health was in its best possible state before embarking on pregnancy. This idea should be more widely spread by all agencies responsible for health education.

The main structure of the organs of the embryo is laid down in the first 8 weeks of pregnancy, and it is during this time that major congenital abnormalities arise. Much effort is devoted to the care of the pregnant woman and her fetus after the 12th week, but there is little medical support for the early embryo. Dietary, smoking and drinking habits may well call for advice from the doctor or midwife, and women of poor physique and those with social or medical problems may need help.

Facilities will hardly permit every woman to seek medical advice before every possible conception, either from her general practitioner or from a hospital clinic, but pre-pregnancy counselling is of especial value for women with problems and anxieties. Such counselling may be the first medical examination that a young woman has had since leaving school, and health problems may be found which would otherwise only have been discovered during pregnancy.

A pre-pregnancy counselling clinic might give advice to women with any of the following conditions:

- women who have had an unsuccessful pregnancy from abortion, perinatal death, or preterm labour, obstetric complications, intercurrent disease or fetal genetic abnormalities
- women with some continuing disease who are anxious to know whether pregnancy would exacerbate this, or whether the child might be harmed. The list of possible conditions is unlimited, but includes heart disease, hypertension and renal disease, diabetes, haemoglobinopathies, epilepsy, thromboembolism, mental disease and treated malignant disease. Some women may be taking drugs for the control of their disease, and possible effects on

the fetus might call for variation in the drugs or their dosage
- women with a family history of disease. Such conditions as diabetes, epilepsy, haemophilia, sickle-cell disease, muscular dystrophy and Huntington's chorea, although some of them are rare, will call for advice about the possibility of inheritance. Genetic problems may require special investigation, including chromosomal studies by experts.

ANTENATAL CARE

The number of times a woman needs to be seen during her pregnancy will vary. The first visit, when she comes to make arrangements, should be as early in pregnancy as possible. In a first pregnancy, she should be seen every 4 weeks until the 28th week, fortnightly until the 36th week, and then weekly until the onset of labour. If complications arise more frequent visits will be necessary. A woman who has had at least one normal pregnancy may not need so many visits.

There are many different models for antenatal care. Shared care with the general practitioner may save the woman travelling, promote good doctor–woman relationships, and reduce over-crowding in hospital clinics. Women with normal pregnancies may have all or most of their antenatal care with their midwife; many midwives now operate in small teams, one of whom will actually deliver the woman. Other community based systems of care exist in which only one or two visits are made to the hospital clinic for scans or consultations. Each antenatal visit should have a clear purpose, and should follow a planned protocol.

BOOKING VISIT

At the first visit the following procedure is followed:

General medical history

It is important to know whether the woman has had any significant illness, including cardiac disease, renal disease, diabetes, rubella or a blood transfusion. Tactful enquiry should be made about possible exposure to HIV infection, whether or not she has taken drugs, her level of alcohol intake and cigarette consumption, as well as a brief survey of her general social and environmental circumstances. Previous surgical treatment, particularly gynaecological operations, may be relevant.

Family history

Any disease with a hereditary tendency, including diabetes and hypertension, is recorded. A family history of twins may be given; their zygosity should be known to assess the full significance.

Past obstetric history

If the woman has been pregnant before, she is questioned about previous pregnancies and labours. For example, a history of repeated abortion or premature labour might suggest cervical incompetence, and intrauterine fetal death might suggest the possibility of hypertension, diabetes or rhesus incompatibility. A history of raised blood pressure might suggest pregnancy induced hypertension, with the possibility of recurrence.

The history of previous labours is a guide to what may be expected in the coming labour. For example, if the woman has had a long labour ending in instrumental delivery, resulting perhaps in the birth of a dead or injured child, it is possible that she has pelvic contraction. A history of postpartum haemorrhage would be a warning of possible recurrence. It is essential to know the birth weights of any previous children. The cause of any stillbirth or neonatal death should be ascertained whenever possible, sometimes by writing to the doctor who was then in charge of the patient.

It is important to know whether the children were breastfed, and of any feeding difficulties which may possibly be overcome by special advice.

History of the present pregnancy

The date of the first day of the last menstrual period must be carefully recorded, with a note of the woman's normal cycle and of any irregularities. If she was using oral contraception this may be relevant, as may a history of subfertility and the form of treatment used.

Any episode of vaginal bleeding or pain in early pregnancy should be noted, and the occurrence of any incidental illness or drug treatment recorded.

The doctor will then examine the woman.

General examination

The woman is weighed, and her height and development are noted, including any abnormal gait or deformity. The breasts are examined to exclude a tumour, and to check the nipples for breastfeeding. The blood pressure is recorded, and the heart and lungs examined. The teeth should be inspected for the presence of gum infection and caries, and dental care encouraged. The legs should be examined for the presence of varicose veins, oedema and any other abnormality.

Abdominal examination

An examination of the abdomen is made, including auscultation of the fetal heart sounds. If the pregnancy has reached 24 weeks a Pinnard stethoscope will suffice, while before this fetal blood flow through the fetal heart can be demonstrated with a hand held Doppler monitor from as early as 12 weeks.

Vaginal examination

In some clinics a vaginal examination is made at the woman's first visit. The position of the uterus (anteverted or retroverted) and its size in relation to the menstrual history are determined, and extrauterine abnormalities such as an ovarian cyst or a fibroid may be discovered. A cervical smear is taken for cytological examination. Some idea of the general shape of the pelvis may be gained. At most antenatal clinics efforts to estimate pelvic capacity are usually postponed until the 36th week, when the tissues are more relaxed. Many obstetricians now forego the pelvic examination at the first visit, since fetal size and the absence of other pelvic masses will be checked at the first ultrasound scan.

Investigations

A chest X-ray is no longer performed as routine but it may be wise in women coming from countries where tuberculosis is still common. A midstream specimen of urine is examined for bacteria and tested for glucose and protein.

Blood is taken for determination of ABO and rhesus blood groups, and is screened for the presence of abnormal antibodies. Haemoglobin concentration is estimated, and for women of

Afro-Caribbean, Asian and Mediterranean origin haemoglobin electrophoresis is performed to screen for haemoglobinopathy. Serological tests for syphilis are carried out, as well as rubella antibody screening, and if the woman is not rubella-immune a note is made to offer vaccination in the puerperium. In many clinics the serum is examined for alpha-fetoprotein at 16–18 weeks, a high level indicating a possible neural tube defect or other congenital abnormalities. A low level in conjunction with oestriol and hCG levels may predict an increased risk of Down's syndrome. The presence in the blood of hepatitis B antigen should also be detected, so that precautions can be taken to avoid infection of attendants during venepuncture or delivery, arrangements made to vaccinate the new born, and advice given to the woman and her family. HIV antibody screening should be done in at risk women. A random blood sugar estimation is often performed.

Ultrasound examination

At most clinics a routine ultrasound scan is performed at 10–16 weeks to confirm the gestational age and to exclude any gross fetal abnormality. A full anomaly scan may be done between the 18th and 20th week.

SUBSEQUENT VISITS

At every subsequent visit the blood pressure is recorded and the urine may be tested. A careful check is kept on the woman's weight. Provided that it is within normal limits at the outset, the increase should not exceed 12 kg during the pregnancy. Weight gain is variable during the first 12 weeks, and there may be weight loss if nausea and vomiting are particularly troublesome. From the 16th to the 28th week the average gain is about 6 kg, and thereafter weight is gained at about 2 kg every 4 weeks. Excessive weight gain could indicate that the woman is developing hypertensive disease, or that she has an excessive intake of carbohydrate which needs to be checked; often it is just an increase in maternal fat. This should be regulated for excess weight is hard to lose after childbirth with a new baby to be looked after.

The haemoglobin concentration is estimated again at around 30 and 36 weeks, and many clinics repeat the blood sugar estimation.

At every visit the height of the fundus is recorded, and from the 32nd week onwards the presentation of the fetus. After the 36th week it is important in a primigravida to determine whether the widest diameter of the fetal head has entered the brim of the pelvis, or if it has not done so, whether it can be made to engage. A coarse estimate of the volume of amniotic fluid is made and the fetal heart may be auscultated although this is not always necessary if there are good fetal movements. Any suspicion that fetal growth is impaired calls for full investigation of the case, usually by ultrasound measurement of the fetus.

What has been described is routine antenatal care. Any abnormality will demand further investigation and treatment, with more frequent visits to the clinic, attendance at a day care fetal unit, or admission to hospital for observation. Some special tests are described in the next section.

ANTENATAL FETAL MONITORING

Measurements of fetal growth

The most reliable indication of fetal well-being is normal growth. Assessment of fetal size by abdominal palpation is not to be undervalued, especially if the observations are made by one observer and carefully recorded. The fundus–symphysis pubis height is often measured, in centimetres, with a tape measure. Repeated ultrasound examination of the biparietal diameter of the fetal head and ultrasound measurement of the abdominal circumference and femur length of the fetus are more precise measurements which may be applied to any case in which there is clinical suspicion of retarded growth, or reason to fear it, for example in cases of hypertension. Recognition of a large fetus (macrosomia) is also important.

Investigation of circulatory and other fetal responses

Apart from asking the mother to keep records of the fetal movements or kick counts, antenatal recording of the fetal heart rate over a period of 20 minutes or so may be made with a cardiotocograph to confirm fetal well-being.

A normal fetus has periods of inactivity during which the heart rate shows little variation, and episodes of active movement, during which the heart rate shows increased beat-to-beat variation and accelerations. Uterine contractions of pregnancy may cause temporary slowing of the rate. A fetus that shows none of the normal variations in heart rate, or one that shows prolonged slowing of the heart rate with a uterine contraction, may be at risk of hypoxia. Blood flow in the placental, umbilical and fetal vessels may be assessed with modern Doppler blood flow machines.

Assay of hormones and enzymes to assess fetoplacental function

The placenta produces several hormones and enzymes and various measurements of these substances in maternal blood or urine have been used as placental function tests. Biophysical tests have largely superseded these, and human placental lactogen (hPL) and oestriol assays are now rarely performed.

Investigations to exclude fetal abnormalities

Ultrasound screening will demonstrate the fetal sac (*see* Fig. 2.8) and the fetus from very early in pregnancy and detectable abnormalities of the fetus are detectable with increasing gestational age. In late pregnancy radiological examination is rarely used if fetal abnormality or fetal death is suspected.

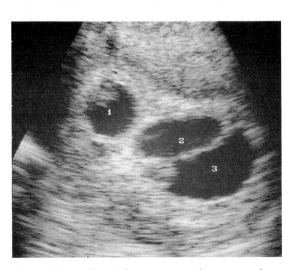

Fig. 2.8 An early triplet pregnancy demonstrated on a transvaginal ultrasound scan

Amniotic fluid may be obtained by abdominal paracentesis so that fetal cells found in it can be grown in tissue culture for chromosomal analysis, for example in cases of Down's syndrome. In some other inherited diseases biochemical study of the fluid is also helpful.

Chorionic biopsy allows earlier examination of fetal cells for chromosomal study. A cannula attached to a syringe is passed through the cervical canal or transabdominally into the uterine cavity and a sample of villi is obtained by aspiration. A greater risk of miscarriage and a risk of fetal limb deformities if the test is done before the 9th week has limited the application of this technique. It is particularly useful in the diagnosis of haemoglobinopathies.

In cases of open neural tube defects alpha-fetoprotein may be found in excess in maternal serum and detailed ultrasound scanning is used to confirm the lesion, or amniocentesis may be performed to estimate the fluid alpha-fetoprotein level.

Estimations of rhesus antibody titres in maternal blood and bilirubin concentrations in amniotic fluid are frequently used in the management of cases of haemolytic disease.

The technique of cordocentesis allows aspiration of blood from the umbilical cord under ultrasound visualization. This provides a sample of fetal white cells for rapid chromosomal analysis; in some centres, blood gas and base balance, and hormonal studies help the assessment of the fetus.

ADVICE DURING PREGNANCY

During pregnancy the woman should be advised to attend the education classes which are now generally available at antenatal clinics. Usually, each woman is given a booklet, which she can read at leisure, with explanations of events in pregnancy and labour, advice about the care of her health and information about services available. With the advent of so many special investigations such as blood tests and ultrasound many clinics also have leaflets to explain these. The woman will also need information about financial grants and social services.

It is important to allay any anxiety about labour. A simple explanation of the stages of labour should be given to the woman so that she will know what to expect. This is a convenient time to discuss analgesia and to help the woman to prepare a birth plan if she so wishes. A visit to the labour ward where she can meet some of the staff will help to familiarize her with a strange environment. If a woman's partner or friend is to be present during labour they should also receive preparatory instruction.

DIET

A number of investigations on the effect of diet on the outcome of pregnancy have been made. A poor quality diet may predispose to preterm labour and increased perinatal mortality, but claims that hypertension can be prevented by modification of the diet during pregnancy have not been substantiated.

There is no need for a large increase in calorie value of the diet; 2400 calories is recommended, but the distribution of its constituents requires consideration. Protein should be increased, and at least two-thirds of the protein should be of animal origin, i.e. meat, milk, eggs, cheese and fish. If these are taken the intake of fats will be adequate. Carbohydrates can be reduced slightly to compensate for the increased calorie value of the protein, and more severely restricted if weight reduction is necessary.

In the latter half of pregnancy there is need for a considerable increase in the intake of calcium, phosphorus and iron, and probably of other trace elements, to supply the needs of the growing fetus and to prepare for lactation. Milk, cheese, eggs, meat and fresh green vegetables are foods rich in mineral salts, and a well-balanced diet will contain sufficient minerals, except perhaps for calcium and iron.

Calcium

The amount of calcium required daily by an adult is 0.5 g; during pregnancy the amount is increased to 1.5 g. Calcium is contained in milk, cheese, some vegetables and bread. It is difficult to be sure that all the calcium taken is absorbed; for example, the phytic acid of bread flour produces an insoluble salt of calcium.

Iron

Many women have poor iron reserves at the beginning of pregnancy, so it is necessary to check the haemoglobin level throughout pregnancy. The daily absorption of iron from an ordinary diet is about 1.2 mg, while the requirement during pregnancy averages 3.5 mg. An iron supplement is therefore often given. The preparation commonly used is ferrous sulphate 200 mg three times daily. This may cause gastric irritation in some women and constipation in others; ferrous fumarate 300 mg daily, or a slow-release preparation, may be used in these women.

During pregnancy megaloblastic anaemia from deficiency of folic acid may occur, and in many clinics combined pills are used, containing iron with a daily dose of 0.5 mg of folic acid. Following the use of folate for women with a previous CNS abnormality, in an effort to reduce a repeat problem, the Department of Health now recommends for all women to take 4.0 or 5.0 mg of folic acid supplement daily prior to conception and for the first 12 weeks of pregnancy.

REST AND EXERCISE

Although violent exercise is imprudent during pregnancy, the woman should be encouraged to continue all ordinary activities. In many clinics women are given instruction in antenatal exercises. These are directed more to posture and general physique than to the muscles especially concerned with childbirth. Many women benefit from instruction in muscular relaxation, so that they can relax voluntarily during the uterine contractions of the first stage of labour. Relaxation may also be achieved by deep breathing during pains. This is the basis of the method of so-called psychoprophylaxis. Other classes give instruction in yoga or active birth. Adequate sleep must be secured, with a sufficient number of hours in bed. Sleeplessness is occasionally troublesome towards the end of pregnancy. It may be treated, if it is severe enough, with sedatives such as promezathine hydrochloride 10 mg, benzodiazepines such as temazepam (Normison) 20 mg or flurazepam (Dalmane) 15 mg but these are rarely required.

REGULATION OF THE BOWELS

Constipation is a troublesome complication in many pregnant women. With a diet containing plenty of fruit, vegetables and bran a daily action of the bowel can usually be ensured; otherwise mild aperients such as senna and lactulose may be prescribed.

COITUS

Women can be reassured that intercourse is not harmful during pregnancy except when there is a threat to miscarry or a previous history of abortion. It should be avoided if there has been antepartum haemorrhage.

PREPARATION FOR LACTATION

The best preparation for lactation is to ensure that the expectant mother is aware of the normal course of events following delivery and is mentally prepared for breastfeeding.

Attention is given during antenatal examination to the nipples. A poorly developed, retracted or inverted nipple cannot be drawn into the infant's mouth, and may be traumatized because the baby cannot fix onto the nipple properly. The nipples should be examined to see whether they are retracted or inverted, and to ascertain whether they will protract. The external appearance of the nipple is not a certain guide and the base of the nipple should be gently pinched to make it protrude. If the nipples are retracted some advocate the mother should wear glass or plastic nipple shells during the day, and at night during the latter part of pregnancy, and this will in some cases correct the abnormality.

There should be no attempt to harden the nipples with spirit; only ordinary washing is necessary. Dry skin on the nipples may be treated with an occasional application of lanoline. Expression of the breasts during the last few weeks of pregnancy has been advocated as a way in which congestion may be prevented, but this requires special instruction to the mothers, and not all of them find it easy.

The breasts should be supported by a well-fitting brassiere which does not press upon the nipples.

VOMITING OF PREGNANCY

From about the 6th to the 12th week of pregnancy nausea or vomiting in the morning or at other times of the day is so common that it is accepted as a symptom of normal pregnancy. It is often retching rather than vomiting. It nearly always stops before the 14th week and does not disturb the woman's health or her pregnancy.

Since morning sickness sometimes occurs when the woman does not know that she is pregnant it seems to be caused by something other than a psychological reaction to pregnancy, and the increased incidence of vomiting in cases of twins and hydatidiform mole has led to the theory that it is the result of higher levels of chorionic gonadotrophin, or an increased sensitivity to it. The vomiting occurs at the time of peak output of this hormone in normal pregnancy, but studies comparing hormone levels and sensitivities in cases of excessive vomiting with those of normal controls have not consistently supported this theory.

If the vomiting is persistent and disturbs the patient's health it is termed *hyperemesis gravidarum*. In these severe cases the only biochemical abnormalities which have been found are those which are secondary to vomiting, starvation and dehydration, namely ketosis, electrolyte imbalance and vitamin deficiency. In the years before the management of electrolyte imbalance was understood hyperemesis was a significant cause of maternal mortality, and in fatal cases severe weight loss, tachycardia and hypotension, oliguria, neurological disorders from vitamin B deficiency and jaundice from hepatic necrosis were seen. In the last 20 years the incidence of hyperemesis in the UK has greatly diminished, and maternal deaths or cases requiring termination of pregnancy are now virtually unknown.

In some of the persistent cases there may be psychological factors. Mere removal to hospital without any other treatment often leads to dramatic and immediate improvement.

Management

In any case of severe or persistent vomiting it is essential to exclude any other possible cause for it such as pyelonephritis, intestinal obstruction, infective hepatitis or cerebral tumour.

Ordinary morning sickness can be simply treated by reassurance and sometimes by giving one of the antiemetics which have been proven to be non-teratogenic such as meclozine 25 mg, cyclizine 50 mg or promezathine 25 mg, up to three times daily.

If the vomiting is severe the woman is admitted to hospital. If it continues, her dehydration, ketosis and electrolyte imbalance require treatment by intravenous fluids and antiemetics, and sometimes with vitamin supplements, commonly in the form of intravenous multivitamin preparations, although it is vitamin B6 which is required. The treatment is regulated by daily studies of the blood chemistry and cessation of vomiting, normal urinary output and weight gain are indications of recovery. Oral feeding is begun as soon as possible, starting with fluids and progressing to semi-solids and eventually to a full diet. The possible need for psychotherapy must be considered.

PTYALISM

An apparent increase in the amount of saliva occurs occasionally during early pregnancy. In extreme cases the woman spits out her saliva instead of swallowing it and lives in an aura of wet handkerchiefs or spitoons. Treatment is unsatisfactory. Psychological factors should be borne in mind.

VARICOSE VEINS

Varicose veins of the legs may cause considerable discomfort during pregnancy, and may be associated with oedema and thrombophlebitis. Fortunately pulmonary embolism from such thrombi is extremely rare. Support by elastic stockings may give some relief. Surgical treatment is not advised during pregnancy because the condition often improves considerably after delivery. The residual lesion should be assessed about 3 months after delivery before deciding on surgical treatment.

Vulval varices

Vulval varices may be uncomfortable during pregnancy, and are an occasional cause of a vulval haematoma during labour. They may be compressed with a sanitary towel.

HAEMORRHOIDS

Haemorrhoids may first appear and may be made worse during pregnancy. Surgical treatment is not advised during pregnancy except to evacuate a painful perianal haematoma (a thrombosed pile) under local anaesthesia. Aperients should be prescribed. The anal region should be carefully washed and dried after defecation and analgesic ointment applied.

PRURITUS VULVAE AND VAGINAL DISCHARGE

An excess of vaginal discharge from any cause can lead to vulval pruritus. The commonest cause during pregnancy is infection with *Candida albicans* which may be associated with the lowered renal threshold for sugar which occurs in many pregnant women. In all cases of pruritus the urine should be tested for glucose, and if there is any reason to suspect the possibility of diabetes a glucose tolerance test will be required. In candidiasis there may be vulvitis with redness of the skin, or vaginitis with masses, or plaques, of white cheesy material lightly adherent to the epithelium. Microscopical examination of a little of the discharge in a drop of normal saline will show mycelial threads, and *Candida* will be grown on culture. A variety of fungicides may be used for treatment, such as clotrimazole (Canesten) pessaries and cream for three nights.

Vaginitis during pregnancy may also be caused by *Trichomonas vaginalis*, which causes a profuse offensive, purulent discharge, in which the organisms can be found on microscopical examination. Metronidazole tablets 200 mg three times daily by mouth for 7 days may be used during pregnancy. No adverse effect on the fetus has been shown with this drug, but it may be wiser to rely on local treatment before the 12th week.

CRAMPS IN THE LEGS

Transient nocturnal painful spasms of the small muscles of the feet or of the muscles of the calves sometimes occur during pregnancy. Such cramps have been attributed to calcium deficiency, but there is no satisfactory evidence to support this. It is more likely that they are due to temporary circulatory insufficiency. The cramp tends to improve spontaneously in late pregnancy, but can be troublesome if it returns during labour.

ACROPARAESTHESIA

This is not uncommon during pregnancy. There is a sensation of pins and needles in the hands with some sensory loss, and sometimes weakness of the small muscles. The woman finds it difficult to use her hands for fine work. Acroparaesthesia has been attributed to oedema in the carpal tunnel involving the median nerve, and this will explain the cases in which the signs have the appropriate distribution, but in many cases the ulnar border of the hand and sometimes even the forearm, are involved, and in these cases pressure on the lowest part of the brachial plexus near the first rib must be considered.

Some, but not all, women obtain relief from diuretics such as chlorothiazide. Splinting the wrists at night is sometimes helpful. Recovery after delivery is the rule.

HARMFUL EFFECTS OF DRUGS ON THE FETUS

After the discovery that when the drug thalidomide was given in early pregnancy, it was associated with gross fetal deformities, all drugs used during pregnancy have come under close scrutiny.

Thalidomide was used as a sedative and antiemetic but was withdrawn in 1962 after it was found to cause gross limb deformities (phocomelia) and other fetal abnormalities if given to the mother between the 30th and 70th days of pregnancy. Yet care is necessary before attributing any solitary case of deformity to a particular drug. In every 1000 viable births there will be about five perinatal deaths from congenital malformation and

more than 10 other infants with malformations of clinical significance. The great majority of these malformations occur as a result of genetic disturbance or pathological events entirely unrelated to any drugs which the mothers may have taken during pregnancy. Properly controlled comparisons are always necessary in studying such relationships.

Most drugs pass the placental barrier, and possible effects on the fetus must always be considered when drugs are prescribed during pregnancy. It is an important principle of teratology that it is not so much the nature of any harmful agent as the time in embryonic development at which it acts that chiefly determines the abnormality produced. In general, the earlier in pregnancy that a teratogenic agent acts, the more severe will be the malformation. Furthermore, there is a short critical period during which each developing structure is particularly vulnerable (*see* Fig. 2.9). In addition, there seem to be many maternal factors which affect the outcome, such as the standard of nutrition and hormone levels. It should be a general principle to avoid the administration of any drug during the early weeks of pregnancy unless it is clearly necessary for the treatment of a maternal condition.

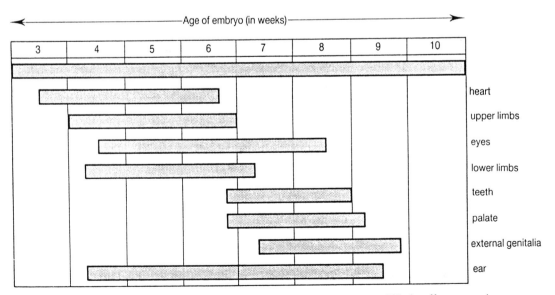

Fig. 2.9 Developmental milestones of the embryo in early pregnancy. Times of likely affect on various organs and limbs are shown

SEDATIVES AND ANALGESICS

Morphine and pethidine

Morphine or pethidine, given within 2 or 3 hours before delivery, will depress the fetal respiratory centre. This effect can be antagonized by an injection of naloxone.

Heroin

If the mother is a heroin addict the baby may show withdrawal symptoms after delivery, with restlessness and failure to feed, and consequent loss of weight.

Aspirin

Non-steroidal anti-inflammatory drugs such as aspirin and indomethacin may inhibit prostaglandin synthesis and produce premature closure of the fetal ductus arteriosis. In very high doses they could theoretically postpone the onset of labour. Low dose (paediatric) aspirin used in prophylaxis of hypertension does not carry this risk.

Diazepam

Diazepam (Valium) administered in large doses before delivery will depress the fetal medullary centres and cause loss of the normal base-line variation of the heart rate (*see* Fig. 2.9); there is hypotonia after delivery.

Phenytoin

Phenytoin (Epanutin) given to control epilepsy is a folic acid antagonist, and if it is used during pregnancy additional folic acid must also be given.

DRUGS AFFECTING THE CARDIOVASCULAR SYSTEM

Digitalis

Digitalis has no harmful effect on the fetus.

Hexamethonium

Hexamethonium may cause ileus in the newborn child.

Reserpine

Reserpine will cause a transitory non-infective nasal discharge in the infant, but neither reserpine nor hexamethonium is used very often.

Propanolol and atenolol

With propanolol and atenolol there may be fetal bradycardia and neonatal circulatory depression.

Antihypertensives

Methyldopa, bethanidine, guanethidine, hydralazine and Nifedapine seem to have no harmful effect on the fetus. Both atropine and scopolamine accelerate the fetal heart rate, but the effect is transitory and not harmful except that it may confuse interpretation of CTG traces.

ANTICOAGULANT DRUGS

Heparin

Heparin does not cross the placenta.

Oral anticoagulants

Oral anticoagulants reach the fetus. Warfarin may cause bony and facial abnormalities if it is administered in early pregnancy, and in late pregnancy there can be retroplacental haemorrhage or bleeding into fetal tissues, with intrauterine death. Intravenous or subcutaneous heparin is therefore preferable, at least during the first trimester and the last 4 weeks of pregnancy.

ANTIBACTERIAL DRUGS

Sulphonamides

Sulphonamides compete with bilirubin for binding sites on serum albumin, and therefore increase any possible risk of kernicterus after birth.

Salicylates

Salicylates may act in a similar way to sulphonamides, so that it is wise to discontinue these drugs before birth if there is any probability of premature birth or if there is rhesus incompatibility.

Co-trimoxazole

Co-trimoxazole is a folate antagonist and therefore a potential teratogen.

Tetracyclines

The children of mothers taking tetracyclines during pregnancy may have greenish-yellow staining of their milk teeth, and there may be interference with enamel growth leading to imperfect dentition.

Streptomycin

With streptomycin there is a theoretical risk of damage to the 8th nerve of the fetus, but with ordinary doses this seems to be very rare.

Chloramphenicol

Chloramphenicol may cause postnatal collapse and hypothermia. The penicillins, erythromycin and most other antibiotics pass into fetal blood and the amniotic fluid, but seem to have no harmful effect.

Metronidazole

Metronidazole has not been proven to be teratogenic, but it is not usually prescribed in the first trimester.

HORMONES

Progestogens

Androgens such as testosterone should not be given to the mother during pregnancy as they may cause virilization of a female fetus. Synthetic progestogens (19α group), metabolized through the androgenic pathways, have also caused virilization.

Allyloestrenol and dydrogesterone are safer than ethisterone or norethisterone, and hydroxyprogesterone hexanoate seems to be without risk.

Diethylstilboestrol should not be given to pregnant women as it may cause vaginal septal defects in female fetuses if given early enough. Its use is associated with vaginal adenosis which may rarely proceed to vaginal carcinoma 15 to 20 years later. Estradiol, given by mouth or transdermal application (in cases of ovarian failure and ovum donation) appears to carry no risk to the fetus.

Adrenal steroids

Adrenal steroids may be given in physiological doses to pregnant women with Addison's disease without risk to the fetus. In a few uncommon diseases, such as lupus erythematosus, polyarteritis nodosa, status asthmaticus and thrombocytopenic purpura, steroids are used in larger pharmacological doses. Animal experiments have suggested that there may be some risk of fetal cleft palate, but human evidence is doubtful, and it is justifiable to use these drugs for severe maternal illness.

RADIOACTIVE SUBSTANCES

Radioactive isotopes of iodine, strontium and phosphorus cross the placenta and become localized in fetal tissues. Neither these substances nor any others that emit gamma rays should be used during pregnancy.

ANTITHYROID DRUGS

Drugs such as thiouracil or large doses of iodine may cause fetal goitre or hypothyroidism. Established treatment of maternal thyrotoxicosis is very important and must continue. Concomitant thyroxine may need to be administered also to go through to the fetus.

CYTOTOXIC AND ALKYLATING DRUGS

All of these, including methotrexate, busulphan, cyclophosphamide, chlorambucil and many others, may harm the fetus. They are unlikely to be called for during pregnancy, but to the mother with neoplastic disease it may be necessary to give such drugs in spite of the fetal risk. This must be fully discussed with the woman.

SMOKING DURING PREGNANCY

Smoking in pregnancy is harmful to the fetus and will cause reduction in birth weight with an increase in perinatal mortality. It has been suggested that raised maternal carbon monoxide levels produce an increase in carboxyhaemoglobin which interferes with oxygen transport, but the ill effects are more likely to be caused by nicotine which has a vasoconstrictive effect on maternal vessels in the placental bed. The adverse effect of smoking is greater if the mother is hypertensive. It has been alleged to be responsible for more neonatal pathology than any other known cause, and its effects may continue to be manifest well into childhood. Despite much educational effort many smokers seem to be unwilling to discontinue in pregnancy.

ALCOHOL DURING PREGNANCY

Children born to mothers who drink heavily have a low birth weight and an increased neonatal and infant mortality. A few will have the fetal alcohol syndrome with a characteristic facial appearance, with a broad base to the nose, epicanthic folds, a long upper lip and a small lower jaw, with mental retardation. There is an increased incidence of other serious malformations. While no harmful effect has been shown from small doses of alcohol, on general principles even these must be undesirable in the early weeks of pregnancy.

OBSTETRICAL EXAMINATION

Any obstetrical examination should be preceded by consideration of the history, including the past medical record, the past obstetric history and the record of the present pregnancy.

ABDOMINAL EXAMINATION

By abdominal examination it is possible to ascertain:

- the size of the uterus, and to note whether it corresponds with the period of amenorrhoea
- the size of the fetus
- the lie, presentation and attitude of the fetus
- the relationship between the brim of the pelvis and the presenting part
- whether the fetus is alive
- the presence of abnormal conditions, such as excess of amniotic fluid, twin pregnancy or abdominal tumours.

Inspection

By inspection, an impression of the general shape and size of the uterus can be gained and fetal movements may be seen. If the occiput is anterior and the back of the fetus is to the front, the abdomen will appear smoothly convex; if the occiput is posterior, a flattening of the abdomen may be seen above the symphysis pubis. If the lie is transverse the uterus will be wider than usual, and the height of the fundus may be a little lower than with a longitudinal lie. If twins are present the uterus also appears wide but the height of the fundus will be higher than normal. It is traditional to observe the presence or absence of striae grav-

idarum and linea nigra, but they are of no importance or significance to the prognosis.

Palpation

(*See* Fig. 2.10.) The obstetrician stands on the woman's right side. A regular routine is followed:

The level of the fundus

The level of the fundus is determined by using the ulnar border of the hand and little finger and moving it downwards from the xiphisternum (*see* Fig. 2.1). Many obstetricians measure the distance between the upper edge of the symphysis pubis and the top of the fundus with a centimetre tape. From 14 to 34 weeks the distance in centimetres approximates to the weeks of gestation.

The presentation

As the majority of fetuses present by the vertex, the lower part of the uterus should be palpated next to establish whether the presentation is cephalic. This is done by placing both hands on the lower abdomen above the pelvic brim in the last weeks of pregnancy (*see* Fig. 2.9). The head is recognized because it is rounder and harder than the breech. Its mobility will depend upon its relation to the pelvic brim. If it is completely above the brim it is freely moveable and the fingers of both hands can be made to meet below it; if the widest diameter has entered or passed through the brim the head is usually fixed and is described as engaged; if the head is deeply sunk in the pelvic

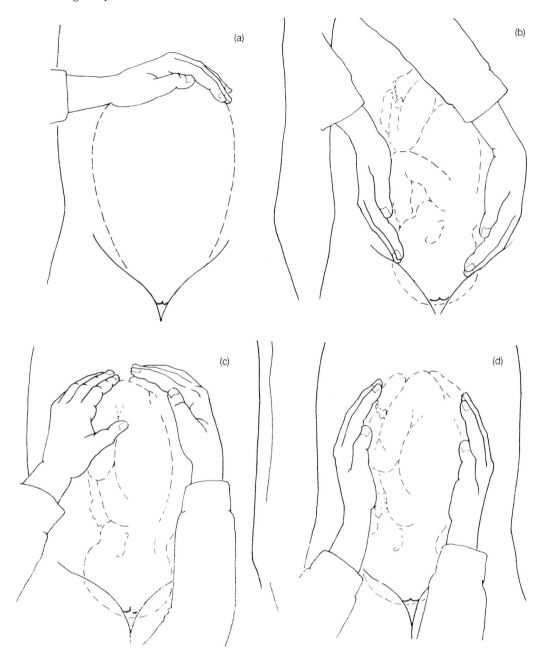

Fig. 2.10 Engagement of the fetal head in the maternal pelvic brim assessed

cavity it may be difficult to feel it except with the finger tips.

The relationship between the head and the pelvis may be described in fifths (*see* Fig. 2.11). A head that is completely free is described as 5/5; one that is beginning to enter the brim as 4/5, and one that has a major part in the brim as 3/5. Once the widest diameter has passed the brim the notation 2/5 is used, and for a head deeply engaged in the pelvis 1/5. The fifths refer to the proportion of the head still palpable above the brim.

The degree of flexion of the head can often be

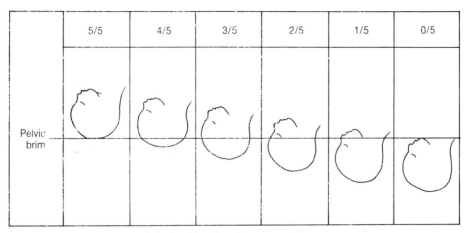

	5/5	4/5	3/5	2/5	1/5	0/5

Pelvic brim

Fig. 2.11 Engagement of the fetal head in the maternal pelvic brim assessed in fifths

made out by abdominal palpation. For example with an LOT position of the occiput the well-defined forehead (sinciput) will be felt on the right, but the occiput is less easily felt on the left as it is at a lower level, indicating that the head is well flexed. In such a case the anterior shoulder will be palpable near the midline (*see* Fig. 2.10(b)).

If the head is above the brim of the pelvis it may be felt more easily by the single-handed grip shown in Fig. 2.12. Gentleness is needed in all these abdominal examinations.

Palpation of the fundus

The fundus should then be palpated with both hands (*see* Fig. 2.10(c)). If the breech occupies the fundus a broad mass will be felt; this is not as round or hard as the head. The breech is continuous with the fetal back, and a foot or knee can often be felt near it and may move under the hand. If the head occupies the fundus it is felt as a harder, rounder, smoother and more mobile mass than the breech. The head can be moved independently of the body, and it can sometimes be balloted between the two hands.

Abdominal palpation

The sides of the abdomen are then palpated to discover on which side the back lies. If the back is directed more to the front than to the back, a broad smooth surface will be felt on one side of the abdomen, and on the other side a number of small knobs, which are the limbs (*see* Fig. 2.10(d)). If these small parts are felt all over the abdomen the

back must be directed posteriorly and to the side, and deep palpation may be necessary to feel it.

If the fetus is lying transversely, the head is felt in one flank and the breech is felt in the other; if it is lying obliquely the head is often felt in one iliac fossa, and the breech higher up on the opposite side (*see* Fig. 2.13).

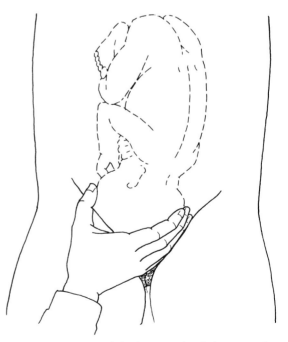

Fig. 2.12 Palpation of the lower pole of the uterus by Pawlik's method

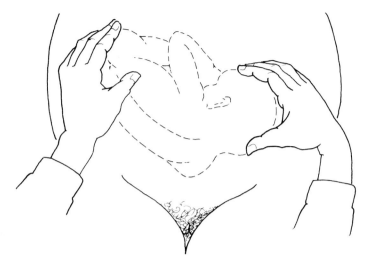

Fig. 2.13 Abdominal palpation of fetus lying transversely

Exclusion of cephalopelvic disproportion

In the last 4 weeks of pregnancy, if the head is presenting but above the brim, an attempt is made to discover if it will engage, and thus to exclude cephalopelvic disproportion developing in labour. This may be done by moderate pressure on the head in a backwards and downwards direction. Another method is to raise the woman's head and shoulders, asking her to take the weight of her trunk on her elbows, and the head will often enter the brim (*see* Fig. 2.14). Perhaps the best method of all to test whether the head will engage is to examine the woman when she is standing.

At each visit the level of the fundus and the size of the fetus must be carefully noted. If the duration of pregnancy is uncertain, repeated observations are most useful, and any suggestion that the fetus is small-for-dates or is not growing normally calls for investigation by ultrasound. If the uterus is

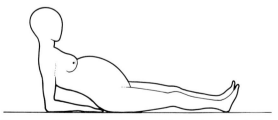

Fig. 2.14 Abdominal palpation. The patient is asked to sit up (or lean on her elbows) to test whether the head will engage

larger than expected there may be an excess of amniotic fluid or a multiple pregnancy. Twins may be diagnosed by palpation of more than two fetal poles, and often two heads are clearly recognized. Ultrasound will confirm the diagnosis.

Auscultation

The fetal heart sounds are listened for. They are best heard over the back of the fetus in vertex and breech presentations; it is only in face presentations that they are heard over the front of the fetus. In occipitoposterior positions, in which the back is directed to one side and posteriorly, the sounds may be heard either in the flank or near the midline.

In breech presentations the heart sounds will often be heard at a higher point in the abdomen than in vertex presentations, at the level of or a little above the umbilicus, unless the breech is deeply engaged, when they will be heard at the usual level.

VAGINAL EXAMINATION

A vaginal examination may be performed at the patient's first attendance at the antenatal clinic to confirm the pregnancy and its duration, to determine the position of the uterus and to exclude other abnormalities such as ovarian cysts. The bony pelvis may also be examined, but the best time to assess the size and shape of the pelvis is at

Table 2.1 Modified Bishop score

Score	0	1	2	3
Dilatation of cervix (cm)	0	1 or 2	3 or 4	5 or more
Consistency of cervix	Firm	Medium	Soft	–
Length of cervical canal (cm)	>2	2–1	1–0.5	<0.5
Position of cervix	Posterior	Central	Anterior	
Station of presenting part (cm above ischial spines)	3	2	1 or 0	Below

about the 36th week, by which time the soft tissues will be more relaxed and examination is easier.

However, most women attending antenatal clinics will be examined by ultrasound, which will confirm the duration of pregnancy, show that the fetus is alive, and exclude other abnormalities. If this is done there is no need to examine the woman vaginally. A few women miscarry in early pregnancy and may blame vaginal examination for this.

Full aseptic precautions are required if a pelvic examination is performed on any woman who may be in labour. During labour the most important observations will be the degree of dilatation of the cervix, whether the membranes are ruptured, the recognition of the presenting part, and the determination of its position and level in the pelvis.

Some obstetricians use the *Bishop score* to record the state of the cervix (Table 2.1). It enables different observers to relate their findings accurately. A score of less than 5 suggests that labour is unlikely to start without ripening of the cervix; a score of 7 indicates that labour should commence easily. This information is valuable if, for any reason, induction is contemplated.

The ischial spines are useful landmarks in determining the level of the presenting part during labour, and the degree of flexion or extension of the head can be found by palpation of the fontanelles. Feeling the posterior fontanelle indicates that the head is well flexed; feeling the anterior fontanelle indicates that the head is not well flexed, and this is usually associated with occipitolateral or occipitoposterior positions.

RADIOLOGICAL EXAMINATION

With the development of ultrasound examination there is now hardly any place for radiological examination of the fetus. Any irradiation of the fetus, particularly in the early weeks, may be associated with fetal abnormalities or cause abnormal mutations in the genes of the sex cells. It has also been suggested that even moderate exposure to irradiation *in utero* may increase the incidence of leukaemia in childhood, although this has not been universally accepted. Radiological pelvimetry still has a limited place. Where this is indicated it should be done late in pregnancy, with as few exposures as possible and may be replaced by computerized tonography (CT) which gives only $\frac{1}{20}$th of the dose.

ULTRASOUND EXAMINATION IN OBSTETRICS

Sound is the orderly transmission of mechanical vibrations through a medium. The number of vibrations that occur in 1 second is known as the frequency and is measured in Hertz (Hz); one Hz is one cycle per second. The human ear can detect sound in the frequency range of 20 Hz to 20 kHz (20 000 Hz). Sound above this range is known as ultrasound. Medical diagnostic instruments use

frequencies in the range of 1–10 megaHertz (MHz) with 1 MHz being 1 million cycles per second.

Ultrasound is generated from crystals that have the piezoelectric property. If the crystal is stimulated with electricity it changes its width and in so doing generates vibrations that travel into the human body. The waves are scattered and reflected back to the crystal by differences in the sound properties between tissues or within tissues. The scattered waves are received at the crystal and converted back into electricity. In practice the ultrasound probe (transducer) contains many crystals which are activated in a predetermined sequence. The returning echoes are then reconstructed by a computer and presented as a picture on a television screen. All ultrasound is now real-time scanning and this is achieved by replacing the static picture at a rate of more than 20 times per second so that the human eye cannot detect the change between static frames.

Measurements can be made using ultrasound by the simple expedient of determining the time taken for the ultrasound to travel to the structure in question and then to return (go-return time).

$$\text{Depth} = \frac{\text{Velocity} \times 2}{\text{Time from transmission to reception}}$$

Accuracy of measurement is limited to about 1 mm with the commonly used obstetric ultrasound frequencies and modern electronic callipers.

The resolution of ultrasound equipment is determined by the transducer frequency; the higher the frequency the better the resolution. Unfortunately, however, as the frequency increased the depth to which the ultrasound can penetrate decreases. Transabdominal obstetric transducers use frequencies in the range of 2.5–5.0 MHz. In early pregnancy resolution has been vastly improved by the introduction of transvaginal transducers. Because this can be placed closer to the uterus and ovaries they use frequencies in the range of 6.0–7.5 MHz.

CLINICAL APPLICATIONS OF ULTRASOUND

Diagnosis of pregnancy

Modern pregnancy tests use a monoclonal antibody against the beta subunit of hCG and will detect pregnancy at conception plus 10 days; that is about 4 days before a missed period.

The recent introduction of transvaginal ultrasound techniques has improved the resolution and the diagnosis of pregnancy can be made much earlier than with abdominal scans (Table 2.2).

Resolution of early pregnancy problems

Women who present in early pregnancy with pain and bleeding and who, on clinical examination, have an open internal cervical os are diagnosed as having an inevitable or incomplete miscarriage and require an evacuation of retained products of conception. Ultrasound only has a role to play if the woman has a previous problem such as treatment for infertility or previous miscarriages.

In those who bleed or have pain and in whom the cervical os is closed, ultrasound will make the definitive diagnosis as follows:

Missed abortion – a fetal pole is seen within an intrauterine gestation sac but no fetal heart activity is detected.

Incomplete abortion – an intrauterine gestation sac is not visualized but there are products of conception (*see* Fig. 2.15).

Anembryonic pregnancy – an intrauterine gestation sac is visualized but a fetal pole is not seen (*see* Fig. 2.16). Care has to be taken with this diagnosis because it may be a normal but early pregnancy. The date of the woman's last menstrual period should not be used as a guide. The volume of the gestational sac should be determined and if it is over 3 ml (equivalent to 7 weeks' gestation) the diagnosis is confirmed. If it is less the scan should be repeated in 1 week when, in normal pregnancy, the gestation sac will have more than doubled in size and the fetal pole becomes visible.

Ectopic pregnancy – prior to the introduction of transvaginal ultrasound the main role of ultrasound in a woman with a suspected ectopic pregnancy was to exclude an intrauterine pregnancy, as a co-existence ectopic or heterologous pregnancy

Table 2.2 A comparison of scanning methods in early pregnancy

| | Scanning technique | |
	Transabdominal	Transvaginal
Visualization of gestation sac	5 weeks	4 weeks
Detection of fetal heart activity	6 weeks	5 weeks
Detection of a fetal pole	7 weeks	6 weeks

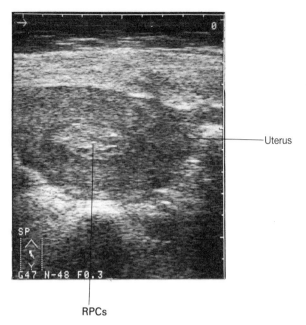

Fig. 2.15 An ultrasound scan of a transverse section of the uterus demonstrating bright echoes from within the uterine cavity due to retained products of conception (RPCs)

is very rare. Only in about 5 per cent of cases could an ectopic gestation be firmly diagnosed by visualizing a fetus in an extrauterine position (*see* Fig. 2.17). With transvaginal ultrasound over 90 per cent of ectopic pregnancies can be confidently diagnosed. The ultrasound could then be used to guide a needle into the pregnancy to allow injection of methotrexate to destroy the trophoblast.

Hydatidiform mole – this has a characteristic appearance that used to be described as a snowstorm in the 1970s using ultrasound scanners but with modern equipment the vesicles can be clearly seen (*see* Fig. 2.18).

Routine ultrasound examination

Almost all pregnant women have an ultrasound examination in the first half of pregnancy if they present to the hospital in time. The purposes of this examination are:

- to confirm or establish the gestational age
- to exclude or diagnose multiple pregnancy
- to exclude structural abnormalities
- to search for minor markers of chromosomal abnormalities
- to localize the placenta
- to increase maternal bonding to the fetus. Seeing the fetus on the ultrasound screen is often the first real contact that a mother has with her fetus.

The timing of routine examinations varies between units but fits to one of three patterns:

- a booking ultrasound examination. The woman is scanned at the booking clinic to confirm gestational age and exclude multiple pregnancy. Exclusion of abnormalities is less effective in early pregnancy. This scan is also offered to women wishing the triple test to exclude Down's syndrome

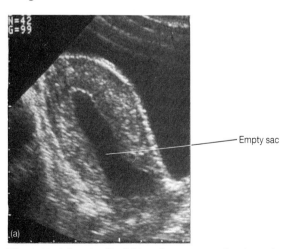

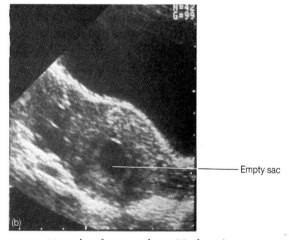

Fig. 2.16 A longitudinal (a) and transverse (b) view of pregnant uterus at 11 weeks of amenorrhoea. No fetus is seen within the gestation sac allowing the diagnosis of anembryonic pregnancy to be made

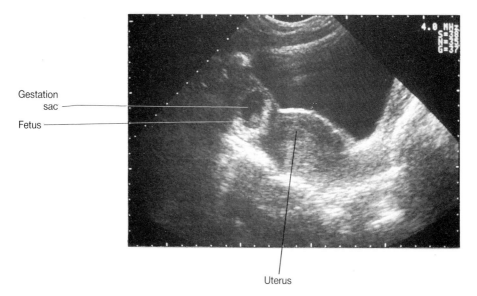

Gestation sac

Fetus

Uterus

Fig. 2.17 A transverse section through the pelvis demonstrating an extra-uterine (ectopic) pregnancy

- a 16–20 week ultrasound examination. This is the most sensitive time to perform all the purposes listed above
- both a booking and a 16–20 week ultrasound.

In addition some units offer a scan at 32–34 weeks' gestation, mostly to determine fetal size but also to check the site of the placenta and to screen for abnormalities that may only be detectable in the last half of pregnancy.

Establishing gestational age

General principle

Up to 40 per cent of women attending antenatal

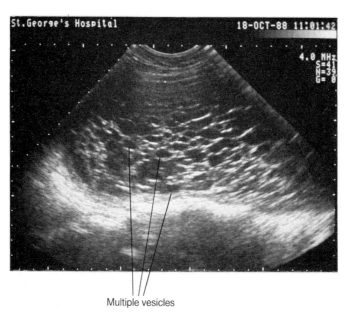

Multiple vesicles

Fig. 2.18 An ultrasound scan demonstrating the appearances of a hydatidiform mole on modern equipment

clinics have an unreliable menstrual history because:

- they do not know the date of their last period
- they have irregular cycles
- they conceived within 2 months of stopping the contraceptive pill
- they have had bleeding in early pregnancy.

The most accurate way to establish gestational age in these women is by ultrasound measurements of the fetus (fetal biometry). In addition, there is some evidence that up to 25 per cent of women with a firm menstrual history may be more than 2 weeks adrift from ultrasound measurements. Although evidence is scant it appears that dating all women by means of fetal biometry reduces those who will exceed 42 weeks' gestation with positive benefits to the fetus.

Charts of ultrasound dimensions are derived by taking women with sure menstrual dates and measuring their fetuses at differing weeks of pregnancy. By regression analysis two types of chart may be plotted (*see* Fig. 2.19(a) and (b)):

A growth chart – in this case the independent variable (plotted on the *x*-axis) is the gestational age and the chart illustrates how the measured fetal parameter increases throughout pregnancy

A size chart – in this case the measured fetal parameter is the independent variable and the chart

allows the gestational age to be determined from that parameter.

In addition to using the correct style of chart, it is preferable for these charts to have been derived from the same geographical population to that on which they will be used.

Early pregnancy

Gestational age can be calculated from measurement of the gestational sac volume but this is rarely done because it is relatively inaccurate and as soon as the fetus is visible the crown–rump length (CRL) can be measured. Figure. 2.20 illustrates how the CRL is measured and Fig. 2.21 shows the growth of the CRL in early pregnancy.

Mid-trimester

Up to about 24 weeks' gestation the gestational age can be accurately established by measurement of the fetal biparietal diameter (BPD) and femur length (FL). Figures 2.21 and 2.23 demonstrate the sections on which the measurements are made and Figs 2.22 and 2.24 demonstrate the growth of the BPD throughout pregnancy and Fig. 2.25 illustrates the chart of the BPD which is used to establish gestational age. Having made the BPD and FL measurements they are interpreted as follows:

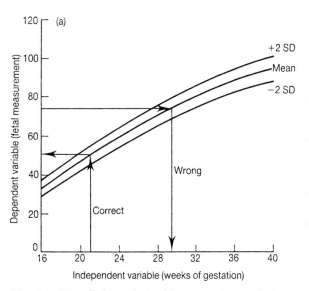

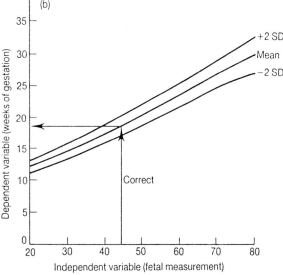

Fig. 2.19 Use of charts derived by regression analysis. (a) Is a chart that demonstrates the growth of the dependent variable (usually a fetal dimension) throughout gestation; (b) is a size chart and is used for establishing gestational age from a measured fetal parameter. Note how the regression lines are different for each of the above charts

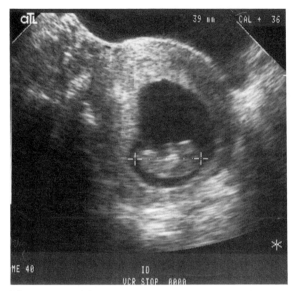

Fig. 2.20 An ultrasound scan demonstrating measurement of the crown rump length

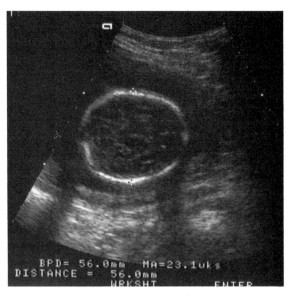

Fig. 2.22 An image of BPD

- if the BPD and FL fall within the data reference range for the postmenstrual age then this establishes gestational age
- if the woman has no menstrual dates or the measurements fall outside the data reference

range then the BPD and FL are plotted on size charts. If the prediction of gestational age from both measurements agrees then the gestational age is established
- if the BPD falls within the data reference range but the FL is shorter then short limbed dwarfism should be excluded
- if the FL falls within the data reference range but the BPD is smaller then microcephaly or a neural tube defect should be excluded.

In cases of continuing discrepancy the woman should be referred for a detailed ultrasound examination and an expert opinion.

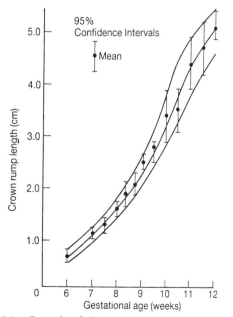

Fig. 2.21 Growth of the crown rump length in early pregnancy

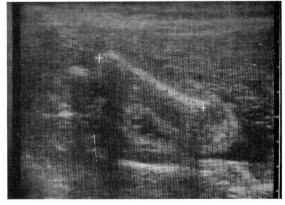

Fig. 2.23 An ultrasound scan demonstrating measurements of the femur length

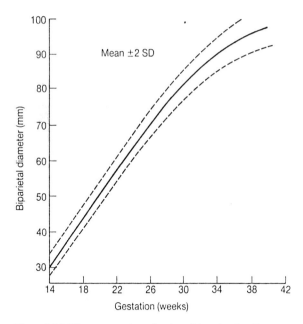

Fig. 2.24 The growth of the biparietal diameter throughout pregnancy

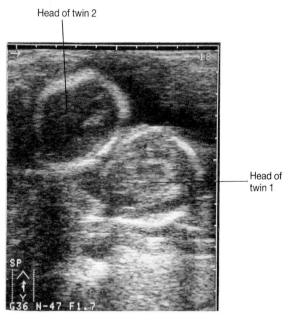

Fig. 2.26 An ultrasound sound scan demonstrating twins

Late bookers

If a woman books after 22 weeks' gestation the BPD and FL should be measured; if they agree with the postmenstrual dates these are accepted. If

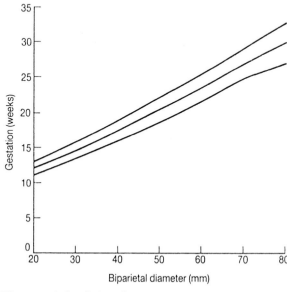

Fig. 2.25 A head-size chart. This is used to determine the gestational age a measurement of the biparietal diameter

the measurements are smaller (or larger) then it is not possible to determine whether the baby is small (or large) for gestational age or if the dates are wrong. The woman should then be scanned serially in order to monitor growth. As a good general rule, fetuses that are growing do not die.

Diagnosis of multiple pregnancy

Careful ultrasound examination should allow all multiple pregnancies to be diagnosed. Figure 2.26 illustrates twins but higher multiples must always be carefully excluded. Gestational age is established as above but using the measurements from the bigger fetus if there is a discrepancy in size.

Congenital abnormalities and all the complications of pregnancy are much more common in monozygous twins (about 30 per cent of twins) so an attempt should be made to establish zygosity as follows:

- number of placentae – two distinct placentae are almost always associated with dizygosity
- fetal sex – opposite sexes means dizygosity
- thickness of the dividing membrane – with good ultrasound equipment this may be demonstrated to be four layers thick (dichorionic diamniotic twins) or to only contain two layers (monochorionic diamniotic).

Twins have a higher incidence of impaired growth than singleton pregnancies and so should undergo serial ultrasound examination to measure growth velocity.

Placental localization

In the first half of pregnancy about 5 per cent of women will have a low lying placenta whereas only 0.5 per cent will have placenta praevia. The term 'placenta praevia' should not be used before 28 weeks because it is defined as a placenta that lies partially or wholly in the lower segment of the uterus; this does not begin to form until 28 weeks' gestation.

Placental localization in the first half of pregnancy is worthwhile because if the placenta is in the fundus it will not move lower in the uterus. Women with low lying placentae do not need admission, restriction of physical activity or avoidance of intercourse but should be advised to report any vaginal bleeding. They should be rescanned at 32–34 weeks when most of them will have a placenta sited in the upper segment. There is controversy about whether those women who still have a low placenta should be admitted if they are asymptomatic. It is probably sensible to admit those in whom the placenta covers the os.

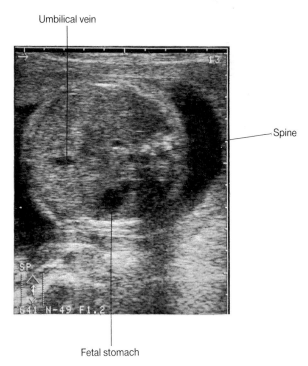

Umbilical vein

Spine

Fetal stomach

Fig. 2.27 An ultrasound scan demonstrating the section on which the fetal abdominal circumference is measured

ASSESSING FETAL GROWTH AND SIZE

Serial ultrasound assessment of growth

The following women are at high risk of carrying a fetus that will show an abnormal pattern of growth in pregnancy:

- reduced growth
  - previous small baby
  - previous preterm labour
  - underweight women (less than 45 kg)
  - maternal vascular disease
  - pre-existing hypertension from any cause
  - pregnancy induced hypertension
  - collagen disorders
  - diabetes mellitus with renal complications
  - heavy smokers
  - alcoholics and drug addicts
  - women with sickle cell disease
  - women with recurrent APH

  - women with a raised serum alpha-fetoprotein level but a structurally normal fetus
- increased growth
  - insulin diabetes mellitus
  - obese women.

In such cases there should be serially examination of fetal growth by measurements of the fetal head circumference and abdominal circumference (*see* Fig. 2.27).

Screening for small-for-gestational-age (SGA) babies

There is no test that can be applied to a newborn infant to determine if it has suffered from intrauterine growth retardation (IUGR). This is best defined as the presence of a pathology which is impairing fetal growth and if removed would allow normal growth. This definition has no clinical usefulness and paediatricians and epidemiologists therefore judge smallness on size at birth. Some arbitrary statistical cut-off point is chosen, com-

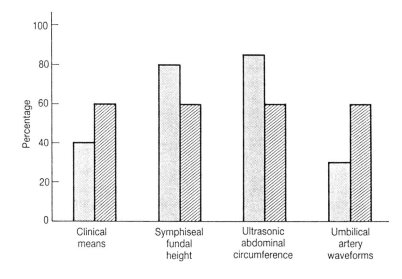

Fig. 2.28 The sensitivity (solid columns) and false positive rate (hatched columns) of current screening tests designed to detect babies with a birth-weight of less than the 10th centile

monly less than the 10th centile for gestational age, and babies below this are defined as SGA. Many of these babies are just normal biological variants and have no evidence of long-term problems.

Figure 2.28 illustrates the value of current tests at predicting a birth weight of less than the 10th centile for gestational age. The most sensitive test is a measurement of the fetal abdominal circumference at 32–34 weeks' gestation.

PATTERNS OF IMPAIRED GROWTH

In broad general terms impaired fetal growth may be divided into two types:

1. Symmetrical smallness (Fig. 2.29(a) and (b) – here there is an impairment in the growth velocity that affects all parts of the fetus equally. The causes are:

- idiopathic (>95 per cent)
- chromosomal abnormalities
- transplacental infections (TORCH organisms)
- heavy maternal smoking
- heroin addiction
- fetal alcohol syndrome
- chronic maternal undernutrition
- sickle cell disease.

2. Asymmetrical smallness (*see* Fig. 2.30(a) and (b) – in this case the growth of the fetal head is protected until late in the course of the disorder. The causes are:

- placental insufficiency
- pregnancy induced hypertension
- severe maternal cardiac or renal disease
- multiple gestation
- idiopathic (about 40 per cent).

Management

Symmetrical SGA

Most cases are the biological lower limits of normal. A careful scan should be performed looking for markers suggestive of chromosomal abnormality and maternal infections should be excluded by blood tests.

Growth should be monitored on a fortnightly basis but tests of fetal well-being are not indicated unless there is:

- an increase in the head/abdomen ratio of the fetus suggesting the superimposition of asymmetry
- associated maternal hypertension
- abnormal uteroplacental or umbilical Doppler waveforms.

If growth continues at a normal velocity induction of labour is not indicated.

Asymmetrical SGA

This group of babies are most at risk of the problems of placental failure which include:

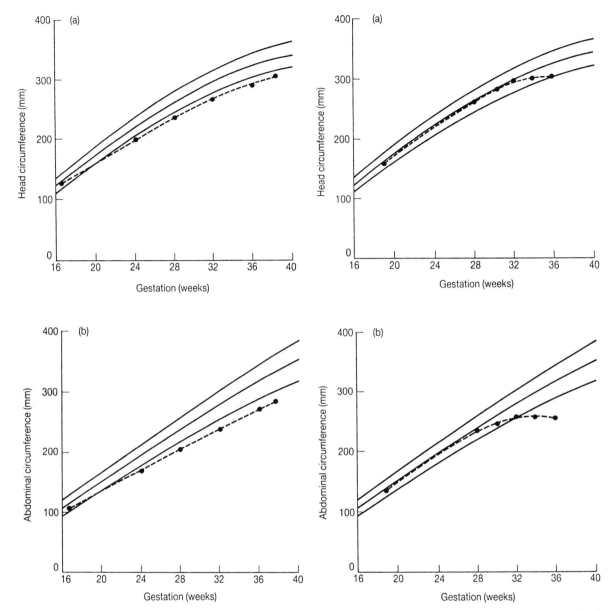

Fig. 2.29 Charts of growth of (a) the fetal head circumference, and (b) the fetal abdominal circumference demonstrating the growth pattern of a symmetrically small fetus

Fig. 2.30 Charts of growth of (a) the fetal head circumference, and (b) the fetal abdominal circumference demonstrating the growth pattern of an asymmetrically small fetus

- stillbirth
- asphyxial handicap leading to cerebral palsy or major mental handicap
- hypoglycaemia, hypothermia, hypocalcaemia
- polycythaemia

- premature delivery
- necrotizing enterocolitis
- haemorrhage into the lungs or the gut.

Appropriately timed delivery may avoid most of these problems.

Management

Growth should be monitored on a fortnightly basis and tests of fetal well-being are indicated in this group of babies. Delivery is indicated for an abnormal test result, cessation of growth, abnormal Doppler waveforms from the fetal circulation (see below) or maturity. Maturity is usually taken to be 34 or more weeks and induction of labour may be attempted if the cervix is favourable.

Tests of fetal well-being

Maternal appreciation of fetal movements

Mothers are asked to record fetal movements either for a specified hour in each day or until their fetus has moved 10 times. A decrease in the number of fetal movements or no fetal movements in a 12-hour period warrants ultrasound examination and a cardiotocograph.

Antenatal cardiotocography (CTG)

This is a record of the fetal heart rate over a 20–40 minute period. The CTG is classified as:

- *Reactive* – more than two accelerations in a 40 minute period; an acceleration being defined as a rise in heart rate above the baseline of >15 bpm lasting for >15 seconds
- Non-reactive
- Deceleratory – showing decelerations (of >15 bpm lasting >15 seconds) usually in the absence of contractions.

The risk of perinatal death within 1 week of a reactive CTG is about 4/1000 births but once the CTG shows a deceleratory pattern the fetus has usually sustained asphyxial brain damage or myocardial ischaemia. In addition, about 20 per cent of deceleratory CTGs are false positives in that the pattern is not seen when labour is induced. However, this is still the most widely used method of assessing fetal well-being.

Biophysical profile

Using ultrasound, fetal movements and breathing movements can be seen and quantified and the biggest vertical column of amniotic fluid may be measured. Fetal tone may be assessed by looking for flexion/extension movements of the fetal body or opening and closing of the fetal hand. All have been used individually as means of assessing fetal well-being but they have been combined with the antenatal CTG to form the biophysical profile. The risk of perinatal death within 1 week of a normal score is 1/1000 births. However, the profile may take up to 90 minutes to complete and it is probable that just as much information can be obtained from the CTG and measurement of amniotic fluid columns as from performing the whole profile.

Doppler waveforms from the fetal circulation

If sound or ultrasound is directed at moving red blood cells then the returning echo will have undergone a shift in frequency that is related to the velocity of the red cells. This is due to the Doppler effect and the waveforms that are obtained can be displayed on a screen (*see* Fig. 2.31).

Simple, continuous wave equipment is relatively

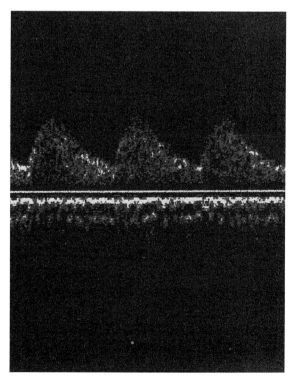

Fig. 2.31 A Doppler sonogram demonstrating the waveform the umbilical artery (upper trace) and vein (lower trace) in a normal pregnancy in the third trimester

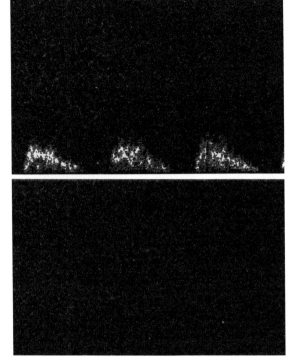

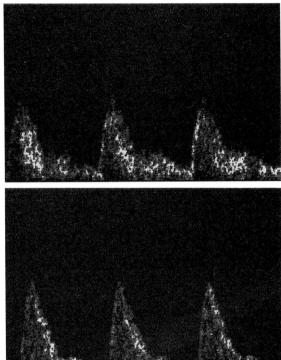

Fig. 2.32 A Doppler sonogram demonstrating loss of end-diastolic frequencies in the umbilical waveforms

Fig. 2.33 Sonograms from the demonstrating fetal aorta obtained by means of pulsed Doppler ultrasound in (top) normal pregnancy, and (below) a pregnancy complicated by growth retardation and central shunting (see text)

cheap and can be used to get waveforms from the umbilical arteries. Normal waveforms (*see* Fig. 2.31) carry a favourable outlook and need to be repeated only once a week; risk of perinatal death in that time is estimated at 1/10 000. Increasing resistance within the fetal circulation in the placenta (due to hypoxia and ischaemia in the intervillous space) leads to loss of frequencies in end-diastole (*see* Fig. 2.32). This is associated with an increased risk of fetal hypoxia and acidosis and if the fetus is viable it should be delivered.

If the fetus is not viable then more complex and expensive ultrasound is needed. Colour coded Doppler ultrasound allows detection of blood flow in vessels that are often too small to be visualized on real-time ultrasound. If the flow can be detected it can be measured by means of pulsed Doppler ultrasound. In this case a Doppler cursor is placed over the area of blood flow and by altering a depth gate waveforms can be obtained from the indicated blood vessel. At present, waveforms are usually obtained from the fetal descending aorta (reflecting blood flow to the fetal body and placenta) and from the middle cerebral artery (MCA) reflecting blood flow to the brain. In normal pregnancies the MCA does not demonstrate recordable frequencies in end-diastole whereas the aorta does (*see* Fig. 2.33(a) and 2.34(a)). In situations of increased placental resistance to blood flow central shunting occurs and the MCA has recordable end-diastolic frequencies whereas the descending aorta loses these frequencies (*see* Fig. 2.34(a) and 2.34(b)).

THE SAFETY OF ULTRASOUND

Ultrasound has been widely used as a diagnostic tool in clinical medicine for more than a quarter of a century. To date no studies have shown any deleterious effects of ultrasound on the fetus or the mother. A good deal of research has been carried out on animals and tissues, but the adverse effects that have been reported from some of these studies have occurred with ultrasound intensities, pulse

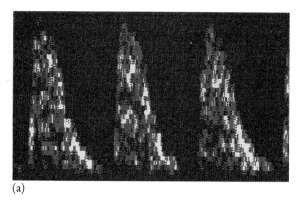

(a)

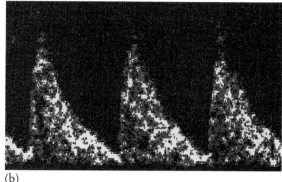

(b)

Fig. 2.34 Doppler waveforms from the middle cerebral artery in (a) normal pregnancy, and (b) a pregnancy complicated by growth retardation with central shunting (see text)

lengths and exposure times far in excess of anything that is currently used, or is likely to be used, in obstetric diagnostic ultrasound. The search for possible hazards will continue, but in the meantime there is no reason to withhold the proven benefits of diagnostic ultrasound.

PRENATAL DIAGNOSIS

This phrase usually includes both the screening tests and the diagnostic tests that are available for congenital abnormalities that affect the fetus. The range of conditions that can be diagnosed has and is increasing rapidly due to techniques of deoxyribonucleic acid (DNA) analysis, particularly with the use of the polymerase chain reaction (PCR).

Screening tests

History

This is really only useful for women who have a positive personal or family history of a particular abnormality or syndrome. Ideally, such women should be seen at a pre-pregnancy clinic by a medical geneticist for counselling. Most common conditions that affect the fetus, however, do so in women without such a history, for example more than 90 per cent of women who give birth to an infant with spina bifida have no relevant history.

Maternal age has long been used as a basis for screening for Down's syndrome and this is considered later under triple testing.

Examination

This rarely helps although the clinical suspicion of impaired fetal growth, oligohydramnios or polyhydramnios may lead to an ultrasound examination.

Routine booking test

Syphilis serology is a useful means of preventing congenital syphilis and rubella immunity protects against congenital rubella infection. Other organisms that cross the placental, such as *Toxoplasma* may be sought in specific situations.

Maternal serum alpha-fetoprotein (MSAFP)

AFP is an oncofetal protein that is produced from the yolk sac and then later from the fetal liver. It is a plasma protein that is excreted into the amniotic fluid via fetal urine. Precisely how AFP crosses the placenta into the maternal circulation is unknown. However, in cases of open neural tube defects MSAFP may be high as it leaks out of neurological tissue. Most neural tube defects (NTDs) are anencephaly or spina bifida.

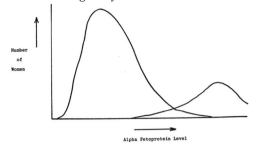

Fig. 2.35 Bimodal distribution of MSAFP

Table 2.3 Risk of chromosomal problems with increasing maternal age

Age (years)	Incidence
35	1 in 250
36	1 in 143
37	1 in 125
38	1 in 111
39	1 in 83
40	1 in 71
41	1 in 63
42	1 in 30

Although MSAFP demonstrates a bimodal distribution (*see* Fig. 2.35) there is a cross-over between normal and affected pregnancies. The level of MSAFP which is chosen as the cut-off point is therefore a trade off between wishing to detect all cases and of not including too many normal pregnancies. In areas where the prevalence of NTDs is 3–4/1000 births than a cut-off level of 2.5 multiple of the median will detect 99 per cent of anencephalics and 85 per cent of open spina bifida with a 2 per cent false positive rate.

Using the MSAFP levels is very dependent upon accurate knowledge of gestational age and therefore all women who are going to have a MSAFP estimation should have a preliminary ultrasound to establish gestational age. This will also exclude other causes of raised MSAFP such as twins, blighted ova and defects of the fetal abdominal wall. It will detect all the cases of anencephaly and in experienced hands it will also detect most cases of spina bifida.

Women with raised MSAFP may undergo amniocentesis in order to allow measurement of amniotic fluid AFP; those with high levels then are offered abortion. Alternatively (and increasingly) they undergo detailed ultrasound examination to exclude or diagnose the NTD.

Women with raised MSAFP and a structurally normal fetus are at high risk of preterm labour and small babies and so should be scanned serially.

Triple testing for Down's syndrome

Until recently the only screening test available for trisomies was maternal age. The incidence of chromosome abnormalities increases with maternal age as demonstrated in Table 2.3.

It has been traditional to offer women over 35 years an amniocentesis for chromosomal analysis of the fetus. Even if all such women accepted only about one-third of Down's syndrome babies would be detected, for although the incidence is less in young women they have many more babies.

Down's fetuses appear to behave biochemically as though they are about 2 weeks less mature than their gestational age. Measurements of MSAFP, unconjugated oestriol (uE3) and hCG can be combined with maternal age by a computer program so that a new risk figure can be given to the woman. Amniocentesis is usually recommended if the risk is greater than 1 in 250 but not otherwise. By this means some older women can avoid amniocentesis and it is projected that the pick rate for Down's syndrome should increase to about 60 per cent.

Ultrasound examination

Ultrasound is in the unique position of being both a screening test and a diagnostic test, depending upon the skill of the operator and the resolution of the equipment. The following are examples of its use as a screening test:

- recognition of the cranial signs associated with spina bifida. These are dilated cerebral ventricles (*see* Fig. 2.36) and the lemon shape of the skull (*see* Fig. 2.37).
- recognition of symmetrically impaired growth, oligohydramnios or polyhydramnios, which should lead to a detailed ultrasound examination
- recognition of the structural markers of chromosomal anomalies. These include cardiac defects, duodenal atresia, omphalocoeles, nuchal fat pads and abnormalities of the position of the hands and feet. It is believed that

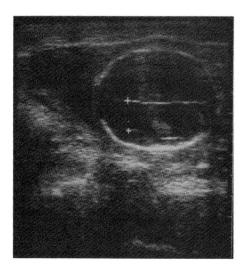

Fig. 2.36 Ultrasound scan of dilated cerebral ventricles

about 85 per cent of Down's fetuses will demonstrate such markers. Their presence should therefore lead to counselling about amniocentesis, a definitive test of the fetal chromosomes.

Diagnostic tests

Apart from ultrasound these are all invasive procedures, performed with ultrasound guidance, and carry a risk of losing a normal pregnancy.

Amniocentesis

About 20 ml of amniotic fluid is removed by puncture of the pregnancy sac by transabdominal insertion of a needle (*see* Fig. 2.37). This fluid contains amniocytes and cells shed from the fetus.

These cells are cultured in the laboratory until they divide; it is on the dividing cells that the chromosomes can be studied. AFP may be measured in the amniotic fluid to aid the diagnosis of a raised MSAFP.

The risk of losing a normal pregnancy from amniocentesis is about 1 in 100. In about 1 in 1000 cases the karyotype is wrong as maternal cells have been obtained (this can never be the case if the karyotype is male).

Amniocentesis is usually performed at 16 weeks' gestation and results are not available until about 19 weeks so abnormalities require a late termination of pregnancy. Recently amniocentesis has been offered as early as 10 weeks' gestation; with new genetics techniques the result is available within a week.

Indications for amniocentesis are:

- *Karyotyping* based on a personal or family history, maternal age, higher risk, triple testing results or the presence of ultrasonic markers (see above).
- *Haemolytic disease.* In cases of rhesus isoimmunization maternal IgG antibodies readily cross the placenta and cause haemolysis of fetal blood. The fetus excretes lysed blood as bilirubin which can be estimated by studying the optical density of amniotic fluid. Direct fetal blood grouping, haemoglobin estimations and haematocrit have largely replaced this technique. The RhD type (Rh positive or negative) can now be diagnosed using PCR DNA techniques on fetal DNA obtained from amniotic fluid. This may be clinically useful where the mother is RhD negative and has anti-D and her partner is heterozygous for RhD. In this clinical context only half of the offspring may be affected by maternal antibody. Identifying the

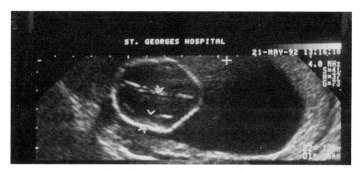

Fig. 2.37 The lemon shaped head seen in Down's syndrome (courtesy of Dr R Patel)

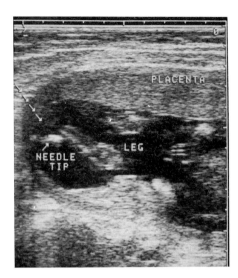

Fig. 2.38 Ultrasound scan used to assist amniocentesis. The needle track is indicated by the arrows. The upper edge of an anterior placenta has been avoided

used to detect neural tube defects in women with raised MSAFP but this has largely been replaced by detailed ultrasound scanning.

- *Inborn errors* of metabolism can be detected by biochemical studies performed either directly on the amniotic fluid or by administering substrate to amniocytes. This technique has now largely been replaced by chorionic villus sampling.

- *Lung maturity.* Measurement of the ratio of lecithin to syringomyelin (L/S ratio) was used to determine lung maturity. A value of more than two indicates that respiratory distress syndrome due to lack of surfactant is unlikely. This test is rarely used now except perhaps with a fetal abnormality that is compatible with a good quality of life but which is worsening.

fetuses at risk will allow appropriate use of fetal blood sampling and intra-uterine transfusion, as necessary, without jeopardizing Rh negative fetuses with invasive investigative and therapeutic procedures.

- *Neural tube defects.* Measurement of amniotic fluid AFP and acetylcholinesterase has been

Chorion villus sampling

This is normally performed at 8–12 weeks but it may be performed transabdominally at any stage of pregnancy. A sample from the edge of the placenta is removed by means of a hollow needle and suction. The needle may either be inserted through the cervix at 8–10 weeks (*see* Fig. 2.39), at 10–12 weeks, or abdominally (*see* Fig. 2.40). The technique has opened the way to diagnosis of many conditions as the material obtained is particularly suitable for DNA analysis.

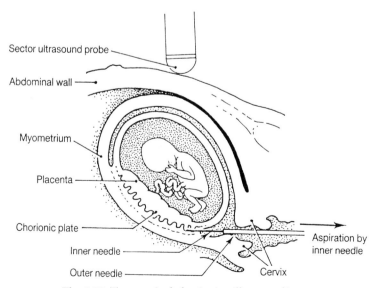

Fig. 2.39 Transvaginal chorionic villus sampling

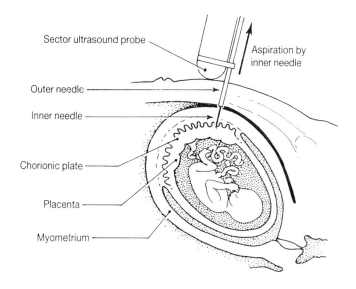

Sector ultrasound probe

Aspiration by inner needle

Outer needle

Inner needle

Chorionic plate

Placenta

Myometrium

Fig. 2.40 Transabdominal chorionic sampling

Chromosome results are available within 2–3 days from the direct preparation so if the result is abnormal a vaginal termination of pregnancy can usually be performed early under general anaesthesia.

The test carries a 4 per cent risk of miscarriage but as about 2 per cent of pregnancies will miscarry at 9–11 weeks the increased risk is only 2 per cent. The result will not reflect the fetal karyotype in about 2 per cent of cases and such women may require a later amniocentesis to confirm the result. Indications:

- *Karyotyping*
  The indications are as for amniocentesis

- *Inborn errors of metabolism.*

These are rare but are usually inherited in an autosomal recessive fashion. They result in an enzyme defect which allows accumulation of a harmful substrate. Examples are Gaucher's disease, an abnormality of glycogen storage and Niemann–Pick disease, an abnormality of lipid metabolism. Direct enzyme assays on the chorionic tissue give a result in 2–3 days.

- *DNA analysis*

As chorionic tissue is growing it is rich in DNA and is highly suitable for gene probing. This may be performed by a direct gene probe or by a more complex method known as restriction length polymorphisms.

Once the gene associated with the condition has been cloned a radioactive probe can be made which will bind to the gene site on the DNA and therefore allow it to be recognized. This technique can be used to identify such conditions as alpha-thalassaemia in which the gene is missing or sickle cell anaemia in which the abnormal gene prevents specific enzymes from cutting up the DNA in the normal way.

Unfortunately direct gene probes are currently only available for a few conditions caused by a single gene mutation. Some conditions, such as beta-thalassaemia, can be caused by multiple mutations. In such cases the affected child and its family has to be studied to determine if the gene is informative. DNA from each family member is broken down into short lengths by specific enzymes and a specific segment (known as a restriction fragment length polymorphism – RFLP) attached to the gene in question is sought. If a RFLP is found in an affected child and one or both of its parents then it can be sought in chorionic tissue in the next pregnancy.

- *Transplacental infections*

Chorionic tissue may be examined for organisms such as rubella, *Toxoplasma* or cytomegalic virus either by electron microscopy or by magnifying the small amount of virus particles that are present by polymerase chain reaction.

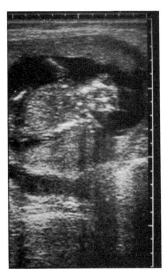

Fig. 2.41 Ultrasound scan of anencephalitic fetus detected early in pregnancy

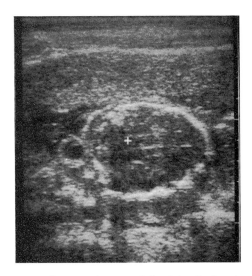

Fig. 2.42 Ultrasound scan of fetal head showing an encephalocoele to the left of the figure

Cordocentesis (transabdominal umbilical blood sample)

This technique allows rapid karyotyping (2–3 days) and involves taking 1–2 ml of fetal blood from the umbilical vein using a fine needle and ultrasound guidance. It carries about a 1 in 100 risk of losing a normal pregnancy as the result of the procedure.

It is most commonly indicated when a structural abnormality in the fetus suggests an underlying chromosomal problem. It is also used to diagnose

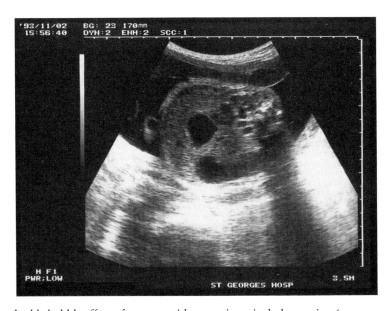

Fig. 2.43 The double bubble effect often seen with upper intestinal obstruction (courtesy of Dr R Patel)

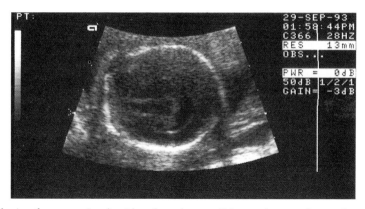

Fig. 2.44 Microcephaly is often associated with other abnormalities such as a distorted ventrical as seen here (courtesy of Dr R Patel)

transplacental infection (by specific IgM) and to diagnose and treat rhesus disease.

Other invasive procedures

Almost any fetal organ can (and has) been biopsied under ultrasound guidance but fetal skin and liver are readily accessible.

Drainage of the fetal bladder in conditions of obstruction may prevent back pressure destruction of fetal kidneys and allow normal levels of amniotic fluid so allowing normal fetal lung development.

Ultrasound examination

In experienced hands ultrasound is a diagnostic test that carries little or no risk to the pregnancy. Ultrasound diagnoses abnormalities by:

- recognizing the absence of a fetal part, for example the absent cranial vault in anencephaly (*see* Fig. 2.41).
- recognition of solid or cystic structures attached to or within the fetus, for example the encephalocoele in Fig. 2.42.
- recognizing the effects of the abnormality on adjacent structures, for example the double bubble caused by duodenal atresia (*see* Fig. 2.43).
- recognizing abnormal growth pattern such as the reduced growth of the fetal head that is seen in microcephaly (*see* Fig. 2.44).

With modern, high resolution equipment most organs in the body can be examined in detail and normality or otherwise assured.

FURTHER READING

Chudleigh P. and Pearce J.M.F. (1993) *Obstetric Ultrasound: How, why and when (2nd edition)*. Churchill Livingstone, Edinburgh.

3

ABNORMAL PREGNANCY

ABNORMALITIES OF THE PELVIC ORGANS

RETROVERSION OF THE UTERUS

Pregnancy often occurs in a retroverted uterus and is nearly always uneventful. In the majority of cases the uterus rises up into the abdomen at about the 12th week. In a small minority the uterus remains retroverted, and as it grows it comes to fill the pelvic cavity completely. The cervix is directed forwards, and upwards (*see* Fig. 3.1). The uterus is then said to be *incarcerated*. As the bladder base is attached to the supravaginal cervix the base of the bladder is distorted and the urethra is elongated. Retention of urine follows and the bladder becomes distended.

The acute retention is extremely painful and if it is not relieved by catheterization overflow incontinence will eventually occur, with frequent escape of small volumes of urine.

Diagnosis

If difficulty in micturition occurs between the 12th and 16th weeks of pregnancy the possibility of incarceration of a retroverted uterus must always be considered. On abdominal examination the distended bladder will be felt; it may contain as much as 1.5 litres of urine, and must not be mistaken for the pregnant uterus. On vaginal examination the posterior vaginal wall is found to be pushed forward by a smooth elastic swelling which occupies the hollow of the sacrum. The cervix is found to be high up behind the symphysis pubis and directed forwards and upwards; this confirms that the swelling in the pelvis is the uterus.

A fibroid in the posterior wall of the uterus might give rise to similar symptoms and signs

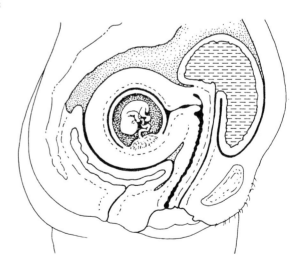

Fig. 3.1 Retroversion of the pregnant uterus. Note the marked elongation and narrowing of the urethra which occurs as a result of upward displacement of the bladder base

during pregnancy. A pelvic ultrasound scan will help to distinguish between the two.

Treatment

If a woman is found to have a pregnancy in a retroverted uterus before the 12th week there is no need to interfere as the uterus will probably right itself and rise up into the abdomen as it enlarges.

In the few cases in which retention of urine occurs a Foley catheter is immediately passed and the bladder is drained continuously. It is worth noting that the catheter will have to be inserted to a much greater distance than usual because of the marked urethral elongation. As the bladder empties the uterus nearly always rises up into the abdomen. When the uterus is palpable abdominally it means that normal bladder anatomy is restored and the woman will again be able to void urine spontaneously.

Only in extremely rare cases in which the uterus remains incarcerated in the pelvis after catheterization will correction by manipulation under anaesthesia be required.

PROLAPSE OF THE PREGNANT UTERUS

Pregnancy may occur in a partially prolapsed uterus. As the uterus grows it eventually becomes too large to sink through the pelvic brim and usually, although not invariably, the cervix is then drawn up and no longer descends. Minor degrees of prolapse merely require support with a ring pessary until the uterus has grown well up into the abdomen.

Pregnancy in a completely prolapsed uterus is very rare. If the cervix remains outside the introitus it may become oedematous and fail to dilate during labour. The woman should be kept in bed for some days before labour, with the cervix within the vagina to allow the oedema to subside.

If a woman becomes pregnant after an operation for prolapse most obstetricians would advise delivery by caesarean section. If the cervix has been amputated it usually dilates more quickly than usual, but occasionally, because of scarring, it either tears or fails to dilate. In such a case caesarean section is carried out.

It is generally agreed that if stress incontinence has been successfully cured by operation, caesarean section is strongly advised.

CONGENITAL ABNORMALITIES OF THE UTERUS AND VAGINA

The more gross uterine abnormalities are often accompanied by infertility, but lesser degrees of malformation do not prevent pregnancy occurring.

A woman with a double uterus may become pregnant in one or both sides. The risk of abortion or premature labour is increased, but labour is often normal. If pregnancy occurs in one uterus it is possible for the other one, which undergoes both myometrial and decidual hypertrophy, to obstruct the descent of the presenting part into the pelvis, and caesarean section may be necessary. The nonpregnant uterus is sometimes mistaken for a pelvic tumour such as an ovarian cyst.

With a bicornuate or septate uterus there may be repeated abortion. In other cases a malpresentation such as a transverse lie or a breech presentation may occur.

A vaginal septum may be present with or without a double uterus. Usually there is no obstruction to delivery; in most cases there is no need for surgery in pregnancy but the septum can be excised or divided during the second stage of labour when the fetal head is pressing on it.

UTERINE FIBROIDS

The incidence of clinically detectable uterine fibroids during pregnancy is about 5 per 1000 in Caucasian women, but is much higher in Negro women.

EFFECTS OF PREGNANCY ON FIBROIDS

During pregnancy the tumours may undergo several pathological changes:

Increase in size and softening

There is hypertrophy of the muscle fibres of the tumour as in the rest of the myometrium, and there is also increased vascularity and oedema, so that it becomes larger and softer.

Necrobiosis (red degeneration)

This change is seldom seen in fibroids except during pregnancy. Rapid degeneration is caused by obstruction to the venous outflow from the tumour as it enlarges. The cut surface of the softened tumour has a reddish-purple colour.

The woman experiences pain at the site, and the fibroids becomes very tender. There is usually pyrexia and sometimes vomiting. This condition must be distinguished from other causes of acute abdominal pain in pregnancy, especially acute appendicitis, concealed placental abruption and torsion of an ovarian cyst which require active treatment – with red degeneration the symptoms and signs subside spontaneously sometimes in a few days. During that time the woman may require treatment with strong analgesic agents and, if vomiting is prominent, intravenous fluids.

Torsion on a pedicle

Torsion on a pedicle of a pedunculated subper-itoneal tumour is a rare accident, but is more common during and just after pregnancy than at other times.

EFFECTS OF FIBROIDS ON PREGNANCY

Pregnancy

Miscarriage is sometimes associated with fibroids. It is only likely to occur with a subendometrial tumour.

Labour

Fibroids occur most commonly in the body of the uterus, and such tumours are drawn up out of the pelvis as the uterus enlarges, so that they do not obstruct labour. Cervical fibroids which remain in the pelvis prevent the head from engaging, may cause malpresentations and obstruct labour.

The third stage of labour may be complicated by postpartum haemorrhage, especially with suben-dometrial tumours and particularly if the placental attachment is over the tumour.

Puerperium

A subendometrial fibroid may become infected during labour, and separation of a necrotic tumour may cause late postpartum haemorrhage. After delivery fibroids regress as part of the general process of involution of the uterus, but they do not totally disappear.

Diagnosis

Fibroids which project are easily felt in the wall of the pregnant uterus, and are distinguished from fetal parts by their fixed position. They become more evident when the uterus contracts, whereas fetal parts become less obvious. They may be very difficult to detect when they have become softened and flattened in the uterine wall; however such tumours are very unlikely to interfere with labour.

If a woman who is known to have a fibroid uterus becomes pregnant the signs of pregnancy are sometimes masked, or confusion may arise about the expected date of delivery because the uterus is larger than expected for the period of amenorrhoea. The tumours may make the swelling feel unlike the normal pregnant uterus in shape or consistency, or make it difficult to feel fetal parts. Confusion may occur if there is amenorrhoea from some cause other than pregnancy, such as the menopause. Conversely bleeding from threatened abortion may be attributed to menstrual bleeding if the signs of pregnancy are obscured.

In cases of doubt a beta unit hCG test for pregnancy is performed, and an ultrasound scan is carried out. This will show the presence in the uterus of a gestation sac containing a fetus, and the characteristic echo of the fibroid.

Treatment

Pregnancy

Most women with fibroids pass through preg-nancy without difficulty.

If necrobiosis (red degeneration) occurs the symptoms may be severe enough to require admis-sion to hospital, but they almost always subside within a week or so without any treatment other than rest in bed and analgesic drugs. Myomectomy is unnecessary and dangerous, carrying a high risk of abortion and also of severe haemorrhage during the operation.

Torsion of a pedunculated subserous fibroid causes severe pain and vomiting and calls for laparotomy and removal of the tumour. Two rare conditions with which torsion of a fibroid may be confused are torsion of the pregnant uterus itself (a complication which usually only occurs in a preg-nant woman with a double uterus) and intra-

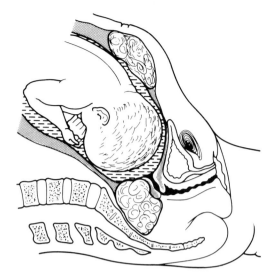

Fig. 3.2 Fibroids complicating pregnancy. The tumor in the anterior wall of the uterus has been drawn up out of the pelvis as the lower segment was formed, but the fibroid arising from the cervix remains in the pelvis and will obstruct labour

peritoneal haemorrhage from a ruptured vein on the surface of a fibroid.

Labour

All pregnant women with fibroids should be delivered in hospital. The great majority can safely be left to deliver vaginally even if the fibroid seems to lie below the fetal head before the onset of labour (*see* Fig. 3.2). Only in very few cases will the fibroid remain in the pelvis as uterine retraction occurs in labour. In such cases, labour becomes obstructed and a caesarean section solves the problem. No attempt should be made to remove fibroids at the time, since this may be hazardous and they diminish greatly in size after delivery. The decision about definitive treatment should be made at least 3 months after delivery when hysterectomy or myomectomy can be safely performed, if indeed any treatment is needed.

OVARIAN CYSTS

Ovarian cysts are not commonly associated with pregnancy, the incidence of clinically obvious cysts being less than 1 in 1000 cases. Any type of ovarian cyst may occur, but luteal cysts, simple serous cysts and benign teratomatous cysts are the most common. It must be remembered that about 10 per cent of cysts in women under 30 years are malignant, and this proportion rises in older women.

Torsion of the pedicle of an ovarian tumour occurs more often during and just after pregnancy than at other times. Rupture of a cyst or intracystic haemorrhage may occur; necrosis from pressure on a cyst during labour, infection and suppuration may follow.

Pregnancy is usually undisturbed by the tumour unless torsion or some other complication occurs. Labour is unaffected unless the tumour lies in the pelvic cavity when obstruction is probable. This is relatively more common with ovarian tumours than with fibroids because the latter usually rise up into the abdomen before labour begins.

Diagnosis

In the early months of pregnancy the uterus is easily distinguished by bimanual examination from the rounded and usually mobile ovarian tumour lying behind it in the rectovaginal pouch.

The chief condition which has to be distinguished from an intrapelvic ovarian cyst associated with pregnancy is retroversion of the gravid uterus. Contrary to expectation an ovarian cyst is not displaced to one side but lies in the midline behind the uterus. It is cystic and tense and a groove may be felt between the cyst and the cervix, which is directed downwards and backwards. A retroverted gravid uterus lies in the rectovaginal pouch, but the body of the uterus is soft and the cervix is displaced upwards and forwards.

A twisted ovarian cyst with an early uterine pregnancy might be mistaken for a tubal pregnancy, but the correct treatment is laparotomy for both conditions. An ovarian cyst that cannot be recognized as a separate structure from the uterus might lead to an erroneous diagnosis of fibroids or a bicornuate uterus. The diagnosis can often be clarified by an ultrasound scan.

Treatment

Because of the possibility of malignancy, torsion, rupture or intracystic haemorrhage, an ovarian cyst discovered during pregnancy should be removed as soon as possible after 12 weeks. This rule does not apply if there is a strong suspicion that the cyst is a luteal retention cyst. Such cysts are characterized by diminishing size on serial ultrasound scans and are best managed conservatively.

It is wise to wait until after the 12th week before operating, as there is a risk of miscarriage if the corpus luteum has to be removed with the ovary before the placenta has taken up its hormonal function. If the tumour proves to be innocent every effort is made to conserve ovarian tissue; the cyst or tumour is enucleated from the rest of the ovary.

If the diagnosis is not made until late in pregnancy, and if the ovarian tumour is not in the pelvic cavity, vaginal delivery is awaited and the tumour is removed in the early puerperium. If the cyst is in the pelvis, it is likely to obstruct delivery and caesarean section with removal of the tumour is necessary.

SOLID OVARIAN TUMOURS

Ovarian tumours which are solid or are cystic with a solid component on an ultrasound scan are much more likely to be malignant than are cysts which are smooth-walled and have no solid elements. Tumours with a solid element must be treated surgically as soon as possible.

CARCINOMA OF THE CERVIX UTERI

CERVICAL INTRAEPITHELIAL NEOPLASIA (CIN)

During pregnancy CIN is much more common than invasive cancer of the cervix as CIN afflicts the younger age group.

It is common practice to take a cervical smear from women at the time of their first visit to the antenatal clinic unless there has been bleeding, the woman is very anxious, does not give her consent or has recently had a cervical smear. If abnormal cells are found the cervix is carefully examined with a speculum to exclude obvious invasive cancer. If the cervix appears normal on ordinary inspection colposcopy should be performed to locate the areas on the cervix from which the abnormal cells originated. Thereafter a colposcop-

ically directed biopsy, taking only a small amount of tissue, can be carried out. If a diagnosis of dysplasia or CIN is established, and invasive cancer excluded, treatment can safely be left until after delivery. Treatment is described in *Gynaecology by Ten Teachers*.

If any woman has not had a cervical smear examination during pregnancy this should be done at the postnatal visit.

INVASIVE CARCINOMA

Invasive cancer of the cervix is a rare complication of pregnancy because the disease most often arises in the later years of menstrual life, when pregnancy is less likely to occur. The disease may cause no symptoms, or may come to light because of bleeding, sometimes after coitus or vaginal examination. It quickly progresses in the same way as in the non-pregnant woman, with free bleeding and purulent discharge from the friable or ulcerated lesion on the cervix. When there is any doubt about diagnosis biopsy is essential.

If the growth is endocervical, delay in diagnosis may occur. Because invasive carcinoma of the cervix can occur in pregnant women, it is very important to consider doing a speculum examination on all women who experience painless vaginal bleeding during pregnancy.

Treatment

This depends on the stage of pregnancy at which the diagnosis is made. In early pregnancy therapeutic termination should be carried out and the cervical lesion should then be treated by radiotherapy or surgery in the usual way. In late pregnancy caesarean section should be performed and may be combined with a radical hysterectomy (caesarean–Wertheim operation).

It is difficult to know what to do when the diagnosis is made in midpregnancy before the fetus is viable. In such cases treatment may sometimes be postponed for a short time to allow the fetus to grow to a size and maturity at which it has hope of survival. Such a postponement should not exceed 4 weeks and can only be agreed to after consultation with the woman and her partner.

PLACENTAL ABNORMALITIES

PLACENTAL SITE, ANATOMY AND ADHERENCE

An abnormal implantation site

In normal circumstances the placenta is attached to the upper segment of the uterus, occupying part of the fundus and passing down onto the anterior or posterior walls. Whereas the precise site of implantation in the upper segment is unimportant, a low lying placenta, with part or all of the placenta in the lower segment of the uterus, can be associated with either bleeding or obstruction of the birth canal during labour. In around 4 cases per 1000 the placenta is partially or wholly implanted in the lower segment – a placenta praevia. The risk of placenta praevia is increased with increasing maternal age, increasing parity and previous caesarean section. It may also be more likely in twin pregnancy. This latter is because the relative increase in placental surface area compared with that of the uterine cavity means the placenta may cover part of the lower segment as well as the fundus and upper segment. When the placenta encroaches significantly on the lower segment of the uterus then, inevitably, there will be haemorrhage before or during labour. Uterine contractions retract the lower segment; this creates a shear force at the site of placental attachment which causes placental separation, significant bleeding, and can present as either antepartum or intrapartum haemorrhage.

Extra lobes of the placenta

Normally the placenta is a single disk, but occasionally it has an extra lobe so that the shape is not completely round. The presence of a bi-lobate or tri-lobate placenta is unimportant, but the presence of a separate succenturate lobe, which is usually attached to the main body of the placenta only by membranes and blood vessels, is clinically significant. This is because a succenturate lobe may become detached when the placenta is removed during the third stage of labour and this may then be retained in the uterus. A retained piece of placental tissue will prevent complete contraction of the uterus allowing postpartum haemorrhage. It

is therefore important that after every delivery the placenta is carefully inspected to make sure that a succenturate lobe has not been left behind in the uterus. In particular, the placental margin should be carefully inspected for signs of blood vessels with an open lumen running right to the edge of the placenta indicating where they have been torn across. This occurrence, or that of a circular defect in the membranes, both suggest a missing succenturate lobe. The uterus must be explored under regional or general anaesthesia and the retained piece of placenta removed manually.

Circumvallate placenta

Where the original area of attachment of the chorionic plate to the uterine wall is small and placental growth has continued beyond its margin, a fibrous ring is seen on the fetal surface of the placenta causing a circumvallate placenta (*see* Fig. 3.3b). This placental variation is not clinically important as the placenta continues to function quite normally.

Hydrophic placenta

In severe cases of isoimmunization the fetus is anaemic and odematous (hydrops fetalis). The placenta shows the same changes, being enlarged, pale and odematous, with a marked increase in placental weight.

Morbid adherence of the placenta

During the second stage of labour the empty uterus contracts down dramatically. Associated shear forces cause placental separation. This occurs through the stratum spongiosum of the decidua, such that the superficial part of the decidua comes away with the placenta while the deeper part remains on the uterine wall. The level of separation is determined by the degree of penetration of the chorionic villi in the first and second trimester. Normally penetration occurs into the deep layers of the decidua but no further.

Morbid adherence of the placenta results from increased penetration of the decidua and myomet-

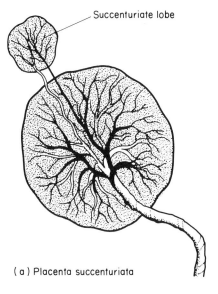

(a) Placenta succenturiata

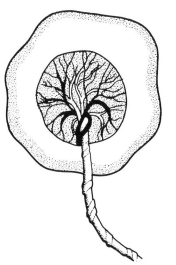

(b) Placenta circumvallata

Fig. 3.3 Placenta abnormalities: (a) a succenturate placenta, i.e. with succenturate life, (b) a circumvallate placenta

rium by the villi. The degree of morbidity is determined by the depth of invasion. The terms used to describe the degree of morbidity are:

- *placenta accreta* – the placenta is partially or completely adherent to the uterus with penetration of villi into the superficial part of the myometrium

- *placenta increta* – the villi penetrate deeply through the decidua into the myometrium
- *placenta percreta* – penetration can even be seen on the serosal surface.

Although placental percreta and increta are indeed very rare, occurring in 1 in 20 000 pregnancies, some degree of morbid adherence will occur in about 1 in 500 pregnancies. Morbid adherence is of great importance clinically because it makes it impossible to remove the placenta completely thus exposing the mother to a severe risk of postpartum haemorrhage (see Chapter 5).

Tumours of the placenta

Apart from choriocarcinoma, tumours of the placenta are rare, and consist of masses of chorionic villi with hypertrophied blood vessels. Such vascular tumours are known as haemangiomas or chorangiomas. The tumour is usually single and the rest of the placenta is normal. Such a tumour is a rare cause of polyhydramnios, due to exudation of fluid from the surface of the tumour.

THE UMBILICAL CORD

Abnormal length

The usual length of the cord is about the same as that of the fetus, 50 cm at term, but considerable variations occur. Excessive length predisposes to descent of the cord below the presenting part and prolapse. The formation of loops round some part of the fetus may cause intrauterine death in the very rare cases in which the cord is pulled tight. During delivery the treatment for a loop round the neck is to slip it over the head after this is born, or if this cannot be done to divide the cord between clamps.

With an abnormally short cord delay in the second stage of labour, premature separation of the placenta or inversion of the uterus are theoretical accidents.

Knots in the cord

These may be formed by fetal movements, the fetus passing through a loop which later forms a knot. Knots are rarely tight enough to obstruct the circulation, but they do occasionally cause intrauterine fetal death. Local protuberances of Wharton's jelly may give an appearance of knotting and have been described as false knots; they are of no importance.

Abnormal insertion of the cord

The cord is usually attached to the centre of the placenta, but sometimes it is attached to the edge of the placenta (squash racket placenta). This is of no importance.

In rare cases the cord is attached to the membranes at some distance from the edge of the placenta, and at this point the vessels may divide into branches which run on the membranes for some distance before reaching the edge of the placenta (velamentous insertion of the cord). This can be dangerous to the fetus if the vessels happen to pass across part of the chorion that lies below the presenting part (vasa praevia), as a branch may be torn when the membranes rupture, leading to fetal blood loss or even exsanguination.

Single umbilical artery

This is an uncommon finding but can be associated with other abnormalities of the fetus. The cut end of the cord should always be inspected after delivery, and if only one artery is seen the child should be carefully examined for other defects notably those of the kidneys, ureters or bladder.

GESTATIONAL TROPHOBLASTIC DISEASE

Gestational trophoblastic disease is a rare complication of pregnancy in which abnormal trophoblastic proliferation occurs, most commonly as a benign hydatidiform mole, but occasionally as a highly invasive tumour – choriocarcinoma. Although the majority of cases of trophoblastic disease present in the first trimester as a uterus filled with abnormal placental tissue, but with no signs of a fetus (*see* Fig. 3.4), choriocarcinoma can occur after normal pregnancy or, less commonly, after a spontaneous miscarriage or an ectopic pregnancy. A full account is given in *Gynaecology by Ten Teachers* (sixteenth edition), but because of the importance of the subject, a shorter version is included here.

AETIOLOGY

An hydatidiform mole occurs because of an abnormal fertilization process. This may be due to either the fertilization of a normal ovum with a duplicated haploid sperm or, more occasionally, the fertilization of an ennucleated ovum by two sperm. In the former case the conception would be chromosomally 46 XX and in the latter 46 XX or occasionally 46 XY. A particular feature of this abnormal fertilization process is that the resultant conception has a nucleus of paternal origin and a cytoplasm of maternal origin in each cell. Cases of partial hydatidiform mole where a fetus is present are usually diploid conceptuses with one maternal and two paternal sets of chromosomes.

BIOLOGICAL BEHAVIOUR AND HISTOLOGICAL CLASSIFICATION

Gestational trophoblastic disease must be regarded as a disease with a spectrum of activity ranging from benign to highly malignant, dependent upon the degree of trophoblastic invasion. All cases have a malignant potential which may not be histologically apparent at the initial diagnosis and so necessitate careful follow-up. Histologically, there may be a partial hydatidiform mole with evidence of a conceptus, a complete hydatidiform mole with well-formed, well-differentiated vesicular structures, a less well-differentiated invasive mole or lastly a frank choriocarcinoma.

In hydatidiform mole the microscopic features are hydropic villi and trophoblastic hyperplasia, which in partial mole affect only the syncytiotrophoblast and in complete mole affect both cyto- and syncytiotrophoblast (*see* Fig. 3.5). Invasive moles have significant trophoblastic hyperplasia and persistence of placental villi with some degree of invasion into the myometrium and occasionally metastases, though without disease progression.

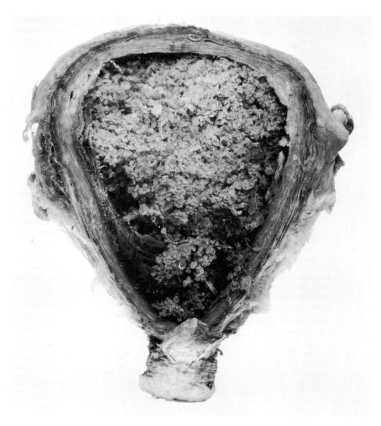

Fig. 3.4 Uterus containing a hydatidiform mole

Choriocarcinoma is a frank carcinoma with both cytotrophoblastic and syncytiotrophoblastic elements. There are in fact two important malignant entities, gestational choriocarcinoma and the rare placental site trophoblastic tumour in which the malignancy arises from the trophoblast of the placental bed.

When making the diagnosis of hydatidiform mole, it is important to note that both the features of hydropic change and trophoblastic proliferation must be present and that hydropic change alone does not indicate the presence of a hydatidiform mole nor the potential for malignant transformation.

RECOGNITION OF MOLAR PREGNANCY

All molar pregnancies have the potential to develop into a potentially fatal carcinoma. All women, therefore, need careful surveillance and follow-up until that risk has been excluded. This presents a practical problem. Because molar pregnancy is relatively rare, occurring in approximately 1.5 births per 1000 in the UK and between 0.2 and 2 births per 1000 worldwide, no individual gynaecologist can build sufficient clinical experience in its management. The risk of molar pregnancy developing into choriocarcinoma is equally small at between 2 and 5 per cent.

In the UK the management problem has been resolved by setting up regional centres for the monitoring of molar pregnancies and the treatment of choriocarcinoma. These centres are at Charing Cross Hospital in London, the Jessop Hospital in Sheffield and Ninewells Hospital in Dundee. They are able to provide up-to-date management and advice and they keep careful cancer registry records. The effectiveness of this system clearly depends first on the recognition of molar pregnancy and then on appropriate referral to one of the specialist centres.

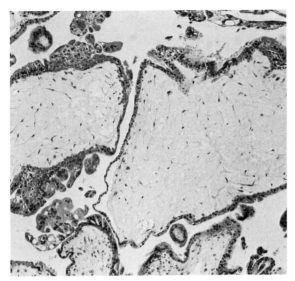

Fig. 3.5 Microscopic section of hydatidiform mole showing otstic spaces due to hydropic change and proliferation of both syncyto and cytotrophoblast

Most cases of molar pregnancy are now recognized by ultrasound, either at a routine early examination, or following referral as a threatened miscarriage with bleeding in early pregnancy. The ultrasound picture may show a typical well-differentiated complete hydatidiform mole with marked vesicular structures and notably no fetus, but alternatively when the ultrasound picture shows the presence of a non-viable fetus, there may be trophoblastic disease present in the placenta, as a partial mole. Finally, there may simply be a picture of an enlarged uterus containing amorphous placental debris, such as is seen in many cases of incomplete miscarriage. Where the ultrasound picture is not diagnostic the woman should be treated as a threatened miscarriage and a diagnosis can only be made if the uterus is evacuated and the retained products of conception are sent for histology. This should be routine practice with all cases of incomplete miscarriage, since even at evacuation the operator may not recognize that trophoblastic disease is present when there are no typical vesicles present. Without histological examination the diagnosis may be missed.

When looking at a complete hydatidiform mole ultrasonically, the multiple vesicles reflect the sound. This gives a snow storm appearance. Better quality pictures can be obtained with transvaginal, rather than with abdominal, ultrasound examination.

MANAGEMENT

A molar pregnancy must be removed by suction evacuation of the uterus. This method will remove the abnormal placental tissue most effectively. It has the added advantage of enhancing uterine contraction, thus reducing the risk of bleeding which is greater with molar pregnancy than with an incomplete miscarriage. Because of this risk of bleeding, syntometrine should also be given during the operation. When the uterus has been evacuated the next stage depends on the results of histological examination of the trophoblastic tissue. Those with gestational trophoblastic tumour will require chemotherapy, those with a partial or complete mole surveillance.

SURVEILLANCE

Surveillance is performed by assay of hCG, one of the first examples of the use of a tumour marker. Normally hCG falls within 4 weeks of evacuation so persistent hCG means that there is persistence of trophoblastic tissue, and that the woman may have a gestational trophoblastic tumour. Follow-up can be by assay of either urine or serum, the later being more accurate at or near the lower limit of detection. Follow-up continues 2-weekly after evacuation until hCG is absent, monthly for the first 12 months and then 3 monthly to a total of 2 years. Further follow-up is advisable for a short period after any subsequent pregnancy when there may be further risk of trophoblastic tumour.

TREATMENT OF GESTATIONAL TROPHOBLASTIC TUMOUR

As with all gynaecological cancers, treatment could theoretically be by surgery, radiotherapy or chemotherapy. Gestational trophoblastic tumours have proved highly sensitive to chemotherapy which is almost invariably the treatment used. Surgery has little place, because of the high vascularity of the tumour leading to a high risk of bleeding and also because of the effectiveness of chemotherapy. Its principle role is where the tumour is resistant to chemotherapy and there are single metastases persisting despite chemotherapy, for example in the

Table 3.1 Criteria for intervention with chemotherapy

- High levels of hCG more than 4 weeks after evacuation
  Serum value greater than 20 000 IU/l; urine greater than 30 000 IU/l
- Increasing hCG values at any time after evacuation as shown by a minimum of three rising values
- Persistent uterine blood loss
- Evidence of metastases in the central nervous system, kidney, liver or gastrointestinal tract or pulmonary metastases greater than 2 cm in diameter or greater than three in number
- Any level of hCG persisting 4–6 months after evacuation

Note: Women in group one are at risk of uterine perforation because of the high level of trophoblastic activity in the uterus.

lungs or brain. Radiotherapy may be effective in this malignancy, but there is little experience of its use.

CHEMOTHERAPY

The criteria for intervention with chemotherapy used in the UK were devised at Charing Cross Hospital in the early 1970s. They have been adopted by the WHO study group on trophoblastic disease and are shown in Table 3.1.

CHOICE OF CHEMOTHERAPY

Initial experience with chemotherapy was achieved using Methotrexate and later with Actinomycin. Combined chemotherapy is now used and the type of therapy is determined by whether the women are thought to be at low or high risk. It must be appreciated that some receiving chemotherapy have merely an invasive mole, whereas others are at a high risk of dying of the disease, because they already have metastatic choriocarcinoma. The cri-

Table 3.2 Criteria for assessing risk in gestational trophoblastic tumour

- Increasing age
- Type of previous pregnancy
- Duration between pregnancy and chemotherapy
- hCG level
- ABO blood group compatibility with sexual partner
- Tumour size
- Size of metastases
- Number of metastases identified
- Success of prior chemotherapy

teria used to determine risk are shown in the table (*see* Table 3.2).

The worse prognosis is in the older woman where trophoblastic tumour follows abortion or term pregnancy rather than hydatidiform mole, when a long duration exists between diagnosis and chemotherapy, where there is a high hCG level, where there is evidence of extensive disease or failure to respond to prior chemotherapy and when the woman and her partner are of opposite ABO blood groups.

Those at low risk are treated with Methotrexate and Folinic acid over an 8 day course with 6 day intervals between courses. High risk women are treated with the EMA/CO regimen (Etoposide, Methotrexate, Actinomycin, Cyclophosphamide, Oncovin (Vincristin)) or the POMB regimen of Cis-Platinum, Oncovin (Vincristin), Methotrexate and Bleomycin. In general the EMA/CO regimen is used first and around 90 per cent of women will survive following treatment with this. In those who do not respond the POMB regimen will achieve remission in around 40 per cent. These are highly toxic treatments with a high incidence of side effects, but are justified by the success rates achieved.

ANTEPARTUM HAEMORRHAGE

The term antepartum haemorrhage is applied to bleeding from the vagina occurring at any time after the 24th week of pregnancy and before the birth of the child.

Antepartum haemorrhage may be described under three headings:

- haemorrhage from a normally situated placenta. This is *placental abruption*

- haemorrhage from partial separation of a placenta abnormally situated on the lower uterine segment. This is termed *placenta praevia*. Haemorrhage is unavoidable when the lower segment becomes stretched in labour
- haemorrhage from a lesion of the cervix or vagina such as an erosion, a polyp or a carcinoma.

PLACENTAL ABRUPTION

Varieties

Owing to the separation of a portion of the placenta from its uterine attachment, part of the wall of the intervillous space is breached and maternal blood escapes from the opened sinuses. This blood may track down between the membranes and the wall of the uterus and escape at the cervix (*revealed haemorrhage*, see Fig. 3.6) or remain inside the uterine cavity (*concealed haemorrhage*, see Fig. 3.7). In fact the distinction is more clinical than anatomical; in nearly all cases there is some external bleeding, but in concealed cases the degree of shock is out of all proportion to the external loss.

Cause and pathology

The reported incidence varies from 1 in 80 to 1 in 200 deliveries. In many cases no cause can be discovered, but about a quarter of the cases are associated with hypertension and proteinuria. It is not certain that the hypertension or proteinuria is the cause of the haemorrhage; proteinuria may follow rather than precede it.

It has been claimed that placental abruption is caused by folic acid deficiency. The disease is more common with advanced parity and in lower income groups. There is also said to be an association with megaloblastic anaemia, which is also associated with folic acid deficiency. Studies of folate levels and of the bone marrow have given some support to the theory. However, placental abruption is not especially frequent in Nigeria, where megaloblastic anaemia is common, and there has been little evidence that folic acid supplements during pregnancy reduce the incidence in this country.

A few cases of placental separation are caused by trauma during external version or a direct blow on the abdomen. There is almost certainly a relationship with cigarette smoking.

The bleeding in cases of placental abruption may be of any degree from a small retroplacental haematoma, which may not affect the fetus and which may only be discovered after the placenta is delivered, to a large collection of blood which distends the uterus and kills the fetus by separating a large part of the placenta.

In severe cases haemorrhages occur into the substance of the uterine wall and may extend to the peritoneal surface which then looks bruised and oedematous.

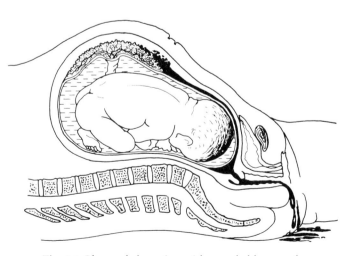

Fig. 3.6 Placental abruption with revealed haemorrhage

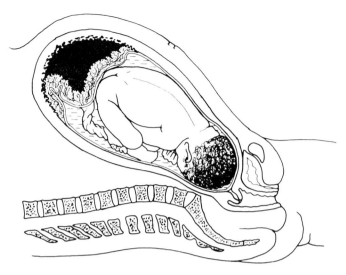

Fig. 3.7 Placental abruption with concealed haemorrhage

It is sometimes stated that such haemorrhage interferes with uterine contraction, and that this explains the continued bleeding which occurs in some, even after delivery of the fetus and placenta. This is improbable as the uterus is tense from spasm, rather than relaxed, before delivery. In concealed cases the whole uterus is tense and tender, but in revealed or less severe cases there is more localized tenderness over the site of haemorrhage. If postpartum haemorrhage occurs after placental abruption it is caused by hypofibrinogenaemia and failure of blood clotting rather than by atony of the uterus. Disseminated intravascular coagulopathy is common after moderate abruption and should always be considered and an investigation be made. This can lead to renal cortical necrosis or lower nephron nephrosis, which result in anuria.

Symptoms and signs

Placental abruption with revealed haemorrhage

With or without any obvious cause the woman notices blood coming from the vagina. The amount of blood lost varies, but is usually not great. There may also be slight abdominal discomfort and tenderness over the placental bed if it is anterior.

The initial diagnosis is between abruption and placenta praevia. Tenderness over the placental site, engagement of the fetal head (in late pregnancy) and fetal death or fetal distress suggest that the symptoms are due to abruption. However, in many slight cases the fetus is not adversely affected and in multiparae the fetal head may not engage until the onset of labour, even in normal women. If the head is not engaged the placental site should be localized by ultrasound scan.

It is dangerous to attempt to exclude the presence of a placenta praevia by vaginal examination and the passage of a finger through the cervix. If the placenta should prove to be low-lying such an examination may cause torrential bleeding. In exceptional cases, in which the diagnosis is in real doubt and the degree of bleeding is such that investigation by ultrasound scan would take too long, the woman should be taken straight to the operating theatre and delivered by caesarean section.

Placental abruption with concealed haemorrhage

The symptoms and signs vary with the severity of the case. In the extreme cases the gravity of the woman's condition may be out of all proportion to the amount of blood effused into the uterus. Shock is due not only to the haemorrhage but also to the painful uterine distension. In spite of severe shock with a low blood pressure) the pulse rate may not be raised, at least for a time, and this can be misleading.

The overdistension of the uterus causes severe

and constant abdominal pain. The uterus may be larger than would be expected for the period of gestation reached and more globular in outline. It is hard, of wooden consistency and is extremely tender. The fetal outlines cannot be made out and the fetal heart cannot be heard.

Should the blood loss be smaller or the degree of myometrial damage be less the symptoms are correspondingly less severe. The woman complains of a sudden attack of abdominal pain and at the same time feels faint and suffers from nausea. She looks ill, her mucous membranes are pale, and the pulse rate is raised. No abnormality may be discovered on abdominal examination, except that palpation of the uterus elicits tenderness, often localized to the placental site; there is difficulty in detecting these signs with a posteriorly implanted placenta if a sufficient area of placenta has been detached or if a large enough area of myometrium is put into spasm the fetal heart sounds are often not present. A correct diagnosis is important as such a woman is likely to have a recurrent and more serious haemorrhage. Protein may be present in the urine.

In many cases the bleeding is partly concealed and partly revealed and the signs and symptoms are mixed. The severity of a case with external haemorrhage must never be judged solely by the amount of blood lost *per vaginam* but by the general condition of the woman. A trifling external loss may be accompanied by serious concealed haemorrhage.

Diagnosis

Revealed haemorrhage

Revealed haemorrhage in placental abruption simulates placenta praevia, and in many cases an immediate distinction is not possible. Placenta praevia is suggested by an absence of pain, a history of recurrent attacks of bleeding, by the absence of hypertension or proteinuria, by a malpresentation or an unduly high presenting part. If the fetal head is engaged or investigation by ultrasound shows that the placenta is certainly situated in the upper uterine segment a firm diagnosis of abruption can be made.

The danger of vaginal examination, except in the operating theatre with all preparations made for caesarean section, has already been emphasized.

After delivery the diagnosis can be confirmed by examining the membranes. The hole through which the child is delivered may be close to the placental edge if the case was one of placenta praevia.

Concealed haemorrhage

The diagnosis must be differentiated from cases of intraperitoneal haemorrhage caused by advanced ectopic gestation, spontaneous rupture of the uterus, or acute polyhydramnios. These, although very rare complications of pregnancy, closely resemble placental abruption of the concealed variety. They must be diagnosed by the history and on the physical signs present in each case. Other conditions complicating pregnancy, such as red degeneration of fibroids, torsion of the pedicle of an ovarian cyst, volvulus, intestinal obstruction, acute appendicitis or peritonitis from any other cause, may have to be considered during the early stages of a concealed haemorrhage. Some vaginal loss of blood is nearly always present and this makes diagnosis easier. Ultrasound examination of the uterus has proved invaluable in the diagnosis of concealed accidental haemorrhage since the retroplacental clot can usually be visualized and the presence of low lying placenta excluded.

Prognosis

The most important factor in the prognosis is the degree of shock and its duration. The amount of blood lost will obviously be important. If the uterus starts to contract rhythmically, instead of remaining in spasm, the prognosis is improved, even if the external bleeding increases temporarily when the woman goes into labour. The sooner the uterus is emptied, the sooner bleeding will be arrested and the risks of disseminated intravascular coagulopathy will be reduced.

In cases of revealed haemorrhage the maternal risk is small, but in severe cases of concealed haemorrhage the maternal mortality may exceed 10 per hundred thousand total.

The prognosis for the child is bad. The perinatal mortality is over 100 per 1000 total births. This is caused by asphyxia from placental separation, sometimes combined with pre-existing fetal growth retardation or preterm delivery.

Treatment

All cases of antepartum haemorrhage should be

admitted to hospital. In severe cases this is essential; in less severe cases the diagnosis is often uncertain and placenta praevia cannot always be excluded. A vaginal examination should not be made before the transfer of the woman to hospital. In cases with severe shock, a blood transfusion or one of the plasma expanding substances should be set up before moving the patient.

Since the term revealed haemorrhage includes cases which vary in severity from a slight loss of blood to a profuse flooding no single method of treatment will be applicable to all cases.

Revealed haemorrhage with slight bleeding

If the amount of bleeding is only slight and placenta praevia has been excluded by an ultrasound scan there is no need to do more than admit the woman to hospital for rest and observation. In many cases no further bleeding occurs and the pregnancy continues to term. The eventual appearance of the placenta will show that part of it has been detached, being brown, shrunken, and more solid than the rest and sometimes having old blood clots adherent to it depending on the stage of gestation.

Revealed haemorrhage with more severe bleeding

This group includes cases in which the amount of bleeding is sufficient to be dangerous or there is a degree of shock. In most severe cases the fetus is already dead and need not be considered. The object of treatment is to empty the uterus, contract and retract it with as little bleeding as possible and without added risk to the mother. If the woman is in labour this is best achieved by allowing the uterus to empty itself. Caesarean section is seldom indicated if the fetus is dead. A blood transfusion of 4 to 6 units should be given if there are any signs of shock. If the woman is not already in labour that should be induced, by low rupture of the membranes, without delay. If regular contractions do not follow an oxytocin drip may be given; only in a few exceptional cases in which profuse bleeding persists should the uterus be emptied by caesarean section.

Many women are already in labour, or labour soon follows induction. The second stage of labour may be shortened with the aid of the forceps or the vacuum extractor. The third stage should be conducted with care and an intravenous injection of oxytocin 5 units or syntometrine 500 μg should be given immediately after the birth of the head.

Caesarean section is not required in cases with slight bleeding unless there is evidence of fetal distress. The operation may be indicated when the fetus is alive, the woman is not in labour and bleeding is severe, but it should only be performed after maternal shock has been treated.

Concealed haemorrhage

Most cases of concealed haemorrhage are very serious, because the woman has collapsed as a result of the loss of blood and the painful distension of the uterus. The first essential is to treat the shock and not to attempt to deliver the child before this has been done. If the woman is first seen at home an emergency obstetric unit (flying squad) or paramedical resuscitation team should be called and resuscitation commenced. Immediate transfer to hospital with an intravenous infusion running is essential. In hospital the condition of the woman is continuously observed. Pain relief will be necessary and can be given liberally if the baby is dead.

In a severe case several litres of blood will be needed. The pulse and blood pressure are not always reliable indices of need; the speed of transfusion is best judged by monitoring the central venous pressure with a catheter inserted through the external jugular vein.

As the state of shock passes off improvement usually occurs. Treatment will be influenced by the parity of the woman and by the duration of the pregnancy. The majority of these women are multiparous women, and as the fetus is small and premature, an easy quick delivery is to be expected if labour is established. After the initial treatment for shock the membranes should be ruptured. It was formerly taught that the membranes should not be ruptured until the uterus had lost its wooden hardness, as it was feared that further loss of blood would occur from an atonic uterus. There is little ground for this fear and continued bleeding is more often due to hypofibrinogenaemia, and the danger of this or of renal necrosis is reduced by early rupture of the membranes and consequent reduction of intrauterine tension.

If the fetus is alive caesarean section is now recommended. The results in the past were poor, because the operation was performed in desperate cases, some of which had received inadequate transfusion, and in some of which DIC was present so that bleeding could not be checked, even by hysterectomy. The section must be performed before irreversible shock is established. Most cases respond rapidly to artificial rupture of the membranes. When the fetus is dead, section should only be considered if the woman's condition is deteriorating in the absence of uterine contractions, especially in the case of a primigravida with a tight cervix.

The possibility of DIC as a result of concealed haemorrhage should be borne in mind. It should be suspected in any case of delayed or absent clotting. A clotting screen including fibrinogen estimation should be made and a critical level is considered to be 1.0 g/l. It can be treated by transfusing fresh frozen plasma which, unlike the blood issued by the transfusion service, contains fibrinogen in addition to all the coagulation factors. If fresh blood is available this will provide approximately 1 g in each 500 ml. However, the fibrinogen level is often less important than the circulatory condition.

Treatment after the labour is over

The fact that the woman has been delivered after placental abruption and that there is no great amount of postpartum haemorrhage does not necessarily mean that she will do well. In the absence of efficient treatment some of these women die a few hours after delivery from heart failure. The women are not safe until recovery from shock is complete, as indicated by the general condition, the pulse rate and blood pressure. An amount of postpartum haemorrhage which would be trifling in the case of a robust woman may be of grave consequence in the case of one who has had severe antepartum haemorrhage, and it is always prudent to obtain a generous amount of cross-matched blood for these women.

It is important to continue with blood transfusion until the woman's condition is restored to normal or until the total blood loss has been made good. Bleeding in the third stage must be controlled by ergometrine intravenously during delivery and intramuscularly afterwards. Many of these women are best nursed for 24 hours after delivery in an intensive care unit.

PLACENTA PRAEVIA

A placenta is described as praevia when it is wholly or partly attached to the lower uterine segment. Haemorrhage is inevitable when labour begins.

DEGREES OF PLACENTA PRAEVIA

The degree of encroachment onto the lower uterine segment is important because both treatment and prognosis are determined by it. Figure 3.8 illustrates the classification which is often used:

- *type I*   the placenta is only partly attached to the lower segment. Its lower margin dips into the lower segment but is at a little distance from the internal cervical os. This is sometimes called a lateral placenta praevia
- *type II*   more of the placenta is attached to the lower segment so that its lower margin reaches down to the internal os
- *type III*   the placenta overlies the undilated internal os, but if a finger was passed through the cervix it would be able to reach the margin of the placenta
- *type IV*   the placenta overlies the undilated internal os, but if a finger was passed through the cervix it could not reach the margin of the placenta. This and type III are often referred to as central or complete placenta praevia.

This classification may be useful for descriptive purposes, but most unfortunately it gives the impression that the diagnosis of placenta praevia and its degree is made by passing a finger through the cervix. This method of diagnosis can be disastrous and is to be avoided.

More recently, a simple classification has allowed the clinical rather than the pathological aspects to be emphasized. Lateral and marginal placenta praevia correspond to the former types I and II. Central placenta replaces old type III and IV for who would wait in labour to tell the differentiation? Treatment is just the same.

Pathology

Placenta praevia occurs about once in 250 pregnancies, and is slightly more common in women who have had several children and in multiple pregnancy because of the large placental area. Those who had a lower segment caesarean section pre-

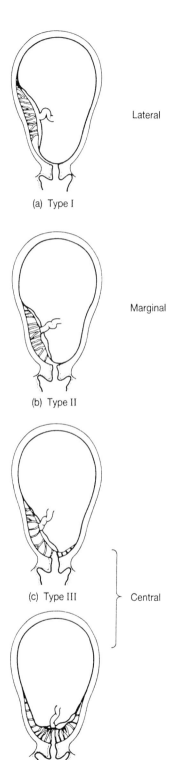

(a) Type I   Lateral

(b) Type II   Marginal

(c) Type III   ⎱ Central

(d) Type IV

Fig. **3.8** Classification of degrees of placenta praevia

viously also have an increased risk of lower segment implantation of the placenta on the scar.

A placenta praevia is often irregular in shape and variable in thickness. It may cover a larger area than normal, and is often pathologically adherent in part to the lower uterine segment. These changes are explained by the comparatively poor blood supply which the placenta obtains from the less vascular lower segment. The cord frequently has a low marginal insertion.

The haemorrhage is from maternal vessels which are opened up by separation of the placenta as the uterine contractions dilate the lower segment. The separation during pregnancy may be slight, but it is inevitably greater during labour, when severe bleeding occurs. Except in cases in which the placenta is torn there is no loss of fetal blood, but in severe cases fetal oxygenation will be impaired because of placental separation or compression during labour, or because maternal haemorrhage causes anaemia and hypotension, with reduced blood flow on the maternal side of the placenta. The cord vessels may be compressed in vasa praevia.

Symptoms and course of labour

During the last 12 weeks of pregnancy (and occasionally earlier) the woman notices slight haemorrhages from the vagina. These occur without evident cause, perhaps during sleep, but they may also follow hard exercise or any local disturbance such as coitus. There are usually repeated slight warning haemorrhages, but occasionally the first bleed may be a severe one, and in a few cases there is no bleeding until labour starts. There is no pain and the fetal movements and heart sounds are usually normal.

During labour severe haemorrhage is inevitable as the cervix dilates.

In the third stage of labour there may be postpartum haemorrhage because the placental site is larger than normal and lies on the lower segment which may not retract efficiently.

Diagnosis

A history of repeated painless losses of blood in late pregnancy, small in amount at first but usually increasing, is strongly suggestive of placenta praevia.

On abdominal examination the fetal head is not engaged. It may be high and freely mobile; the breech may be presenting; or the lie may be oblique, because the placenta occupies the lower segment and prevents the head entering the pelvis. There is no tenderness and the fetal heart sounds are present.

In cases with slight bleeding it is safe to inspect the cervix by gentle passage of a speculum to exclude any incidental cause of bleeding. However, if a cervical erosion is found this does not exclude placenta praevia, and only if any cervical lesion is actually seen to be bleeding should it be accepted as a possible cause of the haemorrhage.

Hypertension or proteinuria are not found, except coincidentally.

Placental localization

Unless the woman is bleeding so profusely that immediate treatment is essential, an attempt should be made to determine the position of the placenta. Several techniques have been tried in recent years but the only method now in common use is ultrasound scanning which is now freely available in all maternity units in Britain. It is without risk to the woman or fetus and is the method of choice. If a scan is performed at any stage of pregnancy for other reasons placental localization should always be undertaken at the same time, thus enabling many cases of placenta praevia to be diagnosed before any antepartum haemorrhage has occurred. A persistent malpresentation in late pregnancy, even in the absence of bleeding, requires ultrasound examination, when a placenta praevia may be discovered.

The characteristic mottled shadow on the screen may not outline the lower edge of the placenta accurately, but repeated scans can safely be performed if necessary. Further a vaginal probe will give an excellent image. The placenta sometimes appears to be low-lying before the 32nd week, but re-examination later in pregnancy after the lower segment has formed may exclude this. A fundal placenta is easy to outline, and this immediately excludes placenta praevia.

Prognosis

The chief causes of death in cases of placenta praevia are haemorrhage and shock; without efficient treatment the danger is great. The amount of bleeding will be least with a type I (lateral) placenta praevia, but progressively greater with the more central types. Both antepartum and postpartum haemorrhage may occur. Some hazards arise from caesarean section, which is the necessary treatment, but in well-equipped units, with properly trained staff, these hazards are minimal.

The essential measure to reduce the risk of placenta praevia is to transfer any woman with slight antepartum haemorrhage to hospital at once, so that accurate diagnosis and preparation for treatment can be made before severe haemorrhage occurs. The maternal mortality in cases managed in this way is very low. The outlook for the fetus is good if active measures are taken to treat maternal haemorrhage and to deliver the fetus early by caesarean section. Prematurity may be a problem.

Treatment

Following an episode of painless blood loss, the woman must be admitted to hospital as soon as possible. With more severe bleeding an intravenous infusion is set up before transferring the woman urgently to hospital. It is most important that no vaginal examination is made.

In all cases with severe bleeding immediate active treatment is required and delay is perilous, but in the most common type of case seen today the woman is admitted when the bleeding has only been slight and there is time for investigation; it is usually possible to make a diagnosis by clinical and ultrasound examination. The chief cause of fetal mortality used to be prematurity, but if the bleeding was only slight and the fetus was some weeks premature, then the risk of keeping the woman in bed while the fetus grew was justifiable, provided that the mother was in a hospital where treatment by blood transfusion and operation was available immediately should severe bleeding occur. This expectant attitude greatly reduced the fetal mortality without increasing the maternal mortality. If severe bleeding starts while the woman is in hospital caesarean section is performed immediately. The exact position of the placenta is of no importance and digital examination is dangerous and unnecessary.

In a few cases the diagnosis is still uncertain after ultrasound examination, and in these *few exceptional cases* when the pregnancy is near term it may be justifiable to make a pelvic examination, provided that it is done in the operating theatre with the woman anaesthetized and the instruments sterilized and ready for caesarean section. A finger is passed very gently through the cervix. If a placenta of type II, III or IV is encountered, or if moderate or severe bleeding is precipitated, then caesarean section is performed. If the placenta is of type I, and particularly if it is situated anteriorly, low rupture of the membranes is done, which allows the presenting part to descend and compress the lower margin of the placenta and so control the bleeding. If no placenta praevia is felt it is still wise to rupture the membranes, as the finger has now been inserted into the cavity of the uterus with a risk of introducing infection.

If the woman is shocked caesarean section should not be performed until this has been corrected by transfusion. Section should still be performed if there is severe bleeding and the fetus is premature or dead; the primary purpose of the operation is to control the bleeding by emptying the uterus and allowing it to retract.

If the placenta lies anteriorly a few obstetricians recommend upper segment caesarean section as there may be severe bleeding from large vessels in the lower segment and from the placenta during a lower segment operation, but it is nearly always found in practice that the lower segment operation can be safely performed, with the advantage of a much more secure uterine scar.

In all cases blood transfusion should be freely employed because, although the woman may recover from the initial haemorrhage, the possibility of further bleeding in the third stage must be anticipated.

VASA PRAEVIA

A rare cause of antepartum haemorrhage occurs when there is a velamentous insertion of the cord and the vessels lie on the membranes covering the internal os in front of the presenting part. When the membranes rupture these vessels may be torn and vaginal bleeding occurs. The blood lost is fetal blood and the fetus may become exsanguinated. The diagnosis is usually missed, but is suspected if the fetal heart rate alters abruptly after rupture of the membranes. Vaginal blood loss can be tested for fetal haemoglobin. If the diagnosis is made the fetus should be delivered as quickly as possible probably by caesarean section.

POLYHYDRAMNIOS AND OLIGOHYDRAMNIOS

POLYHYDRAMNIOS

This is an excess of amniotic fluid. The average volume of fluid at term is 800 ml, but a range of 400 to 1500 ml is accepted as normal. Probably only volumes in excess of 2000 ml would be noticed as abnormal on clinical examination.

In most of the cases the excess fluid accumulates gradually (*chronic polyhydramnios*) and is only noticed after the 30th week. In a few exceptional cases polyhydramnios occurs earlier and more quickly (*acute polyhydramnios*), and many of these cases are associated with uniovular twins. The composition of the fluid in cases of polyhydramnios does not usually differ from normal.

Aetiological factors

Polyhydramnios occurs more often in multiparae than in primigravidae. It may have fetal or maternal causes, although frequently no cause is found (*see* Fig. 3.9).

Fetal causes

Polyhydramnios may occur with *twin pregnancy* of either type. Usually only one sac is distended. The association of acute polyhydramnios with uniovular twins is due to twin-transfusion syndrome.

Polyhydramnios often occurs with *anencephaly*. There may be some exudation from the exposed brain, but a more likely explanation is that the fetus does not swallow normally.

Polyhydramnios can be due to *oesophageal or duodenal atresia*, when inability to swallow is certainly the explanation. In all cases in which there has been an excess of fluid the newborn infant must be examined to exclude intestinal atresia.

Polyhydramnios may also occur with other fetal abnormalities, including spina bifida and a rare cause is *chorioangioma of the placenta*.

There may be excess of fluid in cases of *hydrops fetalis*, as in severe rhesus alloimmunization.

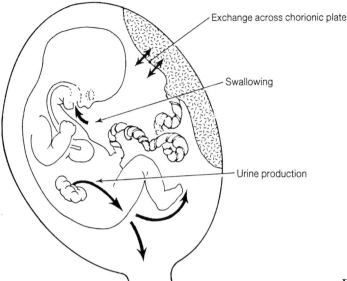

Exchange across chorionic plate

Swallowing

Urine production

Fig. 3.9 Circulation of amniotic fluid

Maternal causes

Polyhydramnios may be associated with maternal diabetes. Not only is there an excess of amniotic fluid, but the placenta and fetus are large. There may be an excess of glucose in the fluid, but this is only found in a proportion of the cases and does not explain the polyhydramnios, which is probably caused by fetal polyuria. The polyuria is secondary to maternal and therefore fetal hyperglycaemia. Polyhydramnios is therefore most often seen when maternal diabetes is badly controlled.

Clinical features

The woman may notice that her abdomen is unduly enlarged and that the fetus is unusually mobile. If the uterus is very much enlarged she may have dyspnoea and indigestion. However, it is extraordinary how tolerant the woman may be of even an enormous accumulation of fluid, provided that it has formed slowly. In the rarer cases of acute polyhydramnios there is abdominal pain and vomiting.

The physical signs depend on the amount of amniotic fluid. The abdomen is larger than expected for the duration of pregnancy, and the abdominal muscles may be stretched. It may be difficult to feel the fetus and the fetal heart sounds may be

muffled or inaudible. A fluid thrill can be elicited. The fetus is unduly mobile and the presentation is unstable. Oedema of the abdominal wall and vulva is sometimes seen. The tightness of the uterus varies, but in cases of acute polyhydramnios the uterus is very tense.

Diagnosis

Chronic polyhydramnios has to be distinguished from multiple pregnancy. This may be difficult, especially as polyhydramnios may complicate multiple pregnancy; in such a case the diagnosis of twins is easily missed. If the twin pregnancy is not complicated by polyhydramnios the essential clinical observations are the discovery of two heads and an unusual number of limbs. An ultrasound scan must be a routine in all cases of polyhydramnios to exclude multiple pregnancy or fetal abnormalities such as anencephaly.

If pregnancy co-exists with a large ovarian cyst the diagnosis from polyhydramnios can be difficult. Ultrasound scan will show two sacs, only one of which contains a fetus.

Acute polyhydramnios may simulate placental abruption with concealed haemorrhage, but in the latter condition the uterus is hard and tense and the fetal heart sounds are absent. In most cases of placental abruption there is at least a little external bleeding. The clinical assessment of amniotic fluid volume is subjective and even with ultrasound it is not possible to obtain a precise measurement of the volume. Ultrasonography is widely used however, to measure the deepest pool of amniotic fluid, free of cord or limbs; if this is more than 8 cm the woman is considered to have polyhydramnios.

Effects on pregnancy and labour

Spontaneous preterm labour may occur. The membranes may rupture suddenly and there is a risk of prolapse of the umbilical cord. Because the fetus is unduly mobile malpresentations may occur. If a large quantity of amniotic fluid escapes suddenly the placental site may diminish in area, and this may lead to separation and antepartum haemorrhage. After delivery there is a risk of postpartum haemorrhage.

The perinatal mortality is greatly increased with polyhydramnios because there may be a fetal abnormality, and because of the possibility of preterm labour, cord prolapse and malpresentation.

Treatment

There is no known method of controlling the production or absorption of amniotic fluid, except that improved control in cases of diabetes may reduce the prevalence of polyhydramnios. Polyhydramnios without symptoms and without any evidence of fetal abnormality requires no treatment.

If ultrasound or radiological examination shows a gross fetal abnormality labour should be induced by rupturing the membranes and setting up a Syntocinon infusion after ensuring that the lie of the fetus is longitudinal. Routine ultrasound examination at 18–20 weeks should ensure that some fetal abnormalities are detected before polyhydramnios develops.

In cases near term in which the woman is in serious discomfort labour should be induced. When there is a great deal of fluid under tension there is some risk of placental separation after rupturing the membranes, and some obstetricians draw off part of the fluid by abdominal amniocentesis before the induction.

Abdominal amniocentesis is particularly suitable in cases in which the pregnancy is not sufficiently advanced for safe induction but the woman is in discomfort and there is no evidence of fetal abnormality. After localizing the placenta with ultrasound, an epidural needle is inserted into the amniotic sac and fluid is withdrawn with an epidural catheter passed through the needle. Up to 2 litres of fluid may be removed, provided that it is only allowed to escape slowly. Although there is some risk of labour starting after amniocentesis the discomfort is relieved for a time. Unfortunately the fluid is often quickly replaced. The procedure may be repeated if necessary. There is always a slight risk of perforating a fetal vessel and causing bleeding into the amniotic sac.

OLIGOHYDRAMNIOS

This means deficiency of amniotic fluid, and is most often associated with poor placental function and fetal growth retardation.

If pregnancy continues beyond term there is a slight fall in the volume of fluid, which is not in itself of serious significance. Ultrasonography is essential to measure the deepest pool of amniotic fluid, which if less than 2 cm, constitutes oligohydramnios.

Severe oligohydramnios is seen with obstructive lesions of the fetal urinary tract and with renal agenesis. In the latter the fetus has a typical facies (Potter's syndrome). The nose is hooked, the lower jaw is underdeveloped and the ears are set low. The amnion is studded with tiny white nodules. Microscopical examination shows these to be islands of degenerate squamous epithelium resting on a bed of flattened amniotic cells. It is thought that the squamous cells have been rubbed off by the dry skin of the fetus. There is usually pulmonary hypoplasia and the fetus invariably dies within 48 hours of birth.

In a few cases of oligohydramnios there is no evidence of renal or any other abnormality. The fetus has little room to move and if it presents by the breech in early pregnancy it will be unable to alter its position and version is impossible. Deformities of the limbs, such as talipes and ankylosis of joints, may be caused by pressure, and amniotic adhesions may form bands which can constrict a limb.

AMNIOTIC ADHESIONS

Amniotic bands may occur between the amnion

and the head, body or limbs of the fetus. They are often associated with deformities, including craniofacial lesions and constriction rings or even amputation of limbs. Possibly some of them result from early rupture of the amniotic sac, so that the fetus lies in a false cavity between the amnion and chorion.

HYPERTENSIVE DISORDERS IN PREGNANCY

The fact that some pregnant women had epileptiform fits was known to Hippocrates, who lived in the fourth century BC. The condition was called *eclampsia*, although the word does not refer to fits; the original Greek word meant flash out, in the sense of a sudden event. Little more was known about eclampsia until, in 1843, Lever (of Guy's Hospital) found that many of the women who had fits also had albumin in their urine. However, it was not until early in this century, when the sphygmomanometer was introduced, that it was recognized that eclampsia was associated with hypertension. The fact that albuminuria and hypertension could precede the onset of fits gave rise to the concept of pre-eclampsia as a clinical condition.

For many years it was postulated that a toxin was liberated from the pregnant uterus, and the disorder became known as toxaemia of pregnancy. All efforts have so far failed to demonstrate any such toxin and the word toxaemia is now avoided. The term pre-eclampsia is criticized because only a small proportion of women develop eclampsia, and the term *pregnancy induced hypertension* is now used. It will be easier to discuss the numerous theories of the aetiology of eclampsia and hypertension during pregnancy after describing their clinical and pathological features.

In these conditions signs precede symptoms. This means that early diagnosis is only possible if the pregnant woman is examined regularly, but it also means that early recognition and treatment of hypertension during pregnancy will almost always prevent eclampsia, which is a serious danger to the life of both mother and fetus.

The classic description of pregnancy induced hypertension is that it is a condition occurring after the 20th week of pregnancy, usually in the third trimester, in which at least two of the three signs (hypertension, proteinuria and oedema) are present. The fact that oedema is found in about 50 per cent of normal pregnant women makes precise diagnosis difficult in some of the milder cases.

Further difficulties are that in some cases hypertension is pre-existing and therefore does not come into the above category, while in other cases the woman's blood pressure before pregnancy is unknown. Confusion has arisen in the past because of failure to classify hypertension during pregnancy accurately, and because of lack of agreement on definitions. The cases may be classified:

- pregnancy induced hypertension. This term includes eclampsia and cases formerly described as pre-eclampsia
- chronic hypertension preceding pregnancy, of any aetiology
- chronic hypertension with superimposed pregnancy induced hypertension.

PREGNANCY INDUCED HYPERTENSION

The signs of this condition usually appear over a period of several days in the following order:

fluid retention (or excessive weight gain)
hypertension
proteinuria.

However, they can appear in any order or all together in less than 24 hours. Some discussion of each of the signs is required.

Weight gain

This is the least specific sign. During normal pregnancy the average weight gain is about 12 kg. Sometimes there is a loss of weight during the first trimester if nausea and vomiting are prominent. After the 12th week there is an average weight gain of about 0.5 kg per week until term, when the weight gain becomes less; indeed there may be a loss of about 0.5 kg in the week before delivery. There are great individual variations and the pattern may be distorted by dieting, overeating and vomiting. The weight gain is made up of the weight of the fetus, placenta and fluid, the increase in size of the uterus and breasts, the increased blood volume and expansion of the extracellular fluid, and fat deposition.

There is no evidence that women who are overweight before pregnancy or have an above-average weight gain throughout pregnancy are more likely to develop hypertension than slimmer women. However, a fat arm may lead to an incorrect diagnosis of hypertension when the pressure is measured with a standard sphygmomanometer.

Oedema during pregnancy

A woman of average weight (55–60 kg) normally increases her extracellular fluid (apart from that in the fetus, placenta and amniotic fluid) by over 2500 ml during the course of pregnancy. Osmotic equilibrium is maintained by the retention of sodium. This degree of fluid retention may cause slight thickening of the skin, rings on the fingers will be tighter and, if the carpal tunnel is restrictive, oedema of the sheath of the median nerve may cause paraesthesiae of the fingers.

Excessive fluid retention eventually gives rise to oedema. This can usually first be detected over the lower subcutaneous surface of the tibia by gentle sustained pressure, but ultimately the feet and ankles are obviously swollen. Oedema of the feet and ankles may also be caused by pressure of the uterus on the pelvic veins, or by associated varicose veins. This non-significant dependent oedema is particularly common at the end of the day, especially when the woman has to spend much time standing, and during warm weather. Dependent oedema of moderate degree is so common that it is usually disregarded unless it is accompanied by other signs. Oedema of the fingers or face is more significant because gravity has little effect upon the accumulation of fluid in these places, but even here it does not establish a diagnosis of pregnancy induced hypertension.

Blood pressure during pregnancy

Most younger women have a resting blood pressure of 110–120 mmHg systolic and 60–70 mmHg diastolic. There is wide variation in healthy women and the response to anxiety adds to the differences found at the first visit to a clinic or to the doctor. Ideally, to assess the changes observed during pregnancy, the pressure before pregnancy should be known, but as this is rarely the case it is necessary to take the readings made during the first trimester as the indicator of the normal state.

The control of blood pressure is affected by normal pregnancy. Pooling of blood in the legs and splanchnic area may result in transient cerebral ischaemia when the pregnant woman stands up suddenly or stands still for a time. The faintness that results is often regarded in folklore as an indication of pregnancy. The pulse pressure is slightly higher during pregnancy than in the non-pregnant state because the diastolic pressure is lower. In the midtrimester some women have a slight fall in both systolic and diastolic pressure, a change that reverts before term. In late pregnancy another factor causes variability; the pressure of the uterus on the large pelvic veins and the inferior vena cava brings about a diminished return of blood to the right side of the heart and induces a low-output hypotension. This most commonly occurs when the woman is lying on her back, especially if she is under an anaesthetic, and is termed the *supine hypotensive syndrome.*

There is some disagreement about what may be regarded as a normal blood pressure during pregnancy. In most clinics 140/90 mm is taken to be the dividing point between physiology and pathology. Some will diagnose hypertension in pregnancy if there is an increase of 30 mm in systolic pressure or 15 mm in diastolic pressure over the baseline readings. The rise should be observed on two readings taken 6 hours apart.

The levels of hypertension used in general medicine are not appropriate in obstetrics. A pregnant woman with a blood pressure of 140/90 mm may be running into danger and at 160/110 she might develop eclampsia. Pregnant women are in a young age group, but during pregnancy the blood pressure may sometimes rise very quickly and the fetus may be at risk with relatively slight hypertension. The physiopathology of eclampsia differs from that of essential hypertension, whether benign or malignant.

If the blood pressure rises acutely, with the diastolic reading above 100 mmHg, the woman may complain of a severe and persistent frontal headache and may vomit. These are warning signals of the possibility of eclampsia. But of course many pregnant women develop headache for other reasons such as migraine, and in the absence of hypertension a headache during pregnancy is not to be taken as a complication of the pregnancy.

The woman who has an abnormally high blood pressure before pregnancy requires special consideration. In her case a rise in the diastolic pressure of 20 mmHg is usually required before a diagnosis of superimposed pregnancy induced hypertension can be made.

Proteinuria during pregnancy

The term proteinuria is more accurate than albuminuria. Although at first the protein that appears in the urine is albumin, the molecules of which are among the smallest of the plasma proteins, as hypertension worsens larger molecules, including globulins, appear in increasing proportions. Proteinuria is most easily detected with test paper strips although older methods, including boiling the urine, remain satisfactory. The test paper strips are very sensitive and indicate a degree of proteinuria which would not be detected by older methods. Nevertheless, proteinuria during pregnancy is always potentially serious and demands investigation. Besides hypertension of pregnancy, the main causes are urinary tract infection and chronic renal disease.

Before accepting that proteinuria is present contamination from vaginal discharge must be excluded by obtaining a clean catch midstream specimen, after cleansing the vulva with sterile water or saline.

Urinary infection can be excluded by *Nephra* dipsticks or examining the urine microscopically, when pus cells and bacteria can be seen. Examination confirms the diagnosis and indicates the most appropriate treatment.

Proteinuria associated with chronic renal disease rarely presents a diagnostic problem because there is usually a history of the disease predating the pregnancy. Even if there is no such history it is likely that the proteinuria will be present throughout pregnancy, while pregnancy induced hypertension is very uncommon before the third trimester.

After excluding these other causes of proteinuria it is reasonable to attribute it to pregnancy induced hypertension. It is then always a sign of serious significance. The risk of intrauterine death of the fetus and of eclampsia is increased many times over when proteinuria occurs, even when there is only a slight increase in blood pressure.

A rarer type of proteinuria is *orthostatic proteinuria*, which occurs after the woman has been on her feet for some time. It is not peculiar to pregnancy, and is not uncommon in adolescents. The diagnosis is most improbable in a fully ambulant pregnant woman. Unexplained proteinuria necessitates further investigation, but in some women the proteinuria promptly disappears; then the diagnosis of orthostatic proteinuria can be made by testing the urine before and after activity.

Sometimes other uncommon causes of proteinuria during pregnancy, such as disseminated lupus erythematosus, may need consideration.

Clinical features of pregnancy induced hypertension

Pregnancy induced hypertension with proteinuria occurs in 2–4 per cent of primigravidae and hypertension without proteinuria in 15–20 per cent. In second and subsequent pregnancies the incidence of hypertension, with or without proteinuria, is only about one-tenth as great as in first pregnancies.

Those multiparae who develop the condition almost always have a history of a similar event in the preceding pregnancy or pregnancies. Unless there is some underlying vascular disease, hypertension usually becomes less severe from one pregnancy to the next.

Pregnancy induced hypertension is more common in women over 35 years, especially in primigravidae. Those who have essential hypertension are more likely to develop a further rise in blood pressure during pregnancy than normotensive women, but they do not invariably do so. However, in women with hypertensive renal disease such a rise in pressure almost always occurs. It is more common in women with diabetes mellitus, and almost invariably occurs when there is diabetic vascular disease.

There is an increased incidence of hypertension in association with multiple pregnancy, with polyhydramnios, with severe rhesus incompatibility and in cases of hydatidiform mole. The last association is of interest because the hypertension is frequently severe, occurs early in pregnancy (usually 16–20 weeks), and in the absence of a fetus.

One other condition deserves to be mentioned, although it is extremely rare. This is abdominal pregnancy, where the placenta has become attached to structures outside the uterus, but has succeeded in maintaining the fetus into the third trimester. In these cases hypertension frequently occurs.

Management of pregnancy induced hypertension

Although the aetiology of pregnancy induced hypertension is still obscure, empirical management has achieved considerable improvement in the prognosis for both mother and baby. It is possible that the severity of the disease has also declined in recent years. As a result the incidence of eclampsia in countries where there is comprehensive antenatal care has fallen from 1 in 1000 to 1 in 5000 births.

The most important factor in management is early diagnosis. Because hypertension becomes more likely as pregnancy nears term, antenatal attendances should be more frequent at this time. At each visit the woman is weighed, oedema is sought over the tibiae and in the fingers, the resting blood pressure is recorded and the urine tested for protein. Excessive weight gain, moderate oedema, and a rise of blood pressure not exceeding 130/80 mmHg may be dealt with by advising rest at home. A check on the effectiveness of this advice must be made within a week.

If the blood pressure reaches 140/90 mm or the diastolic pressure rises 20 mm above the booking pressure the woman needs special supervision, with daily observation of the blood pressure and urine testing. Most obstetricians are now prepared to manage mild non-proteinuric cases at home, provided that the community midwife regularly checks the urine for protein, or by attending a hospital day-assessment unit. Worsening of the hypertension or proteinuria would necessitate immediate hospital admission.

The aim of treatment is to obtain a live baby, as mature as possible, while preventing injury to the mother. It would be ideal to maintain the pregnancy into the 36th week, always providing that the placental exchange remains adequate and the mother's blood pressure is under sufficient control to minimize the risk of eclampsia, cerebral haem-

orrhage and placental abruption. Hypertension will not be cured until the fetus is delivered, and, however well controlled it may appear to be, it is usually necessary to keep the woman in hospital or under close supervision until this time.

The most important component of treatment is rest in bed although there is debate about its efficacy. This will usually result in stabilization of the blood pressure and, because uterine blood flow is relatively greater when the woman is at rest, better fetal growth.

Sedatives are no longer employed as these have no effect upon the blood pressure; and can depress the central nervous system of the baby if it is born prematurely.

Hypotensive agents and diuretics should only be used in special circumstances. Reduction of the blood pressure to normal values can often be attained by the use of potent hypotensive agents, but this may be at the expense of placental function, which in many cases seems to depend on the raised blood pressure, in the same way as a damaged kidney may depend upon an increased blood pressure to maintain a tolerable function.

Routine observations in hospital will include records of the blood pressure 4-hourly, and more frequently if the condition is worsening. Generalized oedema is sought and the mother is weighed on alternate days to check for fluid retention. A record should be kept of fluid intake and urinary output in all but the mildest cases, for reduction in urinary output almost always precedes eclampsia. Each urinary specimen should be tested for protein. Twenty-four hour urine collections are performed at least twice a week for creatinine clearance and protein excretion. Significant proteinuria is 500 mg or more per 24 hours. A falling creatinine clearance and increasing proteinuria indicate early delivery. The blood urea level, platelets and clotting factors should be checked on admission and periodically. Serum urate levels are of value in assessing progress of the disease and the prognosis for the fetus. Levels of 450 mmol/l and above are associated with a poor fetal prognosis.

Regular palpation of the uterus to detect growth failure or diminishing fluid volume is helpful in judging placental function. These observations should be supplemented by ultrasonic measurements of the biparietal diameter and circumference of the fetal head and the abdominal circumference of the fetus. Serial measurements may indicate impairment of fetal growth especially abdominal girth. An estimation of the amount of amniotic fluid should also be obtained.

Doppler blood-flow measurements of the circulation in the placental bed give direct evidence of resistance to flow in the placental bed.

Another indication of fetal well-being is its activity. This can be assessed by the mother counting the number of fetal movements over a given time, a kick count, by recording fetal breathing movements with ultrasound, or by observing the variation in fetal heart rate that accompanies fetal activity or occurs in response to uterine contractions.

In conditions in which the fetus may be compromised, such as pregnancy induced hypertension, it is common practice to make daily simultaneous observations on the fetal heart and uterine contractions with a cardiotocograph an *antenatal CTG* or non-stress test.

Sampling of fetal blood by cord puncture under ultrasound guidance, and estimation of pH or PO_2 has a place in the severely compromised fetus.

A time may come in any pregnancy complicated by hypertension when the fetus would be safer delivered and in a cot, even prematurely, than in the unhealthy environment of the uterus. The purpose of all these investigations is to decide when this moment has arrived. In some cases labour starts prematurely, and in most cases there is a ready response to induction. In any event the woman should not be allowed to pass her expected date of delivery.

The worsening situation – imminent eclampsia

If the diastolic blood pressure remains above 100 mmHg in spite of rest in bed, or if proteinuria persists, and the duration of gestation is more than 36 weeks' labour should be induced without hesitation. During labour the fetal heart rate should be continuously monitored.

If the fetus is very immature, between 28 and 33 weeks, hypotensive agents may be used to prevent further elevation of the blood pressure. However, the diastolic pressure should not be reduced to less than 90 mm, as this can reduce placental perfusion and lead to fetal death. Among the many powerful drugs on the market no one agent seems to be better than another. It is sensible to become familiar with one agent, and methyldopa is often chosen. There has been debate about the safety of beta-blockers in pregnancy, but these are now increasingly used. If control of the blood pressure cannot be achieved, if the amount of protein in the urine is increasing or heavy, or if there is evidence of placental failure, then early delivery is advisable, often by caesarean section.

In cases between 33 and 36 weeks' judgement is exercised. The excellent care available in modern neonatal units has encouraged the decision to deliver the fetus early if there is any doubt about poor placental function or hazard to the mother.

Signs that an eclamptic fit is imminent include a continually rising blood pressure, brisk reflexes, increasing oedema of the face and hands, heavy proteinuria, oliguria, headache and disturbances of vision. Vomiting and epigastric pain, with tenderness over the liver, indicate hepatic haemorrhages or necrosis. Urgent action is required to prevent fits, to reduce the blood pressure, to promote diuresis and to deliver the fetus.

An intravenous injection of diazepam (Valium) 10 mg is given at once, followed by a slow infusion containing 40 mg in a litre of 5 per cent dextrose solution. This acts as a sedative and raises the threshold at which fits occur. Other effective sedatives which are sometimes used intravenously are chlormethiazole, magnesium sulphate and phenytoin.

The blood pressure can be most promptly reduced by intravenous administration of hydralazine (Apresoline). While monitoring the blood pressure closely, hydralazine 10 mg is injected intravenously over a period of 15 minutes. The initial dose is usually followed by a continuous slow infusion of 50 mg of hydralazine in a litre of Hartmann's solution. The rate of infusion is adjusted to keep the blood pressure just above 140/90 mm; if the pressure falls below this the infusion is stopped.

Frusemide (Lasix) 40 mg intravenously will produce a diuresis within minutes. This helps to reduce the blood pressure and appears to protect the kidneys from possible damage caused by a vascular shutdown. An indwelling catheter will avoid the discomfort of a full bladder or the disturbance of frequent micturition.

External stimuli are liable to trigger off a fit, therefore undue noise, bright lights and painful procedures should be avoided.

The method of delivery will depend upon the parity of the woman, the stage of pregnancy and the state of the cervix. In all but the most severe cases, and cases in which placental function is notably poor, vaginal delivery should be the aim. If delivery has to be effected before the 34th week, when induction is less certain, caesarean section may be considered, and it would be advised if there was not an immediate response to induction, or if labour was not progressing quickly.

Because of the possibility of postpartum eclampsia the blood pressure is carefully watched after delivery and the urinary output is recorded. Drugs should be withdrawn gradually over the course of 3 or 4 days, depending on the progress the woman makes.

ECLAMPSIA

Eclampsia remains one of the most serious complications of pregnancy. It may occur before, during or shortly after delivery. The mortality varies with the number of fits, the quality of treatment and the speed with which it is made available. A maternal mortality of 200/100 000 and a perinatal mortality of 150/1000 still occur, even in countries with modern obstetric services.

Eclampsia is characterized by the occurrence of major epileptiform convulsions in women who usually, but not always, have the signs and symptoms described above under imminent eclampsia. Four stages in the fits are described:

- **aura** – a visual phenomena, with flashes of light and spots before the eyes
- **cry** – caused by spasm of the respiratory muscles and larynx
- **tonic phase** – during which the woman loses consciousness, has generalized muscular spasm and becomes cyanosed. The fetus may show signs of hypoxia
- **clonic phase** – during which violent movements occur, the tongue may be bitten, and vomiting and even inhalation of vomit may occur. In severe cases recurrent fits occur and the woman remains deeply unconscious between them. The risks of hypoxia, inhalation of vomit and of cerebral haemorrhage caused by elevation of the blood pressure during the fits are considerable. Placental abruption, disseminated intravascular coagulation and renal necrosis are other serious dangers.

Treatment

The whole emphasis of this chapter has been on prevention of eclampsia by vigilance during pregnancy, during labour and soon after labour. It is especially important to recognize cases of worsening and fulminating hypertension so that further development can be prevented by delivery followed by full sedation for at least 48 hours.

If eclampsia does occur the aim is to prevent further fits. The greater the number of fits the worse is the prognosis for the mother and baby. Any stimulus may precipitate another fit, so the excitability of the central nervous system must be reduced by sedatives or anaesthesia and everything should be done to cut down external stimuli such as noise, bright light and discomfort, arising especially from a full bladder or a strained position in bed.

Heavy sedation is given immediately. Apart from those sedatives already described, magnesium sulphate may be given by slow intravenous infusion, initially 4–5 g over 20 minutes and then 1–3 g per hour. Because the woman is comatose or else heavily sedated there is a risk of both asphyxia and hypostatic pneumonia. A clear airway must be maintained and oxygen may be required. Provision must be made for aspiration of any vomit. False teeth must be removed, and a gag may be placed between the jaws. The woman may need restraint to prevent her injuring herself during the fits. An indwelling catheter will both prevent the stimulus of an overfull bladder and allow accurate observation of the urinary output.

Hypotensive agents, as already described may reduce the risk of cerebral haemorrhage or of placental abruption.

The woman will not be safe from the possibility of further fits until she is delivered. The mode of delivery will depend to some extent on the prognosis for the fetus. If it is dead or very small, caesarean section may not be justified, but with a fetus of reasonable size the risk of labour with a poorly functioning placenta may not be acceptable. If caesarean section is not chosen labour must be rapid and easy. Forceps or the vacuum extractor are often used for delivery, and with any additional obstetric problem section is advised.

Eclampsia may occur up to 48 hours or more after delivery. In this country postpartum eclampsia is common and equal in incidence to antepartum eclampsia, perhaps because of relaxation of vigilance after delivery. It may be very severe and difficult to control. For treatment the woman is deeply sedated until the signs have improved.

These seriously ill women are now often nursed in an intensive care unit with the skill and help of anaesthetics. Deep sedation and positive pressure ventilation may help to reduce the risks of cerebral haemorrhage while alteration to fluid balance prevents renal complications.

Remote prognosis of pregnancy induced hypertension and eclampsia

After pregnancy induced hypertension or eclampsia approximately one-third of women will be found to have residual hypertension. The question is whether these conditions caused the hypertension. At present it is held that these women would eventually have developed hypertension even if they never had a pregnancy. The evidence that supports this view is that the incidence of deaths from cardiovascular disease is the same at all comparable ages in nulliparous women and in women who have borne children.

Some women have hypertension in every successive pregnancy and most of these will later be found to have essential hypertension.

Pathology of pregnancy induced hypertension and eclampsia

It is fortunately uncommon nowadays to have an opportunity to study the postmortem features of these conditions. The available evidence throws scant light upon the aetiology of the disease or of the mechanism by which the hypertension, oedema and proteinuria are produced. Most of the lesions observed in fatal cases result from the hypertension and a disseminate intravascular coagulation rather than provide an explanation for it (*see* Fig. 3.10).

Cerebral lesions

Because of the raised intracranial pressure the brain is oedematous and the convolutions are flattened. There are small haemorrhages scattered throughout its substance which may become confluent. Massive haemorrhage in the brain may be a cause of death. Haemorrhage and oedema may also occur in the retina.

Hepatic lesions

The lesions found in the liver in eclampsia are

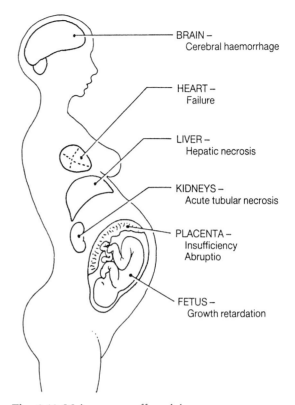

BRAIN –
Cerebral haemorrhage

HEART –
Failure

LIVER –
Hepatic necrosis

KIDNEYS –
Acute tubular necrosis

PLACENTA –
Insufficiency
Abruptio

FETUS –
Growth retardation

Fig. 3.10 Main organs affected by severe pregnancy-induced hypertension

diagnostic of the disease. There is no other disorder which produces similar changes. Macroscopically the liver is enlarged and there are patchy focal red and yellow areas, of which the red are caused by haemorrhage and the yellow by necrosis of the liver cells. The red and yellow patches are visible under the capsule and throughout the cut surfaces. On microscopical section the haemorrhages are mainly grouped around the portal canals, but they may be so extensive that they completely disrupt the liver architecture. By interrupting the blood supply they cause necrosis in the periphery of the lobules with fatty change, which is responsible for the yellow colour. The extravasated blood shows many fibrinous thrombi.

The epigastric pain and liver tenderness which may occur in eclampsia probably arise from distension of the capsule. The interference with liver function is caused by destruction and damage to liver cells, and if this is severe it will result in jaundice. This is therefore a very serious sign. The damage to the liver may be so great that death occurs from hepatic failure.

Renal lesions

The primary renal lesion is in the glomeruli which show swelling of their cells and of the underlying basement membrane. The whole glomerulus appears to be so stuffed with its own swollen cells that it looks as if there is no room for blood to flow through the capillaries, but in fact a greatly diminished flow continues. It is generally held that the reduced renal blood flow is caused by vascular spasm, although the cause of the spasm is uncertain. Beyond the glomerulus the rest of the nephron which is supplied by the afferent arteriole is starved of oxygen. The result is necrosis of the proximal and distal convoluted tubules.

Depending on circumstances, large or small areas of the kidney may be involved. In extreme cases the cortex of the kidney may be destroyed almost entirely. This is less common in eclampsia than in cases of placental abruption. In eclampsia the renal ischaemia is usually less severe and causes areas of patchy necrosis which affect the tubules rather than the glomeruli. Such lower nephron necrosis may be reversible, so that spontaneous recovery occurs. Whether the glomeruli or the tubules are chiefly affected, there is always some degree of renal failure, which in severe cases may progress to anuria.

These pathological changes explain the proteinuria by damage to the glomerular cells, and the oliguria by reduction of glomerular filtration. Sometimes the amount of glomerular filtrate may be very little diminished but the tubules are incapable of concentrating the fluid which reaches them.

Aetiology of pregnancy induced hypertension and eclampsia

There are many theories of the cause of these conditions but none is entirely satisfactory. A group of clinical observations must be fitted into any theory, however complex:

- there is no instance of hypertension of this type occurring outside pregnancy, although a number of other diseases have some similarities to it. It resolves completely when the pregnancy is

over unless some structural damage has been caused by the hypertension
- it occurs more commonly in first than in subsequent pregnancies, with the same partner
- the incidence is high when there is pre-existing vascular disease or long-standing hypertension
- it tends to be less severe in successive pregnancies unless there is pre-existing vascular disease
- it occurs more frequently as pregnancy advances but its progress can often be slowed or even arrested. However, the only cure of the disease is the ending of the pregnancy
- intrauterine death of the fetus is often associated with considerable improvement in the signs of the disease
- it is not necessary for a fetus to be present but trophoblast tissue must be. With a hydatidiform mole severe hypertension may occur early in pregnancy
- it is more common in multiple pregnancy
- it is more common in women who have diabetes mellitus, especially if there is diabetic arterial disease, nephropathy or retinopathy
- it may occur in haemolytic disease with a hydropic placenta and fetus
- the fetus is frequently small for gestational age and shows signs of intrauterine malnutrition which has often preceded the development of the signs of pregnancy hypertension
- hypertension sometimes appears or becomes worse after delivery, even to the stage of eclampsia. The occurrence of this has decreased since the risk of the pressor effect of ergometrine in these women has been recognized.

Current theories of the aetiology of pregnancy induced hypertension centre round three pathophysiological systems:

> immunological mechanisms
> altered vascular reactivity
> coagulation disturbance.

Measurable differences can be observed during pregnancy in all these systems, and attempts have been made to find an explanation for pregnancy induced hypertension based on these observations. It is possible that there are different aetiological factors in different cases, and that we may be dealing with a number of linked conditions which share the physical signs of hypertension, proteinuria and oedema. In the present state of knowledge all that we shall attempt is to describe some of the changes that have been observed during pregnancy and discuss how they may be related to pregnancy induced hypertension. Neither the observations nor the theories are simple.

Immunological mechanisms

Trophoblastic tissue contains HLA, ABO and tissue-specific antigens and immunoglobulins. Since half of these antigens are of paternal origin, development of maternal antibodies which would reject the placenta and fetus might be expected. Trophoblast seems to be capable of only low-grade antigenic activity, because surface coating with protective sialomucin and barrier antibodies occurs. There is also a reduction in maternal immune responsiveness, perhaps an effect of higher levels of hCG, hPL, progesterone and cortisone. Other maternal hormones and pregnancy proteins have been shown to be lymphocytic immunosuppressants.

It is known that from early pregnancy trophoblastic cells escape into the maternal circulation; this may occur to a greater extent in hypertensive pregnancy. Thus an excessive amount of antigen might be released into the maternal circulation and result in the formation of antigen–antibody complexes which are deposited in specific sites such as the renal glomeruli and placenta. Certainly immunofluorescent techniques have demonstrated deposition of immune complexes and IgM in renal glomeruli and spiral arterioles of the placental bed. This deposition might be responsible for the development of the hypertensive process. Trophoblast also invades the spiral arterioles in the first half of pregnancy converting them into low-resistance conduits for maternal blood flowing into the placenta. If this invasion is inadequate, the spiral arterioles have a higher resistance and less blood flow, and it is possible that pregnancy induced hypertension is an attempt to compensate for this (*see* Fig. 3.11).

The increased incidence of hypertension in first pregnancies is not explained.

Altered vascular reactivity

Renin is produced and stored in the juxtaglomerular apparatus of the kidney. It is also produced by the placenta and the fetal kidney. Renin acts on renin substrate to produce angiotensin I

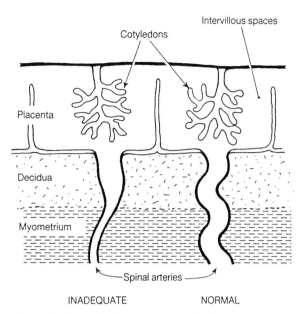

Fig. 3.11 Cytrophoblast invasion of spiral arteries: normal (right); inadequate (left)

which is converted to the potent vasoconstrictor angiotensin II.

In normal pregnancy there is decreased response to the vasoconstrictor effect of angiotensin II, while in hypertensive pregnancy the response is increased and there is an increase in circulating levels of angiotensin II. This reduced response to angiotensin II in normal pregnancy may result from a counterbalancing effect of prostacyclin (prostaglandin PGI_2), a potent vasodilator present in blood vessel walls. Prostacyclin concentrations increase in normal pregnancy. A mild increase in angiotensin II levels appears to cause a rise in blood pressure and an increase in utero placental blood flow, perhaps by stimulating increased production of the local vasodilator prostacyclin.

In severe pregnancy induced hypertension there seems to be a reduced concentration of prostacyclin and its metabolites in maternal blood and in uterine and umbilical vessels. With reduction in prostacyclin activity there is less opposition to the vasoconstrictor action of angiotensin II and the prostanoid thromboxane. Placental blood flow is thus *reduced*. A reduction in renal vascular wall prostaglandins might have a similar effect on renal blood flow, producing hypertension and renal damage.

The increased incidence of hypertension in first pregnancies is not explained, nor the reason for any reduction in prostacyclin activity.

Coagulation disturbance

Reduction in the number of circulating platelets, increase in fibrinogen degradation products and reduction in fibrinolytic activity have been observed in women with severe pregnancy hypertension, suggesting that intravascular coagulation is occurring. Alterations have also been found in levels of factors VIII, IX and X. Fibrin deposition is a well-known histopathological feature of pregnancy hypertension. It has been postulated that the placenta may release thromboplastin, which causes disseminated intravascular coagulation, and that the fibrin deposition in the kidney and placenta results in the development of hypertension and placental insufficiency.

It is difficult to know whether the disseminated intravascular coagulation is cause or effect. Although some studies have suggested that alteration in the coagulation mechanism occurs *before* the development of hypertension, others have failed to confirm this. Early reports that treatment with heparin improves the outcome in early hypertension of pregnancy have not been confirmed, although the coagulation parameters are certainly improved. It is possible that low dose aspirin may be of prophylactic benefit by suppressing thromboxane more than prostacyclin.

CHRONIC HYPERTENSION PRECEDING PREGNANCY

Chronic hypertension preceding pregnancy may result from chronic pyelonephritis, chronic nephritis, polycystic disease of the kidneys, renal artery stenosis, coarctation of the aorta, phaechromocytoma or, most commonly, essential hypertension. Before making a diagnosis of essential hypertension the other possible underlying causes of hypertension must be excluded. For the obstetrician the importance of hypertension is that women starting pregnancy with a raised blood pressure are more likely to develop superadded pregnancy induced hypertension than those who are normotensive. However this only occurs in a proportion of cases, and two out of three women who have a mild or moderate degree of essential hypertension at the start of pregnancy do not have

a further rise of blood pressure and have babies of normal birth weight. However, hypertension caused by renal disease is more likely to progress.

If a pregnant woman is not seen until after the 12th week of pregnancy an accurate base-line pressure reading may not be obtained because of the tendency of the pressure to fall slightly in the middle trimester. This tendency is most marked in those who have mild hypertension of recent origin, and it is of good prognostic significance because the incidence of superadded hypertension is lower in this group of women than in those hypertensive women who do not show such a fall.

Management

A resting blood pressure of 140/90 mmHg or more during the first 20 weeks of pregnancy is usually deemed to be significantly raised, and at that time in pregnancy is unlikely to be due to pregnancy induced hypertension.

If hypertension is found an effort must be made to discover any underlying cause of it, such as those mentioned above. The previous history may indicate the cause. The femoral pulses must be palpated; absence of pulsation would suggest the possibility of aortic coarctation. The urine is examined for protein and casts, and bacteriologically examined if any pus cells are found. The blood urea and serum urate levels may be estimated and other renal function tests may be required. Unfortunately, even if some underlying cause for the hypertension is discovered, specific treatment can seldom be given during pregnancy, and the management described here may be all that is possible until after delivery.

Proteinuria

Proteinuria is of serious prognostic significance when found with hypertension, for it implies that there is both renal and cardiovascular disease. Which of these is primary is not of immediate importance in obstetric practice, and fortunately most cases of essential hypertension have not progressed far enough to cause renal damage during the childbearing years. However, proteinuria may be due to chronic pyelonephritis, chronic nephritis or to relatively rare diseases such as systemic lupus erythematosus involving the kidney. This may be diagnosed by the discovery of LE cells in the blood.

Mild essential hypertension

In all cases of mild essential hypertension (less than 150/100 mmHg) the women must be seen more frequently than usual because of the risk of a further rise in blood pressure during pregnancy. A rise in diastolic pressure of 20 mm, or the appearance of proteinuria, is an indication for admission to hospital. Careful watch must be kept on the growth of the fetus both by clinical observation and with the help of ultrasound.

Moderate or severe hypertension

With moderate or severe hypertension (above 150/100 mmHg) the woman should be admitted to hospital for rest. The effect of rest gives a valuable indication of the prognosis, for if the blood pressure is thereby reduced a favourable outcome can be expected. Such a woman may be sent home, but weekly checks of blood pressure and fetal growth must continue.

If the blood pressure does not fall with rest in bed, thought should be given to the use of hypotensive drugs. In using such drugs the prime consideration is the protection of the mother from the dangers of high blood pressure, such as cerebral haemorrhage and left heart failure. Artificial lowering of the blood pressure will only minimally improve the chance of survival of the fetus; indeed there is a danger that it may diminish the blood flow to the placenta and affect the fetus adversely.

Hypotensive drugs

The preference among the many hypotensive drugs at present is for methyldopa. It is started at a dosage of 250 mg twice daily and then slowly worked up towards a maximum dose of 4 g daily, depending on the response of the blood pressure. It has the special advantage that its effects are not dependent upon the woman standing up; with some other drugs a hypotensive effect is only seen when the woman is on her feet, and then the pressure may fall so precipitously that she feels faint. Several other hypotensive drugs are in use, including labetalol 100–200 mg twice daily and hydralazine 25–100 mg twice daily.

Some women are already taking hypotensive drugs before they become pregnant; these drugs should be continued, adjusting the dose if necessary.

Management

The obstetric management in essential hypertension is exactly the same as for pregnancy induced hypertension. It is essentially that of securing delivery at the best time to avoid serious maternal complications and to prevent fetal death *in utero*. When the diastolic pressure is between 90 and 99 mm the perinatal loss will be about 40 per 1000; if it is between 100 and 109 mm, about 60 per 1000 and if it is above 110 mm, about 120 per 1000. Much of this fetal loss occurs in the last few weeks of pregnancy, so that it is often best to secure delivery before term. The best time for this will depend on the particular case, but in general the higher the pressure the earlier the delivery should be. Renal and placental function tests, ultrasound measurements of fetal growth and fetal heart monitoring may all play a part in determining the optimum time for delivery. Assessment of the maturity of the fetal lung by measurement of the lecithin–sphingomyelin ratio in the amniotic fluid may be helpful in deciding whether to induce labour.

The method of induction or delivery depends on the particular case. In a severe case if there is not a quick response to induction, caesarean section would be performed.

PLACENTAL ABRUPTION AND HYPERTENSION DURING PREGNANCY

In about 25 per cent of cases of placental abruption moderate hypertension or proteinuria (or both) are discovered. It was at one time believed that pregnancy-induced hypertension or essential hypertension were common causes of placental abruption. However, in 90 per cent of cases there is no record of any preceding hypertension. It is certain that in some cases the hypertension and proteinuria follow the bleeding, perhaps as a result of a uterorenal reflex.

It seems likely that both events may occur, i.e. that hypertension may occasionally lead to placental separation, and that placental damage may sometimes cause hypertension.

INTERCURRENT DISEASES DURING PREGNANCY

RENAL DISEASE

Normal changes in renal function during pregnancy

There are marked changes in renal function during pregnancy which include an increase in glomerular filtration rate (GFR) and effective renal plasma flow (ERPF) to values about 50 per cent greater than in the non-pregnant woman (*see* Fig. 3.12). The changes are evident in the first trimester but reach their peak at the end of the second trimester, falling back by about 20 per cent by term. These changes are reflected in parallel changes in blood chemistry, so that mean blood urea falls from an average of 4.3 mmol/l to a low of 3.1 mmol/l at the beginning of the third trimester. Urate behaves in a slightly different fashion, with a fall from a non-pregnant mean value of 0.25 mmol/l to 0.19 mmol/l in the first trimester, and then a rise to 0.21 mmol/l in the second trimester and 0.27 mmol/l in the third (a useful way of remembering the approximate upper limit of normal from 24 weeks to term is to put a 0. in front of the weeks

of gestation, for example 0.24 mmol/l at 24 weeks and 0.36 mmol/l at 36 weeks). The creatinine clearance rises from a mean value of 120 ml/min to 180 ml/min, and this results in a fall in plasma creatinine from 73 μmol/l non-pregnant to a low of 54 μmol/l in the second trimester.

Urinary protein levels are slightly higher in pregnancy, probably due to change in tubular function. Normal non-pregnant urine may contain up to 200 mg protein per litre; up to 300 mg/l is considered normal in pregnancy and up to 500 mg/l may occur on occasions without significant disease. This means that dipstick testing of the urine often shows a trace of protein during pregnancy; the sensitivity is set for the non-pregnant state. Serum albumin falls from 35 g/l to 25 g/l at the end of the second trimester. This is not due to increased loss but to the 50 per cent increase in plasma volume which occurs in pregnancy; the total albumin mass in the vascular compartment increases substantially. It is therefore vital that tests of renal function should be judged against pregnancy specific standards, and it should be appreciated that values of blood chemistry at the upper end of normal for the non-pregnant woman may indicate significant disease in pregnancy.

The impact of renal disease on pregnancy

Women with chronic renal disease but only mild to moderate evidence of renal dysfunction (e.g. plasma creatinine <125 μmol/l, creatinine clearance >60 ml/h) usually have a successful pregnancy with a normal outcome for the baby. This illustrates the large reserve inherent in renal function, and the relatively innocuous nature of excretion products such as urea. Cases are reported where the blood urea was about 30 mmol/l throughout pregnancy and the outcome was normal. Further, pregnancy usually has little effect on the course of most renal diseases. Exceptions are probably lupus nephropathy (secondary to lupus erythematosus), membranoproliferative, IgA and reflux nephropathies. When renal disease progresses to chronic renal failure, hormone excretion becomes disturbed and amenorrhoea or dysfunctional bleeding is common, resulting in infertility. Infertility is also the rule in women on peritoneal

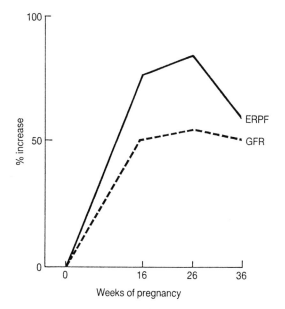

Fig. 3.12 Renal haemodynamic changes in pregnancy. (After Baylis and Davidson in Current Physiology)

dialysis, and in the rare event that conception occurs, the outcome is usually very poor. On the other hand, in women who have a successful renal transplant, pregnancy is usually normal, despite the use of immunosuppressive drugs.

Diagnosis during pregnancy

In most cases of significant disease, the diagnosis will be known before pregnancy. However, if renal disease presents for the first time in pregnancy, it is usual to defer definitive investigation until 3 months postpartum. This is not because X-ray urography and renal biopsy have substantially increased risks for the woman during pregnancy, but if any complications occur they may result in a premature birth. In addition, X-rays during pregnancy increase the risk of childhood leukaemia. This means that careful follow-up arrangements must be made, as often with the stress of a newborn baby to look after, the mother may fail to be investigated until she presents, sometimes years later, with chronic renal failure. Anyone with more than 1 g/l of proteinuria not clearly due to pre-eclampsia (which resolves fully after pregnancy) should be seen at 6 and 12 weeks postpartum. If biochemical renal function remains abnormal at 12 weeks postpartum, an IVU and a renal biopsy are probably indicated.

The general management of pregnancy complicated by renal disease

Despite the generally good prognosis, a successful outcome cannot be guaranteed and therefore close surveillance is justified. Women should be encouraged to plan a pregnancy following pre-pregnancy counselling from an obstetrician and physician with interests in medical disorders in pregnancy. Good diet is important, and advice should encourage non-smoking and no alcohol.

Close monitoring of blood pressure is particularly important, as hypertension is common in women with chronic renal disease. However, a maintained moderate hypertension (not greater than 170/110 mmHg) is compatible with a normal pregnancy outcome unless pre-eclampsia supervenes. Pre-eclampsia is increased in frequency in these women who must therefore be watched closely for any signs of its occurrence. Diagnosis of pre-eclampsia is made difficult if the woman already has a blood pressure above 140/90 mmHg

with proteinuria and a raised blood urea. Accordingly, the diagnosis must rely on detecting changes, notably increases in blood pressure, plasma urate, and urinary protein loss. A flow chart on which the pre-pregnancy levels (if known) of blood urea, urate and creatinine and urinary protein concentration can be entered, with further values being charted every 4 weeks to 28 weeks, every 2 weeks to 36 weeks, and then weekly, is an invaluable aid to management.

An understanding of the normal changes of these levels through pregnancy is essential if unnecessary admission is to be avoided and yet prompt diagnosis achieved. All these measurements are probably satisfactorily made as single measurements so long as they remain normal, but if they become abnormal, 24 hour estimates of creatinine clearance and protein loss should be made. This may require admission to hospital unless the woman is well organized and motivated. Fetal growth is probably best monitored with serial ultrasound scans. Regular urine cultures should be undertaken as any infection can cause deterioration of renal function and should be diagnosed and treated promptly.

Most women can be managed by regular visits to the clinic, but marked deterioration of renal function or increase in blood pressure will require her to be admitted for more detailed monitoring.

SPECIFIC RENAL DISORDERS IN PREGNANCY

Infection

Symptoms of frequency, urgency and dysuria which characterize cystitis due to urinary infection in non-pregnant women are common in pregnancy in women with sterile urine. Thus diagnosis of urinary tract infection (UTI) must depend on proper and prompt investigation if overtreatment is to be avoided. A midstream urine specimen should be obtained, centrifuged, looked at under the microscope, and cultured on appropriate media. The presence of red cells, greater than 40 white cells per high power field, and large numbers of bacteria on microscopy suggests infection, and treatment should be instituted promptly with a broad spectrum antibiotic such as amoxycillin (trimethoprim in women who are allergic to penicillin). The finding of a pure growth of a plausible organism (such as *Escherichia coli*, which causes 80 per cent of infections, or *Streptococcus faecalis*, *Proteus*, or occasionally *Staphylococci* spp.) with a colony count of >10⁵ confirms the diagnosis. The antibiotic should not be changed according to reported sensitivities if therapeutic success is being achieved; rather the culture should be repeated a week after finishing a 5-day course to ensure the urine has been sterilized. If there has been a poor clinical response, then the antibiotic should be changed to one to which the infecting organism is sensitive; preferably use bactericidal rather than bacteriostatic antibiotics in order to clear the renal parenchyma of organisms more effectively. Some antibiotics, notably tetracycline and the sulphonamides, should be avoided because of their potential adverse effects on the fetus, unless their use is clinically imperative.

At one time it was usual to test all pregnant women at booking for asymptomatic bacteriuria, defined as a pure growth of >10⁵ organisms per ml of a pathogenic bacterium on at least two consecutive occasions from a mid-stream specimen of urine (MSU), without symptoms (any organisms found in urine taken by suprapubic aspiration are also significant). However, the cost effectiveness of this procedure has been questioned. Sixty per cent of women with asymptomatic bacteriuria never develop any disease, and may therefore suffer reactions from antibiotic treatment that they did not actually need. In addition about half of all women developing UTIs in pregnancy have no preceding bacteriuria. Finally, most women at risk of infection give a positive history of previous attacks. There are no data which suggest that routine screening reduces complications of pregnancy such as preterm labour, and therefore in many obstetric units in the UK it is no longer carried out.

Acute symptomatic UTI occurs in about 1–2 per cent of pregnancies. Apart from frequency, dysuria and haematuria, pyrexia is a useful diagnostic feature. Abdominal pain is not usually due to UTI, and many cases of preterm labour, placental abruption, and other intra-abdominal catastrophes have been wrongly ascribed to UTI in the first instance, with serious consequences. The diagnosis can only be made with certainty upon the results of investigations (see above), although treatment with antibiotics should be instituted promptly in the presence of pyrexia. *If the result of microscopy and culture is negative, an alternative diagnosis should be sought without delay.* In mild infections,

treatment is usually effective within 5 days, but in women presenting with a pyrexia and acutely unwell, it is probably wise to continue therapy for 2 weeks. Relapses are fairly common, and if they occur, continuing therapy throughout pregnancy may be necessary. In such cases, follow up is important and any recurrence postnatally suggests that an IVU is necessary.

Infection of the renal parenchyma (pyelonephritis) is rare in modern practice, with an incidence of perhaps 1 per 1000 pregnancies. It is a serious condition which can lead to septicaemia and preterm labour. It is characterized by a high temperature (often spiking to 40°C), rigors, and loin pain with tenderness over the renal angle. There is often associated vomiting. Blood and urine cultures should be taken and then vigorous treatment with intravenous antibiotics commenced immediately. This should be continued until the woman is afebrile; oral therapy can then be used. Antibiotic treatment should continue for at least 2 weeks after apparent recovery.

Acute hydronephrosis and hydroureter

The ureters and renal pelves are normally dilated in pregnancy, an effect attributed to progesterone induced muscle relaxation, and compression of the ureters at the pelvic brim by the gravid uterus (*see* Fig. 3.13). However, in rare cases, acute dilatation occurs, causing severe pain. This occurs more commonly on the right side than the left, for reasons which are obscure. Most cases resolve with local heat and bed rest in the lateral position (lying away from the most affected side); if this does not work ureteric catheterization is always effective. The transvesical route is usually successful; if not, a direct ultrasound guided nephrostomy route is available.

Systemic lupus erythematosus

Systemic lupus erythematosus (SLE) is one of the commonest collagen diseases to affect pregnancy, particularly in Chinese women. It often produces serious kidney dysfunction, which can complicate pregnancy management.

The condition may improve, worsen or remain the same in pregnancy. If the disease is in remission at the beginning of pregnancy, the majority of women have a successful outcome. If the disease is active, pre-eclampsia and intrauterine growth

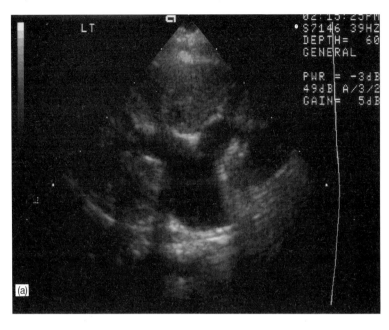

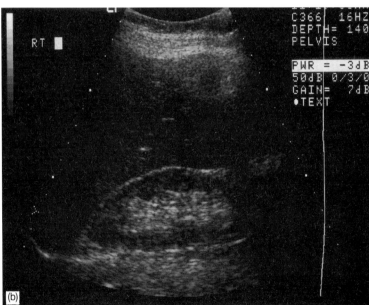

Fig. 3.13 (a) Transverse section of kidney showing dilated collecting system; (b) Longitudinal section of kidney showing collecting system in undilated state

retardation are very common and close monitoring of the disease progression is essential. Treatment with steroids and aspirin should be given if clinically indicated, for usually these drugs have surprisingly little effect on the pregnancy. The onset of renal failure or uncontrollable hypertension may require the pregnancy to be terminated.

About 10 per cent of women with SLE have circulating lupus anticoagulant. This is an autoantibody which often produces a positive Wasserman reaction (historically used as a non-specific test for syphilis). A further group of women have lupus anticoagulant syndrome; about 10 per cent of these subsequently develop SLE. The lupus anticoagulant is so called because it inhibits clotting *in vitro*, but *in vivo* it is associated (paradoxically) with deep vein thrombosis (DVT). Perhaps because of the tendency to thrombosis, these women often have intrauterine death of their fetus, and a history of recurrent miscarriage and DVT should lead one to test for the lupus anticoagulant.

Women with SLE or lupus anticoagulant require specialist monitoring during pregnancy, with regular fetal growth scans and serial monitoring of renal function. Early delivery is often necessary.

Glomerulonephritis

There are a variety of forms of this condition, which can be both acute and chronic. Acute forms are often associated with rapid deterioration of renal function, and are usually treated with high dose steroids. If this fails, delivery of the baby may be necessary. Chronic forms are usually benign in pregnancy unless associated with severe hypertension.

ANAEMIA

Microcytic anaemia

There can be few topics in obstetrics which arouse more controversy than the apparently simple matter of diagnosing anaemia in pregnancy. Nonpregnant women have a lower Hb concentration than men; 11.5–16.5 g/dl compared with 13–18 g/dl. This is ascribed to the effects of menstruation with its resulting blood loss. In pregnancy plasma volume increases by an average of 50 per cent and the red cell mass also increases but by not so much; hence the red cell count and Hb concentration in pregnancy fall. The point at which the fall constitutes anaemia in the pathological sense is vigorously disputed. The World Health Organization has recommended that a Hb concentration of less than 11 g/dl in pregnancy should be called anaemia, and oral iron therapy instituted. This has value in the developing world, where true iron deficiency is common. In developed communities where women have good nutrition, adverse effects in pregnancy have not been

demonstrated until the Hb falls below 9 g/dl. Studies have suggested that no more than 5 per cent of women in well nourished communities have significantly depleted iron stores such that they would benefit from supplementation. Maximum fetal growth is associated with the greatest plasma volume expansion; thus maximum birth weight is seen in women whose Hb levels fall to 9.5–10.5 g/dl in the mid-trimester. Indeed, most fetal morbidity is associated with high, not low, Hb levels. For example the incidence of preeclampsia and intrauterine growth retardation is increased four-fold in women whose Hb never falls below 14 g/dl in pregnancy, as this indicates a poor plasma volume expansion.

Thus most health authorities in the UK now consider that routine iron supplementation in pregnancy is not indicated. The normal daily diet contains about 10 mg of iron, of which about 2 mg is absorbed. This increases to about 4 mg in pregnancy, thus allowing for the developing needs of the fetus.

Some women nevertheless do have a diet deficient in iron, for example, some strict vegans and those with adverse social circumstances with inadequate cooking facilities living on carbohydrate rich diets poor in vitamins and iron. The simplest way to diagnose iron deficiency is by measuring the mean corpuscular volume (MCV). The normal range in pregnancy is 84–99 fl. A value less than 84 fl is termed microcytosis and suggests iron deficiency (unless it is due to thalassaemia). Iron deficiency should be considered in any woman with an Hb less than 9 g/dl, an MCV <84 fl, or an Hb which continues to fall after the mid-trimester – Hb usually falls by about 1.5 g/dl until 30 weeks' gestation, and then rises again by about 1 g/dl towards term (*see* Fig. 3.14).

Serum iron measurements have been widely used in the diagnosis of anaemia in pregnancy in the past, but recent studies show clearly that levels of serum iron correlate poorly with bone marrow iron stores, and therefore this measurement is of little value. Ferritin levels have been proposed as a more informative measurement, but there is no evidence that ferritin values correlate with the outcome of pregnancy. Normal values in pregnancy range down to about a quarter of the nonpregnant, without any association with adverse pregnancy outcome.

Therefore, the appropriate initial response to a diagnosis of anaemia on the criteria described

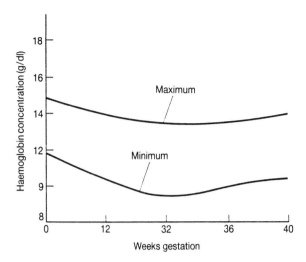

Fig. 3.14 Haemoglobin concentrations during pregnancy

above is to prescribe an oral iron preparation. Current evidence suggests that more than 60 mg daily does not speed the rise in Hb but simply increases the incidence of side effects, which include nausea, gastritis and constipation. Half a 200 mg tablet of ferrous sulphate daily should therefore be adequate treatment for anaemia in otherwise healthy women. There is no place for suggesting two or more tablets daily, as this does not increase the rate of response.

The initial effect of iron supplementation is only seen in the bone marrow, with a marked increase in blastocysts. The Hb level in peripheral blood usually starts to rise after about 10 days, and then increases by about 1 g/dl per week thereafter, until the Hb plateaus at its peak level. There is no increased effect by parenteral administration, which should therefore be avoided unless oral therapy is completely inappropriate (women who refuse to take tablets usually respond to small doses of liquid iron preparations; sometimes changing the formulation of the iron, for example from sulphate to fumarate, also helps). Iron sorbitol citrate (Jectofer) can be given intramuscularly but is a painful injection. Iron dextran (Imferon) can be given intravenously but can cause marked allergic reactions. If it is given, an antihistamine injection and adrenaline must always be available at the bed side. Its use is almost never justified in modern practice.

The very occasional woman with a particularly

low Hb (<6 g/dl) or an Hb of <9 g/dl who fails to respond to oral iron may have a serious underlying cause such as myelofibrosis or even early leukaemia, and should be referred to a haematologist for a bone marrow examination. Usually, however, this simply confirms severe iron deficiency and the woman eventually responds to oral iron therapy. In such women the need for blood transfusion should be considered, particularly if gestation is advanced and there is insufficient time for the bone marrow to compensate. In general, blood transfusion should be avoided in pregnancy because of the risk of a transfusion reaction for, if severe, this may endanger not only the mother, but it also places the fetus at considerable risk because it will interfere with placental perfusion. There is also the risk of infection, for example with hepatitis C or other viruses such as cytomegalovirus. It is usually possible to correct chronic anaemia with iron therapy, and there is no immediate problem with depleted blood volume. However, if blood loss is acute, for example with placenta praevia, then replacement of blood volume is sometimes necessary to prevent shock. Blood transfusion is also sometimes necessary to treat a sickling crisis.

Macrocytic anaemia

The commonest cause of an MCV >99 fl is physiological variation with the second commonest being excessive alcohol intake. Diplomatic enquiries will often confirm the latter cause; some surveys suggest that, during pregnancy, as many as 10 per cent of women consume excessive amounts of alcohol in some parts of the world. Folate deficiency can also cause macrocytosis. Although there is an association between low folate levels and fetal neural tube defect, such as spina bifida, among women with a previous affected baby, or a positive family history, low folate levels have no other known harmful effect in pregnancy, except as part of a generalized severe dietary deficiency. Nevertheless, it is usual to add a small dose of folic acid (350 μg) to oral iron therapy. Higher therapeutic doses (5 mg/day) are usually reserved for prophylaxis against neural tube defect. Vitamin B12 deficiency is often suggested as a possible cause of a macrocytic anaemia, but the possibility can be virtually discounted as significant vitamin B12 deficiency (pernicious anaemia) would almost invariably be associated with infertility. Confusion is sometimes caused because, as with folate, vitamin B12 levels in the blood usually fall to a quarter of the non-pregnant value, which may lead the unwary laboratory technician, or computer, to label it as abnormal. In fact, modern computerized reporting systems label up to 50 per cent of samples from pregnant women as abnormal, because of the change in normal values during pregnancy which are not allowed for by the pathology laboratory.

Symptoms and signs of anaemia

These are usually vague, and simply exaggerations of those normal in pregnancy, that is shortness of breath, tiredness and palpitations. Pallor is an unreliable index of haemoglobin concentration, and most clinicians can tell embarrassing stories of huge discrepancies between their estimates of Hb and the laboratory findings. Diagnosis is by regular Hb estimation, at booking, 26 and 34 weeks, or more often if clinically indicated. The test is cheap, reliable and reasonably specific.

Alpha thalassaemia

Thalassaemia is a disorder of the synthesis of haemoglobin. Haemoglobin consists of two alpha chain proteins and two beta chain proteins (*see* Fig. 3.15). In alpha thalassaemia minor one of the two normal genes (one on each chromosome of the pair) for its production is missing, and the affected individual has a chronically lowered Hb, usually around 9 g/dl, with a microcytosis (usually less than 74 fl). They are physically normal in all other respects, and show almost perfect adaptation to their condition (although they may be more intolerant of sudden blood loss). However, if their partner also has alpha thalassaemia minor, the fetus has a one in four chance of inheriting neither gene for the alpha chain. Severe anaemia results, as Hb without alpha chains is almost totally non-functional. Almost all affected babies develop heart failure with peripheral oedema (hydrops) and die early in intrauterine life. Antenatal diagnosis has therefore little to offer; nature herself removes those severely affected.

Beta thalassaemia

In contrast to the alpha chain, there are a number of different beta chains. The normal adult has mostly HbA_1, with about 1–2 per cent of HbA_2. If

a gene for HbA_1 is missing, the individual has beta thalassaemia minor. This is manifest as a mild anaemia, with an MCV <74 fl. There is usually an increased level of HbA_2, and also fetal Hb, HbF. If both parents are beta thalassaemia minor, the fetus has a one in four chance of having no gene for the production of the normal beta chain for HbA_1, a condition called beta thalassaemia major. *In utero* this is not a problem, because the fetus uses almost entirely HbF. However, following birth, the usual switch to HbA_1 cannot occur and a severe anaemia develops. The individual has to survive on low amounts of HbA_2 and persisting HbF. Frequent blood transfusions are necessary, with resulting iron overload and hepatic damage. Infections may be acquired via the blood transfusion. Life expectancy is greatly reduced.

Beta thalassaemia minor is common in people originating from the Eastern Mediterranean area, particularly Greece and Cyprus. However it can also occur sporadically in many other communities, even the West of England and the East of Ireland, said to be due to the Phoenicians who traded there three millennia ago. All women with an MCV <74 fl should have a haemoglobin electrophoresis; in many centres this is now performed on all samples taken at the first antenatal visit. Electrophoresis can separate the various haemoglobins because of their different electrical charges. Antenatal diagnosis of beta thalassaemia major is available using a specific gene probe on fetal cells obtained at chorionic villus sampling (CVS) or amniocentesis, allowing termination of affected fetuses should the parents request it. This should be explained to all women with beta thalassaemia minor, and their partners offered testing so that antenatal diagnosis can be offered if they also have thalassaemia minor.

Sickle cell disease

This condition is due to an abnormal beta chain, producing HbS. In HbS, the normal glutamic acid in the polypeptide chain is replaced by valine. HbS is quite effective at carrying oxygen, but when the oxygen tension falls, its shape deforms, and the red cell containing it becomes sickle shaped hence the name (*see* Fig. 3.16). The abnormal shape of the sickle cell prevents it passing freely through the capillaries. This leads to interruption of local blood flow, and an increased rate of red cell breakdown. When only one of the two genes for the beta chain

Normal haemoglobin (HbA₁) has two alpha chains and two beta chains:

Fetal haemoglobin (HbF) has two alpha chains and two gamma chains:

In alpha thalassaemia major, alpha chains cannot be made, and therefore the haemoglobin molecule is made up from a tetramer of four beta chains – Hb Barts. This is always fatal.

In beta thalassaemia major, normal beta chains cannot be made, so there is no HbA₁, and all the haemoglobin is either HbF (with gamma chains) or HbA₂ (with delta chains). There is severe anaemia.

Fig. 3.15 Variations in the structure of haemoglobin

is abnormal, the person has sickle trait. They sometimes have a mild anaemia but are otherwise normal although sickling can occur if they become very hypoxic, and special precautions are therefore taken during anaesthesia. The incidence of sickle cell trait is highest in West Africa, where it occurs in 1 in 4 of the population. Its frequency is due to the fact that sickling occurs when the red cell is infected with the malarial parasite, trapping and killing it. This confers a useful protection against the disease, which is common in the area. The incidence falls to 1 in 10 black people in the West Indies, and about 1 in 20 in the UK. If a person inherits both beta chain genes for HbS, they have the disease.

Sickle cell disease is a severe condition, in which sickling occurs spontaneously in many parts of the body where the oxygen tension is relatively low, notably the spleen and bone marrow. Those affected have sickling crises, commonly precipitated by infection, in which the Hb level falls very low. The sickling is associated with microthromboses which often cause severe pain. The mainstay of treatment in a crisis is analgesia, oxygenation, fluid replacement to reduce blood viscosity, and blood transfusion with normal blood to dilute the abnormal cells. Women with sickle disease have major problems in pregnancy, and death can occur from splenic rupture or pulmonary embolism. The fetus is growth retarded in 50 per cent of cases. Ante-

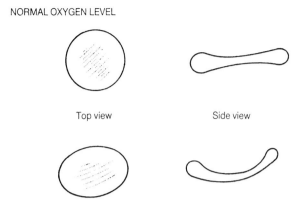

NORMAL OXYGEN LEVEL

Top view   Side view

REDUCED OXYGEN LEVEL producing a sickle cell

Fig. 3.16 Blood cells containing sickle haemoglobin

natal diagnosis by gene probe is available, as for thalassaemia, so if a woman has sickle trait her partner should also be offered a test and antenatal diagnosis discussed if he is positive. At the present time, uptake of screening for sickle disease is much lower than for thalassaemia. This is due to poor knowledge about the condition in the black community, poor uptake of tests by partners (many women in the Afro-Caribbean community in the UK are unmarried) and reluctance to consider termination of pregnancy.

Haemoglobin C disease

This occurs when the glutamic acid of the beta chain is replaced by lysine. It is a less severe condition than sickle disease, but if it occurs in combination with HbS, HbSC disease results, which is almost as severe as sickle cell disease. It is much less common than HbS.

Other haemoglobinopathies

There are over 100 other variants of haemoglobin; many are found in the Indian subcontinent, but they can occur almost anywhere. One specific form of thalassaemia occurs in Portugal, for example. This is a good reason for testing all antenatal women by electrophoresis; in the north-west Thames region all newborn babies are tested for haemoglobinopathies at 6 weeks of age, using the same blood specimen which is used to test for

phenylketonuria and thyroid disorders. Management of these less common haemoglobinopathies is usually in conjunction with a haematologist with a special interest in them.

HEART DISEASE

Cardiovascular changes are among the most obvious physiological adaptations to pregnancy (*see* Fig. 3.17(a), (b) and (c)). There is a steady increase in blood volume from about 8 weeks of gestation, so that by 36 weeks it is 40 per cent greater than in the non-pregnant woman. Following birth, the volume returns to its previous level quite quickly, with a marked diuresis in the first 4–5 days postpartum. Non-pregnant levels are reached by about 4 weeks. Parallel with the increase in volume, there is an increase in cardiac output. This rises from 3 to 5 l/min in the non-pregnant woman to 6 to 7.5 l/min at 36 weeks. Thereafter, there is probably a small fall until delivery, when there is a transient increase after delivery of the placenta and retraction of the uterus. There is then a marked fall over the first 2 weeks of the puerperium.

A major factor stimulating the rise in cardiac output is the reduction in peripheral resistance due to the increased flow of blood through the uterus, and through the skin (to facilitate heat loss, heat being generated by the growth and metabolism of the fetus). There is also increased flow through the kidneys and breasts. The rise in cardiac output is achieved partly by a rise in average heart rate from 65 bpm to 80 bpm, and a comparable increase in stroke volume. Because the increased cardiac output is associated with a decrease in peripheral resistance, blood pressure falls rather than rises. There is usually about a 10 mmHg reduction in both systolic and diastolic blood pressure, which reaches its lowest levels at the end of the second trimester. There is then a gradual rise to the non-pregnant value by term. Thus the incidence of blood pressure equal to or greater than 140/90 mmHg is only about 2 per cent at the booking visit, but rises to 12 to 15 per cent at term. This must be borne in mind when making the diagnosis of pregnancy induced hypertension; a booking blood pressure of 135/85 mmHg is likely to rise above 140/90 mmHg by term due to the natural changes of pregnancy, and does not necessarily therefore indicate a high risk of pre-eclampsia.

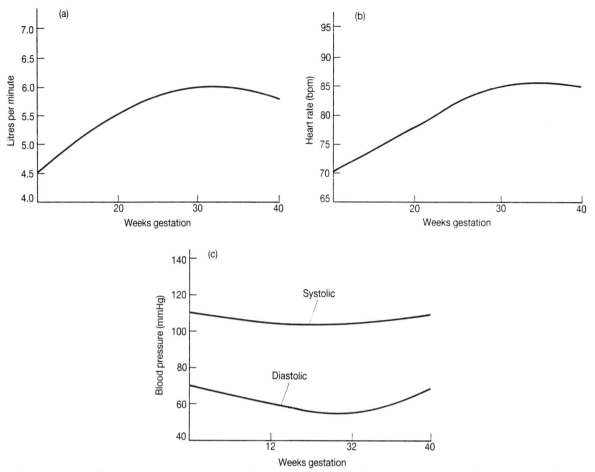

Fig. 3.17 (a) Cardiac output during pregnancy; (b) Resting heart rate during pregnancy; (c) blood pressure during pregnancy

The incidence and aetiology of heart disease in pregnancy

Fifty years ago in the UK, the incidence of heart disease was about 2–3 per cent; now it is only about 0.5–1 per cent. This fall is largely due to the decrease in the incidence of rheumatic fever, an abnormal reaction to a streptococcal infection which causes auto-immune damage to the heart valves. Fifty years ago, rheumatic fever caused about 90 per cent of all cardiac valvular abnormalities, but now the proportion has fallen to less than 50 per cent. However, rheumatic fever is still an important cause of valvular disease in the ethnic minorities, particularly in recent immigrants from the Middle East. It usually causes mitral stenosis, with mitral and aortic regurgitation being less common.

On the other hand, in women presenting with cardiac lesions, the proportion due to congenital defects has increased from about 10 per cent to over 50 per cent. This is because improved medical care has allowed many who previously would have perished to survive infancy and childhood. This is partly due to the introduction of effective surgical techniques allowing total or partial correction of anomalies (hypothermia and heart lung machines, for example), and to the availability of antibiotics and other drugs to treat complications such as bacterial endocarditis. Congenital heart disease presents a bewildering spectrum of disorder, including patent ductus arteriosus, atrial septal defect, ventricular septal defect, coarctation of the aorta, pulmonary stenosis, Fallot's tetralogy, aortic stenosis and Eisenmenger's syndrome. In Eisenmenger's syndrome, there is initially a left to right shunt with greatly increased pulmonary blood flow. As pulmonary hypertension develops, usually in the teens, the shunt reverses and cyanosis occurs. Cardiomyopathy of pregnancy, and disorders of the heart rhythm such as supraventricular tachycardia and Wolf–Parkinson–White syndrome are rare but potentially serious disorders.

Maternal mortality

This is most likely in those conditions which prevent an effective increase in cardiac output in response to pregnancy, notably pulmonary hypertension and mitral stenosis. The normal mitral valve area is 8 cm^2; problems are possible when it falls below 4 cm^2 and likely below 2 cm^2. Values below 1 cm^2 are probably incompatible with life. Eisenmenger's syndrome carries the greatest risk of mortality in pregnancy – about 50 per cent – and is responsible for about a third of all deaths associated with cardiac disease in pregnancy. About half the deaths are related to heart failure and the other half to thromboembolic disease.

Fetal mortality and morbidity

Fetal problems are mainly due to a restriction of maternal cardiac output and hence placental perfusion, resulting in fetal growth retardation (FGR). In many forms of congenital disease, the incidence of FGR is 40–50 per cent. It is also common in rheumatic mitral stenosis if the valve area falls below 2 cm^2. There is also an increased incidence of preterm labour, particularly with FGR.

Principles of management

Ideally, all women with heart disease should have a full cardiological assessment immediately prior to starting their family. This likely to include echocardiography, and in selected cases cardiac catheterization. Precise delineation of the extent of any impairment of cardiac function will allow proper counselling of the women about the risks which the pregnancy poses to her and her child. Such counselling should include:

- the risk of maternal death
- any reduction of maternal life expectancy and how this will affect her family
- the likelihood of prolonged hospital stays during pregnancy
- the risk of recurrence of congenital heart disease in the child (if either parent has a congenital heart defect, the risk of a congenital heart defect in their offspring is about 2.5 per cent)
- the increased risk of preterm labour and FGR.

When the woman is pregnant, she should be encouraged to book in at the antenatal clinic early. Ideally she should be seen at a joint clinic by a cardiologist and an obstetrician. Visits need to be more frequent than in normal pregnancies, often weekly. Continuity of observer is important, as the early signs of cardiac failure may be difficult to appreciate unless the physician is familiar with the woman's previous condition. At the first visit, a very thorough history and examination is essential. Some shortness of breath, ankle swelling, and a soft systolic ejection murmur are all normal in pregnancy.

A detailed medical assessment is necessary at each antenatal clinic check. The woman should be asked *at every visit* about:

- any increase in her shortness of breath, in particular any paroxysmal nocturnal dyspnoea
- any decrease in exercise tolerance
- tachycardia, especially with an irregular rhythm
- any marked increase in tiredness.
- fetal movements.

Examination should always include:

- pulse rate and rhythm
- blood pressure
- jugular venous pressure
- the lung bases listening for the crepitations which can indicate pulmonary oedema
- increasing sacral or ankle oedema
- fundal height measurement to assess fetal growth.

A careful watch should also be kept for any pregnancy complications which can place an extra strain on the heart and precipitate heart failure. These include:

- pre-eclampsia
- twins
- urinary or chest infections

- acute bacterial endocarditis indicated by an increasing murmur, pyrexia, and Osler's nodes
- atrial fibrillation
- anaemia.

Outpatient management is usual, but if there are any signs of incipient heart failure, the woman should be admitted. Bed rest will reduce the load on the heart, and may allow the pregnancy to proceed without the need for additional intervention. Persisting signs such as basal pulmonary crepitations indicate the need for digitalization. Some cardiologists recommend a small dose of a selective beta adrenergic blocker such as atenolol as prophylaxis against arrhythmias. In an acute situation, diuretics should be used to treat heart failure, as in the non-pregnant woman. Many obstetricians are reluctant to administer diuretics long term, as there is some evidence that they increase the risk of FGR by decreasing blood volume. Heart failure should be treated vigorously, as in the non-pregnant, including the use of morphine and oxygen. It has a high mortality in pregnancy. Occasionally, heart failure with mitral stenosis may require emergency surgery such as a closed valvotomy. Regular fetal growth ultrasound scans are advisable, because of the increased risk of FGR.

Delivery represents a time of particular stress for the woman with a cardiac disorder. Ideally, labour should be spontaneous, because this confers the best chance of rapid progress and normal delivery, thus keeping stress to a minimum. Epidural anaesthesia should be recommended, again to reduce the stress of pain. A senior anaesthetist should perform the epidural, to minimize the risks of dural tap, and hypotension which can be a severe problem in someone with a restricted cardiac output. If a normal delivery does not occur readily, then if the second stage is reached an elective forceps or Ventouse delivery is advisable to reduce maternal effort. Caesarean section is often thought, incorrectly, to be a soft option in women with heart disease. In fact, the stress of the surgery, the possible need for general anaesthesia, the increased risk of haemorrhage and post-delivery infections, all increase the risk to the mother. Caesarean section should therefore only be performed for the usual obstetric indications.

Antibiotic prophylaxis is commonly given, to guard against the risk of bacterial endocarditis. In fact there is little evidence that prophylaxis is necessary as a routine, except in cases with previous attacks of endocarditis, or if the woman has a prosthetic heart valve.

It is usual to give only Syntocinon for the active management of the third stage, as ergometrine may cause vasoconstriction, an increase in blood pressure, and precipitate heart failure.

Prosthetic heart valves

Heterografts (e.g. transplanted pig heart valves) are easiest to manage in pregnancy because no special precautions are necessary. However, they only last about 8–12 years, and thus these days most women will have artificial valves, which give a better life expectancy without further operation. They require the woman to be fully anticoagulated, to avoid clot formation on the valve, with resulting embolism. This presents a problem in pregnancy, since the best anticoagulant, warfarin, is embryopathic. Only about 10 per cent of embryos survive the first trimester when the mother is taking warfarin, and they often have congenital abnormalities including punctate dysplasia of the bones. Thus it is usual to transfer the mother to intravenous heparin as soon as the pregnancy is diagnosed (subcutaneous heparin is not sufficient to prevent embolism). Heparin is a large molecule which does not cross the placenta and therefore cannot affect the fetus. Warfarin can be restarted at about 10 weeks, when the risk of teratogenicity has passed. Anticoagulation should be changed back to intravenous heparin at about 36 weeks, because labour may supervene. The intravenous heparin can be stopped an hour or two before delivery, and then restarted after the third stage is complete. If premature labour occurs while the mother is on warfarin, its effects can be reversed within a few hours by injections of vitamin K and the infusion of fresh frozen plasma to replace the missing clotting factors.

DIABETES

Diabetes is the second commonest medical disorder (after hypertension) complicating pregnancy, with an incidence of about 1 per cent.

Normal changes in carbohydrate tolerance in pregnancy

These are essentially a decrease in the fasting

glucose level (from 5 to 4.5 mmol/l) and an increase in the peak glucose level following a carbohydrate challenge, which occurs despite an approximate doubling of insulin levels. These changes are probably due to the large increase in insulin antagonists present during pregnancy, notably human placental lactogen (hPL), sometimes known as chorionic somatomammotrophin. The function of this hormone remains unknown, but it has a structure similar to that of growth hormone, which is also an insulin antagonist.

Despite these changes, which may enhance glucose transfer across the placenta, the mother maintains a remarkably stable glucose level over a 24-hour period. Values of blood glucose usually remain below 5 mmol/l and rarely rise above 6 mmol/l.

Terminology

A number of rather confusing terms are used to describe various disorders of carbohydrate metabolism, which it is useful to summarize here.

Prediabetes

This term is used to describe the situation when parous women, discovered to have diabetes, give a history of previous very large babies (>4.5 kg) or intrauterine death. Although they were not known to have carbohydrate intolerance at the times these events occurred (i.e. when they were prediabetic) it is likely that a glucose tolerance test performed at the time would have been abnormal.

Potential diabetes

This term is applied to women who have features in their personal or family history which put them at increased risk of developing diabetes in pregnancy. These features include:

• diabetes in a first degree relative
• maternal obesity (e.g. >120 per cent of ideal body weight)
• previous large baby (variously considered to be >4 kg or >4.5 kg)
• previous unexplained stillbirth
• previous abnormal glucose tolerance but not diabetic outside pregnancy
• persistent glycosuria
• polyhydramnios.

The concept of the potential diabetic was originally used to define women needing a glucose tolerance test during pregnancy. However, so many women have these features (up to 35 per cent are >120 per cent of ideal body weight in some populations), that performing a full glucose tolerance test on them all is not currently considered a cost effective concept. In addition, screening on such features alone misses half the women with carbohydrate intolerance in pregnancy. There is now a preference for screening the entire population using a simpler test.

Gestational diabetes

This is defined as diabetes that appears in pregnancy and disappears after delivery. However, it is often difficult to be sure that women were not diabetic (as opposed to asymptomatic) before pregnancy, and 50 per cent of women with gestational diabetes will be established diabetics 10 years later. The concept therefore has little value.

Definition of diabetes in pregnancy

The World Health Organization defines diabetes in pregnancy as a fasting glucose ≥7.9 mmol/l, or a value >11 mmol/l 2 hours after a 75 g glucose load (oral meal) (*see* Fig. 3.18). If the 2-hour value is 7.8–11 mmol/l, glucose tolerance is said to be impaired. Impaired glucose tolerance, without any symptoms, is sometimes referred to as *chemical diabetes*. Many authorities have produced their own definitions, sometimes based on a 3-hour

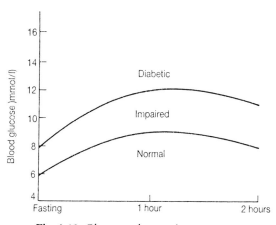

Fig. 3.18 Glucose tolerance in pregnancy

glucose tolerance test (GTT) with seven blood sugar values. There is no evidence that these variations of definition have any great significance.

Diagnosis

Most commonly, women will present as established diabetics, already on injected insulin or oral hypoglycaemic therapy. Occasionally, women present with polydipsia, polyuria, and polyhydramnios. They usually have glucose and ketones in their urine, and treatment is needed urgently if fetal (and even maternal) mortality is to be avoided. A single blood sugar measurement is usually sufficient to confirm the diagnosis; a value over 11 mmol/l is diagnostic without the need to delay treatment while a GTT is performed.

Most authorities now believe that diabetes is most effectively diagnosed before it becomes acute and symptomatic by routine screening of all pregnancies. A common screening method is that proposed by Lind of Newcastle, which is a random blood glucose measurement at booking. If the value is >5.8 mmol/l more than 2 hours after a meal, or >6.2 mmol/l within 2 hours of a meal, a full GTT is performed. The sensitivity of this test is only about 60 per cent; it can be improved by repeating the test at 28 weeks of gestation, when glucose tolerance is under greater stress because of the increased concentration of HPL. About 80 per cent of the women with positive results will turn out to be normal on further testing. The sensitivity of a single glucose measurement can be increased to 80 per cent by taking it 1 hour after a 50 g glucose meal conveniently given as a set volume of a proprietary drink such as Lucozade or Coca-Cola; this method of screening was originally proposed by O'Sullivan and is now quite widely used. If the glucose level is >7.7 mmol/l, a full GTT is performed.

Although widely performed, testing the urine for glucose is not a discriminating test for diabetes in pregnancy. Owing to a decreased renal threshold, glycosuria is common in normal pregnancy, so that glucose spills into the urine even though plasma levels are normal. On testing, 95 per cent of women with glycosuria prove to have normal glucose tolerance. In contrast, diabetes can occur in some women without significant glycosuria, although in severe cases glycosuria almost always occurs. Women with the persistent finding of glucose in the urine should therefore have at least a random blood sugar measurement. A value of >6.2 mmol/l but less than 8 mmol/l is an indication for a 50 g screen, and if this is positive, a full GTT should be performed. If the random sugar is >8 mmol/l, the woman usually has at least impaired glucose tolerance, and it is probably more appropriate to perform a 24-hour profile than a GTT. This involves taking blood after breakfast, lunch, supper, and before going to bed. A normal average value is <5 mmol/l, from 5–6 mmol/l is impaired glucose tolerance which will probably respond to a diet, and >6 mmol/l probably requires active treatment with drugs.

The effect of diabetes on the pregnancy

Maternal complications include an increased frequency of urinary tract infections, candidiasis of the vagina and vulva, and an increased risk of pre-eclampsia (secondary to increased placental bulk) and polyhydramnios (secondary to a glucose induced osmotic diuresis in the fetus).

Fetal complications are even more serious. Poorly controlled diabetes greatly increases the risk of congenital malformations (notably sacral dysgenesis). Fetal macrosomia is common, leading to obstructed labour and shoulder dystocia. It is also associated with sudden intrauterine fetal death late in pregnancy. It has been suggested that this could be due to sudden fluctuations in blood sugar level, leading to a relative hypoglycaemia. The fetus produces more insulin in response to the high sugar levels from the mother, and consequently suffers from hypoglycaemia after birth unless special efforts are made to maintain the blood sugar with intravenous glucose or early feeding. The macrosomia is also associated with polycythaemia, which gives rise to an increased level of bilirubin following birth producing physiological jaundice. The high sugar levels also seem to retard some aspects of organ maturation, so that respiratory distress is more common in infants of diabetic mothers.

The effect of the pregnancy on diabetes

The major effect is the increased insulin resistance, so that the dose of insulin required for insulin dependent diabetics often doubles during pregnancy. The dosage requirement returns to normal within a few hours of the birth.

Management

A 180 g carbohydrate diet may be all that is required in mild cases of carbohydrate intolerance. Such a diet is in any case necessary, even if insulin is required, so that a proper balance can be achieved between the carbohydrate intake and the insulin dose given. The principle of treatment is to normalize the blood sugar as much as possible. To this end, it has been usual to give twice daily doses of a mixture of quick acting and retard insulins, to cover the whole 24 hours. More recently, the novopen system with very fine needles has made it practical to give the insulin in four divided doses each day, giving even tighter control and thus reducing complications. Continuous infusions using mini-syringe drivers have been used experimentally, but conclusive evidence that this gives a better outcome is lacking. The aim is to produce an average blood sugar over the 24-hour profile of <5 mmol/l; a mean value >7 mmol/l would certainly be regarded as unsatisfactory. Most women will now be given their own glucose meters, so that they can check their own profiles on finger prick blood samples at least three times weekly, and often every day. They often need considerable emotional support to achieve this tight control, which is considerably more stringent than the standards thought necessary outside pregnancy.

Diabetic pregnant women are best managed in a joint clinic with a physician and an obstetrician. Most surveillance depends on the values recorded by the woman, although studies show that sometimes poor control is disguised by the woman for fear of censure. For this reason, it is customary to check readings on venous blood from time to time, and to admit women for a 24-hour profile if there is any doubt about the adequacy of their control. During such an admission, it is important to maintain as normal a pattern of physical activity as possible, so as not to disturb the insulin requirements. Another way of checking long-term control is to measure the proportion of HbA that is glycosylated. When the red cell is first produced, it contains mostly HbA, but gradually, with exposure to blood glucose, it becomes glycosylated. The average life of a red cell is 120 days, so that in the presence of a normal level of glucose, the proportion of glycosylated HbA (referred to as HbA_{1C}) plateaus, with the rate of glycosylation being balanced by the destruction of cells containing glycosylated Hb. In the euglycaemic pregnant woman, the percentage of HbA_{1C} ranges from 4 to 6 per cent; values higher than this (especially values >8 per cent) suggest that the cells have been exposed to higher than average levels of blood glucose and therefore that homeostasis has been poor.

Obstetric management is essentially the same as that in a normal pregnancy, with a special watch for complications more common in diabetic pregnancy. In particular, although it is important to make careful measurements of symphysis–fundal height, developing macrosomia can be difficult to detect clinically, and so it is usual to perform serial growth scans of the fetus at for example 19, 26, 32 and 38 weeks. If the fetus is growing particularly rapidly, consideration will be given to delivery at 38 weeks to avoid the complications of sudden intrauterine death, obstructed labour and birth trauma which are more common in this group. If the cervix is favourable, with a Bishop score 6 or more, then induction of labour is an option, otherwise delivery by caesarean section is commonly chosen. When the fetal growth rate appears normal, and there are no other pregnancy complications, it is becoming increasingly common to allow pregnancy to proceed to the due date, or even a little longer, in the hope that spontaneous labour will ensue. However, even in the most liberal centres, about 30 per cent of diabetic women will eventually be delivered by caesarean section. Other methods of antenatal monitoring, such as fetal movement charts and regular non-stress cardiotocograms, have been suggested, but because intrauterine death in the babies of diabetic mothers is usually sudden and unexpected, there is no evidence that these tests are of particular value.

The problem of threatened preterm labour

The incidence of preterm labour in diabetics is increased, some reports suggesting a 12–15 per cent incidence. Part of the increased risk may be due to the polyhydramnios. If a woman is admitted with threatened preterm labour between 26 and 34 weeks, and there are no obvious contraindications such as antepartum haemorrhage, an abnormal fetal heart rate pattern, or a cervix dilated to more than 3 cm, it is usual to suppress contractions with a tocolytic such as ritodrine. Most would administer intramuscular steroids, which halve the incidence of respiratory distress syndrome if the baby goes on to be born. Unfortunately, both these drugs raise the blood sugar, making the diabetic process much more difficult to control. Even women with only carbohydrate intolerance, being treated with a diet, may develop very high blood sugars. Hence it is usually necessary to commence intravenous insulin and glucose infusions, in order to preserve glucose homeostasis, for the duration of therapy.

Management in labour

Once labour is established, either spontaneously or following induction, a continuous infusion of glucose and insulin is set up. The normal daily carbohydrate intake should be given, that is approximately 180 g. Since there are 50 g in each litre of 5 per cent dextrose this can be achieved by giving 1 litre of 5 per cent dextrose every 6 hours (200 g per day), an infusion of 4 litres per 24 hours. While this is not a problem in a fit young woman with normal kidneys, if there is any indication of renal compromise, a fluid overload could result. Some workers therefore advocate adding 25 g glucose to each litre, making a 7.5 per cent solution, and infusing each litre over 8 hours (225 g daily). Whatever the glucose regimen, insulin infusion at at least 1 unit per hour is required so that the cells can use the glucose. The blood glucose should be monitored at regular (e.g. 4-hour) intervals. If the blood glucose rises, then the infusion rate of insulin can be increased. If the blood sugar falls, however, it is a mistake to reduce the insulin infusion rate, as although this will result in a rise in blood sugar, the cells will not be able to use it and keto-acidosis will develop. This has been called *starving in the midst of plenty*. Thus a low blood sugar should be corrected by increasing the glucose infusion rate.

Fetal monitoring is important, and it is usual to recommend continuous electronic monitoring with fetal blood sampling and pH measurements as appropriate.

Contraception

The ideal contraceptive which is fully effective and yet does not accelerate the vascular complications of diabetes, has yet to be invented. Women with diabetes should be advised to persevere with barrier methods if they can. If this is difficult, then a copper loaded IUCD can be used, although there are reports that the failure rate and the risk of infection is increased. The combined oral contraceptive pill is not ideal, because it increases the risk of vascular complications. The progesterone only pill is preferable, if the woman is willing to accept the failure rate and the possibility of menstrual irregularity.

INFECTIONS DURING PREGNANCY

Any maternal infection can have significant effects during pregnancy, for example causing death of the fetus or miscarriage, particularly if it causes a high fever. However, there are certain infections which have specific effects, especially if they can cross the placenta and infect the fetus.

RUBELLA

Rubella is generally a mild illness, which was first recognized in 1941 to have serious fetal and neonatal consequences if the mother contracts it for the first time during pregnancy; reinfection can occur but is very rare and usually has less severe effects on the fetus. Infection in the first 12 weeks of gestation results in fetal infection in over 95 per cent of cases; in 20 per cent of these serious damage to the fetus occurs. The classical triad of injury is deafness, blindness due to cataract, and cardiac malformation. The likelihood of fetal damage is not related to the severity of the illness in the mother, which may be so mild as to pass unnoticed, or at most an influenza-like illness with a rash for 2 or 3 days. The diagnosis is therefore often missed until the baby is born bearing the stigmata of congenital rubella. The diagnosis should be suspected in any small for gestational age baby born with congenital abnormalities, including microcephaly and hepatosplenomegaly in addition to the classical triad as well as many other less common manifestations. The likelihood and severity of damage decreases with advancing gestational age.

Diagnosis is based essentially on the detection of antibodies in the serum of mother or child. It is routine to measure maternal rubella antibodies at the initial antenatal check; the presence of any antibody is evidence of previous infection which is likely to render the mother immune from any further attack by the virus. If the antibody titre is low (<15 IU/ml), a booster vaccination after delivery is recommended. The presence of a very high antibody level may suggest recent infection; detection of rubella specific IgM suggests infection within the last 4 to 6 weeks. The only absolute proof of infection is the detection of antibody in a woman previously shown to be negative. A non-immune woman should be advised to keep away from known cases of rubella while she is pregnant; there is no other specific measure to prevent infection which can be advised. Together with many mild or subclinical cases, about 200 babies a year are born in the UK with congenital rubella syndrome. This is despite an active childhood immunization programme which has reduced the incidence of women susceptible to rubella in pregnancy from 20 per cent to under 5 per cent, although susceptibility rates still exceed 10 per cent in many immigrant groups such as Asians.

All women who are susceptible to rubella should be offered vaccination in the postpartum period. Because the vaccine is a live attenuated virus, there is a theoretical risk of fetal infection if the woman should become pregnant within 3 months and therefore contraception is advised for this length of time. However, recent studies have not shown any increased risk of congenital anomaly in the babies of women vaccinated accidentally in pregnancy and so these women can be reassured that termination of pregnancy is not indicated as the risk of fetal damage appears negligible. Nevertheless, vaccination is not normally performed in pregnancy. Vaccination in childhood is the ultimate preventative measure.

CYTOMEGALOVIRUS

This virus is one of the Herpes group and, like the others possesses the important characteristic of latency. It is spread through the respiratory and genital tracts and anyone infected may subsequently begin to shed the virus. Most primary infections are asymptomatic in the adult, but like rubella, can cause fetal damage in pregnancy. About 50–70 per cent of women are susceptible when they become pregnant; because infection in childhood is less common in the higher socioeconomic classes, they are at the greatest risk. Because there are so few symptoms, there are no definite data on the incidence of infection in pregnancy, but surveys suggest that it may be as high as 1 in 200, with two-thirds of these resulting in fetal infection. Thus there may be as many as 1200 infected babies a year born in the UK. It is not clear whether the timing of infection in preg-

nancy is important, but some data suggest that unlike rubella, it may be the later infections which are more serious. Fortunately, 85 per cent of infected infants are asymptomatic, but this still means about 200 babies a year in the UK are born damaged by the virus. Those worst affected have the stigmata of microcephaly, blindness, deafness, or all three. There are many other manifestations of infection, including liver dysfunction, pneumonitis, chorioretinitis, cerebral calcification and mental retardation. Unfortunately, no vaccine is available and because of the lack of symptoms in women infected, the diagnosis is almost never made before birth.

VARICELLA ZOSTER

Also a member of the Herpes viruses, varicella was widely held to be harmless in pregnancy until the relatively recent description of congenital varicella syndrome. This consists of dermal and skeletal scarring which may produce hypoplasia of the limb bones, muscular atrophy, chorioretinitis, cataracts and cerebral cortical atrophy. However, it occurs so very rarely that women who have varicella during pregnancy should be reassured that the risk to the fetus is very small.

However, infection in late pregnancy can be more of a problem. The baby may be born infected with the virus but before the mother has had time to produce antibodies which can be transferred to the fetus to provide passive immunity. Mothers infected more than 7 days before their baby is born will have had time to develop the antibody and pass it to their baby, but if the rash develops less than 7 days before delivery, progressively fewer babies are born with antibody, and infants born within 3 days have no antibody at all. About one in six such babies will develop neonatal varicella, which carries a 30 per cent mortality rate. However this can be reduced effectively to zero by the administration of anti-varicella-zoster immunoglobulin (VZIG, 200 mg) at birth. Even without administration of VZIG, mortality from neonatal varicella is rare despite the endemic nature of the disease. This is because varicella has an extremely high attack rate in childhood, over 90 per cent of susceptible children being infected in outbreaks. This means a high proportion of pregnant women are immune to systemic disease, although a few will develop Herpes zoster, shingles, shedding the virus from vesicles associated with a single nerve root, usually thoracic. Herpes describes the type of rash and should not be confused with Herpes simplex. Although shingles can start a new cycle of infection in any susceptible children who come into contact with it, babies born to mothers with shingles by definition have passive immunity and neonatal varicella from maternal shingles has not been reported. Because the disease is usually so mild, routine vaccination has never been proposed, although attenuated virus vaccines have been produced.

The symptoms and course of varicella in pregnancy are essentially the same as in the non-pregnant state, but occasionally a rapidly progressive pneumonitis bronchitis and tracheitis develops which can be fatal. Thus any symptoms of cough, shortness of breath or wheeze in a pregnant woman with varicella are grounds for admission to hospital and supportive therapy under barrier nursing conditions. Treatment with hyperimmune globulin and acyclovir may also need to be considered.

HEPATITIS VIRUSES

There are now at least five types of hepatitis virus recognized, labelled A to E. They represent a major cause of morbidity and mortality world-wide.

Hepatitis A

This is the major form of RNA infectious hepatitis transmitted by the faecal and oral routes; it is commonly borne in the water supply. About 50 per cent of the British population have antibodies to hepatitis A, although most will have been unaware of their infection. In pregnancy, it usually causes little problem, although some strains reported from the Indian subcontinent have been associated with fulminating hepatic failure. Congenital defects or infection have not been reported.

Hepatitis B

This is a major cause of mortality in many parts of the world, particularly in Chinese communities such as those in Singapore and Hong Kong, where over half the population are infected with the virus at some time during their life. Transmission of this DNA virus is usually by the sexual route, although it can also occur following blood transfusion or inoculation injury. It causes fulminating hepatic failure in 1 per cent and apparent recovery can be followed by cirrhosis, chronic active hepatitis and hepatocellular carcinoma. The earlier in life the infection occurs, the more likely the person is to become a carrier of the virus; 80 per cent of infants whose mothers were carriers will themselves carry the virus. It is estimated that there are about 200 million carriers of hepatitis B worldwide.

Tests for infection include the antigens HBsAg (a surface antigen) and HBcAg and HBeAg (core antigens), and the antibodies to them. Presence of antigens only in the blood indicate high infectivity; simultaneous presence of appropriate antibody greatly reduces infectivity, and the presence of antibody only indicates immunity (non-infectious). There is a considerable variety of combinations of infectivity, but in principle all babies whose mothers test positive for any of the markers should be treated to prevent vertical transmission. If the mother has only antibodies, then the infant should be vaccinated at birth, at 1 month and at 6 months old. The genetically engineered vaccine Engerix B is now widely available and cheap. If the mother has any antigens in her blood, the baby should also be given passive immunity with 200 IU HBIG (hepatitis B immune globulin) within 24 hours of birth. A further five doses at intervals of 1 month should also be given in cases with high maternal infectivity. Such regimes will prevent carrier status developing in over 95 per cent of infants, and a widespread vaccination programme in Singapore is already having a major impact on the incidence of the disease.

Maternal screening is not recommended for all pregnant women, but must be carried out in all women not born in the UK, and those at increased risk for other reasons, e.g. prostitutes, drug addicts, and anyone with a tattoo. For example, one recent study in immigrants from the West Indies showed a quarter to have been infected with hepatitis B, compared with only one in 40 of a similar ethnic group born in the UK.

Hepatitis C

One of the most recently discovered viruses, this has been isolated by molecular cloning but its full structure remains to be elucidated. Infection is clinically silent in 95 per cent of cases, but it is thought to be a significant cause of chronic hepatitis. Screening has recently been introduced for blood donation, but its significance in relation to pregnancy is not yet established.

Hepatitis D

This is a defective virus which is only functional in the presence of hepatitis B infection. It exacerbates the hepatitic effect of hepatitis B. Its precise incidence is unknown.

Hepatitis E

This virus has so far been identified only in African and Asian populations. It causes acute hepatitis, and is especially fulminating in the pregnant woman. So far, however, it has not been shown to cause chronic hepatitis. In some areas of Iran and Kashmir, pregnancy precipitates hepatitis in 20 per cent of the population, and women affected have an 80 per cent mortality, with an even higher fetal mortality.

OTHER INFECTIONS

Measles and mumps are now rare in pregnancy, and are not thought to be teratogenic. The risks associated with them are very small unless the mother becomes unusually ill. Poliomyelitis can cause paralysis of the newborn if acquired by the mother in the 5 days before labour onset. All three infections are now rare in the UK because of efficient vaccination programmes. Enteroviruses, such as the Echo 9, are not serious in pregnancy, but can cause life threatening infections in the newborn.

TOXOPLASMOSIS

Toxoplasma gondii is a protozoan widely disseminated in the soil. It infects a wide variety of animals including cattle, sheep, pigs and domestic cats. It is most commonly acquired by the human by eating the cystic form in uncooked meat, or the oöcysts from contaminated food such as salads; it can also be transmitted from cat faeces. Maternal infections are usually asymptomatic or associated with a very mild influenza-like illness. The chance of becoming infected during pregnancy is very variable geographically. In parts of France, up to 90 per cent of women are already immune at the beginning of pregnancy, but those remaining susceptible are at increased risk of infection because dietary habits include a significant proportion of uncooked meats. In the UK, only about 10–20 per cent of women are immune when they start pregnancy, but on the other hand their chance of becoming infected is small. If the pregnant woman acquires an infection in pregnancy, the chance of it infecting the fetus varies from 10-25 per cent in the first trimester to 75–90 per cent in the third. On the other hand, the risk of fetal damage from infection decreases from 65 per cent in the first trimester to almost zero for perinatal infections. The classic triad of effects on the fetus is hydrocephaly, retinochoroiditis and intracranial calcification, the latter being a useful diagnostic feature detectable on X-ray or ultrasound. However, other complications include intrauterine growth retardation, hepatosplenomegaly, jaundice, thrombocytopaenia and convulsions. Perinatal mortality is 5–10 per cent and neurological impairment, epilepsy and impaired vision are common in survivors.

It is thought that in the UK about 1 in 500 women acquire a toxoplasma infection in pregnancy each year i.e. 15 000 among 750 000 deliveries; with the variable transmission rate this would imply about 500 fetal infections per year with about 50 babies severely affected. In fact, the incidence of clinically recognized fetal damage is only about one-fifth of this. The reason for the disparity is not understood at present; nevertheless the low risk of damage (about one baby in every 60 000 born) makes it difficult to justify routine screening in pregnancy. Such screening is complicated by the fact that in women shown to be susceptible at booking the only way to demonstrate infection is to repeat antibody screening every 2–4 weeks. If seroconversion occurs, it is difficult to determine whether the fetus has been infected and even testing fetal blood samples from cordocentesis for the presence of the parasite has been shown to be unreliable. If termination of the pregnancy is offered, as many as 9 out of 10 babies lost would have been normal. Treatment of the mother with spiramycin 3 g daily has been recommended to control the infection as an alternative to termination; treatment has to be continued throughout the pregnancy. The safety of the drug in pregnancy has not yet been fully established. The alternative therapy of combined treatment with sulphadiazine, pyrimethamine and folinic acid may be teratogenic in early pregnancy.

Recent infection in women at booking is suggested by the detection of specific IgM; unfortunately the test is only 80 per cent reliable and the antibody can remain in the circulation for up to 2 years (unlike most specific IgMs which only last about 6 weeks). Thus its significance is often difficult to determine.

For all these reasons, routine screening is not currently recommended in the UK, although in many continental countries it is a statutory requirement and women cannot obtain maternity benefits without having it done. However, in France only 10 per cent of women require continued testing through pregnancy, and the chance of a woman needing treatment is therefore low. In the UK, up to 90 per cent of women would require testing and the size of the counselling and treatment problem would be correspondingly greater. There is no vaccine against *Toxoplasma gondii.*

LISTERIA

Listeria monocytogenes is one of seven species of *Listeria*, but the only one pathogenic in the human. It is a slow growing Gram positive rod, which can also occur in a coccoid form. It is present in the bowel of 1 in 10 healthy individuals and is usually harmless. Occasionally, however, it gives rise to a systemic infection in susceptible individuals, and the influenza-like illness which results can produce miscarriage in the first half of pregnancy. Perinatal infection can cause pneumonia, meningitis and septicaemia in the child, and it is an important cause of intrapartum meconium passage in the preterm fetus which otherwise rarely passes meconium in labour. Mortality from established infection is high. The organism is non-sporing but grows best at 4°C and thus is a particular problem in faecally contaminated non-cooked refrigerated foods. High risk foods include coleslaw, dairy products such as soft cheeses, undercooked chicken, hot dogs, and paté. These should all be avoided by pregnant women. Established infections should be treated with ampicillin or penicillin together with gentamycin or kanamycin. The organism is resistant to most cephalosporins.

The incidence of *Listeria* infection is thought to be about 1 per 20 000 pregnancies in the UK. Prevention is currently the only available strategy for its control; screening tests are unreliable and diagnosis relies on the characteristic appearance on microscopy and culturing the organism. Serolog-

ical tests are of no value, as many uninfected women will have antibody present and the titre is no guide to how recent the infection has been. IgM levels are not diagnostic. No vaccine exists.

MALARIA

Worldwide, malaria is the major killer and is rivalled only by tuberculosis for its toll of mortality. It is a protozoan disease transmitted to man by the bite of the *Anopheles* mosquito, prevalent in the tropics. Infection is most serious in non-blacks never previously exposed to the parasite. Sickle cell trait occurs in one in four black people in areas where malaria is prevalent and confers 90 per cent protection, particularly against the fulminant hepatic and cerebral forms of infection. Chronic exposure to the parasite confers substantial, although not complete, protection because of the formation of protective antibodies. However, these antibodies are not life long, and after a period of non-exposure (e.g. in immigrants to the UK), immunity falls. Thus the return of a previous resident to an endemic area is particularly dangerous, as they may neglect proper prophylaxis with anti-malarials such as proguanil, pyrimethamine, trimethoprim, chloroquine, amodiaquine, sulphadoxine, and dapsone. The reason for the profusion of drugs is the steadily increasing resistance of the parasite; some forms now exist which are resistant to all known drugs.

Pregnancy appears to inhibit the immune response in malarial infection, and the disease is therefore even more serious at this time. None of the drugs described above is known to be totally safe in pregnancy, and even though chloroquine and amodiaquine are probably the safest, a small increase in fetal abnormalities such as deafness has been suspected even with chloroquine. Thus it is usual to advise all non-residents to avoid visiting areas where they might be at risk from malaria, especially when they are pregnant. However, if a visit is essential, proper prophylaxis appropriate to the area visited must be taken, as the effects of infection are far worse than the effects of the drugs.

BACTERIAL INFECTIONS

A common genital tract commensal which can also be a major pathogen is the beta haemolytic streptococcus. It is found normally in the genital tract of about 40 per cent of black women, and 20 per cent of other races. It is also more likely to be pathogenic in black women; the reason for this racial difference is not understood. It usually causes no problems unless the pregnancy is otherwise complicated. However, if there is prolonged prelabour rupture of the membranes, it is a common cause of ascending genital sepsis which can progress to cause septicaemia and even maternal death. Septicaemia can also occur in the fetus and newborn, particularly if it is delivered preterm, growth retarded or asphyxiated. A small proportion of normal infants (<1 per cent) colonized in the perinatal period go on to develop secondary disease 7 to 28 days later; meningitis is commonly fatal. There is currently no justification for routine antibiotic prophylaxis unless the mother is known to be a carrier and the pregnancy or birth is complicated, for example by preterm labour. The antibiotic of choice for treatment when indicated because of clinical infection is penicillin; if the mother is allergic to it, then erythromycin can be used. If the mother is a chronic carrier, eradication of the organism from her bowel reservoir is almost impossible, and therefore attempts to treat otherwise normal pregnant women are pointless. However, routine use of chlorhexidine cream to cleanse the vagina during labour may be of some benefit.

MYCOPLASMA

Mycoplasma hominis and T-strain mycoplasma have been associated in some studies with an increased incidence of preterm labour. However, no systematic benefit has been demonstrated to accrue to colonized women treated with erythromycin; the role of these organisms remains an enigma.

SEXUALLY TRANSMITTED DISEASES

CHLAMYDIA

Chlamydia trachomatis is a widespread bacterium causing sexually transmitted disease. Like *Neisseria gonorrhoeae*, it is an obligate intracellular parasite. Reproduction can only occur in the host cell because the organism possesses no ATP of its own. Serotypes A, B and C are among the commonest causes of blindness in the world (ophthalmia neonatorum leading to trachoma), if untreated ophthalmic infection occurs due to perinatal

acquisition of the infection as the fetus passes down the birth canal. It is thought that as many as one in six people in the UK acquire the genital form of the infection, usually with serotypes D–K; they cause non-specific urethritis in the male and are a major cause of tubal infertility secondary to salpingitis in the female. Asymptomatic infection is common, which is one of the factors leading to its high prevalence. A further subgroup of the organism causes lymphogranuloma venereum, rare in the UK.

Chlamydia infection is rarely symptomatic in pregnant women, but in the newborn is an important cause of the common sticky eye. It usually presents with congestion and chemosis (swelling of the conjunctiva) and a purulent discharge. Swabs from the baby's eye, nasopharynx, throat, anorectum and vulva should always be sent for microscopy and culture. If cultures are positive for chlamydia, chlortetracycline eye drops should be administered. Tetracycline should not be used systemically in the pregnant woman or newborn as it will produce yellow discoloration of the developing secondary dentition. Systemic treatment for the baby should therefore be given with oral erythromycin, 40 mg/kg per day for 14 days. The mother should be treated with oral oxytetracycline, 500 mg per day, after meals free of milk products which chelate the drug because of their high levels of calcium, and thus reduce absorption. It is this phenomenon which renders any tetracycline secreted in breast milk harmless to the baby. An alternative is doxycycline 200 mg orally stat, followed by 100 mg daily. Treatment should be continued for at least 7 days, and 14 days in severe cases. If chlamydia infection is detected during pregnancy, treatment should be with oral erythromycin 500 mg four times daily after meals (it can cause gastritis if given on an empty stomach).

Chlamydia psittaci was first identified as an infection of parrots which can cause respiratory infection in the human. However, subsequent to the discovery of strains causing miscarriage in sheep, particularly the TWAR agent (TW for Taiwan, where it was first found, and AR for Acute Respiratory infection), it was found that it could also cause miscarriage in the human. In 1990 the Department of Health in the UK issued an official warning that pregnant women should avoid contact with sheep, particularly their milk, and lambs.

GONORRHOEA

This is much less common in pregnancy than infection with *Chlamydia trachomatis*. Caused by the obligate intracellular diplococcus *Neisseria gonorrhoeae*, colonization is restricted to the lower genital tract in pregnancy, because the presence of the conceptus blocks the passage of the organism into the uterus and Fallopian tubes. If the mother is infected during late pregnancy, the baby can acquire the infection during birth; *Neisseria gonorrhoeae* is also a cause of ophthalmia neonatorum. The mother sometimes presents with a vaginal discharge, although the infection is commonly asymptomatic during pregnancy. Diagnosis depends on positive cultures obtained from cervical and urethral swabs in the mother and eye swabs in the baby. Initial treatment is with penicillin, as most strains found in the UK are still sensitive to this antibiotic. If the strain proves on culture to be resistant, the antibiotic used will be suggested by *in vitro* sensitivities.

SYPHILIS

Infection with *Treponema pallidum* is still found in a significant number of pregnant women in the UK, about 1 in 40 000. The incidence is much higher in immigrant communities, particularly those from Africa and the West Indies. The screening test most commonly used is now the Venereal Diseases Research Laboratory test (VDRL), which has replaced the outdated Wasserman reaction (WR) and Kahn tests. The VDRL test is nonspecific and is also positive in infections with Yaws due to *Treponema pertenue*, for example. False positives can occur in normal pregnancy, in association with a variety of chronic infections including with Herpes simplex, and in drug addicts. It can also be positive in some collagen disorders; in association with systemic lupus erythematosus it should lead to a search for the lupus anticoagulant. If the screening test is positive, the presence of the treponeme must be confirmed by the use of the Fluorescent Treponemal Antibody absorption test (FTA) and the *Treponema pallidum* Haemagglutination Assay (TPHA).

The particular importance of syphilis in pregnancy is that it almost always causes congenital infection. This can produce miscarriage, stillbirth and preterm labour. It can also cause a variety of congenital malformations, notably a flat bridge to the nose, abnormal dentition, and mental subnormality. The newborn may fail to thrive, or present with skin rashes, hepatosplenomegaly, osteitis, choroidoretinitis or meningitis.

Pregnant women with positive serology should be treated with intramuscular penicillin in a dosage regime of 50 000 units per kg body weight, daily, for 15 days. Follow up should be undertaken for at least 2 years, for both mother and baby.

Problems can occur in successive pregnancies, because serology will remain positive even after effective treatment and without re-infection. It is therefore impossible to be sure whether re-infection has occurred or not. The only safe policy is a repeat course of treatment with each pregnancy. If the mother is unhappy about a further course of rather painful intramuscular injections, and it is thought that she has probably not been re-infected a 2-week course with oral amoxycillin is probably an adequate precaution. A low or absent titre of the VDRL test is probably good evidence against recent re-infection but the other tests remain positive for life. All women found to have a sexually transmitted disease during pregnancy should automatically be retested for syphilis.

HERPES SIMPLEX VIRUS

There are two main types of Herpes simplex virus: type I, the characteristic agent producing cold sores around the mouth, and type II, typically associated with sexually transmitted genital infection. However, with the increasing prevalence of oro-genital sex in recent decades, about 40 per cent of genital infections are now with the type I virus. Greater than 90 per cent of adults have been infected orally with type I virus, and so primary infections with this virus are extremely rare in pregnancy. On those rare occasions when a primary infection does occur, it is usually with the type II virus. Such infections can be associated with a severe systemic upset, with fever, rigors, and generalized influenza-like symptoms. The virus seldom infects the fetus, but if it does so it can cause miscarriage or intrauterine death, and is associated with congenital malformations including microophthalmia, choroidoretinitis, and microcephaly.

There is more risk of infection of the fetus during birth from an active recurrent herpetic lesion on the cervix, vagina, or vulva. About 1 in 30 women in the UK have been infected with genital Herpes simplex, although the figure is probably nearer 1 in 5 in the USA. Of these, about one-third never have recurrences, one-third experience recurrences only rarely, and one-third will have recurrences every few weeks or months. Despite this, the neonatal infection rate with herpes is only about 1 case in 33 000 births in the UK, although it is 1 in 5000 in the USA as the higher prevalence of the virus there would suggest. If infection occurs at birth, HSV can cause a disseminated systemic infection involving the liver and CNS as well as the skin. If such a widespread infection occurs, the neonatal mortality is 75 per cent, reduced to 40 per cent if antiviral treatment with acyclovir is given. Milder infections, localized to the skin, are probably much more common, but their incidence is unknown as most will go undiagnosed.

Because of the high risk of serious complication if infection occurs, caesarean section is advised if the mother has active genital lesions when labour starts, provided the membranes are still intact or have not been ruptured for more than 4 hours. At one time, it was common to take cervical and vulval swabs during the last 4 weeks of pregnancy, to see if the virus could be detected. However, as detection requires cell culture inoculations which take up to 7 days to demonstrate infection, by which time the maternal recurrence has usually healed over, this is no longer performed and decisions about caesarean section have to be made on clinical grounds. In fact, because most women experience considerable local pain with a recurrence, and the vesicles of the recurrent eruption are characteristic, diagnosis is usually easy on clinical grounds.

If a woman has frequent recurrences, they can be suppressed by treatment with daily acyclovir. Although not formally approved in pregnancy or the neonatal period, adverse effects from this drug have not been reported. Treatment of the neonate with prophylactic acyclovir should also be considered if the mother declines a caesarean section.

HUMAN IMMUNODEFICIENCY VIRUS

The origin of the HIV virus remains obscure but the most popular theory is that it arose by mutation from a closely related lymphotrophic virus endemic in African Green monkeys. It is a retrovirus of the subfamily Lenteviridae. First identified in homosexual men, it is now clear that the greatest reservoir of infection is in Africa, where the usual route of transmission is heterosexual intercourse.

In many African countries, such as Uganda and Kenya, 30 per cent or more of women attending antenatal clinics test positive for antibodies to HIV. The commonest strain of the virus is known as HIV-I. HIV-II has now made its appearance in some African countries; preliminary experience suggests that it behaves epidemiologically in a similar way to HIV-I.

The virus invades the body by blood contamination of open wounds. Genital trauma during intercourse is thought to be the main route of transmission, but the infection can also be acquired by infusion of contaminated blood products. The virus then infects the T-cell, and macrophages and other cells within the brain. This leads eventually to severe T-helper cell deficiency, with a reversal of the helper/suppressor ratio, a reduction in killer T cells, autoimmune phenomena, and a derangement of cell-mediated immunity. After a latent period, ranging from 18 months to many years, the majority of people infected develop the Acquired Immune Deficiency Syndrome (AIDS). This is commonly manifest as fever, weight loss, diarrhoea and lymphadenopathy. Opportunistic infections include CMV, Herpes simplex, *Mycobacterium tuberculosis*, candidiasis, *Cryptosporidium* and *Pneumocystis carinii*. Tumours such as lymphomas and Kaposi's sarcoma may occur.

At one time it was considered that pregnancy might accelerate progression of the disease, but this is no longer thought to be the case. There is, however, a serious risk of vertical transmission of the virus from mother to baby. Infection is thought to occur by transplacental passage of the virus in about 15 per cent of pregnancies where the mother is HIV positive; fetal infection is not associated with a significantly increased risk of congenital anomaly, intrauterine growth retardation, or preterm delivery, the latter two being more likely if the woman is ill with AIDS. Trauma during delivery can also transmit infection, and so application of scalp electrodes or use of fetal blood sampling is relatively contraindicated in HIV positive women. Recent data suggests that elective Caesarean section may reduce significantly the risk of vertical transmission, as may regular taking of AZT (zidovudine). Breastfeeding is a significant route of transmission, and will double the number of babies infected (i.e. from 15 per cent to 30 per cent). The risk of infecting the child antenatally is probably not increased during the viraemia that accompanies primary infection, but the risk of transmitting the virus by breastfeeding is doubled in these circumstances. It is important therefore that in developed countries, women are counselled against breastfeeding if they are HIV positive. However, in developing and undeveloped countries, the risk of neonatal mortality from gastroenteritis consequent on bottle feeding without a secure sterile water supply may well exceed the risk of HIV infection; in these circumstances breastfeeding is probably the only reasonable option.

It is important to be aware that the screening test for HIV infection is based on the detection of the antibody to the virus. A woman is therefore infectious for a period of about 12 weeks before the test becomes positive. Routine antenatal screening for HIV antibody remains controversial. HIV infection is currently rare in women born in the UK. Women at increased risk of infection include recent immigrants from high risk areas, particularly Africa south of the Sahara, but also the West Indies; infection rates are higher there than those in the UK and other countries, including Spain and Italy. Others at high risk are intravenous drug abusers and women whose partners are bisexual or haemophiliacs.

Because of the serious implications of a positive test, it should not be carried out without full informed consent from the pregnant woman. Such consent requires detailed and sensitive counselling about the implications of a positive test, including difficulties obtaining life insurance, potential problems with employment, and the risk of the child being infected. The baby of an infected mother will carry the mother's antibodies for 6 to 9 months, and it can often take up to a year before testing can establish whether a child is actually infected. A few cases of children showing the antigen, indicating presence of the virus, despite loss of the antibody have now been reported. Termination of pregnancy is currently the only way to guarantee the prevention of the birth of an infected child.

Any woman tested should also receive full information about the nature of the virus, the disease it causes, the treatments available, and the limitations of the testing procedure (latent period and the low incidence of false positives and negatives). Great care must be taken to maintain confidentiality; she should be advised not to tell anyone that she is having the test unless she is sure what their reaction would be if the test was positive, and that they could maintain confidentiality. She should also be counselled about the need for safe sex procedures (e.g. use of condoms) if she is in a high risk group, or if the test is positive. Women at high risk or with a positive result will need detailed counselling and support from an appropriately trained person.

INFLUENZA

Many reports have suggested that fetal damage may be produced by influenza in pregnancy. However, the variability of these reports suggests that the effects are strain specific, and as influenza maintains its virulence by the repeated appearance of new strains, most of which are thought to originate in the domestic pig population of central China, the effect of any given epidemic is impossible to predict. Particularly virulent strains (such as that in the 1918–20 pandemic) can cause miscarriage and preterm labour in up to 25 per cent of cases. Fetal infection is uncommon, congenital malformation rare, and most effects are mediated through systemic disturbance in the mother. Some reports have suggested an increase in subsequent childhood leukaemia, although others have failed to confirm this and the risk was in any case very small.

DISEASES OF THE ALIMENTARY TRACT

DENTAL CARIES

Pregnancy is sometimes associated with rapidly progressive gingivitis (inflammation of the gums). The reason for this may be the gingival hyperplasia commonly seen in pregnancy possibly induced by high levels of oestrogen and progesterone. The presence of gingivitis accelerates caries formation and so proper dental care is particularly important in pregnancy. Dentists sometimes raise anxieties about treating pregnant women, especially in relation to use of X-rays used to determine the extent of disease, but they should be reassured that the effects of normal treatment, including properly conducted dental X-rays and local anaesthetics are essentially nil on the pregnancy. The dentist should be aware of the increased propensity to postural hypotension and fainting, and the caval compression syndrome if the woman is laid flat on her back. If either of these occur, they will resolve spontaneously if the woman is placed in the left lateral recumbent position. If general anaesthesia is used extra precautions should be taken, because of the increased risk of aspiration leading to Mendelson's syndrome.

HEARTBURN

This is due to the relaxation of the cardiac sphincter of the stomach, and the increased intra-abdominal pressure due to the gravid uterus. It sometimes responds to advice for the woman to have small frequent meals, and avoid the supine position; using three or more pillows at night may help. Frequent sips of water can help to wash down regurgitated acid, and in severe cases antacids such as magnesium trisilicate or other proprietary formulations can be beneficial. Excessive use of antacids can however cause diarrhoea and prevent the absorption of some vitamins, and should be discouraged. Regurgitation of alkaline duodenal contents can also occur, in which case sipping lemon juice (a weak acid) may help. In most cases, the suitability of the various remedies can only be assessed by trial and error.

APPENDICITIS

Inflammation of the appendix is no less common during pregnancy than in other groups of young women. It is always feared because when it does occur it is often difficult to diagnose for the pain is less often localized to the right iliac fossa, and tenderness is difficult to elicit if the uterus overlies the inflamed appendix. Pyrexia is often low grade, and symptoms of vomiting and constipation are common in normal pregnancy. Final diagnosis is often only made at exploratory laparotomy, but the operation should be performed promptly if there is a serious suspicion.

INTESTINAL OBSTRUCTION

This is rare but when it occurs is most commonly due to obstruction by a band of adhesions from previous infection or surgery; the band becomes tightened due to displacement of the bowel by the expanding uterus. Other causes include strangulated hernia, volvulus and intussusception. Usually the definitive treatment will be surgical. However,

surgery can precipitate preterm delivery, and in the case of adhesions, conservative therapy by parenteral feeding, for example using a subclavian long line, can buy time to allow the fetus to mature. Induction of labour and delivery of the fetus may then relieve the obstruction, so that surgery can be avoided altogether.

ULCERATIVE COLITIS AND CROHN'S DISEASE

In the absence of acute exacerbations, there is little effect of the disease on pregnancy, and vice versa. If the disease is active, anaemia and maternal weight loss can be a problem leading to intra-uterine growth retardation. Treatment with sulphasalazine or steroids should be given as required.

LIVER DISEASE OTHER THAN INFECTION

Mild jaundice can be caused by severe hyperemesis gravidarum; it usually responds to simple measures such as rehydration.

Intrahepatic cholestasis of pregnancy occurs in about 1 in 1500 pregnancies. The common presenting symptom is itching, which usually comes on in the third trimester. The condition progresses only slowly, and reverses once the baby is born. If the jaundice becomes severe, delivery of the baby is indicated. Liver function tests usually return to normal within 4 to 6 weeks. The condition usually recurs in successive pregnancies, and may also occur if the woman takes the combined oral contraceptive pill.

Fulminant liver failure is also a severe although rare complication of pre-eclampsia. Liver function tests should therefore be a standard screening test in all cases of severe pre-eclampsia.

Acute fatty liver of pregnancy is a rare condition, occurring usually in the third trimester or the puerperium in obese nulliparous women in 1 in 15 000 pregnancies. Initially thought to be almost invariably fatal, improved reporting has now shown that about 80 per cent of women affected survive. Despite this, it accounts for about 2 per cent of maternal deaths in the UK recorded in recent triennial reports. Microdroplets of fat appear in the hepatocytes, and there is cholestasis. The cause of death is often the severe coagulopathy which occurs. Symptoms are non-specific with the acute onset of abdominal pain, nausea and vomiting, headache and jaundice. Treatment is to deliver the baby as soon as possible if undelivered already; subsequent management is supportive in the hope that spontaneous resolution will occur.

NEUROLOGICAL DISEASE

EPILEPSY

This is a relatively common disorder, with an incidence of about 1 in 200 of the population. Petit mal is not a problem in pregnancy, but grand mal seizures can make the woman hypoxic with damaging effects on the fetus. If there have been no recent fits, and the woman is not on therapy, there are unlikely to be any effects of or on the pregnancy. However, if the woman is on anticonvulsants, then fits can become more common. This is because the plasma volume expansion of pregnancy reduces the circulating level of anticonvulsants. Thus, management should include regular monitoring of blood levels of the drugs being used, and increases in dose are commonly required. Any woman having fits who is not on treatment should be referred urgently to a neurologist for investigation and instigation of therapy.

Many women are reluctant to take their medication because they are aware that it is teratogenic in about 2 per cent of cases thus it roughly doubles the background rate of congenital anomaly in pregnancy. However, failure to take medication, with the resulting fits, may put the life of the mother and her fetus at risk and severe hypoxia in a fit may cause damage to the fetus. Therefore, in all cases, women must be advised to remain on an adequate dose of medication. Drugs commonly used are phenytoin, carbemazepine and sodium valproate. The latter is particularly likely to cause spina bifida in the fetus and this was once said to contraindicate its use but the development of high definition ultrasound means that such abnormalities can now be detected, and termination of pregnancy offered if the parents request it. Phenytoin is associated with slightly lowered blood folic acid levels, and it is customary to supplement the diet of the pregnant epileptic on this drug with 500 μg of folic acid per day, although studies suggest that this is unnecessary in most cases. Women taking anticonvulsants may safely breastfeed.

MYASTHENIA GRAVIS

This occurs in about 1 in 20 000 of the population. It is more common in women than men, and the peak incidence is in the childbearing years. It is an auto-immune disorder, caused by circulating antibody to acetylcholine receptor protein. Pregnancy has a variable effect on the disease. The dosage of neostigmine may need to be altered, but pregnancy and labour are usually uneventful. Uterine action is unaffected but forceps delivery or vacuum extraction may be required because of weak voluntary effort. Epidural anaesthesia can be used if appropriate. Postpartum exacerbation of muscle weakness is common, due to concentration of the antibody as plasma volume expansion reverses. Supportive therapy is all that is required. The fetus acquires the antibody before birth, and may exhibit weakness after birth. Occasionally this requires supportive therapy for breathing or feeding. The weakness wears off over about 10 days as the antibody level in the baby declines.

MULTIPLE SCLEROSIS

There is no overall effect of pregnancy on this disorder, although recent data suggest that relapses are less common in the third trimester but more common in the puerperium. The disease does not usually affect the pregnancy unless there is severe disability. Anaesthetists commonly decline to perform spinal or epidural blocks for fear of being blamed for any relapses, although the evidence for this is scanty. Management should otherwise be unaffected.

ENDOCRINE DISORDERS

THYROID DISEASE

The thyroid gland produces two hormones, thyroxine (T4) and tri-iodo-thyronine (T3). These circulate largely bound to plasma proteins. Because the total amount of these proteins increases markedly during pregnancy, measurements of total hormone concentrations increase during pregnancy. However, the amount of free T4 remains unchanged, and the free T3 actually falls. Thus, free T4 measurement is the easiest indicator of thyroid function to measure in pregnancy (the free thyroxine index also remains largely unchanged).

The commonest thyroid disorder worldwide is endemic goitre, due to iodine deficiency. It causes cretinism, in which the baby is born with a large tongue, hoarse cry, dry course skin, a distended abdomen, and mental retardation. This is rare in the UK.

Mild hypothyroidism is relatively common in the white population, affecting as many as 1–5 per cent, but only one-third as many black people are affected. It has been implicated as a cause of postnatal depression. If there is any suspicion of the diagnosis thyroid function tests should be performed. The symptoms are the same as in the non-pregnant woman, with the addition that the fetus may exhibit a marked bradycardia. If the tests confirm hypothyroidism, treatment with oral thyroxine should be given without delay (up to 2 µg per kg body weight may be required).

Thyrotoxicosis occurs in about 1 in 500 pregnancies. It has usually been diagnosed before pregnancy, and the woman established on treatment. It is commonly due to Grave's disease and treatment is commonly with carbimazole but propylthiouracil can also be used in pregnancy. Treatment should be monitored by regular measurement of plasma free T4. A fetal tachycardia is commonly seen if T4 levels rise too high; sequential measurement of the fetal heart rate is a useful clinical index of the efficacy of therapy. Radioactive iodine treatment, which is now probably the treatment of choice in the non-pregnant patients rather than partial thyroidectomy, should never be given in pregnancy because iodine is concentrated 10 times more efficiently in the fetal than the maternal thyroid, and this would result in complete ablation of the fetal thyroid. Partial thyroidectomy should also be avoided because of the risk of rebound crisis following surgery; it may sometimes become necessary if the goitre is very large. Breastfeeding can be allowed so long as the infant's thyroid function is monitored closely, as occasionally enough drug is secreted in the milk to affect neonatal thyroid function.

PITUITARY PROLACTINOMA

Hyperprolactinaemia is an important cause of secondary amenorrhoea and infertility, commonly due to a benign pituitary adenoma. In the days when the amenorrhoea was treated with gonado-trophins to induce ovulation, expansion of the tumour sometimes occurred in pregnancy, causing blindness by pressing on the optic chiasma. Since the introduction of bromocriptine, a dopamine agonist, which causes regression of the tumour, this complication has become rare, despite the fact that bromocriptine is normally discontinued once pregnancy is achieved. Nevertheless, careful monitoring of the visual fields of women with prolactinomas who are pregnant is probably still justified. If there is any evidence of tumour expansion, bromocriptine should be restarted.

ADRENAL DISORDERS

Both Cushing's syndrome and Addison's disease are excessively rare in pregnancy. Phaeochromocytoma is a rare tumour of the adrenal glands which secretes catecholamines. It causes episodic hypertension but its effects can be controlled with combined alpha- and beta-blockers. The definitive treatment is surgery, once the baby has been delivered.

SKIN DISORDERS

The skin becomes more active in pregnancy; the breasts are modified sweat glands which respond to human placental lactogen and to progesterone. Increased pigmentation is common, particularly in the nipples, areolae, vulva, perineum and perianal areas. The abdominal midline usually becomes pigmented to form the linea nigra. Palmar erythema and increased spider naevi counts are also common. Striae gravidarum are common on the breast, thighs and abdomen. It has been suggested that they occur as part of the general softening of collagen under the influence of oestrogen and relaxin and help to allow the expansion of the pelvic and vaginal dimensions during labour. They are idiosyncratic and there is no known way to influence whether they appear except perhaps to limit excessive weight gain. Skin creams have commercial but not medical value.

Pruritis, or itching, of the skin is common in pregnancy. Occasionally it is related to cholestasis of pregnancy, but usually it is just physiological. Similarly, non-specific rashes are common. It is thought they may be due to complement fixing reactions between the skin and fetal antigens. They may respond to treatment with calamine lotion, antihistamines, or simply to reassurance that they are harmless and will disappear when the baby is born.

Pemphigoid (herpes) gestationis is a rare condition which has nothing to do with the Herpes virus. Large, tense bullae appear on the abdomen and legs. It usually responds to high dose oral steroids, and remits when the baby is born.

PSYCHIATRIC DISORDERS IN PREGNANCY AND THE PUERPERIUM

Pregnancy is probably the biggest time of change in any woman's life with obvious physical changes, major physiological changes, and necessary psychological and social adjustments.

In the first trimester, the woman may have to cope with extreme fatigue, nausea and vomiting. Irritability and tearfulness are common. Uncertainty about the viability of the pregnancy has to be tolerated and ambivalence towards the pregnancy is normal at this stage. There may be justified worries about emotional and/or practical support from the baby's father. A few women develop hyperemesis gravidarum. Psychological causes for this are now in doubt. Neurological sequelae may be severe with the development of Wernicke's encephalopathy which can progress to Korsakov's syndrome. Energetic treatment with thiamine supplements is required. During the second trimester, the mother has to adjust to her changing shape, and also may be required to undergo investigations to ascertain the health of the fetus. Ultrasound of the baby has been shown to improve bonding with the newborn child.

During the third trimester, fears about coping with motherhood, the baby's health, encounters with hospitals, and pain in labour are most often expressed. Multigravidae who have had difficult experiences during earlier confinements may worry about the next delivery throughout a subsequent pregnancy. Fatigue, disturbed sleep, decreased concentration, and reduced new learning ability may become significant at the end of the pregnancy.

Libido is often reported as being reduced during pregnancy although some women are more interested in sex, especially in the middle trimester.

In the puerperium, exhilaration is encountered until day 2 or 3 postpartum, with euphoria and restlessness and a decreased need for sleep. Later, 50 per cent of mothers experience the baby blues.

The common symptoms are: brief episodes of weeping (50–70 per cent), irritability (33 per cent), anxiety (50 per cent), forgetfulness (30 per cent), headache (35 per cent), depersonalization (30 per cent), elation (40 per cent), and confusion (35 per cent). The peak incidence of the blues is 4 to 5 days postpartum with an average duration of 2 to 3 days. The syndrome is uncommon after 10 days. An increased length of labour is related to an increased likelihood of the blues, but there is no independent effect of parity. Previous premenstrual tension predicts the blues.

Mothers with special needs include those who have conceived after a period of infertility and those who have lost a child either secondary to miscarriage or stillbirth. Very young unsupported mothers need particular attention. Their own dependency needs may conflict with those of the child. Older mothers may encounter difficulties due to their increased need for autonomy and possibly greater expectations of the professionals they encounter.

EFFECTS OF PREGNANCY ON PSYCHIATRIC DISORDERS

Mood disorders

Many women with a bipolar affective disorder developing early in reproductive years may wish to have children. While pregnancy itself is not a risk factor, the puerperium poses a risk of between 20 and 50 per cent of developing a psychotic illness. Women are generally advised to restart lithium following delivery, but the drug is contraindicated during pregnancy.

Schizophrenia

Schizophrenia is a common psychiatric condition

occurring in 1 per cent of the population. If schizophrenic women become pregnant, they should be encouraged to continue with their usual medication. There are no clear indications that major tranquillizers, either given orally or as a depot injection, have harmful effects upon the fetus.

Eating disorders

While severe untreated anorexia nervosa makes conception impossible, anorectics who become pregnant do not appear to deteriorate during pregnancy. There are reports of later cognitive delay in the infants associated with malnutrition *in utero*. Bulimia nervosa seems to remit during pregnancy. There are no large studies of fetal abnormalities associated with bulimic behaviour during pregnancy.

Mental handicap

The range of mental handicap is wide. The trend toward community rather than institutional care has increased the likelihood of pregnancy in mentally handicapped women. If a mentally handicapped woman becomes pregnant there may be concern about the likelihood of handicap in the offspring, her ability to cope with the pregnancy and labour, and her likely abilities as a mother. The incidence of handicap in the children of disabled mothers is reported as ranging from 2.5 to 9.3 per cent, depending upon whether retardation is present in one or two parents, the IQ level of the parents, cause of handicap, prenatal and socio-economic factors.

The special needs of mentally handicapped mothers in the antenatal clinic reflect difficulties imposed by slowness in learning new information. The mother may not understand the consequences of sexual intercourse, and pregnancy is often diagnosed late.

DRUG DEPENDENCE

The two drugs that cause most concern in antenatal care are alcohol and opiates.

ALCOHOL ABUSE

Currently pregnant women are advised to abstain from alcohol in order to avoid growth retardation in the fetus, increased perinatal mortality and fetal alcohol syndrome (characteristic facies and mental retardation). The amount of alcohol required to cause such problems is contentious.

Alcohol abuse is often concealed in the antenatal clinic, but requires a pragmatic style of management with the use of screening questionnaires and the regular observation of MCV and gamma GT.

OPIATE DEPENDENCE

Addicts not attending drug dependence clinics tend to jeopardize their health because of the necessity to spend all their emotional and financial resources to obtain supplies of heroin. Injecting drugs runs the risk of HIV or hepatitis B infection. The effect on the fetus is potentially serious, with risks of premature labour, growth retardation and neonatal withdrawal symptoms. The latter may occur up to several days after delivery and include a shrill cry, irritability, tremors, respiratory distress and feeding difficulties.

Addicts who are known to the medical services are more likely to take advice to withdraw from opiates slowly during the pregnancy. In later pregnancy even cautious reduction in opiates may cause withdrawal symptoms in mother and fetus, and therefore the lowest dose of methadone which prevents this may have to be continued. Hospital admission may be advisable to assess true opiate needs.

PUERPERAL DISORDERS

POSTNATAL DEPRESSION

This is a common and disabling condition occurring after approximately 1 in 10 live births. The mother often displays excessive anxiety about the baby's health which cannot be diminished by reassurance. She tends to blame herself and believes that she is a bad mother. There may be depressive behaviour, such as tearfulness, irritability and loss of libido, but there are rarely complaints of a depressed mood. The mother may, however, acknowledge that she is not her normal self. She may express concern about rejecting her baby, and be reluctant to feed or handle it. She may fear that the baby is not hers or is seriously deformed in some way. On enquiry, but rarely spontaneously, she may admit to thoughts about harming the baby or suicidal thoughts.

The distinction between postnatal depression and a normal adjustment to having a new baby is often difficult, leading to a delay in diagnosis and treatment. The Edinburgh Postnatal Depression Scale is a useful rating scale which identifies postnatal depression in 80 per cent of cases.

The illness usually remits within 6 weeks if diagnosed and treated, but symptoms may continue for over a year if untreated. Treatment is with antidepressants and psychotherapy. Social interventions are also required. Early treatment is important for the entire family. Infants of mothers with postnatal depression show both cognitive and language delay when assessed at 18 months of age.

Risk factors for postnatal depression include previous postnatal depression, premenstrual tension, history of subfertility, ambivalence about the pregnancy, and early separation from the father or the mother. Possible prophylactic treatments include antidepressants, progesterone injections and suppositories and oestrogen patches. None of these options has yet proved of definite advantage in controlled trials.

PUERPERAL PSYCHOSES

Puerperal psychoses are uncommon illnesses occurring after 0.2 per cent of live births. They usually start within 4 weeks of the baby's birth. The disorders are characterized by delusions, hallucinations and perplexity. A marked variability in mental state throughout the day is common and can lead to diagnostic confusion. The mother may attempt to act on delusions about her infant, believing the infant to be at risk and better off dead. The need to protect the baby may require admission of the mother and baby to a psychiatric facility. Specialized Mother and Baby Units are not available in each region; mothers and their babies may be nursed intensively at home or admitted to a non-specialized facility. Treatment usually involves the use of major tranquillizers, antidepressants, lithium and occasionally ECT.

Risk factors for the development of puerperal psychoses include: primiparity, a past psychiatric history or a family psychiatric history.

PSYCHOTROPIC MEDICATION

Psychotropic medication should be avoided where possible in the first trimester except for schizophrenic mothers who should be maintained on medication. With other drugs, the risks and benefits should be carefully assessed. Most psychotropic medications are safe for the developing fetus. The exception is lithium carbonate which is associated with the production of major cardiac malformations, such as Ebstein's anomaly, in 10 per cent of women taking lithium in the first trimester. Lithium may also cause thyroid abnormalities in the developing fetus if taken on a long-term basis. Benzodiazepines have been associated with cleft palate abnormalities.

During breastfeeding, many psychotropic drugs are completely safe with little drug secreted in the breast milk. Some of the new 5HT re-uptake inhibitors have been shown, however, to be present in substantial quantity in breast milk, but damage to the fetus has not been reported. Lithium carbonate is also secreted in breastmilk in 30–70 per cent of the maternal plasma concentration, and is contraindicated in breastfeeding mothers because of the risk of lithium toxicity in the infant. Major tranquillizers given orally or by depot are safe during the puerperium. In breastfeeding mothers taking major tranquillizers, observation of the infant for drowsiness is important.

THE FETUS AT RISK IN LATE PREGNANCY

During pregnancy and labour the fetus may be at risk of damage or death from many causes. However it has become the custom to talk of the fetus at risk in the more limited sense of at risk from acute or chronic placental insufficiency. This term, imprecise but convenient, covers a variety of pathologies. The predominant one is inadequate maternal blood flow in the utero-placental circulation, but reduced size and efficiency of the surface area available for transport, and poor fetal blood flow in the feto-placental circulation due to arteriolar constriction in the stem villi, are also thought to be the major cause in some cases. Acute placental failure may result from placental separation by haemorrhage (placental abruption) or it may come at the end of a phase of gradually declining placental efficiency. In the latter case acute failure may only become manifest during labour when the uterine contractions interfere with blood flow. This failure is commonly called fetal distress. It is discussed in Chapter 5.

The importance of acute hypoxia, which may cause sudden fetal death during labour, has been appreciated for a long time, but recognition of the long-term damage which may result from chronic intrauterine malnutrition and lead to the birth of a small-for-dates baby is more recent.

PLACENTAL INSUFFICIENCY DURING PREGNANCY

During pregnancy the fetus depends on the placenta and the umbilical vessels for transport of oxygen and nutrients from the maternal blood, and for excretion of carbon dioxide and the products of metabolism. Because of the constancy of the mother's internal environment the composition of the blood reaching the placenta is unlikely to vary much except in desperate conditions of maternal circulatory failure or asphyxia, or in cases of gross metabolic disturbance such as severe diabetes or renal disease. Even if the maternal nutrition is poor the maternal blood will show little alteration, and this explains the fact that a normal sized child may be born in such circumstances, and that the fetus of a mother with poor iron reserves is not anaemic as long as her serum iron concentration is maintained. Fetal ill-health therefore seldom arises because of alteration in the biochemical quality of the maternal blood – the ability of the placenta to act as an organ of transfer is the important consideration.

Restriction of the maternal blood flow to the placental bed can have a serious effect upon fetal growth and development and may happen with or without hypertension. Placental tissue may be lost by infarction or the separation from the uterine wall that occurs when abortion threatens or when there is a small retroplacental haemorrhage. Spasm or thrombosis of small decidual vessels may also have the same effect. In a few women giving birth to babies who have obviously suffered from chronic placental insufficiency the placenta is small for no apparent reason.

If pregnancy is prolonged beyond term the placenta sometimes becomes inadequate for the needs of the large, still growing fetus. Significant structural changes in the placenta are seldom found in such cases. There may be calcification of the decidual plate, but the intervillous circulation is not impaired by this. Although intrauterine death before the onset of labour is uncommon in cases of postmaturity, fetal distress may occur during labour.

Intrauterine death may occur in cases of diabetes but the cause of this is uncertain. The fetus is often very large, but so is the placenta, and placental failure is not an adequate explanation.

In cases of haemolytic disease the placenta is large and oedematous, with persistence of Langhans' layer of cytotrophoblast in the villi, but the fetus probably dies from cardiac failure caused by the severe haemolytic anaemia rather than from placental insufficiency.

INTRAUTERINE GROWTH RETARDATION

Definition

This term (IUGR) implies that the fetus has been subjected to nutritional deprivation or to some cause of growth impairment so that its growth potential is not realized. However, there are small fetuses that are merely constitutionally small and are otherwise perfectly normal. These are called small for gestational age (SGA) and the distinction is very important. SGA is defined by comparing the birth weight of a newborn baby with appropriate birth weight *vs* gestational age charts and several levels of severity may be used, for example below the 10th, 5th or 3rd centiles or alternatively below two standard deviations. It is clear that appropriate charts must be used for ethnic group and sex of baby (*see* Fig. 3.19). It is also vital that the gestational age is known and that the pregnancy has been dated accurately (*see* Chapter 2).

It follows that of SGA babies, some have IUGR and others are normal but small. Conversely, there are some whose weight is above the cut-off percentile, for example on the 20th centile, who should have had a birth weight in the 90th but who have suffered from IUGR. These may look thin and starved and are often called dysmature. Weight alone does not define the pathology, but it is an easy and objective measurement; other endpoints,

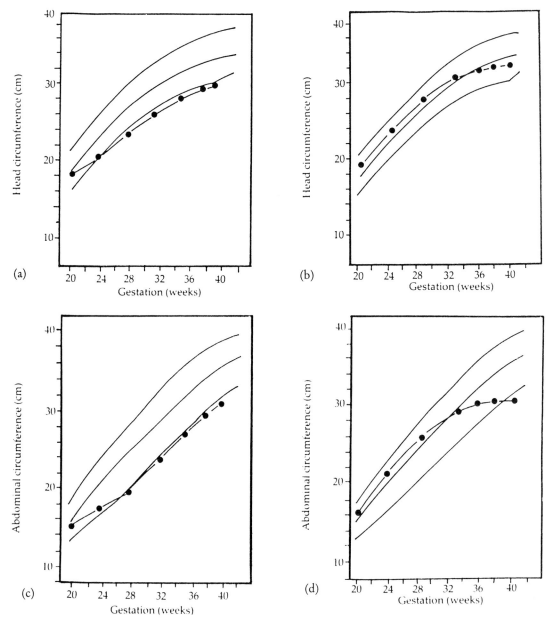

Fig. 3.19 Fetal headgrowth (a and b) and abdominal circumference (c and d) (set against the mean ±250) are plotted in symmetrical (a and b) and asymmetrical (c and d) fetal growth retardation

such as skin-fold thickness, also have their difficulties.

Incidence

A cut-off at the 5th percentile obviously means that 5 per cent of a population are SGA. About two-thirds of these have had IUGR. It has been estimated that 4–6 per cent of all newborn, of any birth weight, are growth retarded. The importance of this group is:

1) their greatly increased perinatal mortality and morbidity

Table 3.3 Factors leading to poor fetal growth

Small parental size
Ethnic group
High altitude
Maternal viral infections (rubella, CMV)
Fetal malformations
Fetal chromosomal abnormalities
Maternal smoking
Maternal malnutrition
Maternal illness (hypertension)
Multiple pregnancy
Placental size
Placental perfusion (maternal and fetal)
Recurrent antepartum haemorrhage

2) the increased incidence of psychomotor problems in childhood
3) the diminished reproductive performance of the females when they grow up and have children themselves
4) a higher incidence of hypertension and non-insulin dependent diabetes in later life.

Aetiology

Fetal weight is influenced by the Y chromosome, parental size, race and altitude. It may be reduced by chromosomal and genetic syndromes, malformations and viral infections (e.g. rubella, cytomegalovirus). Maternal malnutrition may also cause poor fetal growth but the relationship is complex depending on severity, time of onset and duration. It is not clear how much is due to deficiency of calories or of specific nutrient factors such as zinc. Other important causes of poor fetal growth are maternal smoking, a variety of chronic maternal illnesses such as hypertension, lupus and severe anaemia, and multiple gestation. These last causes probably all act by influencing placental size or function.

One of the most important determinants of placental function is utero-placental perfusion. This has been reduced experimentally by various manoeuvres, resulting in IUGR and in human pregnancies with IUGR, vascular abnormalities have been found in the placental bed. It is thought that in normal pregnancies, the cytotrophoblast from the developing placenta invades the spiral arteries in the decidua converting them to wide channels that present very little resistance to maternal blood flow into the placenta (*see* Fig. 3.12). If this invasion is inadequate, the spiral arteries have a higher resistance, placental perfusion is poor and either IUGR or pre-eclampsia can occur.

Another important cause is feto-placental hypoperfusion due to reduced flow in the umbilical circulation. A reduction in the number and diameter of arterioles in the tertiary stem villi has been found.

Prenatal detection of IUGR

Among clinical indicators, a past history of babies with IUGR and a medical history of maternal illness are most significant. Clinical examination includes maternal weight gain, blood pressure, abdominal palpation for fetal size, amniotic fluid quantity, and measurement of the symphysis–fundal height. The sensitivity of clinical methods is poor: at best, 40–50 per cent are detected, i.e. over half of fetuses with IUGR are missed.

To improve the detection rate, various investigations have been employed. By far the most important is ultrasound biometry, which is performed routinely at 32 to 34 weeks in some centres. Various measurements are taken, such as head and abdominal circumferences, biparietal diameter and femur length, and compared to normal charts. The sensitivity is improved to 70–80 per cent, but because the prevalence is fairly low, the positive predictive value is poor. Some further improvement has been achieved more recently by computing an estimated fetal weight from the measurements and constructing normal charts of this parameter against gestational age.

Risks of IUGR

Two types of IUGR have been widely recognized but this is an oversimplification. The symmetrical type is usually a non-growth retarded SGA fetus, although some are genetically abnormal or infected by a virus. In the asymmetric type, the head: abdominal circumference ratio is high, due to slower growth of the latter. It tends to occur fairly late in pregnancy and is thought to be caused by a stimulus such as hypoxia initiating carotid chemoreceptor activity. This produces peripheral vasoconstriction and redistributes the fetal cardiac output away from the kidneys, limbs and gut to the brain, heart and adrenals. Liver glycogen and skin fat stores are low, the fetus becomes thin with

Table 3.4 Risks of IUGR

Antepartum	Slow growth, oligohydramnios, reduced fetal movements, hypoxia, acidaemia, abnormal heart rate, iatrogenic, prematurity, intrauterine death
Intrapartum	Fetal distress, hypoxia, acidaemia, abnormal heart rate, passage of meconium, intrauterine death
Neonatal	hypoglycaemia, hypocalcaemia, polycythaemia, asphyxia, neonatal death, long term handicap (neurological and mental)

a small liver, and it may enter a catabolic state. Amniotic fluid and fetal movements are commonly reduced.

The risks of this condition are shown in Table 3.4.

Management of IUGR

The most important step is to make the diagnosis and then to identify a cause which may require treatment, for example maternal hypertension. The mainstay of management is then serial monitoring with the aim of delivering as mature a fetus as possible, but in good condition, before hypoxic damage has occurred. Bed rest, often in hospital, is advised as this may improve placental blood flow. More specific therapies are not yet available, although there is some evidence that low dose aspirin given from early pregnancy may improve fetal growth.

The methods of monitoring are described in the section on Tests of Placental Function. They are basically clinical (maternal weight gain, uterine size or symphysis–fundal height, and fetal movement counting) and biophysical (cardiotocography or non-stress testing, biophysical profile, amniotic fluid volume, serial ultrasonic fetal biometry, and Doppler studies). Increasingly, Doppler waveform analysis is being used to assess circulatory redistribution by comparing flow in the fetal descending aorta with that in the middle cerebral artery.

Occasionally umbilical vein puncture under ultrasound guidance is used to obtain a fetal blood sample for blood gases and acid–base status.

When the time for delivery is deemed right, a decision for vaginal or caesarean delivery must be made. Frequently, the fetus will not be able to withstand the stresses of labour and an elective caesarean section is the method of choice. If labour is induced, the fetus must be monitored very carefully. Such babies should be delivered in units with the appropriate facilities and with expert neonatal care at hand.

Uterine growth

This is not always easy to assess and the variation between observations may be considerable. Nevertheless simple measurements of the height of the fundus above the symphysis pubis have more to recommend them than the common practice of assessing fundal height in relation to the umbilicus or xiphisternum. The fundal height should increase by about 1 cm weekly from the 16th week of pregnancy, and with an average sized fetus should equal the number of weeks of gestation plus or minus 2 cm. Measurements of abdominal girth are much less satisfactory indicators of uterine growth. In the case of both measurements the variability is lessened appreciably if the same person sees the woman at each antenatal visit. If the uterus seems to be growing slowly this is another reason to perform other placental function tests.

Fetal growth

The most reliable measurements of the fetus are obtained with ultrasound. Serial measurements of the biparietal diameter of the fetal head are comparatively easy to obtain. It should be remembered that the head of the fetus is a privileged area in so far as the greatest part of the cardiac output with the most highly oxygenated blood is diverted there. It is therefore the last part of the fetus to suffer from malnutrition. If earlier signs of poor growth are to be sought it is better to measure the abdominal circumference at the level of the liver and to relate this to the size of the head.

Fetal activity

The most important indicator of placental function is the well-being of the fetus. A valuable sign of a healthy fetus is vigorous activity. The mother may be able to give some indication of this by keeping a kick count. She can be asked to note how frequently the baby moves in a given period, perhaps 30 minutes. Alternatively she can be asked to note how long it takes for the baby to move 10 times.

A more sophisticated way of assessing fetal activity is to produce a continuous record of the fetal heart rate over a period of 30 minutes or more. This is done with a cardiotocograph and is often called a non-stress test. An acceleration in rate with fetal activity and variability of the rate from beat to beat are associated with fetal health. The heart rate may change with a uterine contraction but the normal rate is restored as soon as the contraction passes off.

A still more elaborate method of fetal assessment is the biophysical profile. This is a score based on real-time ultrasound observations of fetal breathing and body movements, tone, quantity of amniotic fluid, and on a non-stress test. Lack of movements and amniotic fluid suggest the fetus is in danger of hypoxia, but inactivity in the fetus may also be physiological. Prolonged observation over more than 30 minutes may be necessary and even then there is a high false-positive rate. Irradiation, the more predictive parts of the score, are often marked by the less discriminating. This test has therefore not been widely adopted in the UK although its components are used.

Doppler studies

The reflection of ultrasound by blood moving along a blood vessel can be detected and compared with the energy output of the source. This provides a measure of the speed of passage of the pulse wave. If the vessel diameter is measurable then the flow of the volume of blood per minute may be calculated. As this is of poor reproducibility the shape of the Doppler wave form, which should show forward flow during diastole as well as systole, is analysed. High diastolic flow indicates low resistance and impaired diastolic flow indicates high down-stream resistance. This is the basis of Doppler blood flow studies of the arcuate arteries in the placental bed and the umbilical arteries in the cord. The former is a reflection of the afferent supply of oxygen and nutrients from the mother to the fetus; the latter is a measure of fetal cardiac output and so of well-being.

Placental function tests

Attempts have been made to judge the functional activity of the placenta by measuring one or more of its hormone or enzyme products in maternal blood or urine. Blood levels of oestrogens and progesterone vary with placental function, but diurnal variations and variations due to activity and posture are so great that they are unreliable indicators of fetal well-being. The excretion of oestrogens in the maternal urine during a 24-hour period gives a more consistent indication of placental function. The normal range of values for excretion is considerable, and isolated observations are of little value, but repeated observations may show if there is any of an obvious trend.

Other tests which are even less frequently employed include serum levels of human placental lactogen (hPL) and of heat-stable alkaline phosphatase. These biochemical tests have fallen out of use.

POSTMATURITY

Postmaturity means the prolongation of pregnancy beyond its normal duration. It is important because surveys of perinatal deaths have shown that the risk of intrauterine death increases, particularly in pregnancies continuing more than 2 weeks beyond term. However, there is no agreement on the normal limits of duration of pregnancy. The difficulty in making any definition is that the precise date of conception in any particular pregnancy is unknown, and even with regular menstrual cycles of normal length the date of ovulation is only approximately known. In women with irregular or prolonged cycles calculations based on the date of the last menstrual period are bound to be inaccurate.

Further, each fetus is an individual, growing and maturing at a slightly different rate so that it is improbable that all fetuses will mature in precisely the same number of days. The cause of the onset of labour at term is uncertain.

Diagnosis

General statistical statements based on large numbers of cases can easily be made, showing that delivery takes place over 2 weeks after the expected date of delivery in between 5 and 10 per cent of cases but in any particular pregnancy the diagnosis of postmaturity is often uncertain.

In cases in which the date of the last menstrual period is absolutely certain, and in which the previous menstrual cycles were of normal length, and when this has been confirmed by an early ultrasound scan, the diagnosis of maturity can reasonably be made. If the menstrual history is uncertain an attempt can be made to assess maturity by scrutiny of the antenatal records to discover whether the size of the uterus was determined by bimanual examination at an early visit by an experienced obstetrician. Between the 8th and the 14th weeks an accurate assessment of the uterine size can generally be made. Ultrasound measurements of the crown–rump length of the fetus up to about the 14th week and of the biparietal diameter of the fetal head up to about the 28th week, give a reliable indication of the duration of gestation.

Other tests for the assessment of maturity, including the date of first felt fetal movements, radiological examination of fetal ossific centres and chemical and cytological examination of amniotic fluid, have now been superseded by ultrasound observations.

Clinical significance

The delayed onset of labour might in theory have disadvantages:

- after term, as the uterine blood flow diminishes and degenerative changes progress, the placenta and placental bed function may become inadequate for a large fetus that is still growing
- there may be mechanical difficulties during labour because of the increased size of the fetus
- a uterus which is slow to begin labour may also prove to be inefficient during labour.

The risk of the fetus dying in the uterus from hypoxia before the onset of labour has probably been exaggerated, although this does occasionally occur. However, there is good evidence that the risk of fetal distress and fetal death during labour is greater in postmature than in normal cases. In part this is due to more difficult labour because of the larger size of the fetus and incoordinate uterine action, and in part to the more frequent occurrence of oligohydramnios. The skull of the postmature fetus is more ossified, so that moulding is less easy. A particular hazard for the new born baby is meconium aspiration.

The risk of hypoxia in postmature cases is increased if there is also hypertension, and perhaps in the case of a relatively older mother.

Management

Postmaturity is often a cause of worry, and sometimes of expense and inconvenience, to the mother. Pressure is often put upon the obstetrician to induce labour. While this is often justifiable in cases in which the menstrual history is certain, in other cases there is a risk that the fetus may, after induction, prove to be premature rather than postmature. The clinical evidence must be carefully reviewed before deciding to induce labour. Uncertainty over gestational age is eliminated by early ultrasound dating and is a very strong argument in favour of routine scanning of all pregnant women in the second trimester.

Many women and their relatives tend to worry when the date given to them as the expected date of delivery has been passed; it is most desirable that every mother should be told that the calculated date is only an approximation, and that normal labour may start up to 2 weeks later.

Each case of suspected postmaturity should be dealt with according to its special circumstances. There is no justification for making a rule that all cases must be induced at some stated week of pregnancy; the risks of indiscriminate induction might well exceed those of postmaturity. There will be more anxiety in the case of older primigravidae, or women with hypertension. Induction might well be recommended if the fetus is evidently large, and for women in whom there is a risk of disproportion if the fetus continues to grow. There can be little justification for induction until the woman is at least 2 weeks overdue except in cases with hypertension or some other reason to suspect placental insufficiency. The cervix should always be checked for ripeness (length, softness, position and dilatation) before a decision is made.

Labour may be induced in cases of postmaturity by insertion of vaginal prostaglandin pessaries or gel. If labour does not quickly follow, amniotomy is performed and intravenous syntocin given. When the fetal head is well down in the pelvic cavity and the cervix is soft and taken up, labour follows induction without delay. Very careful monitoring of the progress of the labour and of the condition of the fetus is required. Particular attention is paid to whether or not amniotic fluid is draining, and to thick meconium. If there is cardiotocographic evidence of fetal hypoxia during the first stage of labour, preferably confirmed by examination of a sample of fetal blood obtained by scalp puncture, caesarean section may be necessary. In the second stage clinical signs of fetal distress call for forceps delivery without delay.

INTRAUTERINE DEATH OF THE FETUS

In addition to cases in which the fetus dies during delivery as a result of asphyxia or difficult labour, others are seen in which it dies *in utero* before labour starts. This is usually followed by expulsion of the fetus from the uterus within a few days. However, in exceptional cases the dead fetus is not expelled from the uterus at once, but is retained for several weeks.

Causes

In some cases the cause of death is obvious; but in others it is obscure, because autolysis of the tissues of the fetus and placenta occurs, and as a result post-mortem findings are difficult to interpret. The causes of intrauterine fetal death may be classified under the following headings:

Pregnancy induced hypertension, essential hypertension and chronic renal disease

Fetal hypoxia is rendered more likely by lack of mid-trimester trophoblast invasion of the mouths of the spinal arteries so they remain narrow. In addition it may be produced by reduction in the maternal blood supply to the placenta because of spasm, and sometimes thrombosis, of the maternal vessels. Added to this there may be separation of the placenta (placental abruption) or extensive clotting of maternal blood around the chorionic villi.

Diabetes mellitus

If maternal diabetes is poorly controlled fetal death *in utero* often occurs.

Postmaturity

This is an uncommon cause of fetal death before labour.

Placenta abruption

Cord accidents

True knots in the cord or constriction of the cord round a limb are very rare causes of fetal death.

Haemolytic disease

In such instances the fetus is usually hydropic.

Unexplained placental insufficiency

Apart from the cases of hypertension and renal disease already mentioned, in a few unexplained placental insufficiency occurs in successive pregnancies. Further investigation may reveal anticardiolipin antibodies in maternal blood. The fetus does not grow at the normal rate and intrauterine death may occur. The placenta is found to be small but appears to be normal in other respects. In the absence of any explanation for some of these cases, the only advice which can be offered is intensive fetal monitoring and rest in bed during most of the pregnancy, with the hope that this will increase the uterine blood flow. Delivery before the time at which previous deaths occurred is wise, either by induction of labour or caesarean section. Tests of fetal well-being may give some guidance in deciding when to advise delivery.

Fetal malformation

With gross malformation intrauterine death sometimes occurs.

Infective diseases

Any disease that causes high fever and toxic illness may cause fetal death.

Untreated syphilis and rare fetal infection with Herpes virus or other viruses may cause death. Severe rubella is sometimes fatal to the fetus.

Pathological anatomy

The fetus is usually born in a macerated condition.

Its skin is peeling and stained reddish-brown by absorption of blood pigments. The whole body is softened and toneless; the cranial bones are loosened and easily moveable on one another. The amniotic fluid and the fluid in all the serous cavities contain blood pigments. Maceration occurs rapidly, and may be advanced within 24 hours of fetal death. Autopsy, although frequently unrewarding, is advised in all cases.

Symptoms and diagnosis

The woman may notice that the fetal movements have not been present for several days, and the breasts may diminish in size. In cases of hypertension the blood pressure sometimes falls. The following signs may be found:

After the 24th week the fetal heart sounds can normally be heard with a stethoscope; failure to hear them on careful auscultation will be strong presumptive evidence of fetal death. Ultrasound can detect the fetal heart beat as early as the 10th week of pregnancy, and if a careful and repeated search shows no evidence of cardiac activity, fetal death is almost certain.

The uterus may be found to be smaller than the duration of pregnancy would warrant. A more accurate sign is to note how much alteration takes place in the size of the uterus during a period of observation. For this the bladder must be empty and the level of the fundus noted accurately. The woman is examined week by week. In some cases the uterus not only ceases to grow but gets smaller because of absorption of the amniotic fluid.

Sometimes secretion of colostrum from the breasts occurs a few days after the death of the fetus.

Ultrasonic examination or a radiograph will show overlapping and disalignment of the skull bones and occasionally the presence of gas in the fetal heart and great vessels.

It will be seen that fetal death is difficult to diagnose from a single clinical examination of the patient, and urgent sonography is usually requested. Very little importance should be attached to the patient's statements about fetal movements; false hope will lead to confusion of intestinal with fetal movement.

Management

Imparting the diagnosis to the parents must be handled sensitively. In the majority of cases labour soon follows death of the fetus, but sometimes labour does not occur for several weeks. In these cases there may be no urgent call for interference, unless the complications of disseminated intravascular encephalopathy occur. However, most mothers want to proceed to induction of labour and delivery as soon as possible. Amniotomy is unwise because of the risk of anaerobic uterine infection from growth of bacteria in the dead placental and fetal tissues if labour does not follow quickly.

Labour can usually be induced with vaginal prostaglandin pessaries or gel or extra-amniotic injections of prostaglandins. It is also possible to induce labour by the intra-amniotic injection of prostaglandins and a hypertonic solution of urea. Such women should have bereavement counselling and, where appropriate, genetic counselling.

MULTIPLE PREGNANCY

In Caucasian women twin conceptions occur spontaneously about once in 90 pregnancies, triplets about once in 10 000 and quadruplets about once in 500 000. The spontaneous occurrence of the higher multiples up to octuplets has been reliably recorded.

The spontaneous incidence of twins is greater in the Negro race, reaching 1 in 30 in some West African tribes, and lower in the Mongol race (1 in 150). A tendency to multiple pregnancy is inherited, and it occurs more frequently in certain families. A woman who has given birth to twins is 10 times more likely to have a multiple conception in a subsequent pregnancy than a woman who has not previously had twins.

The incidence of twin pregnancies rises slightly with increasing maternal age up to 40, and (independently) with increasing parity. All these factors of race, familial tendency, parity and age affect binovular rather than uniovular twins.

Multiple pregnancies have frequently occurred after induction of ovulation, including septuplet and octuplet pregnancies. The risk is slight with clomiphene but considerable with ovarian stimulation with gonadotrophins. These women must be monitored carefully by measuring the oestrogen response and by scanning the ovaries with ultrasound. An injection of hCG to release the oöcytes is only given when it is certain that there are only one or two ripe follicles. Multiple pregnancy may also follow assisted conception when multiple oöcytes or fertilized ova are replaced.

In spontaneous multiple pregnancies the normal sex ratio at birth, with a slight preponderance of males, is altered to a female preponderance.

VARIETIES OF TWINS

Twins may be binovular (dizygous) or uniovular (monozygous) (*see* Fig. 3.20).

Binovular twins

Binovular twins are developed from two separate ova which may or may not even come from the same ovary. This variety of twins is four times more common than the uniovular variety. The children may be of the same or of different sex, and are no more alike than is usual with members of the same family. As they are developed from separate ova and separate spermatozoa their genetic material will differ. They have separate and distinct placentae. Sometimes these are loosely joined at their margins but there is never any anastomosis between their blood vessels. Each fetus has its own amnion and chorion.

Uniovular twins

Uniovular twins are developed from a single ovum, which after fertilization has undergone division to form two embryos. Division may occur in the early stage of segmentation, or later when two germinal areas are formed in one blastodermic vesicle. Uniovular twins are always of the same sex and are often remarkably alike in their physical and mental characters. The arrangement of genes in the chromosomes is identical, and inherited characteristics such as blood groups are necessarily the same. Uniovular twins which arise from very early division each have a complete set of membranes (chorion and amnion); those which arise by later

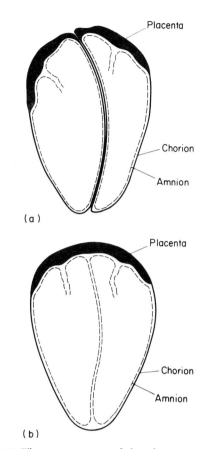

Fig. 3.20 The arrangement of the placentae and membranes in twin pregnancy: (a) probably binovular twins. The placenta are separated and each fetus has its own amnion and chorion; (b) probably uniovular twins. Each fetus has its own amnion, but the chorions are fused and the placenta are anastomosed. (Depending on the race, there is an 80–94 per cent chance that these are binovular)

division have only one chorion but usually have separate amniotic sacs.

Monoamniotic twins are rare; this type is associated with a higher fetal loss from cord entanglement. The umbilical cords are usually separate, but the fetal circulations often communicate by anastomoses in the placentae. This communication may cause unequal development of the fetuses, and occasionally death of one fetus. If one fetus perishes early in pregnancy it is retained until term, when the small fetus is discovered compressed flat on the membranes (*fetus papyraceous*).

When the process of division of a single germi-

Fig. 3.21 Omphalo-thoracophagus twins, joined at chest and abdomen, each infant having two legs as well as two arms. Woodcut from Jacob Rueff's *De conceptu*, 1580

nal area is incomplete some form of *conjoined twins* may occur (*see* Fig. 3.21). Many varieties have been described; they may be joined by the sternum, the pelvis or the head, and the degree of union varies from fusion of skin and soft tissues to formation of a double monster in which head, trunk, viscera or limbs may be shared or duplicated.

DIAGNOSIS OF TWINS

With twins all the early symptoms of pregnancy such as morning sickness may be more pronounced, and pregnancy induced hypertension is common. If the woman has already borne children she may notice an unusual degree of abdominal enlargement and excessive fetal movements. In late pregnancy she may have discomfort and shortness of breath because of the large size of the uterus.

The woman is likely to become anaemic. Apart from the fact that there is a greater increase in plasma volume there is a double fetal demand for iron. As well as iron-deficiency anaemia, megaloblastic anaemia is common with twins. It is important to ensure that the woman has an adequate diet and supplements of iron and folic acid. Oedema of the legs is common and any tendency to haemorrhoids or varicose veins of the legs is accentuated.

On examination the uterus is found to be larger than expected from the duration of gestation. Polyhydramnios may occur with twins, adding to the size and confusing the diagnosis. The diagnosis is simple if two fetal heads can be felt. Both backs

and both breeches may be identified, and an unusual number of small parts. If heart movements are detected with the ultrasound apparatus in two separate areas or directions the diagnosis of twins is almost certain.

By ultrasound scan in early pregnancy two separate gestation sacs can be identified from about the 7th week or sooner. From about the 8th week separate fetal bodies can be detected and from about the 12th week two heads can be distinguished. If routine scanning of all women is carried out at 16 weeks twins should rarely be missed.

If antenatal supervision is poor the diagnosis of twins is sometimes missed until after the birth of the first twin. The high position of the fundus, abdominal palpation of the second fetus and discovery of a second bag of membranes and fetal parts on vaginal examination will make the diagnosis clear. The risk to the mother and to the fetuses is increased because of the following complications.

COMPLICATIONS OF PREGNANCY

Preterm labour

This is the most important factor in causing the increased perinatal mortality, which may reach 500 per thousand. Not only may labour begin before term, but the rate of growth of a twin after the 28th week is slower than that of a single fetus. The reduced rate of growth begins when the total weight of the twins is in the region of 3500 g. There may be a substantial difference in the rate of growth of the twins. The median duration of non-induced twin pregnancy is 37 weeks. About 20 per cent of twin pregnancies do not reach the 36th week.

Discordant fetal growth

The placental mass per fetus is less in twins than in singleton pregnancy, and not uncommonly there is discordant growth, both in monozygous and dizygous twins. The smaller twin has an increased perinatal mortality and an increase in long-term morbidity.

Pregnancy induced hypertension

Pregnancy induced hypertension, including eclampsia, occurs at least three times more frequently than in single pregnancies, and this increases both the maternal and fetal risk.

Anaemia

Because of the increased fetal demand for iron and folic acid maternal anaemia is common in twin pregnancies unless adequate dietary supplements are given.

Polyhydramnios

There will obviously be more fluid in two sacs, but over and above this, one sac may contain an abnormally large amount of fluid.

Congenital malformations

Congenital malformations occur about twice as often in twin than in singleton pregnancies.

COMPLICATIONS OF LABOUR

Malpresentations

Malpresentations are common with twins. In about 45 per cent of cases both twins present by the head (*see* Fig. 3.22); in about 35 per cent one fetus presents by the head and the other by the breech (*see* Fig. 3.23); in about 10 per cent both present by the breech and in about 10 per cent a transverse lie is associated with a cephalic presentation or a breech. It is very rare for both fetuses to lie transversely. Malpresentations increase both the maternal and the fetal risk.

Postpartum haemorrhage

The risk of haemorrhage after delivery is increased because of the large size of the placental site. In the past a general anaesthetic was frequently given to deal with complications of labour and this caused uterine relaxation; with epidural anaesthesia this risk is reduced. If the mother is anaemic at the start of labour any acute blood loss may be serious.

Contrary to earlier teaching, the overdistended uterus usually contracts well during labour, and the duration of labour is not increased.

Cord prolapse

This may occur in association with malpresenta-

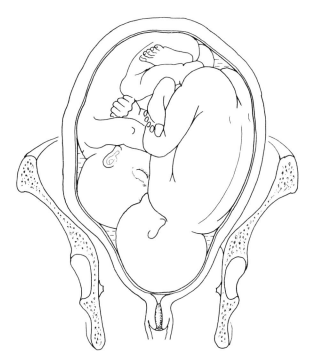

Fig. 3.22 Twin pregnancy. Both presenting by the vertex. Fetuses lying alongside one another. Both heads could be felt, one at the brim and the other above it in the iliac fossa. One back and an unusual number of small fetal parts would be recognized and possibly two breeches in the fundus

tion, especially with the second twin at the time when its membranes rupture.

Locked twins

This very rare complication may prevent spontaneous delivery.

Communal fetal circulations

On rare occasions the circulation of twins may communicate by anastomoses in the placenta. This may give rise to unequal growth or death of one or both fetuses in pregnancy or in labour.

MANAGEMENT OF TWIN PREGNANCY

The antenatal care of a woman with a twin pregnancy should be intensified so that the complications listed above can be detected early. The woman is seen more often than usual from mid-

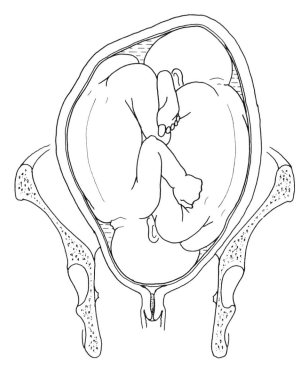

Fig. 3.23 Twin pregnancy. First presenting by the vertex, second by breech. Fetuses lying alongside one another. This offers the easiest diagnosis; one head would be felt at the brim and one at the fundus; two backs with the small parts in the groove between would be palpable and two points of maximal intensity of fetal heart sounds would be made out, one on right and below the umbilicus and the other on the left and above

pregnancy onwards. It is important that she should have adequate rest. The practice of admitting women with twins to hospital from the 30th to the 35th week is now less common, but it may be advisable if the home conditions are poor. It is by no means certain that rest will prevent premature labour, but it may increase the placental blood flow and so improve fetal growth. There is no convincing evidence that the prophylactic use of tocolytic drugs or cervical suture will reduce the incidence of preterm labour in twin pregnancy. In addition many women are uncomfortable when carrying twins and need more rest on that account.

Some obstetricians recommend that labour should be induced at the 40th week to avoid the theoretical risk of placental insufficiency. Induc-

tion may, however, be indicated for hypertension or growth retardation of one or both fetuses.

MANAGEMENT OF TWIN LABOUR

All preparations should have been made for the resuscitation and special care of babies of low birth weight. Because this can only be effectively done in hospital, and because other complications of labour may occur, all cases of twin labour should be in hospital.

The first stage

The first stage of labour is managed in the ordinary way; labour is not often prolonged. Epidural analgesia is very suitable for these cases and an intravenous glucose drip should also be set up. The former will allow any necessary operative intervention without delay and the latter will permit an oxytocic infusion to be started at any time. Because of the changing relationships of the fetuses it may be difficult to hear both fetal hearts, but an effort must be made to locate them. Ultrasound detection of each heart can be performed using two different frequencies to produce two CTG traces. This allows surveillance of each fetus. If either trace becomes abnormal it is usual to move to caesarean section although fetal blood sampling could be performed if it is certain the trace is that of the first or leading fetus. Caesarean section might be performed for prolonged delay in the first stage which did not respond to a careful trial of oxytocin, and in cases in which both twins were breech presentations or were lying transversely or the twins were conjoined.

The second stage

The second stage is also managed in the usual way unless some complication arises. As in other cases the indications for delivery with the forceps or vacuum extractor, or for breech extraction, are undue delay, fetal distress or maternal distress. Unless the perineum is very lax, an episiotomy should be performed routinely, under local anaesthesia if an epidural injection has not been given.

It is after the birth of the first twin that any problems usually arise, and it is the second twin who is mainly at risk. It may have to be delivered without delay and an anaesthetist must be ready in the labour ward during all twin deliveries. The advantages of epidural analgesia have already been pointed out. A doctor who is able to carry out fetal resuscitation should be available in addition to the obstetrician conducting the delivery.

Usually both twins are delivered before the placentae, but rarely with binovular twins the placenta of the first twin will come away before delivery of the second fetus.

Immediately after the delivery of the first twin the abdomen is palpated to determine the lie of the second fetus. If it is oblique or transverse external version is performed to bring one pole of the fetus over the cervix. It does not matter if this is the breech or the head; if extraction of the second twin should become necessary this is often easier by bringing down the legs of a breech than by application of forceps or the vacuum extractor to a very high head.

The fetal heart rate is monitored continuously, preferably by ultrasound, for if the fetus is distressed immediate delivery is required. Distress may occur because the volume of the uterine cavity is reduced after delivery of the first twin, and there may be separation of the placenta on which the second twin depends. It may also be caused by prolapse of the cord, and a vaginal examination to exclude this should always be performed when the second sac of membranes ruptures.

Uterine contractions are often in abeyance for a few minutes after delivery of the first twin. When they start again delivery of the second twin is usually rapid as the birth canal has already been fully dilated. Delivery is conducted in the ordinary way, whether the head or the breech comes first.

If the uterine contractions do not return within about 5 minutes after the delivery of the first twin, Syntocinon (2 units in 500 ml) is added to the glucose solution in the intravenous drip which should already be in place, and the second sac of membranes is ruptured. If this interval is too long the cervix sometimes partly closes. Fewer babies are lost after immediate rupture of the membranes of the second sac, which is usually followed by rapid delivery of the second twin.

If the second twin shows signs of distress or its cord prolapses, it must be delivered reasonably quickly. The head may be found presenting but lying high above the pelvic brim, and application of forceps at that level can be difficult. If forceps delivery is attempted the head should be pressed down as far as possible by abdominal pressure from an assistant, or vacuum extraction may be performed. Internal version and breech extraction is another possibility which should seldom be necessary, but if the breech already presents it is comparatively easy to bring down a leg (or both legs) and extract the baby.

Locked twins

This is a very rare complication occurring in less than 1 in 1000 twin deliveries. If there is unexplained delay in the second stage or difficulty in extracting a first breech it should be considered, and the possibility of conjoined twins should not be forgotten. The diagnosis is made by examination under anaesthesia. General anaesthesia will be necessary if an epidural injection has not already been given.

Locking may occur when the first twin presents as a breech and the second as a vertex. The aftercoming head of the first twin is caught above the chin of the second twin (*see* Fig. 3.24) and if disengagement is not quickly achieved the first twin dies. Caesarean section and delivery of the second twin from above is probably the easiest course, but alternatively under deep general anaesthesia with fluothane the neck of the dead fetus is divided from below and its head pushed up into the uterus to free the head of the second twin. The decapitated head of the first is delivered later with forceps.

Fig. 3.24 Aftercoming head of the first twin locked with the forthcoming head of the second twin

Third stage of labour

There is an increased risk of postpartum haemorrhage because of the large size of the placental site, which may encroach on the lower segment where the contraction of the muscle is relatively ineffective in closing blood sinuses. A prophylactic injection of Syntocinon or syntometrine should be given with the birth of the second twin and the placenta delivered in the usual way.

Caesarean section

Elective caesarean section has a place in twin delivery if the pregnancy is complicated by severe hypertension, or if the woman is older or has a bad previous obstetric history. It may also be considered in cases of preterm labour before 34 weeks if the leading twin presents by the breech.

Triplets and higher multiple births

The increasing use of gonadotrophins and assisted reproduction techniques in infertility has resulted in an increased incidence of triplets and higher multiple births. The exact diagnosis is only likely to be made by ultrasound or radiological examination. All the fetal hazards already described are accentuated, and some of the babies are likely to be very small. The perinatal mortality is high and delivery of triplets or higher multiples is generally best managed by caesarean section. Obviously special arrangements for the provision of adequate expert paediatric help must be made. The impact on neonatal units is considerable, to say nothing of the difficulties of parenthood.

COUNSELLING IN OBSTETRICS

Pregnancy loss at any stage, whether through miscarriage, stillbirth, perinatal death or termination due to abnormality, should alert both medical and midwifery staff to the need of parents to be offered counselling. This crucial area of care from obstetrician and other hospital staff must not be dismissed. If parents are offered clear explanations concerning the events and issues surrounding their pregnancy loss they are more likely to be able to begin to resolve their emotional conflicts. Information will probably need repeating during the days and weeks to follow. If well coordinated, such support could reduce possible litigation. If parents are left in confusion, they may become angry and suspicious of the hospital care they received. Negative repercussions can be avoided if the obstetrician, as the expert source of information during pregnancy, does not prematurely curtail postnatal involvement, simply because the baby has died. If there is no coordinated response, it is likely the woman will become the victim of fragmented care or unwittingly ignored.

There is increasing evidence to suggest that the value of supportive counselling following pregnancy loss is far-reaching, reducing the possibility of chronic grief reactions and other psychiatric sequelae. In the main it is the clergy, social workers and some specialist bereavement nurses who have borne the burden of providing emotional support to parents. All too often, however, it is the sympathetic obstetrician and paediatrician to whom the parents wish to return and from whom they seek comfort as well as answers to their questions.

DIAGNOSIS OF INTRAUTERINE DEATH OR FETAL ABNORMALITY

As soon as a problem is identified by ultrasound scan, parents must be included in the discussion and decision making. Although there are universal responses in shock and grief, each individual and each couple will react uniquely. If there has been previous infertility, a poor obstetric history, or need in the past for psychiatric intervention, the obstetrician should be alerted to these added complications to the grief reaction.

Wherever possible, bad news should be given to a couple when they are together. Relevant information from support organizations is best made available at an early stage as this will not only assist parents in their decision making but also provide additional support, and demonstrate that they are not alone in their tragedy.

Other than when there is an obvious obstetric

indication to expedite the delivery, the woman does not need to be hurried in her considerations. Counselling should include giving clear unbiased information on all possible options, and allowing the woman or couple to make their own choices. The obstetrician's sensitive treatment of the woman at this time will be of utmost importance in her long-term recovery.

INTRAUTERINE DEATH OR TERMINATION DUE TO ABNORMALITY

Until delivery the woman may alternate between denial and acceptance of the reality of her loss. Many people who work specifically in the field of bereavement have suggested that appropriate interventions at the time a baby dies can help prevent future pathological family reactions. Often staff can feel awkward and uncomfortable, but being there to console is as important as what is said, although what is actually said and done during this period will leave lasting impressions. This is not the time for explanations, and information need only be given if specifically sought by parents. Acknowledgement that there will be future questions, and ensuring an appointment is made within 2 weeks is often enough for parents to realize that the medical staff are available and prepared to answer among others the three most often asked questions: What happened? Why did it happen? Is it likely to recur?

The first 24 hours following delivery can be a critical time for the parents psychologically. It is important that the most senior obstetrician visits them as soon as possible. Any known information as to the possible causes and circumstances of the death needs to be made available at this stage. Parents, and especially mothers, can be laden with guilt, and very often it is only the obstetrician who can supply the details necessary to relieve some of those feelings. Asking permission for post-mortem should ideally be delayed until the initial shock has subsided in order to avoid overwhelming the parents.

If in the light of preliminary post-mortem results the obstetrician and paediatrician offer follow up in a planned way, they can play a pivotal role in the parents' recovery. An early postnatal appointment when the obstetrician can provide medical facts to clarify events and decisions prior to the baby's death will help to avoid unnecessary confusion and possible bitterness.

One or both parents may want to talk, even while still in hospital, of the possibilities of another pregnancy. It is usually best to wait for one of them to broach the subject, as it is not unusual for the woman particularly to say she never wants another baby. If advice is given that a pregnancy should not be attempted for 2 to 4 menstrual cycles, good medical information should also be given to substantiate this recommendation. This length of time generally allows for normal menstruation to return, but it needs to be pointed out to parents that if they embark on a pregnancy in 3 months, the expected date of delivery will be close to the first anniversary of the previous loss. Since this can be a difficult time anyway, it may be wise to advise parents of this potential confusion of feelings, so that they can make informed choices. Prior to discharge it is relevant to discuss contraception, together with problems of lactation and methods of suppression.

A couple can also be greatly helped at this early stage if they understand that their reactions are appropriate and consistent with those of other couples in similar circumstances. It may be possible for staff gently to encourage ongoing dialogue between the man and the woman at a time when shock, loneliness and despair can create a distance between them.

LONGER TERM FOLLOW UP

If there has been an efficient communication network within the hospital and into the community, all the relevant support should be mobilized, including general practitioner, health visitor and self help organizations. The next critical time for parents is often around 3 to 6 months following the loss, when an appointment to see the bereavement counsellor or social worker can be of great benefit.

Before embarking on another pregnancy a couple may look to the obstetrician for advice. It is particularly important to listen for specific anxieties which can emerge at this point, and allow time for discussion. This may reveal possible sexual or emotional difficulties which could prevent conception. These can be successfully treated with relative ease if the couple are referred for specialist counselling. Some hospitals do have a counsellor who can help the couple through this difficult time; there may be a need to refer to a sexual dysfunction clinic.

THE NEXT PREGNANCY

Ambivalence is a common feature when a pregnancy is confirmed following a loss. The woman particularly may experience a variety of feelings, principally anxiety, especially around the time the previous loss occurred. She may want to pursue every possible test to ensure the normality of the pregnancy and baby.

Other women experience a feeling of detachment and may not prepare at all for the baby's arrival. If either of these responses becomes extreme, some form of counselling or support may need to be offered.

Towards the end of the pregnancy old feelings of anger, failure and resentment may resurface at a time when most professionals and friends think the couple will be relieved and looking forward to a happy outcome. There can be additional stresses in the couple's relationship as often it is the man who wants to put the previous experience behind him and the woman who may still have a need to talk about the events and feelings surrounding their baby's death. This third trimester of pregnancy can be a time when some parents will value opportunity to talk both about the previous outcome as well as discuss possible management of the labour and delivery for this pregnancy.

Even after the birth of a normal healthy baby, parents who have experienced a previous loss often have very confusing feelings: joy, grief, guilt, and fear that the previous baby may now be forgotten. Bonding may be affected, but professionals should take care not to be too quick to label responses as abnormal. It may simply mean a little extra time and support should be provided for these women and their families. Obstetrician, paediatrician and midwife can all offer this in different ways. Future attachment of mother and baby may well be positively affected by appropriate intervention at this crucial time.

FETAL LOSS IN EARLY PREGNANCY

Vaginal bleeding in early pregnancy is always a cause for concern. It may occur in cases of abortion or ectopic pregnancy, or with a cervical lesion such as a polyp or carcinoma. On many occasions no certain cause of early bleeding may be found. If recurrent, the possibility that there has been chorionic separation remains and fetal well-being be monitored carefully in the later weeks of pregnancy.

ABORTION OR MISCARRIAGE

The terms abortion and miscarriage are synonymous and denote the expulsion of the conceptus before the 24th week of pregnancy. There is no sharp demarcation between late abortion and early premature labour; the division is merely of descriptive convenience.

The obstetrician must understand the diagnosis and treatment of abortion, as it is a common complication of pregnancy, but many of the cases are treated in gynaecological wards rather than obstetric units. A fuller description of the condition and of the law relating to termination of pregnancy will be found in *Gynaecology by Ten Teachers*.

Observation suggests that 10 to 15 per cent of clinically recognized pregnancies end in early abortion, but the loss of very early embryos is certainly greater than this. The most common time for clinically evident abortion to occur is between 7 and 13 weeks.

Causes of abortion

Despite a long list of aetiological factors, the cause of a particular abortion is often uncertain. The known causes include the following.

Malformation of the zygote

A common cause of abortion is an abnormality of the fetus severe enough to cause fetal death. About 70 per cent of these are caused by chromosomal abnormalities, for which either parent may be responsible, although they often arise from spontaneous, unexplained mutation in the zygote itself. Other chromosome disorders which are compatible with life (e.g. Down's syndrome) are also associated with an above average abortion rate. Most abortions of this type are not recurrent, so the prognosis in later pregnancies is good unless several abortions of identical pattern have already occurred.

When there is vaginal bleeding in early pregnancy ultrasound scanning may reveal an empty chorionic sac containing no fetus at all, or one in which the fetal echo is much too small for the dates. This is a blighted ovum, in which abortion is bound to happen. Sometimes layers of blood clot collect round the sac as it collapses.

IMMUNOLOGICAL REJECTION OF THE FETUS

Investigations of the maternal immune response to her pregnancy have now focused on the complex interaction of immune and endocrine factors acting at the endometrial level. Hypotheses on immuno-dystrophism in which cytokines are being evaluated in the context of implantation and embryonic growth will examine the relationships between migrating trophoblast and the large granular lymphocytes. These lymphocytes have natural killer cell activity and seem to be under endocrine control. Earlier views and strategies based on the relevance of maternal–fraternal HLA dissimilarity now rapidly losing support are likely to be put to rest following the conclusion of a multi-centre study failing to show any value from paternal lymphocyte immunization of the mother.

DISEASES OF THE MOTHER

Pregnancy will often continue in spite of maternal disease, but any maternal illness may cause abortion if it is sufficiently severe, especially acute fevers. Maternal infection may involve the fetus, particularly rubella and syphilis, but rarely also malaria, brucellosis, toxoplasmosis, cytomegalic inclusion disease, vaccinia and listeriosis.

Abortion occurs in a few cases of *rubella*, but more often the infected fetus is born alive with congenital abnormalities. *Syphilis* does not cause early abortion and it is an uncommon cause of late abortion in the UK; it is more likely to cause intrauterine death after the 24th week.

The abortion rate is above average in *diabetes* if the disease is not adequately controlled. Intra-

uterine death may occur with *hypertension* and *renal disease*, sometimes before the 24th week.

Severe *malnutrition* is associated with abortion, but it has to be of a degree which is unlikely to be seen in Britain. Although deficiency of vitamin E will cause abortion in experimental animals there is no evidence that it causes abortion in women, and this substance is usually present in adequate amounts in the diet.

UTERINE ABNORMALITIES

The incidence of abortion is increased if the uterus is double or septate, although in many such cases pregnancy is uneventful.

Retroversion of the uterus is not a cause of miscarriage, except in rare instances in which incarceration of the uterus within the pelvis occurs and is left untreated. A fibroid which is closely related to the cavity and the uterus may cause abortion, but other fibroids will not do so.

Previous lacerations of the cervix involving the internal os may result in abortion in the middle trimester, or in premature labour. Very rarely the cervical weakness is of congenital origin; usually it is the result of obstetric laceration, of over-vigorous surgical dilatation of the cervix or core biopsy. During pregnancy the unsupported membranes bulge through the cervix and rupture; miscarriage follows.

HORMONAL INSUFFICIENCY

It has been claimed that insufficient production of progesterone by the corpus luteum before the placenta is fully formed will lead to inadequate development of the decidua and abortion. The evidence for this is poor.

Both thyroid deficiency and hyperthyroidism may be contributory factors in abortion but are unusual.

PATHOLOGICAL ANATOMY

In the first trimester of pregnancy the attachment of the chorion to the decidua is so delicate that separation may follow strong uterine contractions produced by any cause. The resulting haemorrhage into the choriodecidual space leads to further separation. In other cases fetal death precedes uterine contractions, which may occur some days later. The decidua basalis remains in the uterus, and in the majority of cases the embryo, with its membranes and most of the decidua capsularis, is expelled. In some cases the gestation sac is retained in the uterus for days or weeks as a missed abortion. The embryo is reabsorbed and layers of blood clot collect around it to form a carneous mole.

By the 12th week, the placenta is a definite structure, and if it happens after this time the process of abortion is similar to that of labour. Bleeding and painful contractions are followed by dilatation of the cervix, rupture of the membranes and expulsion of the fetus and placenta. If all the conceptus is expelled normal uterine involution follows, but frequently part of the placenta is retained with some blood clot.

CLINICAL VARIETIES OF ABORTION

The following terms are used to describe varieties of abortion:

    threatened
    inevitable
    complete
    incomplete
    septic
    missed
    recurrent.

Threatened abortion

In these cases vaginal bleeding occurs without dilatation of the cervix and with very little or no pain. The clinical distinction from inevitable abortion is based on the relatively slight degree of bleeding and the absence of cervical dilatation (*see* Fig. 3.25). If the bleeding is heavy, or increases, the prognosis is bad, but the abortion should not be regarded as inevitable until the cervix begins to dilate. There may be repeated short episodes of bleeding without the abortion becoming inevitable, and if a slight red loss is followed for some days by old brown altered blood this may have little significance.

The woman is put to bed, and, provided an ultrasound scan confirms that pregnancy is progressing, she rests until some days after all red loss has ceased. A gentle vaginal examination and the passage of a speculum will exclude any other unsuspected cause for the bleeding, such as a cervical polyp; it may also reveal any dilatation of the cervix.

There is no specific treatment and the essential task is to establish that the abortion is threatened and is not becoming inevitable. Ultrasound examination can determine the size of the fetus and show if the fetal heart is beating. It can also give some clue to dilatation of the cervix. In normal pregnancy a gestation sac of about 8 mm diameter

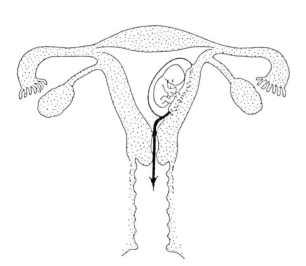

Fig. 3.25 Threatened abortion

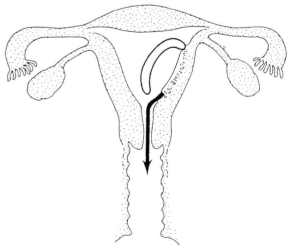

Fig. 3.26 Inevitable abortion

can be identified by the 6th week, and during the 8th week the fetal crown–rump length increases from 15 to 21 mm. The fetal heart action can be detected by 8 weeks and later the abdominal circumference. Repeated scanning is without risk to the fetus and can establish continuing growth. By the 14th week the crown–rump length averages 90 mm; thereafter growth can be monitored by measuring the biparietal diameter of the fetal head. If ultrasound shows an empty pregnancy sac it is probable that there is serious chromosomal abnormality and abortion will occur. Attempts to conserve such a pregnancy are of no value. Ultrasound will also help in the differential diagnosis; if the abortion becomes inevitable dilatation of the cervix and descent of the gestation sac into the cervical canal may be seen.

The measurement of serum hormones and proteins of ovarian trophoblastic, fetal or endometrial origin to predict the subsequent course of pregnancy is not helpful at present. An evaluation of serum measurements of the free beta subunit of hCG and certain placental hormones, in particular pregnancy associated placental protein A (PAPP-A) is currently under way and may well be prognostic (*see* Fig. 3.27(a) and (b)).

If the pregnancy continues the mother may be anxious about the possibility that the fetus is abnormal; she can truthfully be told that this is very unlikely, especially if a later ultrasound scan is normal at 20–24 weeks.

Inevitable abortion

The process is now irreversible. The cervix is open, there is more bleeding and rhythmical and painful uterine contractions continue.

The uterus usually expels its contents unaided. All examinations are carried out with careful aseptic technique. Analgesics such as pethidine may be required. If haemorrhage becomes severe or the abortion is not quickly completed ergometrine or syntometrine is required and the uterus should be evacuated under anaesthesia.

In all cases of inevitable abortion anti-D gammaglobulin 500 μg is injected intramuscularly unless the woman is known to be rhesus positive.

Complete abortion

When all the uterine contents have been expelled spontaneously there is cessation of pain, scanty blood loss and a firmly contracted uterus. If there is no more active bleeding, or if an ultrasound scan shows an empty uterine cavity, no further treatment is required. Many women who have a complete abortion do not require hospital admission.

Incomplete abortion

This term means that some of the products of conception, usually chorionic or placental tissue, are retained. Bleeding continues and may be severe. There is a danger of shock and of sepsis. If blood loss is heavy or continues for more than a few days after an abortion it must be assumed that there are retained products, particularly if the uterus is still found to be enlarged and the cervix is open. The products can be seen on ultrasound scanning. The chief dangers are haemorrhage and sepsis, and they continue until the uterine cavity is empty.

Severe bleeding associated with shock may necessitate blood transfusion by a flying squad before the woman is moved from home to hospital. Intramuscular ergometrine, 500 μg, is given immediately and any placental debris in the cervical canal is removed, under direct vision, using a speculum and ring forceps. Transfusion is continued until the blood pressure reaches a level safe enough for transfer of the woman to hospital. There the uterus is evacuated under anaesthesia. A postoperative prophylactic antibiotic such as ampicillin is wise.

In some cases of incomplete abortion the bleeding is not severe but it continues intermittently for several weeks and the uterus remains enlarged. Surgical evacuation of the uterus is then essential, with histological examination of the products. Some of these cases are due to a placental polyp.

Septic abortion

Infection may occur during spontaneous abortion but it more often occurs after one that has been induced. Blood clot and necrotic debris in the uterus form an excellent culture medium. Infection may spread rapidly to surrounding structures, causing pelvic cellulitis and salpingitis, sometimes with septicaemia and pelvic or generalized peritonitis. There is fever and a raised pulse rate, and there may be lower abdominal pain. Permanent blockage of the Fallopian tubes may follow salpingitis.

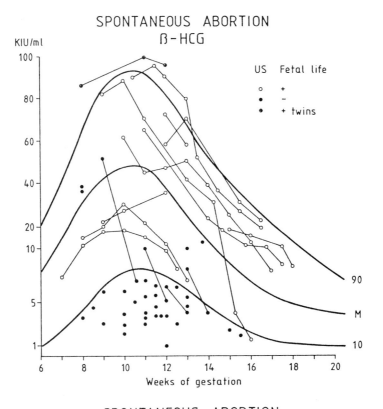

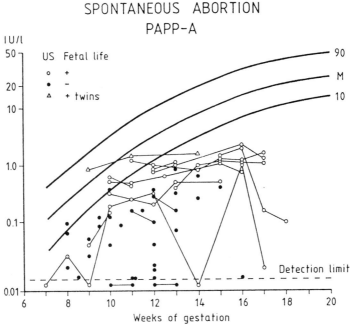

Fig. 3.27 Serum levels of (a) human chorionic gonadotrophin; and (b) pregnancy-associated plasma protein A in 42 women with spontaneous abortion. Heavy lines represent the 10th, 50th and 90th centiles of the normal range. □ = No ultrasonic evidence of fetal heart action; ○ = ultrasonic evidence of heart action; △ = twin pregnancy with live fetuses. WHO reference material for pregnancy proteins 78/610. From Westergaard *et al.* (1985)

The most common septic abortion infecting organisms are *Staphylococcus aureus*, coliform bacteria, *Bacteriodes* organisms and *Clostridium welchii*. Of these the most dangerous are the Gram negative and anaerobic organisms which produce endotoxic shock. The potentially lethal cases of infection with the β-haemolytic streptococcus group A are, fortunately, seldom seen.

The woman must be admitted to hospital and isolated from other obstetric and surgical women. High vaginal swabs and blood specimens are sent for bacteriological culture, while treatment is started with a wide-spectrum antibiotic. There is much debate about the best choice; one that has been recommended is cephradine, 250–500 mg every 6 hours together with metronidazole 500 mg. Both of these can be given intravenously if the woman is vomiting. When the bacteriological report is received the antibiotic treatment is modified according to the sensitivities of the organisms discovered.

Anaemia may occur from haemolysis as well as from haemorrhage, and the haemoglobin level must be determined. Blood transfusion may be necessary.

The uterus must be emptied, but if the bleeding is not too serious; evacuation is best postponed until 24 hours after the antibiotic treatment has taken hold but if there is serious bleeding, it cannot be deferred. In most cases evacuation is performed under anaesthesia with a suction curette or ring forceps. In cases of more than 14 weeks' gestation in which a dead fetus is retained its expulsion may be achieved by oxytocin infusion and vaginal use of prostaglandins.

Some women are gravely ill, especially if anaerobic infection has occurred, with high swinging fever, anaemia and sometimes haemolytic jaundice. Endotoxic shock may be superimposed on hypovolaemic shock, with circulatory failure due to peripheral vasodilation caused by endotoxins released from coliform organisms which have invaded the bloodstream. Before surgical intervention massive doses of intravenous penicillin and metronidazole are used. Both are usually given by mouth, but in severely ill women the intravenous route may be preferable. Intravenous hydrocortisone is sometimes helpful if restoration of the blood volume does not quickly restore the blood pressure. The urinary output must be watched carefully, since oliguria may indicate renal cortical or tubular necrosis.

Missed abortion

This occurs when the embryo dies or fails to develop, and the gestation sac is retained in the uterus for weeks or months. Haemorrhage occurs into the choriodecidual space and extends around the gestation sac. The amnion remains intact and becomes surrounded by hillocks of blood clot with a fleshy appearance, hence the term *carneous mole*. Mild symptoms like those of threatened abortion are followed by absence of the usual signs of progress of the pregnancy. The uterine size remains stationary and the cervix is often tightly closed. Serum placental hormone and protein measurements are low, and if repeated are shown to be falling. The diagnosis, if ever in doubt on initial ultrasound scan, on repeat scan will show no growth of the fetal crown–rump measurement and absence of fetal heart activity.

All missed abortions would probably be expelled spontaneously in the long term, but there may be a delay of weeks or months and many women become distressed once the diagnosis is made, so that active treatment is often chosen. In a few late cases there is a risk of hypofibrinogenaemia after retention of a dead fetus for some weeks, probably caused by thromboplastins from the chorionic tissue entering the maternal circulation.

Successful evacuation of the uterus is usually achieved in late cases with a combination of intravaginal prostaglandins and an intravenous Syntocinon infusion. In early cases the uterus may be emptied surgically with a suction curette after dilating the cervix.

Recurrent abortion

By convention this term refers to any woman who has had three or more consecutive spontaneous abortions. Unless each successive abortion has occurred at about the same time and in similar fashion it should not be assumed that there is an underlying and recurrent cause. Repeated miscarriages can occur by unlucky chance from different causes each time.

Early recurrent abortion can no longer be attributed to progesterone deficiency following inadequate luteal function. The practice of empirical hormone administration in these women cannot be justified scientifically. The realization that luteal function is influenced by events prior to ovulation has led to the practice of follicular phase luteinizing hormone measurements. Studies demonstrating high follicular phase LH levels in association with a high miscarriage rate (65 per cent versus 12 per cent in women with normal LH levels) have led to treatments directed at normalizing LH secretion with gonadotrophin releasing hormone agonists and menopausal gonadotrophin ovulation induction therapy. In some women, typically with polycystic ovarian disease, ovarian electrocautery achieves the same effect.

Repeated late midtrimester abortions may result from incompetence of the internal os of the cervix, which is usually caused by previous obstetric trauma, injudicious surgical dilatation, cone biopsy or cervical amputation. In the non-pregnant state this condition may be diagnosed by a hysterogram or finding that a dilator 3 cm in diameter can be easily passed. During pregnancy it is suspected from a history of repeated, almost painless, abortions after the 16th week, or sometimes by observing the membranes bulging through the partly dilated cervix.

Recurrent abortion is treated by insertion of a purse-string suture of non-absorbable material in the thickness of the wall of the cervix at the level of the internal os before the 16th week (Shirodkhar operation). This is removed shortly before term.

While there have been some adverse reports about the need for or success of this operation, the criticism may perhaps be directed at the choice of women selected rather than at the procedure itself.

When a definite cause for recurrent abortion cannot be established simple general advice is given. Any defects of general health are rectified if possible. Coitus is discouraged at the time of the previous abortion, and the woman may be admitted to hospital for rest in bed, especially at the time at which previous abortions occurred. In all cases she will need constant encouragement and support.

Trials to assess the value of injecting the partner's lymphocytes into the mother in the hope of stimulating her to produce blocking antibodies which prevent cell-mediated rejection of the pregnancy are likely to demonstrate no beneficial effects. Furthermore the use of corticosteroids, heparin or aspirin in women with autoimmune disease such as systemic lupus erythematosus is still to be evaluated systematically.

ECTOPIC PREGNANCY

An ectopic or extrauterine pregnancy occurs when the fertilized ovum embeds in some site other than the uterine decidua. Nearly all ectopic implantations are in the Fallopian tube, typically in the ampullary region (*see* Fig. 3.28), but very rarely the cervix, the ovary or the peritoneal cavity is the site of implantation. In Britain the incidence of ectopic pregnancy is about 1 case to 150 mature intrauterine pregnancies, but much higher incidences are found in countries where pelvic infection is common, e.g. 1 in 28 in the West Indies.

Most of the cases are treated in gynaecological wards, and a fuller account of the condition is given in *Gynaecology by Ten Teachers*.

AETIOLOGY

Fertilization of the ovum normally occurs in the ampulla of the Fallopian tube, and the ovum takes about 5 days to reach the uterine cavity where it implants at the blastocyst stage in the prepared secretory endometrium.

Delay or arrest in transit along the tube may be caused by a number of factors. Previous salpingitis and peritoneal adhesions may kink or compress the outside of the tube. Rarely a congenital diverticulum may be a causative factor. There is a higher incidence of ectopic pregnancy in women using an intrauterine contraceptive device, possibly caused by interference with ciliary function, or normal tubal peristalsis. Tubal surgery as well as assisted reproduction techniques are also an important cause and women who have had such treatments should be warned of the risks of tubal pregnancy, in particular of heterotopic pregnancy, i.e. combined intrauterine and extrauterine pregnancy.

PATHOLOGICAL ANATOMY

The ampulla is the commonest ectopic site of

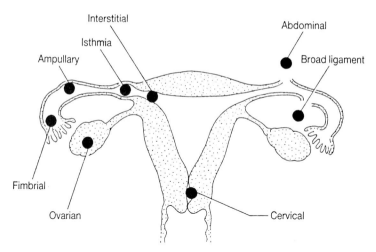

Fig. 3.28 Sites of implantation of ectopic pregnancies; among these, tubal (isthmial or ampullary) make up 95 per cent

implantation of the fertilized ovum. In less than a quarter of the cases the embryo is lodged in the isthmus; the other sites are very rare. The process of embedding is similar to that in an intrauterine pregnancy but, because the tube has no decidua, after the epithelium is penetrated by the tropho-blast the embryo burrows directly into the thin muscular wall, where it grows and distends the tube (*see* Fig. 3.29). Tubal blood vessels are eroded with resulting haemorrhage around the embryo

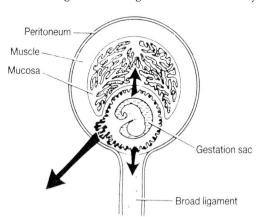

Fig. 3.29 Tubal pregnancy. Diagram to show that the embryo becomes embedded in the muscular wall of the tube. The gestation sac may then rupture into the lumen of the tube, through the wall of the tube into the peritoneal cavity, or between the layers of the broad ligament. Of these, the last is the least common

which then bursts either into the lumen of the tube (*intratubal rupture*) or through the wall (*extra-tubal rupture*) into the peritoneal cavity or occasionally between the layers of the broad ligament.

The uterus also hypertrophies and the endometrium undergoes normal decidual changes. After death of the embryo the decidua is shed either in fragments or as a decidual cast, with some external bleeding.

CLINICAL COURSE AND MANAGEMENT

Symptoms most commonly arise after one menstrual period is missed, although they may occasionally begin before this. However, it is rare for ectopic pregnancy to advance beyond 8 weeks without the occurrence of pain or bleeding, a point which is sometimes helpful in differentiating between the abortion of an intrauterine pregnancy and an ectopic pregnancy.

The dominant feature is low abdominal pain. This often originates on the side of the ectopic implantation and can be acute and severe. This is followed by slight irregular bleeding from the uterus. Profuse intraperitoneal haemorrhage and severe pain will result in sudden shock and collapse of the woman. This may be fatal unless immediate transfusion and laparotomy are undertaken.

A tubal pregnancy may terminate in the following ways.

Tubal mole

Bleeding around the embryo results in embryonic death, the embryo being retained in the tube surrounded by clot. The woman complains of pelvic pain, sometimes unilateral, and slight dark vaginal blood loss. On bimanual examination a very tender swelling is palpable on the side beside a bulky uterus. The diagnosis from salpingitis or torsion of a small ovarian cyst is difficult. An immunological test for pregnancy is often positive for a time. Ultrasound scan and laparoscopy may be helpful, but if there is a pelvic mass exploratory laparotomy must be performed, and the damaged tube is removed.

Tubal abortion

As the name implies, this occurs when separation of an ampullary ectopic gestation is followed by its expulsion through the ostium of the tube into the peritoneal cavity. Blood escapes from the tube to collect in the rectovaginal pouch as a *pelvic haematocoele*. This blood becomes walled off by adhesions, producing a tender cystic swelling behind the uterus. If there is a large haematocoele, the cervix is pushed forwards and upwards, and occasionally retention of urine may occur. Later on, when absorption of the blood has begun, an untreated haematocoele may have a lumpy and uneven consistence. There is often slight fever.

Surgical removal of the damaged tube and the haematocoele may be required as a late secondary procedure.

Tubal pregnancy

Given improvements in the diagnosis of ectopic pregnancy in industrialized societies where sensitive hCG tests and transvaginal ultrasound are readily available, tubal rupture is a much less frequent event, permitting the diagnosis of up to 90 per cent of ectopic pregnancies prior to rupture. Treatment is then directed at removal or destruction of trophoblast tissue rather than removal of the affected tube. Eradication of trophoblastic tissue is effected by endoscopic excision or direct or systematic administration of cytotoxic agents such as methotrexate. The aim is to maintain or restore tubal patency in particular if the woman wishes to become pregnant again.

However, if the diagnosis is delayed, tubal rup-ture may occur. Such a pregnancy usually follows implantation in the isthmus of the tube. In this narrow part of the tube early rupture occurs into the peritoneal cavity. The woman rapidly develops signs of severe intraperitoneal bleeding, with pain, fainting or collapse, pallor, rapid pulse and low blood pressure and signs of an acute abdomen. The whole lower abdomen is extremely tender and there may be some distention. When the woman lies down the blood tracks up to reach the diaphragm and causes referred shoulder-tip pain. On pelvic examination there is exquisite tenderness, but no mass may be felt as the ruptured tube is now empty.

Immediate blood transfusion is required, followed without delay by laparotomy to clamp the bleeding points and remove the damaged tube and the blood clot.

In rare instances the rupture occurs downwards between the layers of the broad ligament to form an *intraligamentous haematoma*.

Secondary abdominal pregnancy

Intraperitoneal rupture, as described above, is occasionally accompanied by relatively little bleeding. On very rare occasions the embryo may then be partially extruded into the peritoneal cavity and may continue to grow, developing partial placental attachment to surrounding structures. If the amnion remains intact the pregnancy may progress. Towards the end of the pregnancy the woman may experience a mock labour which if not observed by a trained midwife or doctor is followed by death of the fetus and thrombosis of vessels going to the placental site. This very rare condition is known as *secondary abdominal pregnancy*; it can be treated successfully if diagnosed in time. A laparotomy after 36 weeks of gestation usually results in a live viable baby. The placenta should be left alone wherever it may be implanted. If the fetus should die, ultimately mummification of the fetus occurs with subsequent calcification, forming a *lithopaedion*. In modern practice the diagnosis of intra-abdominal pregnancy is usually made by ultrasound. It may be suspected because the fetus is felt very easily through the abdominal wall, the lie is oblique and there is a soft mass, the uterus, occupying the pelvic cavity.

Interstitial (cornual) pregnancy

The site of implantation is in that part of the tube which lies in the uterine wall. The pregnancy may continue until about the 12th week, but sooner or later there is extensive rupture of the uterine cornu with very free bleeding from the vascular myometrium. This will necessitate enucleation of the gestation sac and wedge resection of the cornu, or even hysterectomy if there is uncontrollable bleeding. If the diagnosis is made sufficiently early, surgery may be avoided by eradicating the pregnancy by administering systemic methotrexate.

Ovarian pregnancy

This is exceedingly rare and clinically indistinguishable from a ruptured tubal pregnancy. The diagnosis can only be proved by careful histological examination of the gestation sac and ovary after removal.

Rupture of a pregnant rudimentary horn

Rupture of a pregnant rudimentary horn of a bicornuate uterus is not strictly an extrauterine pregnancy, but it is a serious accident with intraperitoneal bleeding. This occurs later than rupture of a tubal pregnancy, often at about the 14th week. The diagnosis is usually made at laparotomy, when it may be necessary to remove the horn.

Cervical pregnancy

Very rarely the embryo implants in the cervical canal. Because of the restriction in its growth the pregnancy seldom progresses beyond the 6th week. There is exceedingly free bleeding and there may be pain from uterine contractions. Evacuation of the pregnancy by curettage does not always stop the bleeding because there is little contractile muscle in the cervix, and deep sutures may have to be inserted to arrest the haemorrhage. Medical treatment may be appropriate in some cases (*see* Interstitial pregnancy).

OTHER CAUSES OF HAEMORRHAGE DURING EARLY PREGNANCY

Bleeding may come from lesions of the cervix:

Cervical erosion

Cervical erosion is very common during pregnancy, when the high level of oestrogens causes proliferation of the columnar epithelium of the cervical canal so that this extends outwards on the vaginal aspect of the cervix, forming a velvety red area, with well-defined edges, around the external os. Such erosions very occasionally cause a slight intermittent blood-stained discharge, but with any bleeding the possibility of malignancy must be considered. If there is any doubt, a biopsy should be done, but in most cases as long as a cervical smear is examined and found to be normal, no treatment is required during pregnancy. Most erosions resolve after delivery.

Cervical adenomatous polypi

These may be found during pregnancy as small, soft, red tumours, attached by a stalk to the cervix near the external os. If they are not causing more than a slight blood-stained discharge, and a smear arouses no suspicion of malignant disease, they may be left alone during pregnancy; otherwise they are easily removed for biopsy.

Carcinoma of the cervix

See Chapter 3.

FURTHER READING

de Swiet M. (1994) *Medical Disorders in Obstetric Practice.* Blackwell Scientific Publications, Oxford.

Bryan E. (1992) *Twins and High Multiple Births.* Edward Arnold, London.

Oates M. (1992) *Psychological Aspects of Obstetrics and Gynaecology.* Ballière Tindall, London.

Oglethorpe R.J.L. (1989) Parenting after perinatal bereavement – A review of the literature. *Journal of Reproductive and Infant Psychology* 7; 227–244.

4

NORMAL LABOUR

THE ONSET OF LABOUR

The uterus when stretched contracts strongly to expel any foreign body from its cavity. In discussing the cause of the onset of labour the problem is not to discover why the uterus starts to contract at term, but to find out why it usually remains quiescent during pregnancy. The uterus clearly has the power of expelling its contents before term, as in miscarriage, premature labour and when labour is induced before term.

It has been suggested that progesterone inhibits the uterine muscle during pregnancy. In rabbits the onset of labour can be postponed by giving large doses of progesterone, but this is not the case in women. Nor has it been shown that the blood concentration of progesterone falls significantly before term.

It has also been suggested that the rising levels of oestrogen during pregnancy sensitize the uterine muscle, so that it eventually responds more easily to stimuli or to oxytocin. There is no increase in secretion of oxytocin at term and labour starts normally in hypophysectomized animals. A prostaglandin pessary (PGE_2) placed in the vaginal vault near term induces labour.

There is some evidence that the fetal adrenal gland plays a part in initiating labour. Anencephalic fetuses may have defective adrenal cortices and with some of these fetuses pregnancy is greatly prolonged unless labour is induced artificially.

The possible part played by prostaglandins in the onset of labour has yet to be fully investigated. Mechanical stimulation of the cervix by the insertion of a finger and separation of the membranes leads to local secretion of prostaglandins. Prostaglandins are present in the decidua and membranes of late pregnancy. It is possible that premature labour may be caused by increased prostaglandin production following rupture of the membranes or ascending infection.

During normal pregnancy the growth of the uterus keeps pace with that of its contents and the limit of stretch is probably not reached even at term; the intrauterine pressure does not rise. However, in cases of polyhydramnios or twins, premature labour is common, so that in abnormal cases over-stretching of the uterus may play some part in the onset of labour.

Labour follows intrauterine death of the fetus, usually after an interval of several days.

During normal pregnancy the uterus contracts intermittently, but these contractions are not strong enough to overcome the resistance of the normal cervix. However, if the internal os of the cervix is damaged or incompetent even these weak contractions may dilate the cervix and labour will follow.

THE UTERINE SEGMENTS

In describing the phenomena of labour the uterus may be divided into two functional segments.

The upper part of the uterus (*upper segment*) contracts strongly, and with each successive contraction the smooth muscle fibres comprising it become shorter and thicker. This powerful segment draws the weaker, thinner and more passive lower part of the uterus up over its contents, and in so doing pulls up and then dilates the cervix.

The *lower segment*, consisting of the lower part of the body of the uterus and the cervix, can contract but is relatively passive compared with the upper segment.

The upper and lower segments are not fully formed until the end of the first stage of labour when they can be clearly seen and the transition between them is quite abrupt (*see* Fig. 4.1). In the non-pregnant uterus and during early pregnancy it is not possible to define the limits of the eventual lower segment, but at the end of pregnancy the lower segment is recognizable; in front it is covered by the loosely attached peritoneum.

During labour, as the cervix dilates and the lower segment is drawn up, its shape changes from a hemisphere to a cylinder. If there is obstruction to delivery the retraction of the upper segment is even more pronounced, and the junction between the two segments forms a distinct muscular ring known as the *retraction ring of Bandl*. In extreme cases this may be palpable and visible per abdomen.

In labour the lower uterine segment, cervix, vagina, pelvic floor and vulval outlet are dilated until there is one continuous birth canal. The forces which bring about this dilatation and expel the fetus are supplied mostly by the muscle of the upper uterine segment, with some assistance in the second stage from the abdominal muscles, including the diaphragm.

The muscle fibres of the upper segment of the uterus not only contract but also retract. When contracting the fibres become shorter and thicker. When the active contraction passes off the fibres lengthen again, but not to their original length. If contraction was followed by complete relaxation no progress would be made. In retraction some of the shortening of the fibres is maintained. Each successive contraction starts not quite at the point where its predecessor ended and the uterine cavity becomes progressively smaller with each contrac-

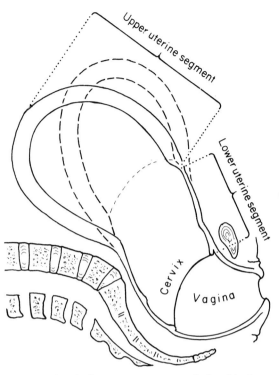

Fig. 4.1 The thick upper segment and the thin lower segment of the uterus at the end of the first stage of labour. The dotted lines indicate the position assumed by the uterus during contraction

tion. Retraction is a property which, though not peculiar to uterine muscle, is more marked in the uterus than any other organ. Later in labour when the placenta is expelled, retraction enables the uterine walls to come together so that there is no more than a potential cavity as it was before pregnancy.

THE STAGES OF LABOUR

Labour is divided into three stages:

- *first stage*, or stage of dilatation, lasts from the onset of true labour until the cervix is fully dilated. Two phases can be identified, the *latent phase* and the *active phase*. In the *latent phase*, cervical dilatation is under 3 cm and contractions may be infrequent, are usually not more than moderately strong and are quite well tolerated even without analgesia
- *second stage* lasts from full dilatation of the cervix until the fetus is born

- *third stage* lasts from the birth of the child until the placenta and membranes are delivered and the uterus has retracted firmly to compress the uterine blood sinuses.

Premonitory symptoms

In most primigravidae the presenting part sinks into the pelvis during the last 3 or 4 weeks of pregnancy, and in lay terms this is spoken of as lightening because the descent of the fundus of the uterus, together with the reduction in the amount of amniotic fluid in late pregnancy, reduces the upper abdominal distension, making the woman more comfortable.

In many multiparae the presenting part does not engage in the pelvis until labour begins. Not infrequently, multiparae experience uterine contractions which are strong enough to be painful, some days, or even weeks, before real labour starts. Such false pains differ from labour pains only in that they are less regular and are ineffective in dilating the cervix.

Symptoms and signs of the onset of labour

These are:

- painful uterine contractions
- a show
- rupture of the membranes
- shortening and dilatation of the cervix.

The contractions

The uterus contracts irregularly and painlessly throughout pregnancy (Braxton Hicks contractions). Labour is recognized by the changes in the contractions when they become regular and painful enough to distract the woman from her usual activities and cause the cervix to be taken up and dilated. The uterus can be felt to harden during each contraction, which begins gradually, works up to a period of maximum intensity and then dies away, the whole lasting about 45 seconds.

At the onset of labour the interval between contractions may be variable and can be as long as 20 minutes. However, it is quite common for the interval between contractions to be as short as 5 minutes right from the onset.

Contractions are often preceded by backache and tend to increase in frequency and duration, gradually becoming more painful. By the end of the first stage of labour the contractions may come every 2 minutes and may last as long as a minute.

The contractions are not under conscious control and occur even when the woman is unconscious, although they may be lessened in frequency or temporarily abolished by emotional disturbance or by distension of the bladder. They may be increased in strength and frequency by such stimuli as a purgative or enema, stretching of the cervix or pelvic floor by the presenting part, by prostaglandins or by injection of oxytocin.

The pain of labour has the same character as that of spasmodic dysmenorrhoea and probably has the same cause – ischaemia of the uterine muscle from compression of the blood vessels in the wall of the uterus. It is analogous to the myocardial pain which occurs when the blood flow in the coronary arteries is restricted. The intermittent nature of the contractions is of great importance to both the fetus and the mother. During a contraction the circulation to the placental bed through the uterine wall is stopped; if the uterus contracted continuously the fetus would die from lack of oxygen. The intervals between contractions allow the placental circulation to be re-established and give the mother time to recover from the fatiguing effect of the contraction. The uterus is a very large muscle and contractions use up a lot of energy; if continued too long this would produce maternal exhaustion.

Electrical traces of the pattern of uterine contractions show that in normal labour each contraction wave starts near one or other uterine cornu. The contraction spreads as a wave in the myometrium, taking 10 to 30 seconds to spread over the whole uterus. As each point is reached by the wave, contraction starts and takes about another 30 seconds to reach its peak.

The upper part of the uterus contracts more strongly than the lower part, and the duration of the contraction is longer in the upper than in the lower segment. This dominance of the upper segment leads to the stretching and thinning of the lower segment and to dilatation of the cervix.

If the wave pattern is abnormal, with the lower part of the uterus contracting first or as strongly as the upper part, no progress in labour will be made. Sometimes the wave spreads erratically in the myometrium and the contractions are uncoordinated.

The duration and strength of each contraction

are relevant both to their efficiency and to the pain felt by the patient. The resting tone between contractions in early labour is about 1.3 kPa (10 mmHg); the intrauterine pressure at which a contraction can be felt by the hand of an observer is about 2.7 kPa (20 mmHg); and the pressure at which the woman feels pain is about 3.3 kPa (25 mmHg). Efficient first stage contractions reach 6.7 kPa (50 mmHg), and they may reach 10 kPa (75 mmHg) in the second stage. The pain threshold of women varies, and may vary in one woman during the course of labour. It is the anxious worried woman, a long time in labour, who may feel pain at a pressure as low as 2 kPa (15 mmHg).

The show

The mucus operculum or plug is expelled from the cervix as it is pulled up at the onset of labour. This is followed by a mucous discharge from the cervix, mixed with a little blood. As the internal os is drawn up (*see* Fig. 4.2) the membranes are separated from the lower uterine segment, and a variable amount of oozing of blood results. Thus the *mucous show* and the *blood show* classically described as signals of the onset of labour can occur at the time of taking up of the cervical canal which can happen before true labour.

Rupture of the membranes

The membranes may rupture at any time during labour, although this usually occurs towards the end of the first stage. When the membranes rupture spontaneously near term it is probable that labour will begin within a short time, although sometimes the onset is still delayed. Early rupture of the membranes is more likely to occur if the presenting part is not engaged or if there is a malpresentation, but it also occurs in many normal cases.

Shortening and dilatation of the cervix

At the beginning of labour the cervix of a nulliparous woman is usually a thick-walled canal, of at least 2 cm in length. However, in other cases the cervix may be found to be shortened, taken up, and partly dilated in the later weeks of pregnancy.

When labour begins the contraction and retraction of the upper uterine segment stretches the lower segment and the upper part of the cervix; the lower part of the cervical canal remains at first unaltered. As the internal os is pulled open, the cervix is dilated from above downwards, becoming shorter, until no projection into the vagina is felt, but only a more-or-less thick rim at the external os, the whole cervix being taken up and its cavity made one with that of the body of the uterus. Without true shortening of the cervix it is unwise to diagnose that a primigravida is in labour (*see* Fig. 4.2).

In women who have delivered children before the external os will often admit a finger before labour has begun and the finger-tip can sometimes be passed through the internal os. Very often the cervix has been taken up. In this case the projection of a small bag of membranes (*see* Fig. 4.2(d)) during a contraction will show that labour has begun.

THE FIRST STAGE OF LABOUR

The contractions of the uterus dilate the cervix. The dilatation of the internal os causes separation of the chorion from the decidua closest to it. Thus a small bag of membranes is formed and is forced into the internal os by the intrauterine pressure. At the beginning of each contraction a little more amniotic fluid is forced into the bag of membranes, the head then comes down like a ball valve and

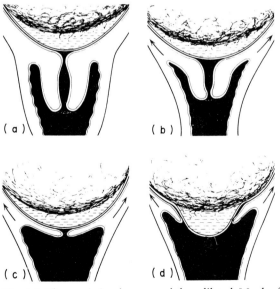

Fig. 4.2 The cervix is taken up and then dilated. Much of this happens in labour but in some women the earlier parts take place during pregnancy

separates the amniotic fluid which is above it from that in the bag, called respectively the hind and forewaters. The bag of membranes may remain intact until nearly the end of the first stage, but even if the membranes rupture early the cervix will normally still become dilated as it is drawn up over the presenting part by the retraction of the upper segment.

During the first stage the fetus does not move downwards to any great degree. When a certain amount of fluid has left the uterus after the membranes have ruptured a new form of pressure comes into play, namely the fetal axis pressure. The upper pole of the fetus, normally the breech, is pressed on by the fundus of the uterus, while the lower pole is pressed down onto the lower segment and cervix. When the membranes rupture early the fetal axis pressure will operate at an early stage, and in modern practice the membranes are often deliberately ruptured during labour because this is believed to encourage more efficient uterine action and shorten labour.

The duration of the *latent phase* of labour need not be defined too accurately. Dilatation of the cervix from 0 to 3 cm can take 6 hours, but slower progress may be normal and is perfectly acceptable provided that the woman is comfortable and in no way distressed. Between 3 and 10 cm dilatation (that is in the *active phase* of labour) the cervix should dilate at a rate of about 1 cm per hour giving a theoretical duration of 7 hours for this phase of labour in both primiparous and multiparous women. During the early part of the latent phase the pains may not be very severe, but towards the end of the active phase they are often very distressing, constituting the most painful part of labour. Vomiting and reflex shivering are not uncommon at the end of the active phase of the first stage of labour.

In the past much has been written about the disadvantage of early rupture of the membranes, but it is now agreed that in normal vertex presentations early rupture is of no consequence. As long as the bag of membranes is intact, the intrauterine pressure is distributed equally over all parts of the fetus. This is still true in a normal case after the membranes have ruptured except for the small part of the fetus which is related to the cervix, because the well-fitting presenting part prevents much amniotic fluid from draining away. If, however, there is a malpresentation or disproportion, early rupture of the membranes may be followed by loss

of nearly all the fluid and the uterus becomes closely applied to the fetus. If labour is prolonged the placenta and cord may be unduly compressed by the retraction of the uterus, fetal hypoxia ensues, and in extreme circumstances the fetus may die.

If the membranes remain unruptured when the cervix is fully dilated the onset of the expulsive stage may be delayed, the cervix not receiving the pressure of the head which should stimulate the uterus to increased activity. If the membranes remain intact after full dilatation they should be ruptured with toothed forceps or a sterile plastic amnihook during a contraction.

THE SECOND STAGE OF LABOUR

There may be very little descent of the fetus during the first stage. However, in the second stage the resistance offered by the lower uterine segment and the cervix has been overcome and the presenting part can be pushed down onto the pelvic floor. The resistance of the pelvic floor then has to be overcome by the uterine contractions, aided by the action of the voluntary muscles of the abdominal wall and the diaphragm.

In the absence of an effective epidural block, full dilatation of the cervix is accompanied by a bearing down sensation during contractions and women are then usually encouraged to push. As the contraction comes on the woman takes a deep breath, then holds it and subsequently bears down with all the force of her abdominal muscles. These partly voluntary, partly reflex expulsive efforts place the fetus under additional stress and pushing should therefore not be allowed to continue for more than one hour. If delivery is then not imminent, assistance in the form of Ventouse extraction, forceps delivery or even a caesarean section may be necessary.

With effective epidural analgesia, the reflex desire to push is lost. In those circumstances, women can be left for one or more hours at full dilatation without pushing, the aim being to allow uterine contractions to cause descent of the fetal head. After that the woman can be encouraged to push, but once again, the duration of pushing should not exceed one hour.

During a contraction the fetal heart rate is often slowed, but it regains its normal rate as soon as the contraction has passed. Such transient bradycardia is of little significance, but if bradycardia is prolonged after each contraction this is a sign of fetal distress.

With each contraction the presenting part is forced down onto the pelvic floor. During the intervals between the contractions, however, the pelvic floor at first pushes the presenting part up again. Retraction now plays an important part, so that the progress made by each contraction is not completely lost during the succeeding interval. Eventually, after being pushed down many times by the contractions and slipping back a little in the intervals between them, a time comes when the presenting part is stationary at the end of a contraction. After this, with each contraction and expulsive effort the head slowly moves down in a forwards direction, becoming more visible. In a primigravida the head may be visible for some time at the vulva before it can emerge. When the widest diameter of the head distends the vulva it is said to be crowned.

In women without an effective epidural block the stretching pain may be very severe as the head passes through the vulva, and will probably cause the woman to cry out, and so to cease from bearing down. To some extent this saves the perineum from damage, as it is likely to be torn if the woman bears down hard while the head is passing through the vulval orifice. It is at this stage that episiotomy may be necessary.

The body of the child is generally born at the next contraction if not by the contraction which expelled the head, and is followed by a gush of fluid.

The caput succedaneum

That part of the head which is most in advance is free from pressure during labour, while the rest of the head is pressed upon by the cervix and lower segment. As a result of venous congestion serum is exuded and an oedematous swelling forms on the scalp, superficial to the periosteum of the cranial bones and not limited by them. This is the caput succedaneum. After delivery, over a few hours or days, it gradually disappears.

If some other part presents, e.g. the face or breech, a comparable oedematous swelling will be formed over the part most in advance.

Moulding

The change in the shape of the head during labour is moulding. The bones of the base of the skull are incompressible, and are joined to each other in such a way that movement is not possible between them. The bones of the vault of the skull *are* compressible, and the sutures allow some movement between the individual bones. The parietal bones and the tabular portions of the occipital and frontal bones can be shaped by pressure, and when forcibly compressed the parietal bones can override the occipital and frontal bones, and one parietal bone can override its fellow. By moulding and overlap the biparietal diameter can be reduced by as much as 1 cm, but excessive moulding may result in intracranial damage.

THE THIRD STAGE OF LABOUR

At the end of the second stage of labour the uterus contracts down to follow the body of the fetus as it is being born. As the cavity of the uterus becomes smaller the area of the placental site is diminished so that the placenta is shorn off the spongy layer of the decidua basalis. Further uterine contractions now expel the placenta from the upper segment into the lower segment and vaginal vault. This process, whereby the placenta leaves the upper segment and occupies the lower segment and vagina, is referred to as *separation and descent* of the placenta.

Immediately after the birth of the child the normal uterus is quiescent for a few minutes. Uterine contractions then begin again, but are not usually painful. When the placenta has separated it presses on the pelvic floor, causing the woman to have an involuntary desire to bear down. The placenta is expelled from the vagina, followed by the membranes and any retroplacental blood clot. There is generally an escape of less than 200 ml of blood as the placenta is delivered. If the uterus does not retract well there is further bleeding, but in the great majority of cases the strong retraction of the uterine muscle compresses the uterine sinuses so effectively that there is little further loss. The uterus can then be felt as a hard round ball about 10 cm in diameter, with the top of the uterus just below the level of the umbilicus.

There are two ways in which the placenta may pass through the vulva. In the Schultz method the placenta presents by its centre, and it delivers inside-out dragging the membranes behind it.

In the Matthews Duncan method, the placenta presents by an edge and slips out of the vulva sideways. It is erroneous to suppose that the way in which the placenta leaves the vagina necessarily implies that that was the way in which it separated from the upper into the lower segment.

Normally the third stage lasts 10 to 20 minutes; with modern active management this can be shorter.

THE MECHANISM OF NORMAL LABOUR

The following terms are used to describe the position of the fetus in relation to the uterus and maternal pelvis.

Lie

This means the relation which the fetus bears to the long axis of the uterus. The lie may be longitudinal, oblique or transverse.

Presentation

The presenting part of the fetus is that part which is in or over the pelvic brim and in relation to the cervix. When the head occupies the lower segment of the uterus the presentation is termed *cephalic*. If the head is flexed on the spine the *vertex* presents.

If the head is fully extended on the spine there is a *face presentation*, and if it is partly extended a *brow presentation*.

If the breech occupies the lower segment the presentation is termed *podalic*.

If the fetus lies obliquely the shoulder generally lies over the cervix and this is called a *shoulder presentation*.

Any presentation other than a vertex presentation is described as a malpresentation.

Position

Position describes the relationship which some selected part (the denominator) of the fetus bears to the maternal pelvis. The denominator varies according to the presentation being described. With a vertex presentation the denominator is the occiput, while with a face presentation it is the chin (mentum) and with a breech presentation, the sacrum.

It is conventional to describe four positions for each presentation. For example with a vertex presentation the occiput could be related to:

- left iliopectineal eminence – left occipito-anterior position (LOA) (*see* Fig. 4.3)
- right iliopectineal eminence – right occipito-anterior position (ROA)
- right sacroiliac joint – right occipitoposterior position (ROP) (*see* Fig. 4.4)
- left sacroiliac joint – left occipitoposterior position (LOP).

However, during late pregnancy and in the first stage of labour the occiput most commonly lies in the transverse diameter of the pelvic brim, and the terms left occipitotransverse position (LOT) and right occipitotransverse position (ROT) are useful.

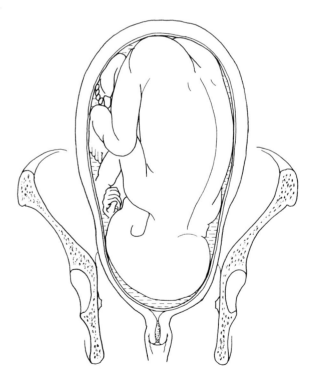

Fig. 4.3 Left occipitoanterior position

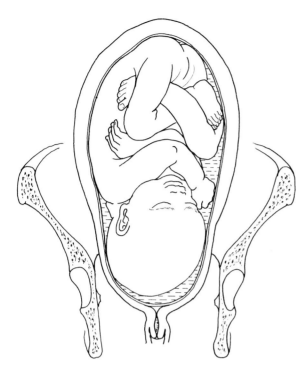

Fig. 4.4 Right occipitoposterior position

Attitude (flexion or extension)

This term refers to the relation of the different parts of the fetus to one another. Normally the head, back and limbs of the fetus are flexed. In some abnormal presentations, which will be described in later chapters, the head or limbs may be extended.

In 96 per cent of cases at term the fetus lies longitudinally, with the head presenting. The reason for this is that the fetus adapts itself by its movements to the shape of the uterus. In the early months of pregnancy the amniotic fluid is comparatively more abundant, and the fetus can move freely; but as pregnancy advances the fetus rapidly increases in size and the volume of fluid becomes comparatively less, so that the fetus is constrained to fit the shape of the uterus. When the attitude is one of complete flexion the buttocks, together with the adjacent parts of the thighs and the feet, constitute a mass which is larger than the head. The cavity of the uterus at term is pear-shaped, with the wider end uppermost; therefore the fetus fits into it best when the breech lies in the upper part of the uterus and the head in the lower part.

If the head cannot readily enter the brim of the pelvis a malpresentation may occur, for instance when the pelvic brim is severely contracted, or when a low-lying placenta or a pelvic tumour reduces the available space in the lower segment.

If the tone of the uterine and abdominal muscles is poor, as may be the case in a woman who has had many children, the factors which normally constrain the fetus to lie longitudinally are absent, and there may be a transverse or oblique lie. If the fetus is dead it may lie abnormally because it does not move and lacks muscular tone.

MECHANISM OF LABOUR WITH VERTEX PRESENTATIONS

The term 'mechanism' refers to the series of changes in position and attitude which the fetus undergoes during its passage through the birth canal. These should be studied with the help of a fetal model and a pelvis, as well as by observation of women in labour.

The head is more or less oval, and fits fairly tightly into the birth canal through which it is pushed. The largest diameter of the pelvis is transverse at the inlet and anteroposterior at the outlet. The head, which normally enters the brim in the transverse or one of the oblique diameters, undergoes rotation and also some change in its attitude as it passes through the pelvic cavity. If the head and pelvis are both of normal size the mechanism of labour is determined by the soft parts rather than the bony pelvis.

Although four oblique positions of the occiput are conventionally described, in most cases the fetal head enters the brim in the transverse diameter. In less than 15 per cent of cases the occiput lies in relation to one of the sacroiliac joints at the onset of labour; it is never in direct relation to the promontory of the sacrum.

The degree of flexion or extension of the head is a most important factor in determining the mechanism of labour and therefore its outcome (*see* Fig. 4.5).

MECHANISM WITH THE OCCIPUT IN THE TRANSVERSE OR ANTERIOR POSITION

For convenience of description the mechanism will be described for the LOT or LOA positions. (For the ROT or ROA positions the same description

Posterior fontanelle

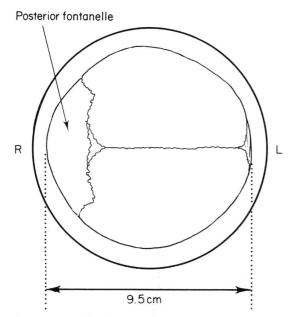

Fig. 4.5 Vaginal palpation of the head in right occipitolateral position. The circle represents the pelvic cavity with a diameter of 12 cm. The head is well-fixed and only the sutures round the posterior fontanelle are felt

applies, but with substitution of right for left and vice versa throughout.)

The descent of the head can be divided into five levels. This analysis is for better understanding only for the five components merge into each other in reality. They are:

> flexion
> internal rotation
> extension
> restitution
> external rotation.

Flexion

The head is often well flexed before labour starts, but if flexion is incomplete when labour starts it becomes complete as the uterine contractions drive the head down into the lower uterine segment. This is because:

* any ovoid body being pressed through a tube tends to adapt its long diameter to the long axis of the tube
* of the so-called head lever (*see* Fig. 4.6).

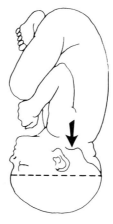

Fig. 4.6 Flexion of the head during labour. The arrow shows the direction of fetal axis pressure and the dotted line indicates the reduction in diameter of the head when it flexes

The occipitospinal joint is nearer to the occiput than to the sinciput (forehead), so the head can be regarded as a lever with a long anterior and a short posterior arm. When the breech is pressed on by the uterine fundus the fetus is subjected to axial pressure and the lever comes into play. The long anterior arm meets with more resistance than the short posterior arm and the head flexes.

Flexion has the advantage of bringing the shortest suboccipitobregmatic diameter of the head into engagement, when the posterior fontanelle of the skull will be at a lower level than the anterior fontanelle.

Internal rotation

In the second stage of labour, the forces propel the fetus progressively down the birth canal. When the head meets the resistance of the pelvic floor the occiput rotates forward from the LOT or LOA position to lie under the subpubic arch with the sagittal suture in the anteroposterior diameter of the pelvic outlet (*see* Fig. 4.7). This internal rotation of the head occurs because with a well-flexed head the occiput is leading and meets the sloping gutter of the levators ani muscles which, by their shape, direct it anteriorly.

Extension

Further advance of the head leads to its passage through the vulva by a process of extension. Once

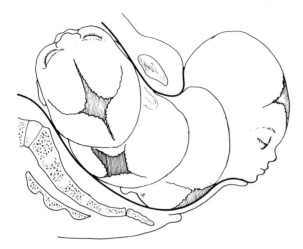

Fig. 4.7 Descent and flexion of the head followed by internal rotation and ending in birth of the head by extension

the occiput has escaped from under the symphysis pubis the head extends, with the nape of the neck pressed firmly against the pubic arch. This extension of the head causes the anterior part to stretch the perineum gradually, until the moment of crowning when the greatest diameter slips through the vulva. Further extension allows the forehead, face and chin successively to escape over the perineum (*see* Fig. 4.7).

Restitution

As the head descends with its suboccipitofrontal diameter in the transverse or right oblique diameter of the pelvis the shoulders enter the pelvic brim in the anteroposterior or the left oblique diameter. When internal rotation of the head takes place the head is twisted a little on the shoulders. As soon as it is completely born it resumes its natural position with regard to the shoulders, the occiput turning towards the woman's left thigh. This movement, which sometimes occurs almost with a jerk, is called restitution, because by it the neck becomes untwisted and the head is restored to its natural relation to the shoulders.

External rotation

As the shoulders descend the right and anterior shoulder is lower and meets the resistance of the pelvic floor before the left shoulder. The right shoulder rotates to the space in front, as did the

occiput, and the shoulders now occupy the anteroposterior diameter of the pelvis. As they rotate, the head, which has already been born, rotates with them and may make a further movement towards the woman's left thigh. The head now lies with the face to the right and the occiput to the left (*see* Fig. 4.8).

Delivery of the body

The shoulders then emerge, the right one escaping under the pubic arch, while the left slides over the perineum. The rest of the body is usually born without any difficulty as its diameters are less than those of the head or the shoulders. The arms are usually folded on the chest, with the hands under the chin.

MECHANISM WITH THE OCCIPUT IN THE POSTERIOR POSITION

The ROP position is more common than the LOP position, and will be described. (For the LOP position the same description applies but with the interchange of left for right and vice versa throughout.)

The mechanism in the occipitoposterior positions depends on whether the head is well flexed or incompletely flexed.

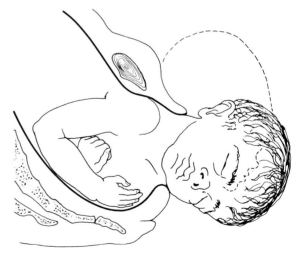

Fig. 4.8 External rotation of the head after delivery as the anterior shoulder rotates forward to pass under the subpubic arch

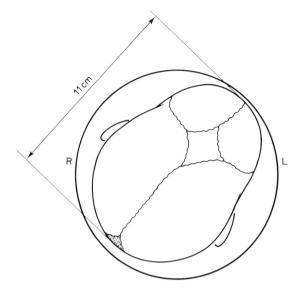

Fig. 4.9 Vaginal palpation of head in the right occipito-posterior position. The circle represents the pelvic cavity with a diameter of 12 cm. The head is poorly flexed, so that the anterior fontanelle presents. Moulding is seen

The well flexed head

If the head is well flexed, the occiput is in advance when the head meets the resistance of the pelvic floor. The occiput slides down the gutter formed by the levator muscles, undergoing long rotation through three-eighths of a circle, to reach the free space under the pubic arch.

When the occiput lies behind and to the right (ROP) it rotates along the right side of the pelvis to reach the front, the shoulders rotating with the head from the left oblique diameter into the anteroposterior diameter. From this point the mechanism is the same as that of the ROA position, with birth of the head by extension.

Delay, which occurs in some cases of occipitoposterior position, is not because of this additional long rotation. If the head is fully flexed, as it must be for long rotation to occur, there is no delay in the labour and no difficulty. The cause of delay, if it occurs, is incomplete flexion, so that normal long rotation does not occur.

The incompletely flexed head

When the occiput occupies the posterior part of one of the oblique diameters of the pelvis the biparietal diameter lies in the bay to one side of the promontory. When the head is pushed down into the pelvis in this position the biparietal diameter is hindered in descending if the pelvis is small or the head is large, and so the forepart of the head descends more easily than the occiput and the head enters the pelvis incompletely flexed (*see* Fig. 4.9).

If the head is incompletely flexed the larger occipitofrontal diameter of the head, which measures 11.5 cm, has to pass through the pelvis instead of the suboccipitobregmatic diameter, which measures 9.5 cm. It is this, and the fact that sometimes neither the occiput nor the sinciput is sufficiently in advance of the other to influence rotation, that explains why some cases of occipitoposterior positions cause difficult and prolonged labour. However, in some cases in which long rotation of the occiput does not occur spontaneously delivery takes place by an alternative mechanism, as follows.

If the head is incompletely flexed with an occipitoposterior position the forehead is as low as the occiput, and being at the anterior end of the oblique diameter of the pelvis it meets the resistance of the pelvic floor before the occiput. The forehead rotates to the front to the free space under the pubic arch, turning through one-eighth of a circle while the occiput rotates backwards into the hollow of the sacrum. The head may now be

Fig. 4.10 Delivery of the head in the face-to-pubes position. 1, shows the head being born by flexion, this is followed by extension, shown in 2

born with the face towards the posterior surface of the symphysis pubis. The root of the nose is pressed against the bone, and the head flexes about this fixed point. The vertex is born by *flexion* and followed by the occiput. As soon as the occiput is born the head extends, the face and chin emerging from under the pubic arch. The vulval orifice is stretched by the occipitofrontal instead of the suboccipitofrontal diameter, with a difference in size of 2 cm, and a severe perineal tear may result (*see* Fig. 4.10).

DEEP TRANSVERSE ARREST OF THE HEAD

In some cases the head becomes arrested with its long axis in the transverse diameter of the pelvis, the degree of extension being such that neither the occiput nor the forehead is sufficiently in advance to influence rotation (*see* Fig. 4.11).

This is described as deep transverse arrest of the head and calls for assistance. Some of these cases are the result of incomplete forward rotation from an occipitoposterior position. Others, perhaps the majority, are the result of the descent of a head which originally lay in the occipitotransverse position and which has failed to rotate anteriorly.

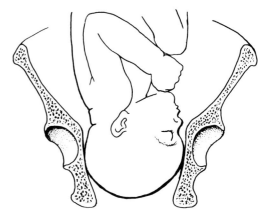

Fig. 4.11 Deep transverse arrest of the head

MANAGEMENT OF NORMAL LABOUR

In Britain today the majority of women are confined in hospital simply because obstetric emergencies such as fetal hypoxia or postpartum haemorrhage may suddenly arise in apparently normal cases. The expertise and facilities to deal with an emergency are immediately available in hospital whereas there is inevitable delay if the crisis occurs in the woman's home, a delay which may be the deciding factor between life and death for the woman or her baby. Many women have the impression that childbirth is an entirely physiological event without risk, but this is not so small in a proportion of cases.

As a compromise for those who prefer home delivery many hospitals arrange for delivery in hospital and early discharge home after 24 or 48 hours, provided that there is a domiciliary midwife and a doctor willing to be responsible when the women returns home. Some units have facilities for a woman to be looked after by her own midwife and general practitioner while she is in hospital and some have rooms which are furnished like ordinary bedrooms in order to make the woman feel more at home. The presence of the child's father, or of a close relative, is welcomed throughout labour. Should an emergency arise in such a birth room the woman can be transferred to the main part of the hospital immediately.

Modern management of labour is carried out by a team consisting of midwives and obstetricians.

The role of the midwife is of the utmost importance. A midwife is the senior professional person present in at least 70 per cent of normal deliveries in the UK. Midwives are normally responsible for the care of women in labour, observing the progress and the condition of the woman and her fetus, alleviating pain, preventing infection and supervising safe delivery. They look for abnormalities in labour and procure medical assistance when necessary, being ready to take emergency measures (e.g. urgent breech delivery or manual removal of the placenta) in the absence of medical help.

No labour can be said to be normal until the third stage is safely concluded. Danger, especially to the fetus, can arise suddenly and unexpectedly, and to secure the greatest safety of the mother and baby, labour is best managed by intensive care techniques. There is no reason why modern methods of monitoring during labour should be psychologically harmful or interfere with normal labour; indeed there is some evidence that they may prevent unnecessary intervention. If their purpose is explained to the woman they should be reassuring to her.

Although the dangers of infection have been much reduced by the use of antibiotics, it is still very important to minimize the risk of introducing infection into the genital tract during labour. Slowly, it is being realized that the important organisms that cause infection in the genital tract after childbirth come mostly from the skin flora of the woman herself and, to a lesser extent from the doctor or midwife. Hence careful antiseptic techniques of cleaning the woman's perineal skin and aseptic washing of the attendant's hands must be adhered to and gloves should be worn. Of less importance is the wearing of masks to guard against nose and mouth organisms, gowns to protect from bacteria and fungi on the clothing or boots to prevent spores and fungi from the outside world. In many labour wards, no masks, gowns or boots are worn, just well-washed gloved hands and a protective apron of light material. Such sterile precautions must be taken by those delivering the woman.

The first stage of labour is proceeding normally if the cervix is progressively dilating and the fetal condition is satisfactory. The second stage is normal when there is progressive descent of the head and the fetus is in good condition. These statements may seem very obvious, but they are essential as the basis of most observations made during labour.

Because of the increased risk to mother and fetus during prolonged labour, labours are no longer advised to continue for great lengths of time. Delay in the first stage is overcome by active management with oxytocics, and delay in the second stage by the use of forceps or the vacuum extractor.

MANAGEMENT OF THE FIRST STAGE

On admission, the woman's antenatal record is reviewed to discover whether there have been any abnormalities during her pregnancy. Records of ante-natal care should be available from hospital record departments or hand held notes. In a few instances there will have been no antenatal care; in such circumstances a complete history must be taken.

In every case the woman's general condition is assessed, her pulse rate and blood pressure are recorded, and her urine is tested for protein and sugar. By abdominal examination the presentation and position of the fetus, and the relation of the presenting part to the brim of the pelvis, are determined. Abdominal examination will also show the frequency and strength of the uterine contractions. The fetal heart rate is checked and any abnormality of rate or rhythm is noted. A vaginal examination will show the degree of dilatation of the cervix, whether the membranes are intact or ruptured and the level and position of the presenting part.

Much of the apprehension from which many women suffer during labour may be removed by adequate explanations beforehand. The best time to give this is in antenatal classes to which the woman may go with her partner. She will find it a great comfort to hear what labour entails in her partner's presence, knowing that he will be at her side all through the confinement. If it is explained to them that the first stage of labour may last up to 10 hours, during which it may be difficult for her to appreciate that there is much progress, it will be much easier to reassure her at the time that all is well. She should, however, always be kept informed about dilatation of the cervix and of the condition of her baby. If it is decided that the best way of checking the fetal heart is with a fetal heart rate monitor its purpose must be explained. The reason for any intervention must also be discussed fully with the woman and her partner. If there is free communication between attendants and the woman there will almost always be full cooperation. At no time in labour should the woman be left alone. The partner should be with her all the time, and the midwife as much as possible.

It is unnecessary to give an enema or to clip or shave the vulval hair. These practices are of no particular benefit and are generally disliked. A warm bath or shower, however, is both hygenic and pleasant.

There is no need for the woman to remain in bed during early labour. If she is up and about, the weight of the amniotic fluid and the fetus helps to dilate the cervix and pressure on the lower segment stimulates the uterus to contract. If the presenting part is not engaged vaginal examination is advisable to exclude prolapse of the umbilical cord when the membranes rupture.

Discomfort during the early part of the first

stage may not be severe, although in primiparae it may cause distress. Towards the end of the first stage the pains become more severe. If epidural analgesia is not employed, drugs such as pethidine 100–150 mg intramuscularly may be given when labour is established if the woman is distressed, but a woman who has been appropriately prepared for labour should be allowed to decide for herself whether, and when, she wants analgesia. The knowledge that help is available gives her confidence and she may prefer to defer its use for a time. As pethidine sometimes causes nausea, metoclopramide (Maxolon) 10 mg may also be given intramuscularly. Some women obtain relief by means of transcutaneous electrical nerve stimulation (TENS) but the equipment is not universally available. Pain relief provided by TENS is best in early labour and if properly used by a trained mother, TENS may postpone the need for stronger analgesia by some hours.

There may be a frequent desire to pass urine during the first stage. When the head is deep in the pelvis the woman may be unable to pass urine and the bladder rises up into the abdomen where it can be seen and felt as a suprapubic swelling. A soft catheter should be passed, as a full bladder has an inhibiting effect on the uterine contractions.

During labour there is delay in the emptying time of the stomach and food or fluids may remain there for several hours. If a general anaesthetic has to be given for any reason there is a risk of vomit being inhaled and the acid contents of the stomach may cause bronchiolar spasm (Mendelson's syndrome). Alkali given by mouth may reduce the severity of this complication, but it is better to withhold solid food during labour. Use Ranitidine 50 mg by intramuscular injection every 6 hours to women who do not have an epidural block and are therefore more likely than those with epidurals to require general anaesthesia if they have a complication requiring urgent operative intervention. Ranitidine effectively reduces the acidity of gastric contents. Feeding traditionally was discouraged in labour, but light easily-digested foods are probably of more help than harm. An example might be chicken soup.

Dextrose solutions given intravenously during labour must be properly controlled. There have been instances of injudicious administration of large volumes of fluid containing no sodium, resulting in maternal and fetal hyponatraemia. If large volumes of intravenous fluid are given for any reason during labour, physiologically balanced infusions such as Hartmann's solution must be employed. Modern drip counters allow strict control of the volume of fluid used.

Large volumes of water are retained during pregnancy, chiefly in the maternal extracellular compartment and unless there has been excessive vomiting significant dehydration or ketosis are unlikely in a labour lasting less than 24 hours.

Towards the end of the first stage, if epidural analgesia is not used, administration of nitrous oxide and oxygen (Entonox) may be started with the onset of each contraction.

Partogram

Once labour has become established, or the membranes have ruptured, all events during labour should be noted on a partogram – a most useful graphical record of the course of labour. Routine observations of the woman's pulse rate and blood pressure, with an assessment of the strength of the uterine contractions are entered on it. Records of the findings at successive vaginal examinations are plotted on a graph, showing the dilatation of the cervix in centimetres against the time in hours. The curve obtained is compared with an average curve for normal primigravidae or multigravidae as may be appropriate in any given population. If the woman's progress is normal her curve will correspond with the normal curve, or lie to the left of it.

Friedman, who introduced the idea of the partogram, has described two phases of labour:

- *latent phase*, from the onset of labour until the cervix is about 3 cm dilated, which may last 3 to 7 hours in a primigravida
- *active phase*, during which dilatation from 3 to 10 cm is more rapid, taking 3 to 7 hours, so that the slope of the partogram curve will be steeper in this phase.

If labour is not progressing normally in the active phase, dilatation of the cervix will become slower or may cease and the woman's partogram will be to the right of the normal curve. A vaginal examination should be made about every 3 hours, and if there is delay the membranes should be ruptured and uterine action augmented by administration of an oxytocic infusion if there is no progress after the membranes are ruptured.

All this information about the strength and

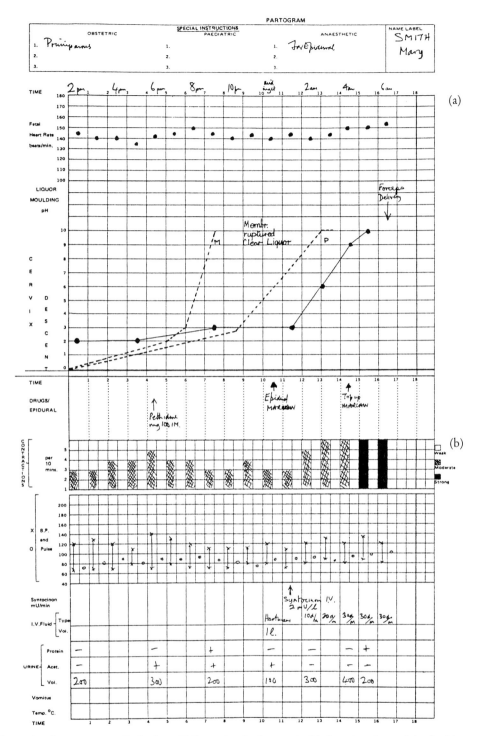

Fig. 4.12 Example of a partogram. Only the middle section is shown. At (a) there is a chart of the fetal heart rate and at (b) there is a chart of the maternal blood pressure and pulse rate. In the illustrative case shown there was a prolonged latent phase. The membranes were artificially ruptured with little effect on the contractions. With intravenous oxytocin (2 units/500 ml) at 10 and then 20 drops per minute the uterine contractions improved and the active phase soon followed. At the point marked M, meconium was seen in the liquor and there were some decelerations in the fetal heart rate, but a fetal blood sample showed a pH of 7.3 and a base deficit of only 4 m eq/l so labour was allowed to continue, ending in an easy forceps delivery

frequency of the uterine contractions, the dilatation of the cervix and (later in labour) descent of the head, and the state of the mother and fetus can be shown on the partogram. Drugs that are given should also be recorded (*see* Fig. 4.12).

Fetal monitoring

The other important observations which must be made during labour relate to the fetus. If simple clinical observations are all that are possible these must be made and recorded regularly. The fetal heart rate is counted with a stethoscope at half-hourly intervals in early labour and at 10 minute intervals in the active phase of labour. The normal rate is between 120 and 160 beats per minute and there is no change of rate, or only a very transient slowing, with the uterine contractions.

Most hospitals now employ fetal monitoring during labour. The uterine contractions can be recorded with a strain gauge strapped to the mother's abdomen, and the fetal heart can be monitored with an ultrasonic device attached in the same way (*see* Fig. 4.13). The ultrasound record of the fetal heart rate is not always satisfactory, although this simple method can be tried

first. After the membranes have ruptured and the cervix has started to dilate a more reliable record can be obtained by attaching a clip to the fetal scalp through the cervix from which a lead passes to a machine which calculates the heart rate continuously by measuring the intervals between R waves in the fetal electrocardiographic cycles.

In normal labour the basal heart rate between contractions is between 120 and 160 beats per minute, with a continuous slight beat-to-beat variation of the order of five beats per minute. In normal labour the heart rate may slow with each uterine contraction, but the slowing is neither profound nor prolonged. Prolonged deceleration trachecardia or loss of baseline variability may be sinister signs.

Where resources are limited monitoring may only be possible for high-risk cases and those in whom clinical signs suggesting fetal distress appear, but to secure the best possible surveillance of the fetus the ideal would be to monitor every labour, because half the instances of hypoxia during labour occur without evident preceding high risk factors. However, some women object to the routine use of monitoring. They dislike any apparent interference in a normal labour and the limitation of free movement caused by the electrical leads to the recording machine; they may also fear fetal injury from the scalp electrode. Restriction of mobility may in future be overcome by radio; telemetry connects the transmitter carried by the mother to the recording machine, so that leads are not required. Particularly in high risk cases, the patient's cooperation can usually be obtained if the purpose and advantage of monitoring is explained.

If the monitor or clinical observations suggest that there may be fetal distress a sample of fetal blood is taken from the fetal scalp to determine the pH and base deficit levels and thus, indirectly, whether there is fetal hypoxia.

These technical methods of checking that all is well with the mother and baby must never be allowed to take the place of close contact between doctors, midwives and the patient. Clinical assessment of well-being must not be omitted or disregarded.

Active management of labour

Most obstetricians believe that with the help of modern technology, they can ensure that childbirth is a safer and happier experience for the

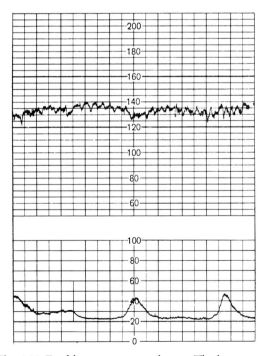

Fig. 4.13 Fetal heart rate, normal trace. The lower trace records the uterine contractions

mother and her baby. In the first stage of labour this includes effective analgesia including epidural block; monitoring the fetal heart rate and the uterine contractions; recording cervical dilatation to ensure progress is being made and the occasional use of intravenous oxytocins. Provided the woman understands the reasons for these measures she will usually accept them in the knowledge that they are undertaken for her and her baby's benefit. It is stressed that they are to be used selectively, not routinely, and in particular that it is only necessary to accelerate labour with an oxytocic drip if progress is abnormally slow.

MANAGEMENT OF THE SECOND STAGE

Without epidural block

During the second stage the woman should be close to her bed and the midwife or doctor should stay with her.

Early in the second stage it does not matter what position the woman adopts, but if she is well propped up with her head upright and her hands behind her knees she will be in a comfortable position to push effectively, with some assistance from gravity. Occasionally, because of supine hypotension, slowing or other changes in the fetal heart rate occur if the woman has slumped into the dorsal position. If such changes occur the woman is turned onto her side and hitched up into a virtually sitting position.

As each uterine contraction pushes the fetal head down onto the pelvic floor, the expulsive reflex comes into play and the woman will generally take a deep breath, hold it, and strain down. In a first labour the woman needs to be encouraged to relax the muscles of the pelvic floor at the time of the contraction. The progress of the descent of the head can be judged by watching the perineum. At first there is a slight general bulge as the woman strains. When the head stretches the perineum the anus will begin to open, and soon after this the caput will be seen at the vulva at the height of each contraction. Between contractions the elastic tone of the perineal muscles will push the head back into the cavity of the pelvis. The perineal body and vulval outlet become more and more stretched until eventually the head is low enough to pass forwards under the subpubic arch. When the head no longer recedes between contractions this indicates that it has passed through the pelvic floor and that delivery is imminent.

When the head begins to appear at the vulva it must be decided whether the woman is to be delivered in a semi-sitting position or on her side. If little help is available the left lateral position may be used but when assistance is available most women are delivered in the dorsal position, as described above.

Different positions may be adopted in the second stage according to individual preference. There has been a recent tendency for a minority of women to ask to give birth on all fours, squatting, kneeling or standing upright, claiming that such positions make delivery easier. Birth chairs have been tried in some hospitals, and usually given up. A few babies are delivered under water. There is little evidence that delivery in any of these positions is more rapid or comfortable, and they have the disadvantage that they make proper observation of the fetal heart rate and control of analgesia very difficult, and care of the perineum hardly possible. However, the right of every woman to deliver her baby in the manner she wishes must be considered sympathetically.

With epidural block

About a third of women have an effective epidural block during the first and second stages of labour; this makes a difference to the management of the second stage of labour.

Such women do not experience a reflex desire to bear down, and once full dilatation is diagnosed, they can be left to continue in labour without pushing for 1 or 2 hours provided that there is no evidence of fetal distress. This gives the fetal head a chance to descend deeper into the pelvis in response to uterine contractions.

When women with an epidural block do start bearing down or pushing, they will not usually know when they are having contractions, nor will they usually experience a reflex urge to bear down during contractions. They will need extra instruction and encouragement from their attendants. Unusual birth positions (all fours, kneeling, squatting or standing upright) cannot be adopted with an effective epidural block.

DELIVERY OF THE BABY

If delivery is left entirely to nature, laceration of the perineum often occurs during birth of the head. By the time the head begins to appear at the vulva some form of analgesia is usually desirable. An epidural or pudendal block or local infiltration may be used but alternatively inhalation of a mixture of nitrous oxide and oxygen (Entonox) is often effective.

At this stage the midwife or doctor must control the head to prevent its being born suddenly and it must be kept flexed until the largest diameter has passed the vulval outlet. Once the head is crowned the woman should be discouraged from bearing down by telling her to take rapid shallow breaths. The head may now be delivered carefully by pressure through the perineum onto the fore part of the head by means of a finger and thumb placed on either side of the anus, pushing the head forwards slowly before it is allowed to extend and complete its delivery and controlling the rate of escape with the other hand (*see* Fig. 4.14).

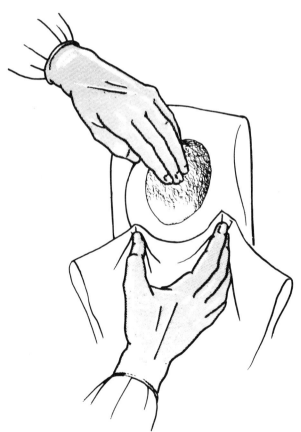

Fig. 4.14 Care of the perineum. The left hand is preventing sudden expulsion of the head, while the fingers and thumb of the right hand are gently helping the head forwards by pressure on each side of the anus

If extension of the head begins before the biparietal diameter has passed through the vulval orifice a larger diameter than the suboccipitofrontal will distend the vulva and a tear may result. Even if the head has become crowned gradually, perineal rupture may occur if the head is allowed to be expelled suddenly and rapidly; it is important that the head should be born slowly and in an interval between contractions.

Episiotomy, or incision of the perineal body, is necessary in some cases and a clean incision is always preferable to an irregular laceration, or even to a grossly overstretched perineal body. It is important to explain to the woman beforehand that it is sometimes necessary, and to assure her that it will not be performed just as routine.

Directly the head is born a finger is inserted to feel whether a loop of cord is round the neck. Such a loop should be slipped over the head; if this cannot be done the cord is clamped with two pairs of artery forceps and divided between them.

The shoulders usually follow with the contraction following the birth of the head, the anterior shoulder being delivered before the posterior. Even if the head has been born without perineal laceration, the shoulders can cause damage unless they are carefully delivered.

If the shoulders do not descend after the birth of the head the mother should be exhorted to bear down. If the shoulders still do not move and the baby's head is becoming cyanosed birth must be assisted. If the shoulders have not rotated into the anteroposterior diameter of the pelvis they must be rotated by digital pressure. If there is still delay attempts must then be made to bring the anterior shoulder under the subpubic arch by bending the neck laterally towards the anus. Once the anterior shoulder has passed the symphysis pubis the posterior shoulder can usually be delivered after pulling the head forwards. As little force as possible should be used, for fear of injury to the brachial plexus. An assistant can help by pressing on the fundus of the uterus at the same time.

In cases of extreme difficulty a finger may be passed up to the anterior axilla of the fetus and the shoulder pulled down; or else a hand may be passed behind the posterior shoulder into the hollow of the sacrum and the posterior arm brought down over the perineum by flexing it at the shoulder and elbow. This makes more room and the anterior shoulder is then easily delivered.

After delivery of the shoulders the rest of the

body quickly follows. As soon as the child is delivered it is held with its head downwards so that any fluid or mucus in the mouth can run out. The mouth and pharynx are sucked clear with a mucus extractor. A healthy baby breathes and cries very soon after it is born; if it fails to do so the baby needs active resuscitation.

Normally the cord should not be clamped until the child has cried vigorously and pulsation in the cord has ceased. If it is clamped immediately the baby is deprived of about 20 ml of blood which would be drawn out of the placenta by the expansion of the lungs. It is best to keep the baby at the same level as the placenta or a little below it. If the baby is held high above the placenta (which may be done inadvertently at caesarean section) blood may run back into the placenta with the risk of subsequent anaemia in the baby. For this reason the practice of placing the baby on the mother's abdomen immediately after birth and before the cord is clamped is of doubtful wisdom; the unwrapped baby may also suffer heat loss. Once the baby is breathing normally and the cord has been divided it should be wrapped up and handed to the mother to hold, cuddle and enjoy.

At first the cord is divided between two artery forceps; later a plastic crushing clamp is placed on the cord 1 to 2 cm from the umbilicus and the cord is cut again 1 cm beyond the clamp. When this is done the cut end of the cord is examined to make sure that both umbilical arteries are present.

If spontaneous respiration is not established soon after birth, resuscitation is the immediate priority and the baby is taken to the resuscitation table directly after the cord is divided.

MANAGEMENT OF THE THIRD STAGE

In a normal delivery, if oxytocic drugs are not injected, the uterus will generally remain quiescent for a few minutes after the delivery of the baby. Regular contractions then begin again. These detach the placenta and push it down into the lower uterine segment and vagina (*see* Fig. 4.15). The mother will become aware of its presence on the pelvic floor, and by straining expel it through the vagina.

The following signs indicate that the placenta has separated and been expelled from the upper uterine segment into the lower segment or even into the vagina:

- the cord moves down and appears to lengthen. It may be difficult to be sure of this, and in case of doubt the fundus of the uterus may be gently pressed upward by a hand placed just above the pubis. If the placenta is still in the upper segment the cord will be drawn up with the uterus
- the uterus rises up because it is now perched on the lower segment which contains the placenta, and when it contracts the empty upper segment feels hard, round and movable from side to side (*see* Fig. 4.15)
- there is often a small gush of blood when the placenta leaves the uterus
- the placenta can be felt with a finger inserted into the vagina.

The third stage of labour is a natural process which can be managed by simple observation and without interference unless bleeding or delay occurs, but even with such conservative management postpartum haemorrhage sometimes occurs if the uterus relaxes. However, nearly all obstetricians now advocate an alternative and more active method of management of the third stage because this has been found to be safer.

Active management of the third stage

After 1935 ergometrine became available as an effective and non-toxic oxytocic agent. It was at first used for the treatment of postpartum haemorrhage after the uterus had been emptied but it was soon found that by giving it at the time of birth, or immediately afterwards, the number of cases of excessive bleeding (defined as a loss of more than 500 ml) was reduced. Ergometrine causes a prolonged contraction of the uterus without periods of relaxation, and while the uterus is contracting there is not likely to be any bleeding. It is therefore a most valuable drug given in a dose of 0.5 mg by intramuscular or intravenous injection in cases of excessive blood loss.

However, ergometrine can cause nausea and vomiting and there is usually a significant rise in blood pressure. Because of these disadvantages there is a tendency for obstetricians to prefer Syntocinon, 5 units, for routine management of the third stage of labour. Syntocinon, an isomer of oxytocin, is given by intramuscular or intravenous injection with delivery of the anterior shoulder in a primigravida or while the head is crowning in a

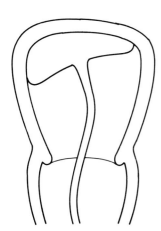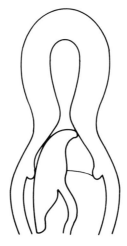

Fig. 4.15 Signs of descent of the placenta. After separation the contracted upper segment is at a higher level and feels more rounded

multipara. With this method the third stage of labour is shortened and blood loss is reduced. As soon as the signs of separation are present, showing that the placenta is in the lower uterine segment and vagina, the placenta is delivered by the Brandt–Andrews' method, as described below.

By injecting an oxytocic drug before the placenta has separated there is a small risk that the placenta may be grasped by the contracting upper segment and become retained, with a slight increase in the number of cases in which manual removal of the placenta becomes necessary. This disadvantage is small compared with the advantage of reduction in incidence of postpartum haemorrhage.

Syntometrine is a preparation containing 500 μg of ergometrine and 5 units of Syntocinon per ml. It can be given as an alternative to Syntocinon alone.

Brandt–Andrews' method of delivering the placenta

With the woman lying on her back the obstetrician places his left hand over the anterior surface of the uterus just above the symphysis pubis, at the presumed level of the junction of the upper and lower segments. An artery forceps is placed on the umbilical cord, which is held just taut but without strong traction with the right hand (*see* Fig. 4.16). The uterus is pushed gently upwards with the left hand and if this can be done satisfactorily it means that the placenta is below the level of the lifting hand and is in the lower segment or vagina. Lifting is now discontinued and, with the uterus contracted, pressure is made with the same (left) hand

in a downward direction while the cord is held taut until the placenta is seen at the introitus. After the placenta has been expelled the uterus is lifted out of the pelvis. It will be noted that the principle of the method is not cord traction but rather elevation of the uterus, which prevents acute inversion of the uterus as a complication.

The membranes generally slip out after the placenta. However, if they do not come away with gentle traction on the placenta they should be held with artery forceps and gently pulled in a rotational fashion, when they will usually be extracted.

Examination of the placenta, membranes and cord

These must always be examined carefully as soon as possible. The maternal surface of the placenta will be seen to be divided into cotyledons, but these should all fit together when the maternal surface is made concave. If any part is missing a gap will be seen.

The membranes should form a complete bag except for the hole through which the fetus passed. The amnion and chorion can be separately examined after peeling the amnion off the chorion. An important but rare abnormality is that in which blood vessels run off the edge of the placenta to a small detached island of placental tissue, a *succenturiate lobe*. If this has come away with the main placenta and membranes it will be plainly seen, but if it is retained there will be a hole in the chorion corresponding to it, to which the vessels pass.

The cut end of the cord should be examined. A

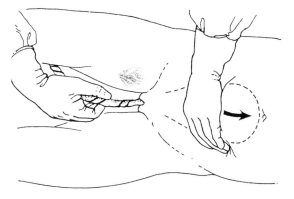

Fig. 4.16 Brandt–Andrews method of delivery of the placenta

rare abnormality is absence of one umbilical artery. This is important as it may be associated with other congenital abnormalities.

If a piece of placenta is retained it is almost certain to cause postpartum haemorrhage, therefore it should be removed as soon as the diagnosis is made. In cases of doubt an ultrasound scan will show whether the uterus is empty. A piece of membrane retained within the uterus will probably not cause complications and will come away in 2 or 3 days; uterine exploration is not necessary.

Examination of the perineum

After the placenta is delivered, the vulval outlet must be examined carefully for lacerations after separating the labia. Any tear other than a minute one should be sutured very promptly.

THE RELIEF OF PAIN IN LABOUR

The amount of pain experienced during labour appears to vary enormously; very few women find that labour is almost painless, but the majority have pain that they describe as severe. Some will manage with minimal doses of analgesic drugs while others demand much more.

During the first stage of labour pain is felt with each uterine contraction. The pressure in the uterus between contractions is of the order of 1.33 kPa (10 mmHg). During a first stage contraction the pressure is about 6.7 kPa (50 mmHg). Most women do not feel pain until the pressure reaches 3.2 kPa (25 mmHg); the beginning of the contraction can be felt with the hand or recorded with a tocograph before the woman feels pain. The pain is believed to be caused by ischaemia of the myometrium, which occurs when the blood flow is arrested or impeded by the contraction.

The nerve pathway for the pain of uterine contractions is the hypogastric plexus and then the pre-aortic plexus, entering the cord as high as the 11th and 12th dorsal segments via the posterior nerve roots.

Pain is also caused by dilatation of the cervix, and this is severe at the end of the first stage. Sensory impulses from the cervix probably enter the cord via the sacral roots; pain towards the end of the first stage is often referred to the sacral region.

The upper vagina distends easily and this does not seem to cause much pain, but during the second stage, when the head is stretching the vulval orifice, severe pain is felt which is different from that of uterine contraction. Pain impulses from the vulva and perineum are carried by the pudendal nerves, and to a small extent by the ilioinguinal, genitofemoral and posterior femoral cutaneous nerves.

During the antenatal period women should be told about the stages of labour in simple terms, avoiding technical words and the jargon of the labour ward. It is particularly important to explain that the first stage of labour is long compared to the second, and that during the first stage the cervix is being opened up, and that there will be no sensation of progress or descent of the baby. In a normal second stage she will be encouraged by feeling the descent of the baby and the realization that the end of labour is not far away. The severity of pain may not be altered by explanation, but fear of the unknown and fear that the labour is not progressing normally can greatly add to mental distress.

There are those who believe that if the woman is tense and anxious this will impede the progress of labour, and that if she is relaxed labour will be both quicker and less painful. While a tense pelvic floor might delay delivery, there is nothing to show that dilatation of the cervix will be affected by the woman's state of mind. However, it is probable that the woman who relaxes completely and rests between pains is conserving her energy, and will be able to make stronger voluntary efforts in the second stage. Moreover, labour will be a less unpleasant experience for her.

Various courses of antenatal exercises have been recommended. If they give the woman confidence they are useful and helpful, although there is little or no scientific evidence that they have any other effect on the course of labour. It is wrong to tell women that if only they follow some pattern of exercises or relaxation that they will have little pain and that labour will be quick and normal; when this sometimes proves to be untrue disappointment may turn into anxiety and recrimination.

Mothers-to-be should be told about the various means of relieving pain, and that these will be available on demand. Secure in the knowledge that help is available many women will not ask for any analgesic until the first stage is well advanced. Drugs should never be given in a routine fashion, but always with respect for the woman's wishes and need. During antenatal instruction classes the women should visit the labour ward and if possible meet the staff who will look after them there, and be given a demonstration of such equipment as the gas and oxygen machine.

During labour women should never be allowed to feel that they have been left alone and deserted. A sensible and affectionate partner who shares the labour is most helpful for he can give much support and comfort, provided that he has also been given a little instruction and told what to expect. The knowledge that a trusted doctor or midwife is present or immediately available will make labour more tolerable.

THE IDEAL ANALGESIC

An ideal analgesic for labour will not harm or endanger the mother or the fetus. In particular it should not:

interfere with uterine action
lead to more operative intervention
depress the respiratory centre of the newborn infant.

In addition to its analgesic effect it should be:

easy to administer

foolproof
predictable and constant in its effects.

Three methods are in common use during labour:

- drugs which are given by intramuscular injection. This method is simple and convenient. Oral administration of drugs during labour is unsatisfactory because absorption is unreliable
- an agent which is inhaled. This method is not suitable for use over long periods, and is therefore only used in the latter part of the first stage; it is more appropriate for the second stage
- an epidural, one caudal block. These give complete relief of pain, but requires skill in injection, and has a few hazards.

ANALGESIC DRUGS

Over the years many drugs have been used, but *pethidine* is now in almost universal use. It is a synthetic drug which is less effective than morphine in relieving pain, but has less of a depressant effect on the respiratory centre of the newborn infant. If it has been given 2 hours of delivery, preparation should be ready to reverse any neonatal respiratory depression with naloxone. The usual dose is 100–150 mg intramuscularly and 50–100 mg can be repeated after 2 hours. In an average labour the total dose should not exceed 400 mg. Pethidine sometimes has an emetic effect; this can be counteracted with metoclopramide (Maxolon) 10 mg intramuscularly.

Morphine 15 mg, sometimes combined with hyoscine 0.4 mg, is seldom used today for normal labour because of its effect on the infant's respiratory centre, but in cases in which the fetus is dead or grossly abnormal (e.g. anencephalic) such drugs will give good pain relief. Another side effect of morphine is the delay it can cause in gastric emptying. This could increase the risk of aspiration if a general anaesthetic were to be needed.

Respiratory depression in the newborn from pethidine or morphine can be counteracted by injecting naloxone (Narcan neonatal) into the umbilical vein. The preparation contains 20 μg/ml, and the dose is 5 μg/kg.

TRANSCUTANEOUS ELECTRICAL NERVE STIMULATION (TENS)

This is a self-administered method for relieving

pain based on the gate control theory of pain. It supposes that excitation of large myelinated afferent nerve fibres reduces the pain impulses conducted by small myelinated and non-myelinated fibres. Part of the analgesic effect may be due to the production of endogenous opioids. A battery-powered stimulator is connected by wires to electrodes placed on either side of the spine in the dermatome corresponding to the pain. It is a harmless method which gives a measure of relief from pain in the first stage of labour in about half of those who try it although in retrospect it is not a highly noted method.

INHALATION ANALGESIA

Nitrous oxide and oxygen

Nitrous oxide and oxygen can be given as the woman requires using Entonox gas cylinders containing a mixture of nitrous oxide and oxygen in equal proportions. The cylinders must not be kept at much below room temperature, because at lower temperatures the gases separate out. Self-administration depends on the woman fitting the mask accurately to her nose and mouth, and closing the valve with her finger. She should take a deep breath as soon as the pain starts and continue breathing throughout the contraction. This gives good and safe pain relief to most.

EPIDURAL BLOCK

This is usually started in the first stage, and it may be continued throughout labour. Now many regional blocks used in labour are combined spinal/epidural blocks.

The anaesthetic is injected into the epidural space through a Tuohy needle which is usually inserted between the 1st and 2nd lumbar spines. Lignocaine (Xylocaine) or Marcain are used depending on the need. A polythene catheter is threaded through the needle and left in the epidural space so that further injections of the anaesthetic can be given as required. The injection calls for some skill and experience of the method, and whoever gives it must maintain continuous supervision and be prepared to deal with immediate complications such as hypotension or temporary respiratory paralysis from accidental intrathecal injection. Long-term neurological sequelae such as

weakness or paraesthesia of the legs or bladder disturbances are very rare.

Existing epidural analgesia blocks can be extended for caesarean section during labour. Problems may arise with dislodged catheters and women at risk of operative delivery require especially close monitoring of the block by the anaesthetist to assess whether the anaesthetic is bilateral in spread and can reliably be extended with top up. Speed of providing anaesthesia for caesarean section can be gained in a good working block by giving a large top up such as 20 ml 2 per cent lignocaine with 1 in 200 000 adrenaline. This can be prepared in advance and the woman ready for caesarean section in less than 10 minutes.

Features of any regional block that are necessary to give proper anaesthesia for caesarean section are a bilateral sympathetic block with dry warm feet, a dense motor block with ability to only move the ankle or toes and a dense sensory block from T4 (nipple line) to S5 (perineum). Sensory block is tested with pinprick, ethyl chloride spray or ice. By whatever means it is tested the desired level of density must be such that the woman feels little or nothing from the nipple line to the perineum as this is necessary to block peritoneal pain.

Neural blockade that is insufficiently extensive or insufficiently dense will result in pain at caesarean section, especially with pulling on the peritoneum. The anaesthetist must be able to assess the extent of the anaesthesia before the start of the surgery if pain at caesarean section is to be avoided. Epidural fentanyl is widely used to decrease the risk of pain and give some central effect. The woman at awake caesarean section requires the support of her partner, her midwife and her anaesthetist. As anaesthetic experience with regional block increases so the occurrence of pain at operation decreases.

SPINAL ANAESTHESIA

The subarachnoid injection of a small dose of local anaesthetic provides a very rapid extensive and dense regional block. This injection can be done as a single dose which has the disadvantage of not being able to prolong the anaesthetic if necessary. A subarachnoid or epidural catheter may be inserted to provide maintenance of anaesthesia. The combined spinal epidural (CSE) uses the Tuohy needle as an introducer for a fine spinal needle and the epidural catheter. This is a commonly used anaesthetic for elective caesarean section and emergency where the woman has no functioning block. The problems of spinal block are the rapidity of onset causing rapid and severe hypotension. This is to be avoided as CSE reduces not only maternal cerebral perfusion but also placental perfusion causing neonatal acidosis at delivery. Hypotension is prophylactically treated with infusion of a litre of intravenous crystalloid such as Hartmanns solution or saline and the use of ephedrine either as infusion or in small bolus doses at the time of the vasodilatation.

Spinal anaesthesia is considered as the method of choice for caesarean section in many units. With the advent of new, less traumatic fine spinal needles the incidence of dural tap headache is very low.

CAUDAL BLOCK

This is an alternative method of introducing an extradural block. A malleable needle is inserted through the sacral hiatus into the sacral canal and a catheter may be left in the extradural space to continue the injection. The lumbar route is now much preferred in developed countries.

PUDENDAL BLOCK

This simple method can be used for analgesia at an operative delivery, including repair of an episiotomy, for forceps delivery or vacuum extraction, for breech and twin delivery. The pudendal nerve is derived from the 2nd, 3rd and 4th sacral nerves. These roots unite above the level of the ischial spine. The nerve passes out of the greater sciatic foramen posteriorly to the ischial spine and re-enters the pelvis through the lesser sciatic foramen. It then enters the pudendal canal where the vessels lie lateral to the nerve. The nerve divides into:

- the inferior haemorrhoidal nerve, giving branches to the anal sphincter and the skin around the anus
- the perineal branch, supplying the perineal muscles and the skin of the perineum and labia majora
- the dorsal nerve of the clitoris, supplying the clitoris and labia minora.

Fibres from the posterior femoral cutaneous nerve and from the ilioinguinal nerve also reach the perineum.

Lignocaine hydrochloride 0.5 per cent is used throughout (without adrenaline). The total amount injected should not exceed 50 ml. A skin weal is raised half-way between the anus and the ischial tuberosity. The index finger of the left hand is inserted in the vagina and the left ischial spine is located. A 20 cm 20 gauge needle is passed through the weal and directed towards the ischial spine with the guidance of the vaginal finger (*see* Fig. 4.17). The needle is directed just posteriorly to the inferior tip of the spine. The plunger of the syringe is withdrawn to make sure that the needle is not in a vein and about 5 ml of lignocaine solution is injected. The needle is inserted a further centimetre and, after testing for intravasation by withdrawing the plunger again, another 5 ml of lignocaine is injected. The process is repeated on the right side.

Some prefer to insert the needle through the vaginal wall rather than the perineal skin.

Pudendal block is usually combined with local infiltration of the vulva, a weal being raised at the fourchette and lignocaine injected here and on each side, extending well forward in both labia majora. If the local infiltration is carried out before the pudendal block it will reduce the discomfort of the manipulations required for the block.

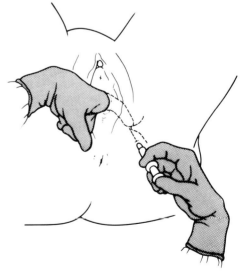

Fig. 4.17 Pudental block. The index finger of the left hand helps to direct the needle to the correct position

Perineal infiltration

If an episiotomy is required in advanced labour with the presenting part well down, direct infiltration of the line of incision with 10 ml of lignocaine 1 per cent is employed.

ANAESTHESIA FOR CAESAREAN SECTION

Epidural and spinal anaesthesia is the method of choice for both elective and emergency caesarean sections. Rates of 80–90 per cent of all caesarean sections are now regularly found in many hospitals both in the UK and USA; the major reason for the decline in usage of general anaesthesia in obstetrics is the increased maternal risk. From 1967 to 1987 there were 195 reported anaesthetic maternal deaths as a result of general anaesthesia mostly given in labour and 10 as a result of epidural block. Features of the woman that increase the hazards of general anaesthesia are:

- failure of endotracheal intubation with reported rates of 1 in 300 cases
- increased risk of aspiration of gastric contents due to poor lower oesophageal tone
- delayed gastric emptying the result of narcotic analgesia during labour.

Further undesirable effects of general anaesthesia are the reduction in placental perfusion, which occurs with the decreased maternal cardiac output at injection of the anaesthetic induction agent, and changes in maternal blood gases that occur with positive pressure ventilation. A reduced placental perfusion and thus possibly reduced oxygenation of the fetus rather than the neonatal effects of anaesthetic drugs may contribute to long-term problems in spite of adequate neonatal resuscitation especially where the fetus is already hypoxic.

Advantages of regional anaesthesia are not only an apparent increase in maternal safety but also:

- an improved fetal outcome if maternal hypotension is avoided
- improved maternal psyche as the experience of childbirth is denied her if unconscious, with a possible benefit of improved maternal infant bonding
- improved maternal cardiovascular stability in severe pre-eclampsia
- reduced maternal postoperative morbidity and depression

- greatly improved postoperative analgesia
- volatile anaesthetic agents are avoided
- less postpartum haemorrhage.

There is possibly a reduced risk of pulmonary embolism in the caesarean section woman as has been shown to occur in women after orthopaedic surgery with epidural blocks.

It is essential that the anaesthetist be experienced with regional anaesthesia having the knowledge and ability to make a full assessment of the anaesthetic blockade before surgery commences. Pain at caesarean section must be avoided and any discomfort treated immediately; a dense neural blockade of dermatomes of T4–S5 is necessary if peritoneal ennervation is to be adequately blocked. Hypotension can occur from either vasodilation in the lower part of the body or aortocaval compression or both; this must be avoided if possible or treated straight away with a vasoconstrictor such as ephedrine and increased lateral tilting. Placental perfusion is compromised by a reduction in maternal cardiac output and hypotension. Complications such as a total spinal block or cardiotoxicity associated with an inadvertent intravenous injection of bupivacaine require full resuscitative skills and the facilities of a fully equipped theatre. A full range of minimally invasive monitoring, blood-pressure, ECG and pulse oximetry is required as for any other anaesthetic.

Early insertion of epidural catheter during labour and proper maintenance of the block in any labour where operative delivery is more than just a possibility will allow between 70 and 80 per cent of all emergency caesarean sections to be performed under regional block. The need for hurried unplanned and unexpected general anaesthesia in labour occurs very rarely. High rates of general anaesthesia for caesarean section during labour indicate poor obstetric and anaesthetic practice. Where there is a risk that general anaesthesia might be required the anaesthetist should be informed so that the woman can be assessed preoperatively. Few if any mothers should be subjected to hurried general anaesthesia without preoperative assessment of their upper airways to avoid possible difficulty with endotracheal intubation.

GENERAL ANAESTHESIA IN OBSTETRIC PRACTICE

Although the advantages of conduction anaesthesia for the mother and infant are numerous there remains a specific role for general anaesthesia. In elective caesarean section the mother's wish to be unconscious for the delivery should be granted unless there is a specific contraindication to this. Elective caesarean section under general anaesthesia is not associated with such a high maternal mortality rate as occurs in emergency caesarean section but still has a coincidental increased postpartum haemorrhage rate. Elective operation on placenta praevia is therefore likely to bleed less under regional than general anaesthesia. Women with a known or suspected placenta accreta or percreta, either where ultrasound evidence of the penetration of the placental villi can be demonstrated or where the woman has had one or more caesarean sections have a relative indication for general anaesthesia so that invasive cardiac monitoring and preparation for replacement of massive blood loss can be made.

Emergency caesarean sections may require general anaesthesia when severe antepartum haemorrhage has occurred, where disseminated intravascular coagulopathy is known or suspected as a result of placental abruption. Unexpected severe fetal bradycardia occurs in less than 1 per cent of a reported obstetric population and 10 per cent of women having an emergency caesarean section. While awaiting general anaesthetic, it is advisable to attempt to reduce fetal hypoxia by ensuring that the mother is in a fully lateral position, increasing her inspired oxygen content, decreasing or stopping contractions by switching off Syntocinon infusion or giving slow diluted increments of salbutamol, especially where hypertonic uterine contractions are present. Eclampsia and severe pre-eclampsia with thrombocytopaenia and coagulopathy, pulmonary oedema or severe liver involvement is likely to require general anaesthesia for caesarean section. It is necessary to avoid or diminish the hypertensive response to intubation that occurs in these women by treating them preoperatively with antihypertensives and giving a preinduction dose of an antihypertensive and an opiate such as fentanyl. Heart conditions such as pulmonary hypertension, right or left shunt, dissecting aortic aneurysm, severe aortic stenosis or coarctation may require intensive monitoring and general anaesthesia; such women are not suitable for rapid sequence obstetric anaesthesia and should have a slow cardiac type induction preferably in a unit where their cardiac condition can be fully

monitored and they can receive immediate surgical attention if necessary. Those with cardiac disease considered suitable for conduction anaesthesia should have an epidural block that is slowly extended so that haemodynamic stability is maintained.

ANAESTHESIA ASSOCIATED MATERNAL MORTALITY

Anaesthesia is an avoidable cause of maternal mortality. It has been among the top three causes after pulmonary embolism and pre-eclampsia from 1972 to 1984. In the 1985–87 report on maternal mortality, it is the sixth commonest cause. The reason for this dramatic decline in the numbers of anaesthetic deaths is not known but may be the result of having a considerable increase in the number of consultant anaesthetic sessions for the labour ward or more experienced trainee anaesthetists on duty for the labour ward. It may also reflect more awareness of the importance of preoperative assessment of the ease of intubation or it may result from an increase in the number of women having caesarean section under regional block.

Women likely to be at risk of difficult intubation are those with small or receding chins, the obese, women who are unable to fully flex and extend their necks or move their lower teeth in front of the upper teeth – indicating a problem with their tempero-mandibular joint. Others are women who cannot open their mouths to the width of two fingers and those in whom, with their mouths fully opened, the faucial pillars and uvula are not clearly seen. No woman should be anaesthetized until an assessment of her upper airway has been performed. Failure to intubate accounts for most of the anaesthetic maternal deaths.

Avoiding aspiration of gastric contents is achieved initially by a decrease in gastric contents by allowing only clear fluids to drink in labour and reducing the number of women who receive intramuscular opiates as gastric emptying is delayed by these drugs. Avoiding general anaesthesia further reduces the number of women at risk of aspiration of gastric contents. When general anaesthesia is used, preoxygenation and rapid sequence induction minimize the time till intubation and dispense with the need to inflate the lungs and possibly inflate the stomach. Cricoid pressure applied before or as soon after unconsciousness as possible by a trained assistant reduces the passive regurgitation of gastric fluid. Cricoid pressure is only released when the trachea has certainly been intubated and the cuff inflated. This can only reliably be shown to have occurred with the use of an end tidal CO_2 monitor. Ranitidine reduces the acidity and volume of gastric contents. It reduces the risk of aspiration of gastric contents with a pH of less than 2.5. The use of 30 ml 0.3 per cent under sodium citrate before induction of anaesthesia can neutralize the gastric contents already in the stomach.

THE USE OF OXYTOCIC DRUGS

OXYTOCIN (SYNTOCINON)

In 1906, Dale found that extracts of posterior pituitary gland had an oxytocic action. These extracts were first used in obstetrics by Blair Bell, founder of the Royal College of Obstetricians and Gynaecologists 3 years later. They contained:

antidiuretic hormone (vasopressin)
oxytocin.

Nowadays the unwanted antidiuretic effect of vasopressin has been eliminated by using a synthetic preparation of oxytocin – Syntocinon.

Oxytocin is an octapeptide. It causes contraction of the myometrium and also of the myoepithelial cells of the breast. The response of the myometrium to oxytocin (Syntocinon) is relatively slight until late pregnancy when, in response to physiological doses, strong but rhythmical contractions occur (unlike the prolonged spasm produced by ergometrine). However, abnormally large doses of Syntocinon will cause sustained contraction, which can greatly reduce the placental blood flow and cause fetal hypoxia or even death.

An increase in neonatal hyperbilirubinaemia has been reported after the use of prolonged high dose Syntocinon to induce labour. The reason for this is uncertain, but it is possible that oxytocin causes osmotic swelling of erythrocytes, reducing their plasticity so that they are more easily haemolysed.

Oxytocin is destroyed in the gastrointestinal tract and Syntocinon is therefore administered by intravenous infusion. The dose is measured in milli-units (mU) per minute based on a standard preparation.

CLINICAL USES OF OXYTOCIN

To induce labour

Syntocinon and low amniotomy can be used but the majority of inductions are now performed with prostaglandins.

To augment slow labour

A Syntocinon infusion may be used to accelerate labour if there is delay from inadequate uterine action but care must be taken to exclude mechanical obstruction as a cause of the delay.

In the third stage of labour

Syntocinon may be used for the preventative treatment of postpartum haemorrhage, given intramuscularly as an injection of 5 or 10 units, or by intravenous infusion at a level of 100 mU/min.

During therapeutic abortion

Syntocinon is sometimes used to enhance uterine contractions during therapeutic abortion induced with prostaglandins.

Method of administration for induction or augmentation of labour

Any woman receiving intravenous Syntocinon must be under continuous supervision, ideally by means of a cardiotograph, which records simultaneously the fetal heart rate and uterine contractions. Infusion pumps regulate the flow of solution better than gravity drips. By adding 10 units of Syntocinon to 500 ml of isotonic saline, one drop of the infusion will contain approximately 1 mU of Syntocinon (assuming 20 drops/ml of infusion).

The starting dose is 2 mU/min, increasing at intervals of 30 minutes according to the strength and frequency of the uterine contractions to a maximum of 32 mU/min. This may be achieved by manual control, or if an intrauterine pressure catheter has been inserted, by a monitor which has an automated feedback mechanism. Once contractions are occurring regularly the rate of infusion can often be decreased, but the infusion should be kept running until the third stage of labour is complete. If at any time there is evidence of fetal distress or hypertonic uterine contractions the infusion is stopped immediately.

SYNOMETRINE

Ergometrine has an almost specific action on the myometrium, but it could also cause a more general vasoconstrictor action, which may cause a rapid rise in blood pressure in women who are already hypertensive. It may also induce nausea and vomiting. Ergometrine maleate may be injected intravenously, intramuscularly or given by mouth. After intravenous injection of 500 μg a strong uterine contraction occurs within 40 seconds, and persists for 30 minutes. After intramuscular injection the time before the uterus contracts is about 6 minutes, and even if hyalase is added to the injection the time will be 4 minutes. Therefore if ergometrine is to be used for postpartum haemorrhage it should be given intravenously, or by direct injection through the anterior abdominal wall into the uterus if peripheral vasoconstriction has rendered the intravenous route difficult. As an alternative for preventing postpartum haemorrhage a mixture of Syntocinon 5 units and ergometrine 500 μg (Syntometrine) may be given intramuscularly. The Syntocinon will act in about 2 minutes, its action will be followed and maintained by that of the ergometrine component.

If the uterus remains flaccid after one dose of 500 μg of ergometrine a second, similar dose may be given. No more than two doses should be given, as occasional cases of severe peripheral vasoconstriction have been reported. If oxytocic action is still inadequate a Syntocinon infusion can be used.

The risk of causing a rise of blood pressure, particularly in a woman who is already hypertensive, has been mentioned. Because of this Syntocinon alone is now often preferred for active management of the third stage of labour, in these cases ergometrine is only used if haemorrhage occurs. It has also been suggested that ergometrine should be withheld in cases of cardiac disease, but there is little evidence of any risk.

Ergometrine should never be given to expedite the delivery of a living child, as the uterine spasm which it produces will stop the placental blood flow, and there is also a risk of uterine rupture. Ergometrine is used in the treatment of abortion and at caesarean section. It is also sometimes used in puerperium if the loss is unduly heavy, but it is futile to give it with the hope of increasing the rate of uterine involution; it causes myometrial contraction, not involution.

FURTHER READING

Chamberlain G. and Stear, P. (1996) ABC of Labour Care. *British Medical Journal*, London.

5

ABNORMAL LABOUR

PROLONGED LABOUR

When obstetricians speak of prolonged labour they invariably mean prolongation of its first stage, a condition which occurs most commonly in primiparae. Prolongation of the second stage of labour is referred to as delay and is dealt with in Chapter 4, which details those procedures that can hasten delivery of the infant through a fully dilated cervix. The physiology of the first stage of labour and the proper management of normal labour have been described in Chapter 4.

Progress in labour is judged by:

- dilatation of the cervix, measured from 0–10 cm
- descent of the fetal head, measured in fifths of the fetal head palpable per abdomen or in centimetres above or below the ischial spines on vaginal examination.

Additional assessments made during the first stage of labour are:

- the condition of the woman, i.e. her pulse rate, temperature, blood pressure, urine output, urinary protein and ketones and psychological state
- the condition of the fetus as judged by such methods as auscultation of the fetal heart rate at regular intervals and looking for meconium in the amniotic fluid; or by cardiotocography, measurement of fetal scalp blood pH or the fetal electrocardiagram

- the size of caput and the extent of moulding, the presence of either or both indicating a tight fit between the fetal head and the maternal pelvis.

THE PARTOGRAM

In all but the most rapid labours progress is usually charted on a partogram (*see* Chapter 4). That portion of the partogram which plots cervical dilation (cm) against time (h) is called the cervimetric graph (or cervicogram) and is the focus of attention in the recognition and classification of prolonged labour. The cervicogram is usually started on admission when labour becomes established (sometimes a difficult diagnosis) or at the time when an amniotomy is done as part of the surgical induction of labour.

The first stage of labour is divided into the latent and active phases. The latent phase is the time during which the cervix is effacing while dilatation is minimal (rarely exceeding 3 cm) and can take 6 hours. The active phase is from 3 to 10 cm dilatation and, at just under 1 cm dilatation per hour, also takes about 6 hours, making a total of about 12 hours for the acceptable normal duration of the first stage of labour in a primipara; multiparae do not take this long as they efface and dilate their cervix at the same time.

Types of prolonged labour

When the first stage of labour exceeds the time limits given above the patient's cervimetric graph moves to lie to the right of the accepted norm (*see* Fig. 5.1) and labour is prolonged. Three types of prolonged labour are recognized:

1. prolonged latent phase, lasting for more than 6 hours
2. primary dysfunctional labour, when uterine activity is either inert or incoordinate from the start of the active phase
3. secondary arrest, when the rate of cervical dilatation, which is normal at first, slows down or stops.

The characteristic cervimetric graphs for each of these three types of prolonged labour are shown in Fig. 5.1 and data about the outcome of normal labour and the three types of prolonged labour are given in Table 5.1.

Prolongation of the first stage of labour occurs in about one in three primiparae and in about one in eight multiparae. In both types of woman prolonged labour is associated with a much higher (10- to 16-fold) caesarean section rate and a roughly four-fold increase of low 5 minute Apgar scores (6 or less) over those that are found in women whose pattern of labour is normal. It is small wonder that prolonged labour is regarded as a high risk condition requiring the best hospital facilities and specialist skills.

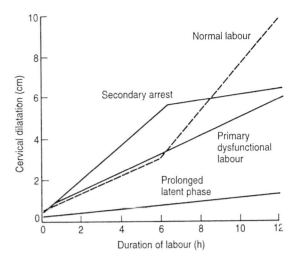

Fig. 5.1 Cervicographs of normal labour (dotted line) and the three types of prolonged labour. See Table 5.1 for information about fetal and maternal outcome in each group

The causes of prolonged labour

Cephalopelvic disproportion

This diagnosis is made whenever the presenting fetal head seems to be too big for the birth canal that it has to pass through; it is by far the most important cause of prolonged labour.

The causes of cephalopelvic disproportion are:

Table 5.1 Incidence and outcome of various types of labour in primiparae and multiparae who go into labour spontaneously

	Cervimetric pattern	Incidence (%)	Caesarean section rate in each group (%)	Apgar scores of 6 or less at 5 min (%)
Primipara (n = 684)	Normal	63.9	1.6	2.6
	Prolonged latent phase	3.5	16.7	8.3
	Primary dysfunctional labour	26.3	20.0	5.5
	Secondary arrest	6.3	27.9	2.5
Multipara (n = 847)	Normal	88.4	0.5	2.4
	Prolonged latent phase	1.4	8.3	8.3
	Primary dysfunctional labour	8.1	8.8	4.3
	Secondary arrest	2.1	5.9	2.2

Fetal
malpositions (Chapter 4)
malformation (e.g. hydrocephaly)
macrosomia or a large baby
Maternal
contracted pelvis
A combination of maternal and fetal
Abnormalities
Abnormalities which can cause prolonged labour in the absence of cephalopelvic disproportion are:

Fetal malpresentation brow, shoulder, face or breech
Maternal abnormalities
pelvic tumour
stenosis or scarring of the cervix
septae or stenosis of the vagina
primary uterine dysfunction (i.e. uterine dysfunction or inertia which is not secondary to disproportion).

The investigation of prolonged labour

The first thing that must be considered before making a diagnosis of prolonged labour is the possibility that there has been uncertainty about timing the onset of labour. Next it is important to assess and examine the woman in order to identify the cause and to determine the condition of mother and baby. To do this it is necessary to take a history, assess the contractions, review cardiotocographic recordings and also to make general, abdominal and vaginal examinations.

The treatment of prolonged labour

There are two alternative courses of action:

- to allow the labour to continue
- to undertake an operative delivery.

A prolonged labour should never be allowed to degenerate into an obstructed labour which has a disastrous outcome.

Allowing the labour to continue

In the absence of fetal distress, maternal distress or severe cephalopelvic disproportion the labour may be allowed to continue in the hope of achieving a vaginal delivery. If the labour is allowed to continue it is important to ensure:

- *adequate analgesia* (often the only way of achieving this is by an epidural block)
- *good maternal fluid balance* with treatment of dehydration by adequate volumes of normal saline or dextrose/saline given intravenously
- *rupture of membranes* if they are still intact
- *good uterine activity*, which often means augmentation of labour with a Syntocinon infusion.

Syntocinon infusions for augmenting labour are usually started at a dose of 2 mU/min and the dose is gradually increased until optimum uterine activity is obtained (but not beyond 32 mU/min). In these circumstances it is important to assess uterine contractions or activity with care and accuracy. The best method available is an intrauterine transducer, particularly if it can be linked to a monitor which automatically measures and prints out uterine activity above the resting baseline in kiloPascalseconds (kPa), a normal value being about 1500 kPas/15 min.

In the absence of such sophisticated equipment fluid-filled intrauterine catheter systems or external sensors which give qualitative recordings of uterine activity are available. In the units where there is no monitoring equipment, the timing and assessment of the strength of uterine contractions by hand becomes one of the functions of the attending midwife, who must keep a careful record, preferably on a partogram.

If prolonged labour is being managed conservatively it is vital to establish good communications with the woman and her partner, relatives or friends, who are invariably anxious and impatient. Detailed and careful explanations of the rationale and the nature of all aspects of management should be offered. Doctors and midwives supervising such women should recognize this responsibility and when they go off duty they should take care to introduce the woman to the new team to whom the responsibility is being transferred.

Operative delivery

In the presence of fetal distress, maternal distress, evidence of arrest of cervical dilatation despite good contractions or frank cephalopelvic disproportion the baby will have to be delivered; the usual method is by caesarean section. In rare circumstances it might be possible to achieve a Ventouse extraction before full dilatation of the cervix but this should only be attempted if there is no caput or moulding, the head is well down in a roomy pelvis, there is no fetal distress and the sole reason for prolongation of labour seems to be ineffective uterine activity.

In Third World countries, where there is poor access to hospitals and traditional attitudes are against it, there may be a reluctance to do caesarean sections because of possible rupture of the uterine scar in subsequent pregnancies and labour. Under these circumstances prolonged labour due to frank cephalopelvic disproportion may, in rare and care-fully selected instances, be treated by symphy-siotomy. However, this procedure should only be undertaken by the experienced and should only be attempted when there are good uterine contrac-tions, when the leading part of the moulded head is deep in the pelvis and when the cervix is more than 7 cm dilated. The immediate complication of sym-physiotomy is damage to the urethra and bladder; long-term problems are instability of the symphy-seal joint with a painful waddling gait and stress incontinence of urine.

FETAL MALPOSITION AND MALPRESENTATION

A malposition is one where the fetal head is presenting but not as a well flexed vertex with the occiput in the anterior quadrant. A malposition includes positions where the head is not the pre-senting part.

OCCIPITOPOSTERIOR POSITIONS

The mechanism of labour in occipitoposterior positions of the vertex is described in Chapter 4. In most cases in which the occiput lies posteriorly at the onset of labour normal vaginal delivery occurs. With good flexion of the head, the occiput (being the first part of the head to meet the pelvic floor) usually undergoes long rotation forward through three-eighths of a circle and is thus directed to the space below the pubic arch. In about 20 per cent of women this does not occur and the malposition persists. A short backward rotation of one-eighth of a circle may then take place which results in the occiput being directed to lie in the hollow of the sacrum. This may occur because the head is poorly flexed, so that the first part of the fetal head to meet the pelvic floor is the sinciput, and this part of the head then rotates forward. In women who have epidural anaesthesia, the musculature of the pelvic floor is lax. This lack of resistance may increase the incidence of failure to rotate when a vertex in the occipitoposterior position meets the pelvic floor.

The direction of rotation of the fetal skull may also be influenced by the shape of the cavity and outlet of the pelvis. In the android type of pelvis there is narrowing of the pelvic cavity from side to side, and the size of the cavity tends to diminish in the lower straits. The pubic arch is narrower and the ischial spines project more into the birth canal. There therefore tends to be less room in the anterior part of the pelvis and a posterior position of the occiput may persist even with good flexion of the head.

With an occipitoposterior position the head is pushed downwards and forwards against the back of the symphysis pubis rather than directly down-wards onto the cervix. Thus some of the effective-ness of the uterine contractions is lost, cervical dilatation tends to be slow and labour is pro-longed. The cervix may be compressed between the head and the pubis, so that progressive oedema of the anterior lip of the cervix occurs.

Diagnosis

Diagnosis during pregnancy is of no importance except that the occipitoposterior position must be recognized as a cause of non-engagement of the head before the onset of labour. During labour a lack of flexion of the head may be suspected if there is early rupture of the membranes with a poorly engaged head.

Abdominal examination

During the intervals between uterine contractions slight flattening of the lower abdomen may be observed, and the limbs are easily felt. It may be difficult to define the back or to hear the fetal heart. The head usually descends through the pelvic brim as labour proceeds, but descent may be slow because with poor flexion a wider diameter presents.

Vaginal examination

Early in labour it may be difficult to reach the presenting part and the membranes may rupture early. When the head has entered the pelvic cavity the most striking feature is the ease with which the anterior fontanelle can be felt behind the pubis. The anterior fontanelle is more easily felt because the head is less well flexed, and also because it lies well forward when the head is in the occipitoposterior position. An attempt should be made to assess the degree of flexion of the head, as the well-flexed head is more likely to rotate. If only the anterior fontanelle can be felt the head is poorly flexed; it is less poorly flexed if both the anterior and posterior fontanelles can be felt; it is well flexed if only the posterior fontanelle is felt (*see* Fig. 5.2).

Although the diagnosis should be made early in labour, it frequently happens that the position is unrecognized until there is delay in the second stage of labour. Diagnosis by vaginal examination

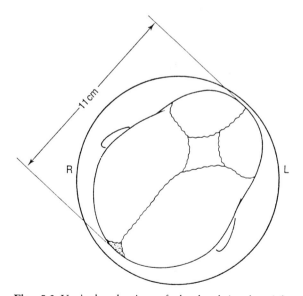

Fig. 5.2 Vaginal palpation of the head in the right occipitoposterior position. The circle represents the pelvic cavity with a diameter of 12 cm. The head is poorly flexed, so that the anterior fontanelle is easily felt

may then be difficult owing to the formation of a caput succedaneum over the presenting part. Before forceps delivery, when it is essential to have accurate knowledge of the direction of the occiput, the fingers may be passed higher to feel the sutures above the caput or the free margin of an ear, which will point to the occiput. An ultrasound in labour may help.

The course of labour in occipitoposterior positions

In about 70 per cent of cases spontaneous rotation of the occiput to the anterior position occurs, and in about another 10 per cent of cases the occiput undergoes short rotation so that delivery in the directly occipitoposterior position (face-to-pubes) can occur. In the remainder assisted rotation will be required.

During labour the uterine contractions may be ineffective because the poorly flexed head fails to press down upon the cervix and provide the reflex stimulation of the lower segment that is produced by the more pointed vertex with full flexion. A long first stage is likely to be followed by a long second stage because the woman is tired and the uterus may be less capable of further strong contractions.

Moulding of the head

When the fetal head descends through the birth canal and flexion of the head is poor the skull will be compressed in the occipitofrontal diameter (*see* Fig. 5.3). If the alteration in shape is extreme the structures of the skull may not adapt well to the relatively sudden change. Great compression along the occipitofrontal diameter causes tension in the posterior vertical part of the falx cerebri. This elevates the tentorium cerebelli, which may ultimately tear at its free margin. The upward dislocation of the tentorium may result in rupture of the great cerebral vein (vein of Galen) or a tributary of it, and fetal damage or death from intracranial haemorrhage.

Management of the first stage of labour

The first stage is managed as in a normal case. Nothing can be done to correct the malposition or to influence the rotation of the head at this stage.

The frequency, duration and strength of the

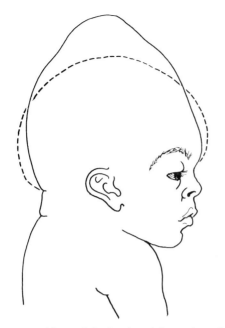

Fig. 5.3 Moulding of the head and formation of caput succedameum is persistent occipitoposterior position

uterine contractions, the dilatation of the cervix and the fetal heart rate are observed in the ordinary way and recorded on a partogram. A continuous epidural block will not only give pain relief but will allow time for spontaneous rotation of the occiput into the anterior position. If progressive cervical dilatation does not occur augmentation of the uterine action with a Syntocinon drip may be tried. If this does not result in better progress in a few hours caesarean section is performed. Caesarean section will also be required if fetal distress occurs.

Management of the second stage of labour

A mistaken diagnosis of full dilatation of the cervix is not uncommon in these cases, when the woman complains of rectal discomfort and a desire to bear down. Careful vaginal examination is essential to establish that the second stage has been reached. The degree of flexion of the head and its position are determined by palpation of the fontanelles. Continued deflexion, a large caput succedaneum or marked over-riding of the skull bones suggest that spontaneous rotation may not occur.

In most cases, provided that the uterine contractions are strong and the woman is able to make good expulsive efforts the occiput rotates forward

and normal delivery takes place. In other cases the baby may be delivered face-to-pubes without any difficulty, although there is a greater risk of a perineal tear.

The indications for interference in these cases are:

- failure of the presenting part to descend
- fetal distress
- maternal distress.

It is desirable for the head to be in the occipito-anterior position, because it presents a smaller and more favourable diameter and therefore the first step in assisting delivery is rotation of the fetal head. This can be performed:

- by manual rotation and forceps delivery
- with Kjelland's forceps
- with the vacuum extractor.

Manual rotation and forceps delivery

Unless an epidural anaesthetic has already been given, a pudendal block or general anaesthesia will be required. After careful diagnostic examination the head is rotated with the fingers in the appropriate direction. Thus, if the fetus is in the right occipitoposterior position the head is rotated so that the occiput travels round the right wall of the pelvis until it is directly anterior; the opposite direction of rotation is followed for a left occipitoposterior position. The shoulder girdle of the fetus should be rotated at the same time as the fetal head. If this is not done the head will tend to slip back after rotation into an oblique or transverse diameter of the pelvis. Rotation of the shoulder girdle may be achieved by pressure through the abdominal wall with an external hand (*see* Fig. 5.4).

It is undesirable to displace the head upwards and this should be avoided, as it will make subsequent application of the forceps more difficult. It is often possible to achieve rotation with the half hand. Rotation is effected by tangential pressure on the side of the head with the fingers only, without grasping the head with fingers and thumb. After rotation to the occipitoanterior position has been achieved the fingers are kept in place to hold the head in position until the obstetric forceps are applied to complete the delivery.

Difficulties arise if the forceps are applied to the head in any position other than with the sagittal suture in the anteroposterior diameter of the pelvis. If this position is not achieved the forceps will

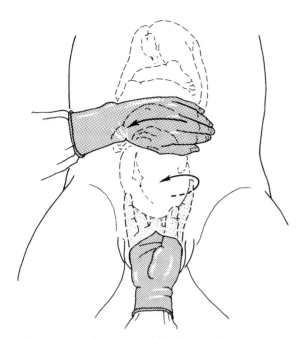

Fig. 5.4 Manual rotation of the fetus in the occipitoposterior position. The right hand passes between the pelvic wall and the fetal head and is about to rotate the head. The left hand placed on the abdomen will assist this rotation by pressure on the shoulder

not lock properly and the handles will not lie together. Squeezing the handles will only compress the head and may cause fetal injury, and when traction is applied the blades may slip off and also cause maternal injury.

Kjelland's forceps

This instrument is of relatively light construction and is designed so that it can be used for rotation of the fetal head, in addition to traction. The pelvic curve of the shank has been eliminated and the lock allows one shank to slide upwards or downwards on the other (*see* Fig. 5.4) Without a pelvic curve the circle described by the blades during rotation is a small one, thus avoiding damage to maternal tissues, bladder and rectum. In experienced hands the instrument is safe and most satisfactory, but it is potentially dangerous in the hands of anyone not trained in its use. The technique of application of Kjelland's forceps differs from that for ordinary forceps in that the instrument is applied with reference to the fetal head in whatever position it lies, and not in relation to the

pelvis. These forceps are used before traction is applied to rotate the head until the occiput lies anteriorly. If the instrument is incorrectly used damage to both maternal tissues and the fetal head may occur.

Vacuum extraction

If the extractor is applied as near to the occipital end of the vertex as possible, and traction is applied, forward rotation of the head often occurs.

Arrest at the pelvic outlet

If it is found that arrest has occurred when the head is so low in the pelvic cavity that with each contraction of the uterus the fetal scalp is easily visible, it is probable that further progress is being prevented by the muscles of the pelvic floor. To reach this level the head will already have undergone a considerable degree of moulding into the shape described for the persistent posterior position of the occiput. It may be found easier to perform an episiotomy and assist the delivery of the baby in the unrotated occipitoposterior position. The large occipitofrontal diameter, albeit shortened by moulding, will have to pass through the pelvic floor and therefore the episiotomy should be adequate. Traction with the obstetric forceps must be careful and only moderate force should be used. The instrument was not designed to fit the head in this position and may slip off. However, only moderate traction is necessary to complete the delivery of a fetal head which is arrested by the perineal muscles at this low level. The vacuum extractor can be used instead of the forceps.

Trial of forceps with an occipitoposterior position

There could be difficulty in rotating and delivering a head in the occipitoposterior position if the baby seems unduly large, if two-fifths of the head is still abdominally palpable, if there is marked caput or moulding, if the ischial spines are unduly prominent, if there is any suggestion of outlet contraction (a rare condition) or if the head is not visible at the height of a second stage contraction. Under these circumstances it is appropriate to conduct a *trial of forceps*. This means embarking on a forceps delivery in the operating theatre with scrubbed nursing staff and everything prepared for caesarean section and with the woman under general or effective epidural anaesthesia. Should there be any difficulty in applying the blades, or if it appears that too much force is going to be needed to extract the baby's head, it is then easy and perfectly proper to abandon the attempt at operative vaginal delivery and to resort to caesarean section without delay and without undue risk to mother or baby.

DEEP TRANSVERSE ARREST OF THE HEAD

This is an arrest in labour when the fetal head has descended to the level of the ischial spines and the sagittal suture lies in the transverse diameter of the pelvis. The occiput is on one side of the pelvis and the sinciput on the other with the head badly flexed. The condition is only diagnosed during the second stage of labour. If the head is firmly fixed in the transverse position obstructed labour will occur.

The occiput may have been obliquely posterior at the onset of labour and only partly rotated forward, or it may have descended from an initial transverse position.

In an android pelvis the anterior surface of the sacrum may be flattened and the ischial spines may be prominent because the side walls of the pelvis are convergent. The head fails to descend to the pelvic floor, where rotation of the head normally occurs.

Diagnosis

Because the progress of labour has ceased the diagnosis rests on vaginal examination made during the second stage of labour. The head will be found to be arrested at the level of the ischial spines, with the sagittal suture lying in the transverse diameter of the pelvis. Both fontanelles are usually palpable.

Management

The head must be rotated so that the occiput is brought to the front and then, and only then, is traction applied. Rotation may be achieved manually, with Kjelland's forceps or using the vacuum extractor. Exactly the same procedure is followed, and exactly the same precautions are taken as have already been described for rotation of the head from occipitoposterior position.

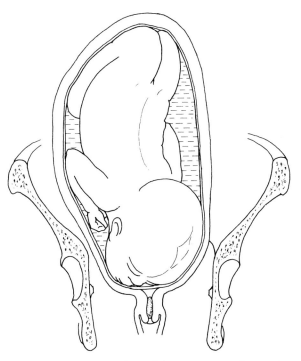

Fig. 5.5 Face presentation, right mentoposterior position. The face from the chin to the bregma presents the submentobregmatic diameter occupying the right oblique diameter of the pelvic inlet. The back faces forwards and to the left, but is extended instead of flexed as in vertex presentation, so that the breech is more prominent and is more easily palpated

FACE PRESENTATION

Face presentation, in which the head is fully extended, occurs about once in 300 labours (*see* Fig. 5.5).

Causes

The most common type of case is one in which a normal fetus actively holds its head extended; even after delivery the infant may keep its head in an attitude of exaggerated extension for some days. In spite of this the head can be flexed on to the chest, showing that there is no spinal or muscular abnormality.

In anencephaly the abnormal head of the fetus is set on the shoulders with the face directed downwards, and the face may present in labour. With congenital tumours of the neck the head may also be extended.

Rarely, with a contracted pelvis, the biparietal diameter of the head is too large to enter the brim of the pelvis, the occiput is held up and the sinciput descends, so that the head becomes extended.

Mechanisms

The chin is the denominator and four positions are conventionally described, analogous to the corresponding positions of the vertex from which they may be said to arise:

right mentoposterior
left mentoposterior
left mentoanterior
right mentoanterior.

The mentoanterior positions are relatively more frequent (*see* Fig. 5.6).

In a mentoanterior position the head engages and descends with increasing extension, so that the submentobregmatic diameter (9.5 cm) comes through the cervix. When the chin reaches the pelvic floor it undergoes internal rotation through one-eighth of a circle and the submental region comes to lie under the subpubic arch. The head is then born by a movement of flexion. Restitution occurs and is followed by external rotation as in vertex presentations.

In a mentoposterior position a similar mechanism occurs, except that the chin has to undergo internal rotation through three-eighths of a circle (*see* Fig. 5.7).

Backwards short rotation of the chin sometimes occurs, but a fetus in such a persistent mentoposterior position cannot be delivered unless it is very small. This is because the head is already fully extended and so further extension to deliver the head is impossible. The head and thorax become impacted in the pelvis and obstructed labour occurs unless assistance is given.

Moulding

In a face presentation the submentovertical diameter of the head is compressed, causing elongation of the occipitofrontal diameter (*see* Fig. 5.8). This shape of the head is called dolichocephaly.

Diagnosis

Abdominal examination

With a mentoposterior position the cephalic prominence is very easily felt; it appears to overlap the symphysis and is felt on the same side as the back, from which it is separated by a deep sulcus. It may

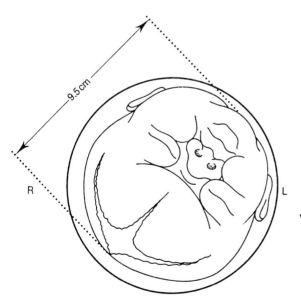

Fig. 5.6 Vaginal examination in the left mentoanterior position. The circle represents the pelvic cavity with a diameter of 12 cm

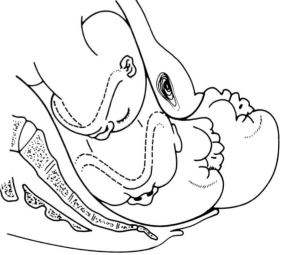

Fig. 5.7 The mechanism of labour with a face presentation. The head descends with increasing extension. The chin reaches the pelvic floor and undergoes forward rotation. The head is born by flexion

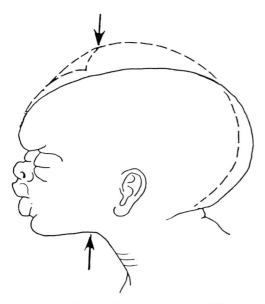

Fig. 5.8 Moulding in face presentation. The arrows indicate the direction of the pressure, which shortens the diameter between the submentovertical and submentobregmatic diameters. The splitting of the skin in the front of the neck and swelling of the face are also shown

be difficult to locate and hear the fetal heart sounds.

With a mentoanterior position the cephalic prominence is again felt on the same side as the back but, being posterior, it is difficult to feel and may be confused with the prominent chest. The fetal heart is easily heard over the chest, and small parts may be felt on the same side. An ultrasound in the labour ward is often helpful.

Vaginal examination

As the face fits less well than the vertex the membranes may rupture early. When the presenting part has engaged in the brim and can be felt, the supraorbital ridges, the bridge of the nose and the alveolar margins within the mouth are recognized. If the face is oedematous it can be mistaken for the breech.

Prognosis

Many face presentations are delivered naturally without difficulty. Face presentations are less favourable than vertex presentations because the

face is a less efficient dilator of the cervix, and because spontaneous rotation of the mentoposterior positions occurs late in the second stage of labour. The emerging diameter, the submentovertical (11 cm), is larger than that with a normal vertex presentation.

Management of labour and delivery

The woman is kept in bed during the first stage and a vaginal examination is made as soon as the membranes rupture to exclude prolapse of the cord. An epidural block or infiltration of the perineum with local anaesthetic and an episiotomy are advisable in primigravidae if there is any delay when the cervix becomes fully dilated and the face reaches the pelvic floor. With a mentoanterior position spontaneous delivery is to be expected. If there is delay in the second stage from inadequate expulsive forces an experienced operator will have no difficulty in applying the forceps.

With a mentoposterior position time should be allowed for spontaneous rotation, which will only occur late in the second stage. If spontaneous rotation does not occur manual rotation of the head to the mentoanterior position is attempted under epidural block or general anaesthesia, after rotation delivery is completed with the forceps. The expert may prefer to use Kjelland's forceps, with which the chin is rotated forwards while traction is applied. When undertaking rotation and delivery for a mentoposterior position the obstetrician should conduct the delivery as a trial of forceps. Vacuum extraction is obviously totally contraindicated with a face presentation.

Caesarean section becomes necessary when there is some complication, such as a contracted pelvis, prolapse of the cord, when the presenting part fails to descend or when there is difficulty with rotation.

The face is always somewhat swollen and discoloured after a face delivery, and the parents should be warned that it may be temporarily unsightly, but that complete recovery is to be expected soon.

BROW PRESENTATION

The causes of a primary brow presentation (*see* Fig. 5.9) include those of a face presentation, but often the reason for partial extension of the head is not evident.

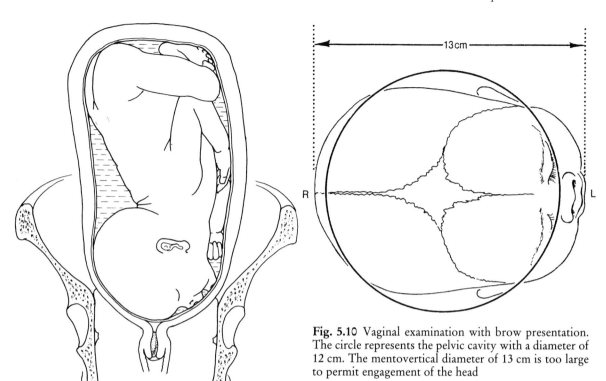

Fig. 5.10 Vaginal examination with brow presentation. The circle represents the pelvic cavity with a diameter of 12 cm. The mentovertical diameter of 13 cm is too large to permit engagement of the head

Fig. 5.9 Brow presentation. The head is above the brim and not engaged. The mentovertical diameter of the head is trying to engage in the transverse diameter at the brim

Extension of the head before labour may be termed primary, and extension during labour secondary.

Two types of secondary extension occur, one at the level of the pelvic brim and the other at a lower level in the pelvis. If the pelvic brim is contracted, descent of the wide biparietal diameter of the head may be impossible and the narrower anterior part of the head may descend a little into the pelvis, so that the head becomes partially extended.

Similarly, in some cases of occipitoposterior position, the wider occipital end of the head may be held up in the sacral bay of the pelvis, while the sinciput descends and extension of the head occurs in the cavity of the pelvis. This can only occur with a small fetus.

A persistent brow presentation is fortunately rare. If a head of normal size lies with its longest diameter of 13 cm across the brim of a normal pelvis it cannot engage (*see* Fig. 5.10). If this occurred in a fetus of average size obstructed labour would result. However, when the fetal head is quite small in proportion to the pelvis it may be driven down into the pelvic cavity and be born as brow presentation.

With a brow presentation the head becomes very much moulded, with compression of the mentovertical diameter and lengthening of the occipitofrontal diameter.

Diagnosis

In cases in which extension of the head occurs early in labour the diagnosis may be difficult. On abdominal examination the head is above the brim, with some overlap, and the cephalic prominence is on the same side as the back. This malpresentation should always be suspected when non-engagement of the head is noted, particularly after the membranes rupture in a woman who has had previous easy deliveries.

As a rule the membranes rupture early in labour, and there is some risk of prolapse of the cord. On vaginal examination, except in the case of extension of a head lying in the occipitoposterior position in the pelvic cavity, the presenting part will be high. The examining finger encounters the forehead, with the orbital ridges and bridge of the nose in front and the anterior fontanelle behind. An abdominal ultrasound may help.

Management

In a few cases a brow presentation is discovered by ultrasound or radiology in the antenatal period. If there is no evidence of disproportion or any other abnormality nothing should be done, as in most cases the head will flex when labour starts, and spontaneous delivery will occur.

If the head is discovered to be partially extended in early labour and there is no evidence of severe disproportion a short trial of labour is permitted, and this may result in further extension of the head to a face presentation and engagement in the pelvic brim. If the head fails to engage or if there is evidence of disproportion, a caesarean section is performed. With a brow presentation above the brim caesarean section is the usual outcome.

If the head has entered the pelvic cavity in an occipitoposterior position and has undergone further extension there will be a brow presentation with the chin directly anterior. The cervix becomes fully dilated but the head is arrested. If the head is now rotated anteriorly so that the occiput becomes anterior flexion of the head occurs and vaginal delivery with forceps is usually easy. Rotation may be performed manually or with Kjelland's forceps. Once again the trial of forceps philosophy is the correct one and if this is any difficulty a caesarean section should be done.

BREECH PRESENTATION

Breech presentation occurs in about 3 per cent of labours near term, but is more common than this before term. It is found in about 25 per cent of pregnancies at the 30th week, and so its incidence is higher in premature births. Spontaneous version occurs in more cases by the 36th week.

There are two main varieties of breech presentation. In the most common variety, especially in primigravidae, both hips are fully flexed but the knees are extended (*see* Fig. 5.11). This is called an *extended* or *frank breech*. If the legs are fully flexed at both hips and knees (*see* Fig. 5.12) the term *flexed breech* is applied.

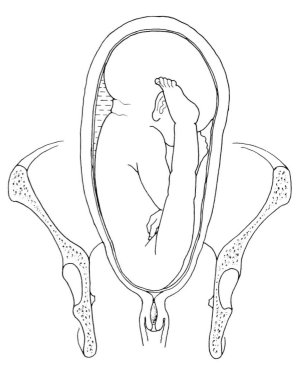

Fig. 5.11 Frank breech (also known as extended breech) presentation with extension of the legs

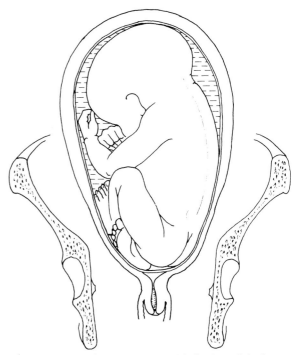

Fig. 5.12 Breech presentation with flexion of the legs

In the course of obstetric manipulations one leg of a breech may be drawn down through the cervix while the other leg remains extended at the knee and flexed at the hip, this is known as *half breech*.

With the breech with extended legs (frank breech) the buttocks accurately fit the lower segment and cervix and prolapse of the cord is an uncommon complication, whereas it may sometimes occur with a flexed breech.

Causes

In most instances breech presentation occurs by chance in an early labour. Before the 30th week of pregnancy the uterine cavity is more or less spherical and the long axis of the fetus may lie in any direction. In late pregnancy the cavity is ovoid, with the fundus wider than the lower pole. Fetal movements normally turn the fetus until the flexed legs occupy the more spacious upper part of the cavity and the head fits into the narrower, lower part. If kicking movements are ineffective because the legs are extended, or the frank breech becomes engaged in the pelvic brim, then a breech presentation will persist. Further fetal growth makes the free space in the uterus relatively less, so that spontaneous version becomes less likely as term is approached. Extension of the legs is probably the most common reason for persistent breech presentation.

If there is an excess of fluids free fetal movement may continue, and a breech presentation at term may occur by chance. For the same reasons malpresentations are more common in multiparae with lax uterine and abdominal musculature.

With twin pregnancy either fetus may prevent the head of the other from engaging, and will also prevent free movement in the uterus, so that malpresentations are common.

In a small proportion of cases some underlying abnormality will be discovered which prevents the head from entering the pelvis, and in some of these cases a breech presentation may occur. Such abnormalities include placenta praevia, contraction of the pelvic brim and pelvic tumours; in the same way breech presentation may occur in cases of hydrocephalus. If pregnancy occurs in one horn of a double uterus, breech presentation may occur because the horn tends to be narrow with the wider pole at the top of the uterus.

Diagnosis

During pregnancy

During pregnancy the diagnosis is made by abdominal examination. The hard, spherical and ballotable fetal head is felt in the fundal region of the uterus. The woman may have noticed discomfort in this region. Admittedly the breech of a mature fetus may feel firm, but it is never so hard or ballotable as the head. If a considerable amount of fluid is present the head can be balloted from side to side between the two hands.

The fetal heart sounds are heard at a higher level than in the case of a cephalic presentation; at term the point of maximal intensity is at the level of or slightly above the maternal umbilicus.

If the diagnosis is uncertain in late pregnancy vaginal examination may resolve it, but if doubt still remains an ultrasound examination will confirm.

During labour

After rupture of the membranes additional information can be obtained by vaginal examination. In the case of a flexed breech a foot may present and the projection of the heel will distinguish it from a hand. With an extended breech the rounded buttocks superficially resemble a fetal head, but the hardness of bone and sutures is absent, and the anus, sacrum and scrotum (in a male) should be identifiable.

Mechanism of labour

Four positions of the breech are conventionally described:

>   left sacroanterior
>   right sacroanterior
>   right sacroposterior
>   left sacroposterior.

However, in most cases the bistrochanteric diameter (10 cm) engages in the transverse diameter of the pelvic brim, with the back to the front. During labour the breech descends into the pelvic cavity and internal rotation brings the bistrochanteric diameter into the anteroposterior diameter of the pelvic outlet. The breech is born by lateral flexion of the trunk, the anterior buttock appearing first (*see* Fig. 5.13). This movement of lateral flexion is

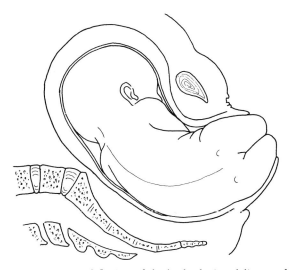

Fig. 5.13 Lateral flexion of the body during delivery of the breech. The anterior (right) buttock is under the pubic arch and the posterior (left) is escaping over the perineum. The child's sacrum looks directly to the mother's right side, and the body is flexed round the symphysis. The shoulders are entering the pelvis in the oblique diameter so that there is also a slight twist on the blyd

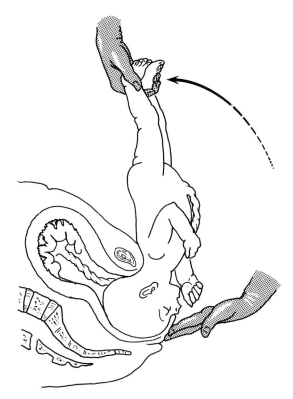

Fig. 5.14 Birth of the aftercoming head. The nape of the neck is under the symphysis, the chin is escaping over the perineum. In delivery the child's body is supported by holding the ankles. Sudden expulsion of the head may be prevented by gentle pressure on the brow but delivery of the head is best controlled with the forceps as in Fig. 5.18

determined by the curve of the birth canal, and more flexion is necessary with a rigid perineum. At one time it was thought that if the legs of the fetus were extended they splinted the trunk and prevented lateral flexion. This is not the case, and the frank breech usually descends easily. Its conical shape makes it a better dilator of the cervix than the rounded head.

The rest of the trunk is born by further descent, together with the arms which normally remain flexed in front of it. The anterior shoulder emerges first under the pubic arch and is quickly followed by the posterior shoulder. The flexed head engages in the transverse diameter of the pelvic brim. Forward rotation of the back occurs as the head descends into the pelvic cavity and then undergoes internal rotation, with the occiput coming to lie behind the symphysis pubis. The neck rests against the pubic arch as the head is born by the face sweeping over the perineum (*see* Fig. 5.14) and the baby is born by a backward somersault onto the maternal abdomen. Posterior rotation of the occiput occurs infrequently, and then the head is born face-to-pubes. This mechanism is less favourable as

the larger occipitofrontal diameter of the head distends the vulva.

Prognosis

The perinatal mortality in primiparae and multiparae is higher in breech than in vertex delivery, even if complications such as contracted pelvis, placenta praevia, fetal abnormality or prematurity are excluded. In mature, uncomplicated cases it has been reported as being between 15 and 30 per 1000. The increased mortality in such cases is due to:

intracranial injury
hypoxia.

Intracranial injury

The risk of a tentorial tear and intracranial haemorrhage is greater with an aftercoming head. Less time is available for moulding, and rapid compression and subsequent decompression of the head is particularly likely to produce this injury.

Fetal hypoxia

Interference with the placental circulation may occur from compression of the umbilical cord by the trunk during delivery, from uterine retraction which may separate the placenta before birth of the head or from delay in the delivery of the head for more than 10 minutes after the body has been born. Even if the delay is not as long as this there is some risk to the fetus from premature inspiration with the aftercoming head undelivered, when mucus may be inhaled into the air passages and obstruct them after delivery.

Fetal hypoxia may also be caused by prolapse of the umbilical cord. This is rare with a frank breech, but more likely with a flexed breech.

Management

Breech deliveries must be under the management of at least a registrar with good obstetric experience. All such cases must be delivered in hospital.

External cephalic version

External version means changing the fetal presentation by manual pressure through the woman's abdominal wall. It is desirable to correct a breech presentation by version so long as the operation does not itself increase the risk to the fetus. Version should not be attempted if there is:

- gross pelvic contraction, when caesarean section is necessary. There is no purpose in attempting version but in a case of doubtful disproportion, version may be useful so that a trial labour can subsequently be conducted with the vertex presenting
- placenta praevia or other cause of antepartum haemorrhage, because of the risk of causing further placental separation
- a caesarean or hysterectomy scar in the uterus
- a bad obstetric history
- twin pregnancy, when version is prevented by the second twin.

The theoretical risks of external version are:

- placental separation
- cord entanglement
- premature rupture of the membranes
- precipitation of premature labour
- uterine scar dehiscence
- transplacental haemorrhage with rhesus sensitization if the mother is rhesus-negative and her partner, and thus her baby, is rhesus-positive.

External version need not be attempted before the 32nd week in a primigravida and 34th week in a multipara, because in a large proportion of cases spontaneous cephalic version will occur. However, after the 38th week version is unlikely to succeed because of the size of the fetus and the relative decrease in the volume of amniotic fluid. Factors which make version difficult are abdominal fat, failure of the woman to relax, uterine irritability, deep engagement of the breech and extension of the legs.

Before version is attempted the exact position of the fetus must be determined by palpation, and the fetal heart auscultated. Relaxation of the abdominal muscles may be easier if the woman bends her knees slightly. More recently, tocolytic drugs, for example Salbutamol have been used to relax the uterus with good results from those who are familiar with these techniques. The first step is to disengage the breech from the pelvic brim; if this cannot be done no further attempt at version should be made. The breech is pushed upwards with the fingers of both hands, then pressed upwards and laterally with one hand while the other hand presses on the head, displacing it in the direction which will increase flexion (*see* Fig. 5.15). Steady pressure is more likely to be effective than jerky movements. After any attempt at version, successful or otherwise, the fetal heart rate should be counted. Immediately after version the head is often high and not engaged, and this should not be taken as evidence of disproportion.

If, by 36 weeks, version had not been successful it used to be common practice to attempt version under anaesthesia but this is rarely performed nowadays since it may increase the risk of placental separation or premature labour.

Planning the method of delivery

If a breech presentation persists the obstetrician must check that the pelvis is normal in shape and

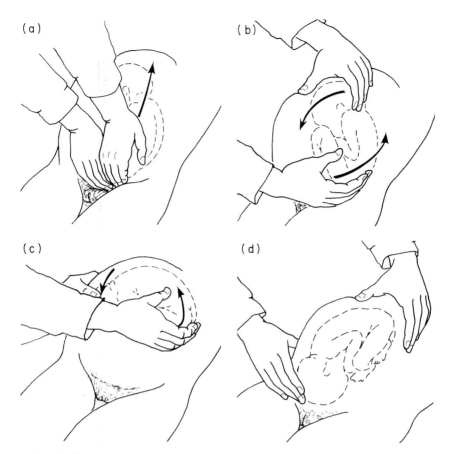

Fig. 5.15 External cephalic version. (a) The breech is disengaged from the pelvic inlet; (b) version is usually performed in the direction which increases flexion of the fetus and makes it do a forward somersault; (c) on completion of version the head is often not engaged for a time. The fetal heart rate should be checked after the external version has been completed

size; this should be done in all women, regardless of parity. The history of previous confinements and clinical examination of the bony pelvis are of importance. Most obstetricians recommend a lateral X-ray or CAT scan of the pelvis. Those against lateral X-ray pelvimetry cite the danger of irradiating the fetus and the inaccuracy of measurements on X-rays. The latter can be reduced by the obstetricians themselves reading the films and doing the measurements. A CAT scan involves much less irradiation and produces very clear images (*see* Fig. 5.16). Its disadvantage is that it must be performed with the woman lying down; this removes any effect of gravity on engagement and also prevents measurement of the angle of inclination of the pelvic brim.

An attempt should also be made to assess the

size of the baby. Clinical assessment should not be overlooked but the best predictions of fetal weight are made from ultrasound measurements of fetal abdominal circumference. Any baby whose estimated birth weight is over 3500 g is best delivered by caesarean section if it is presenting by the breech. The same is true in any circumstances where there is reason to believe that the shape or size of the pelvis are abnormal; breech birth is very dangerous in the presence of even slight disparity between the size of the baby and the size of the pelvis, especially a straight sacrum. Finally, it is essential to exclude extension of the fetal neck, as this will result in the mentovertical diameter of the head (13 cm) meeting the pelvic brim once the trunk has been delivered. The subsequent obstruction to the head may well cause fetal hypoxia. This

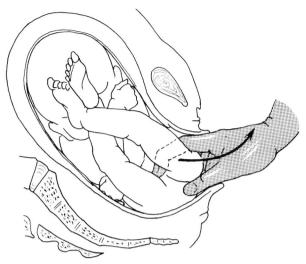

Fig. 5.16 Groin traction during delivery of breech with extended legs after episiotomy

relatively unusual but potentially lethal abnormality may be detected on ultrasound scan at the time of estimating the fetal weight.

The role of induction of labour

It was once argued that, ideally, breech delivery should occur at around 38 weeks, a time when the fetus is mature but is smaller than it would be at 40 weeks and supported by an optimally functioning placenta. The view that this would constitute a good indication for inducing labour at 38 weeks no longer stands. Indeed, many believe that vaginal delivery should only be contemplated if labour is of spontaneous onset, caesarean section being preferable to induction of labour, whatever the indication.

The place of caesarean section in breech delivery

Before discussing the management of vaginal breech delivery it must be asked whether, in view of the fetal risk, caesarean section should not always be preferred. The maternal risk is increased slightly, if the delivery is by section, yet in Britain about 70 per cent of women with breech presentations have a caesarean section.

With vaginal breech delivery it is possible that fetal death may occur, but even if the fetus is born alive there is some risk of it being injured. Many

women know this and request delivery by caesarean section. Many obstetricians agree with them, and also point out that even if labour is allowed to begin, many cases will end in section for delay or fetal distress.

Even if this view is not wholly accepted, it is generally agreed that any additional complication during pregnancy or labour, or anything untoward in the patient's history, is an indication for section. It would be advised for any primigravida over 35 years of age, and for any woman who has previously been infertile or has had stillbirths. It is now generally held that caesarean section is preferable to vaginal delivery with babies whose gestational age is less than 34 weeks, the very preterm baby being unduly susceptible to intracranial bleeding caused by trauma or asphyxia.

Pelvic contraction, even if it is only slight, is an indication for section. An android pelvis is particularly unfavourable. Section would be performed in cases associated with placenta praevia, and in cases of prolapse of the cord before full dilatation of the cervix.

Management of vaginal breech delivery

The conduct of the first stage of labour is exactly as described for a vertex presentation. Early rupture of the membranes may occur as the presenting part, particularly of a flexed breech, may not fit the brim well. As soon as the membranes rupture a vaginal examination should be made to exclude prolapse of the umbilical cord. Throughout this stage of labour a watch is kept, as always, for failure to progress or any evidence of maternal or fetal distress and any of these complications are best treated by caesarean section.

In the second stage a vaginal examination must always be made to confirm full dilatation before the woman is allowed to bear down. Spontaneous descent of the breech onto the pelvic floor usually occurs, whether the legs are extended or not. When the breech reaches the pelvic floor the woman should be placed in the lithotomy position. An episiotomy is performed as soon as the baby's anus becomes visible at the height of a contraction so that the intact perineum will provide no obstacle to the delivery of the breech, and to facilitate the ultimate delivery of the head. Epidural analgesia will provide total pain relief and good relaxation of the pelvic floor, but may interfere with bearing down efforts. As an alternative, pudendal block or local infiltration with lignocaine is necessary for the episiotomy.

In the case of a flexed breech the feet and legs present and as they appear they may be eased out. If the legs are extended no assistance should be given to the birth, which must be allowed to take place as the result of maternal effort and the force of uterine contractions. First the anterior, and then the posterior buttock is born by a process of lateral flexion, so that the buttocks appear to climb the maternal perineum. After the buttocks have been born the baby's feet may need to be released by flexing first the anterior and then the posterior leg at the knee.

Delivery of the trunk and head

As soon as the umbilicus is born pulsation of the cord can be observed. Compression of the cord by the side walls of the birth canal may sometimes cause spasm of the umbilical vessels and obliterate this fetal pulse. Absence of pulsations in the cord during delivery is only an indication for haste if the fetal trunk is pale and limp and the danger of asphyxia seems greater than rapid extraction of the fetus with its risk of intracranial haemorrhage.

A finger should be inserted into the vagina to make certain that the arms are lying folded on the chest. As soon as the body descends so that the inferior angle of the anterior scapula becomes visible at the introitus, the anterior arm can be gently hooked out after which the posterior arm can be released. The baby's body is now allowed to hang down to exert slight traction in the direction of the pelvic axis. As soon as the head has descended onto the pelvic floor and the nape of the neck can be seen below the pubic arch, the baby's legs are held just above the ankles, and by exerting slight traction its body is lifted to the horizontal position. The face will then appear at the vulva and, as the baby can now breathe, the head can be delivered slowly and carefully. Lifting the baby's body upwards before the head has passed completely through the brim will not assist delivery, and may injure the cervical spine. In most units forceps are used routinely for the delivery of the head.

Easy delivery may not occur if the arms are extended. This complication must be dealt with at once by Løvset's manoeuvre. This depends on the fact that the subpubic arch is the shallowest part of the pelvis. By downward traction the anterior

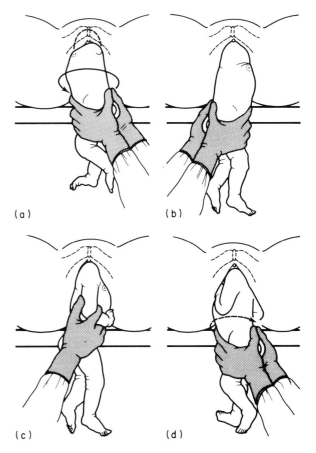

(a)   (b)   (c)   (d)

Fig. 5.17 Løveset's manoeuvre

shoulder is brought to lie behind the symphysis pubis so that the inferior angle of the scapula can be seen. When this is done the posterior shoulder will lie below the pelvic brim. The pelvic girdle and thighs are then firmly held with both hands and the fetus is turned through 180° with the back upwards, while moderate traction is maintained (*see* Fig. 5.17(a)). By this means the posterior arm is brought to the front and inevitably appears under the pubic arch (*see* Fig. 5.17(b)). The arm may be delivered spontaneously; otherwise it is easily hooked out with a finger (*see* Fig. 5.17(c)). The other shoulder now lies in the hollow of the sacrum. The fetus, therefore, is again rotated through 180° in the opposite direction, the fetal back again being kept upwards (*see* Fig. 5.17(d)). The remaining arm will appear under the pubic arch and can easily be delivered.

Difficulty with the aftercoming head

After the birth of the shoulders the flexed head normally enters the pelvis in the transverse diameter of the brim. As descent occurs the head undergoes internal rotation so that the occiput comes to lie behind the symphysis pubis. The neck rests under the pubic arch and normally the head is born by a movement of flexion. In most cases the head will enter the pelvis if the fetal trunk is allowed to hang downwards for not more than 1 minute. Any failure of descent or difficulty in the delivery of the head calls for immediate delivery. The body of the baby having been born, the placental circulation is impaired by retraction of the uterus and pressure on the cord, and respiratory efforts will certainly be made. If the birth is not completed within the next few minutes death from hypoxia is likely. The obstetric forceps must be available for immediate use at every breech delivery whenever the descent of the head is arrested. The body of the baby is lifted upward by an assistant and the blades are applied beneath it. The fingers of the right hand steer the left blade of the forceps to the left side of the pelvis, and then the right blade is applied. Only moderate traction should be required to complete delivery as episiotomy has already been performed (*see* Fig. 5.18).

Very rarely the fetal head cannot be made to enter the pelvic brim. This means that an error of judgement has been made and pelvic contraction

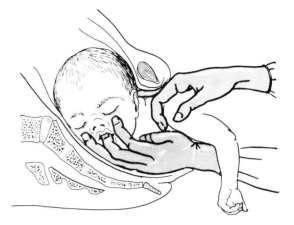

Fig. 5.19 Jaw and shoulder traction applied to a fetus with the head at the level of the pelvic brim. Forceps could not be applied at this level

has been overlooked (or very rarely that hydrocephalus is present). Apart from the fact that the head may be lying transversely, flexion will almost certainly be poor, and the application of forceps at this level would be difficult and dangerous. The best hope of saving a desperate situation is to pass one hand up in front of the fetal thorax, where one finger presses on the jaw to produce flexion. The other hand is passed along the back until the index and middle fingers can curve over the shoulders to exert traction (*see* Fig. 5.19). The head is drawn down by traction on the shoulders, and if necessary rotated until the neck lies under the pubic arch. Suprapubic pressure by an assistant may also help the head pass through the pelvic brim. Caesarean section at this late stage is perilous to the baby for the body has been born to the shoulders. Delivery may then be completed with forceps as already described. This sort of traction may cause injury to the cervical spine or to the brachial plexus.

Difficulty at the pelvic outlet may occur if it is contracted, or if the occiput has rotated posteriorly; delivery is achieved with forceps. If the occiput lies posteriorly and cannot be rotated the forceps are applied in front of the baby's body, which is supported by an assistant.

Most of the difficulties encountered in breech delivery can be avoided if:

- vaginal delivery is only attempted if the pelvis is capacious, there is no hint of delay in the first stage of labour and the breech readily descends

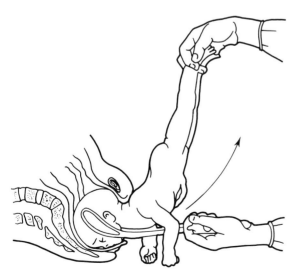

Fig. 5.18 Delivery of the aftercoming head with forceps

to the pelvic floor in the second stage of labour
- full dilatation of the cervix is confirmed before bearing down efforts are encouraged
- an episiotomy is performed under adequate analgesia
- Løvset's manoeuvre is used if the arms are extended
- forceps are used for slow delivery of the after-coming head.

Spontaneous expulsion of the buttocks, rapid and minimally assisted delivery of the trunk and shoulders and slow forceps delivery of the head are the ideal.

TRANSVERSE AND OBLIQUE LIE

The baby may lie with its long axis transverse or oblique in the uterus, when the point of the shoulder is usually the presenting part (*see* Fig. 5.20). In the absence of antenatal care shoulder presentations occur about once in 500 labours, but in many cases can be corrected by external version before labour.

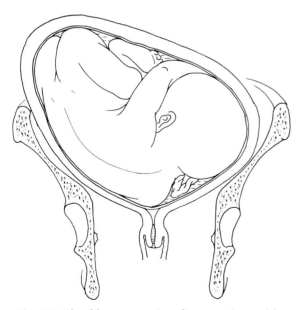

Fig. 5.20 Shoulder presentation, dorsoanterior position. The head is in the iliac fossa, the back lies anteriorly and obliquely and the breech is on the right above the right iliac fossa

Cause

By far the most common cause of an oblique lie is multiparity associated with a lax uterus and abdominal wall. It is not infrequently found on antenatal examination before 36 weeks and it is therefore more common in cases in which labour starts prematurely. An oblique lie may be found with polyhydramnios or multiple pregnancy, and may be caused by anything which prevents engagement of the fetal head, such as contracted pelvis, placenta praevia or fibroids. If a transverse lie occurs in a primigravida or recurs in successive pregnancies the possibility that it is caused by a uterine malformation (arcuate or subseptate uterus) must be considered.

Positions

The fetus may lie with the head in either iliac fossa, with the back sloping obliquely across the pelvic brim, while the breech usually occupies a somewhat higher position than the head, on the opposite side of the abdomen. Only two positions are described, dorsoanterior and dorsoposterior, the former being the more common.

Diagnosis

On abdominal examination the uterus appears asymmetrical, and is broader than usual with the fundus lower than expected for the duration of pregnancy. On palpation the hard round head is felt in one iliac fossa, with the softer breech on the opposite side. No presenting part is felt over the brim. In the centre of the abdomen the back will be felt in dorsoanterior positions, and small parts in dorsoposterior positions (*see* Fig. 5.21).

On vaginal examination at the beginning of labour the presenting part is too high to be felt. During labour the membranes usually rupture early. When the cervix becomes dilated an arm or a loop of cord may prolapse. Diagnosis of a shoulder presentation depends on recognition of the acromion process, scapula and adjacent ribs. A prolapsed arm must be distinguished from a leg; the elbow is sharper than the knee, and absence of a heel and abduction of the thumb will distinguish a hand from a foot.

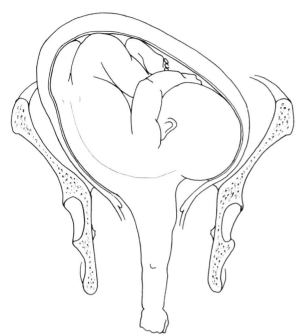

Fig. 5.21 Shoulder presentation with prolapse of the arm. The shoulder and arm, forming the apex of the wedge, are driven into the brim; the head and neck on the side and the trunk on the other form the sides of the wedge

Course of labour

A fetus lying obliquely cannot be born vaginally unless it is macerated or very premature. There is no true mechanism of labour, and an untreated case will end in obstructed labour and fetal death.

Neglected or unrecognized shoulder presentations are extremely serious, since all the risks of obstructed labour are present, including rupture of the uterus. Fetal death is usual from interference with the placental circulation or prolapse of the cord.

Treatment

During early labour

If an oblique lie is discovered early in labour it may still be corrected by external version if the membranes are intact. Once the lie has been corrected the membranes should be ruptured and the uterine contractions will usually maintain a longitudinal lie. If an oblique or transverse lie persists in labour

caesarean section is performed, being safer than the dangerous operation of internal podalic version, in which a leg of the fetus is pulled down and it is then delivered as a breech.

Late in labour when the shoulder has become impacted

In neglected cases, when the uterus is tonically retracted, any form of version is extremely dangerous, owing to the very great risk of rupturing the thinned out lower uterine segment. Even if the fetus is dead caesarean section is probably the safest procedure. In this case a low vertical incision in the uterus is preferable to the usual low transverse incision.

UNSTABLE LIE

Unstable lie is a self-descriptive term referring to a fetus which frequently changes its axis from transverse to longitudinal to oblique. The condition may be associated with polyhydramnios, placenta praevia, pelvic tumour or uterine contraction but usually occurs in multiparae with a lax uterus and abdominal wall. Gentle external version should be used to correct the malpresentation whenever the woman is examined. Such women should be warned to come to hospital immediately if they rupture their membranes or have any symptoms suggestive of the onset of labour. Those women who live at some distance from the hospital should be admitted from 38 weeks onwards to await the spontaneous onset of labour by which time the malpresentation has usually corrected itself.

The place of induction

If it is not practical to await the spontaneous onset of labour or there are obstetric indications for delivering the baby, induction can be contemplated. With an unstable lie it is best, after correcting the fetal lie to longitudinal, to start with a Syntocinon infusion and to delay amniotomy until there are uterine contractions and the presenting part has settled into the pelvic brim. Care should be taken to keep the presenting part in the pelvis when the forewaters are ruptured and to be on the lookout for cord prolapse at that time. Controlled release of amniotic fluid is desirable and may be achieved by using a Drew Smythe Catheter to perform a high puncture of the membranes rather than an amniotomy hook to puncture the membranes. If there is polyhydramnios 500 ml or more of fluid may be drawn off each abdomen by amniocentesis before the forewaters are ruptured. This will lessen the risk of cord prolapse. Whenever an induction is done with an unstable lie the labour ward staff should be informed so that everything is ready for caesarean section if the cord prolapses or if there is recurrence of the malpresentation during labour.

Many obstetricians would bypass all these problems if there is a persistent oblique lie after 40 weeks and perform a caesarean section selectively.

FETAL MALFORMATIONS THAT CAUSE DIFFICULTY IN LABOUR

These are mentioned in this chapter to complete the discussion of fetal causes of difficult labour.

HYDROCEPHALY

This may be discovered during pregnancy by chance at an ultrasound examination for some other reason, if the head is noticed to be unduly large, or during the course of investigation of suspected disproportion. During labour the wide separation of the cranial bones at the sutures may be recognized on vaginal examination. As the head cannot enter the pelvis a breech presentation may occur.

Obstruction during labour is treated by perforation of the head. It is often sufficient to pass a widebore needle into the head and draw off the cerebrospinal fluid. This may be done through the maternal abdominal wall or through the cervix when it is 5 cm dilated. After withdrawal of the fluid the head may collapse sufficiently for spontaneous delivery to occur.

If the fetus presents by the breech the after-coming head can be collapsed after the trunk has been delivered by passing a metal cannula up through the spinal canal (via a spina bifida if one is present).

If spontaneous delivery does not follow tapping, traction can be applied to the scalp with Willett's toothed forceps.

ANENCEPHALY

In some cases of anencephaly difficulty in delivery of the shoulders may occur, because they are often wide and have not been preceded by a head of normal size. There should be no hesitation in dividing the clavicles (cleidotomy) with strong scissors and applying strong traction in the axilla.

CONJOINED TWINS

Conjunction of twins may be suspected during pregnancy if ultrasound or radiological examination shows that the twins constantly maintain the same relative positions, especially if they are facing one another. In other cases the diagnosis is made during twin labour when there is delay and a hand is passed into the uterus to determine the cause. Surprisingly, vaginal delivery has not infrequently been reported when the twins have been small and the junction between them has been fairly pliable, but in most cases caesarean section is unavoidable, and is preferable to embryotomy.

PELVIC ABNORMALITIES AND DISPROPORTION

PELVIC ABNORMALITIES

The size of the female pelvis is of primary importance; if it is large enough there will be no difficulty in the passage of an average fetal head through it. The shape of the pelvis is of secondary importance, and only has a bearing on the mechanism of labour and the ease of delivery if the pelvic dimensions are small in relation to those of the head.

Certain terms are used to describe variations in the shape of the pelvis which were discovered by anatomical and radiological studies. Very few pelves fit the criteria exactly, and any pelvis may be a mixture of the types about to be described.

Gynaecoid pelvis

The gynaecoid pelvis conforms to the accepted female type in that the brim is rounded, with the

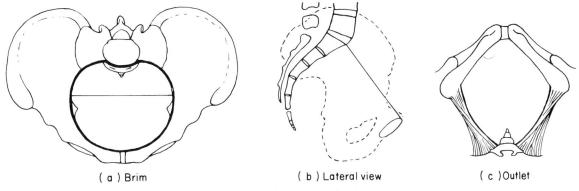

Fig. 5.22 The gynaecoid pelvis

widest transverse diameter slightly behind its centre (*see* Fig. 5.22). The subpubic arch is rounded, with an angle of at least 90 degrees.

Android pelvis

The android pelvis has many of the characteristics of the male pelvis. The brim is heart-shaped, so that the widest transverse diameter is much nearer to the sacrum than it is in the gynaecoid pelvis (*see* Fig. 5.23). The side walls tend to converge, the ischial spines are prominent, the sacrum is straight and the subpubic arch is generally narrow, with an angle of 70 degrees or less. Both the anteroposterior and transverse diameters of the outlet tend to be reduced. This type of pelvis is funnel-shaped, with diameters which decrease from above downwards, and disproportion thus becomes worse as labour proceeds.

Anthropoid pelvis

The anteroposterior diameter of the brim exceeds the transverse diameter (*see* Fig. 5.24). The pelvis tends to be deep and the sacrum often has six segments instead of five; this is known as a high assimilation pelvis. Very often the sacrum and the axis of the pelvic cavity are less curved than in the gynaecoid pelvis and the subpubic arch may be a little narrow, but the sacrosciatic notches are wide and the anteroposterior diameter of the outlet is large, so there is no difficulty.

Platypelloid pelvis

The platypelloid pelvis is described as the simple (non-rachitic) flat pelvis. The brim is elliptical with a wide transverse diameter (*see* Fig. 5.25). The subpubic arch is wide and rounded.

It is emphasized that, except in the case of the android pelvis, these variations in shape have little

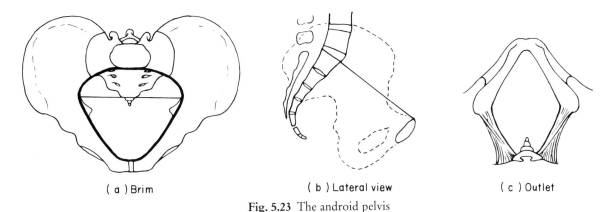

Fig. 5.23 The android pelvis

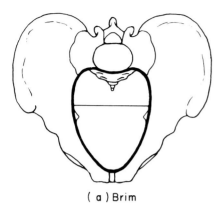 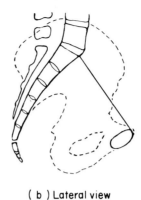 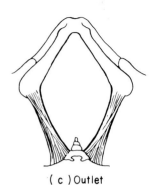

(a) Brim   (b) Lateral view   (c) Outlet

Fig. 5.24 The anthropoid pelvis

effect on the normal mechanism of labour unless there is considerable reduction in the size of the pelvis. The android pelvis is unfavourable because of the tendency to contraction of the outlet.

ABNORMALITIES OF THE PELVIS

Abnormalities of the pelvis may be classified into three groups:

- those caused by developmental abnormalities of the pelvic bones
- those caused by disease or injury of the pelvic bones
- those caused by abnormalities of the spine, hip joints or lower limbs.

Developmental abnormalities of the pelvic bones

Developmental variations in shape have already been described and the android pelvis is the only one of these that commonly affects labour adversely. However, even a round gynaecoid pelvis may be small in size; it is then called a *generally contracted pelvis*.

The presence of a sixth segment in the sacrum, which may be found with an anthropoid pelvis, has given rise to the term *high assimilation pelvis*, but this is not of great significance.

Very rarely malformation of the pelvis may result from defective development of one side of the sacrum, which is fused with the ilium (Naegele pelvis, *see* Fig. 5.26).

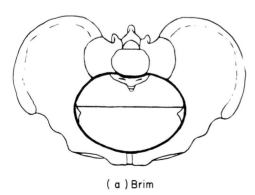 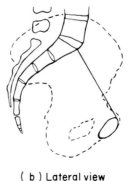 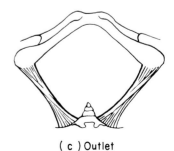

(a) Brim   (b) Lateral view   (c) Outlet

Fig. 5.25 The platypelloid pelvis

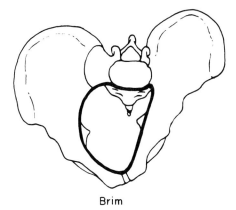

Fig. 5.26 Asymmetrical Naegele pelvis

Fig. 5.28 Kyphosis with forward rotation of the lower part of the sacrum and reduction of the anteroposterior diameter of the pelvic outlet

Disease or injury of the pelvic bones

The most important abnormality in this group is rickets, which causes softening of the bones in early childhood, with flattening of the pelvic brim (*see* Fig. 5.27). Fortunately rickets is uncommon in Britain. A similar disease, osteomalacia, which may occur in adults, can cause very severe distortion of the pelvis; it, too, is almost unknown in Britain but is still seen in Africa and Asia. In achondroplasia the pelvis is often small.

Pelvic deformity may be caused by malunited fractures, or by new growths arising from the pelvic bones.

Abnormalities of the spine and lower limbs

The shape and inclination of the pelvis may be affected by kyphosis of the spine (*see* Fig. 5.28) or by spondylolisthesis, in which the 5th lumbar vertebra slips forward on the sacrum and obstructs the pelvic inlet.

With congenital dislocation of a hip joint there may be oblique distortion of the pelvis, but this rarely causes difficulty.

LABOUR IN THE PRESENCE OF PELVIC ABNORMALITY

When the size of the fetal head exceeds the size of the pelvis even after moulding has taken place,

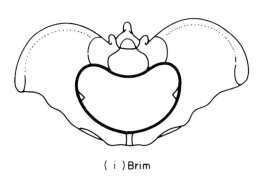

(i) Brim

(ii) Lateral view
(Showing forward displacement of the sacral promontory)

Fig. 5.27 Rachitic pelvis

normal vaginal delivery will clearly be impossible. If, however, there is not absolute disproportion vaginal delivery may occur if the uterine contractions are sufficiently strong to overcome the relatively increased resistance.

In practice the type of deformity is less important than the size of the pelvis relative to that of the head. The mechanism of the passage of the head through the pelvis will differ according to the shape of the pelvis, but in general there is a tendency with most pelvic abnormalities for the forepelvis to be small, thus increasing the risk of malpositions of the fetal head.

CEPHALOPELVIC DISPROPORTION

Disproportion between the size of the fetal head and that of the maternal pelvis will cause difficult labour and danger to the fetus. The size of the head at term is genetically related to the size of the maternal pelvis – small women tend to have small babies – so that in the absence of pelvic deformity the majority of labours proceed without mechanical difficulty. However, the obstetrician must always be on the lookout for possible disproportion, both in the antenatal clinic and during labour.

Disproportion in the Western world is usually found at the pelvic inlet. If the widest diameter of the head will pass through the inlet it will usually pass through the pelvic cavity and outlet. Occasionally the pelvis is to some degree funnel-shaped with a relatively narrower outlet than inlet, particularly in parts of the world where infant rickets and osteomalacia are found. Even in this case the head will often pass because by the time it has reached the outlet its diameters have been reduced by moulding. In every case of disproportion the efficiency of the uterine contractions plays a major part in determining whether or not vaginal delivery will occur. Equally important is the attitude and position of the fetal head. A deflexed occipito-posterior position is most unfavourable in the presence of even a minor degree of cephalo-pelvic disproportion, because of the wider diameters that need to pass through the pelvis.

The diagnosis of disproportion

The following are the steps in the antenatal diagnosis of disproportion:

PAST ORTHOPAEDIC AND OBSTETRIC HISTORY

A history of disease or fracture of the spine, hips or pelvis should alert the practitioner to the possibility of pelvic deformity. A distinction is made between primigravidae and multigravidae.

In a multigravida the pelvis has been tested up to the size of the largest baby previously delivered, but in a primigravida the size of the pelvis is unknown. Disproportion in a multigravida is usually because she has a fetus larger than any she has previously borne. In dealing with a multigravida, therefore, it is important to obtain details of her previous labours, including for each the duration, the mode of delivery, the outcome and the birth weight and subsequent progress of the child.

GENERAL EXAMINATION

Although it is true that small women may have small babies, it is also true that a small woman has a small pelvis, and a greater chance of disproportion. Any woman whose height is less than 150 cm comes into this category.

General disease of the skeleton or disease of the lower limbs or spine should also be noted in view of any possible effect on the shape of the pelvis, which itself may have been deformed by rickets or previous fractures.

Abdominal examination

Abdominal examination in late pregnancy should indicate whether or not the widest diameter of the fetal head has passed through the pelvic inlet. In primigravidae the head normally engages between the 36th and 38th weeks. In multigravidae engagement often does not occur until labour starts.

In African women the pelvic inlet has a higher angle of inclination than is seen in Caucasian and Asiatic women and because of this, in African women engagement of the head is uncommon before the onset of labour, even in primigravidae.

If the head is not engaged before term an attempt is made to discover if it will engage, and thus exclude inlet disproportion. This is most simply done by moderate pressure on the head in a backwards and downwards direction.

Pelvic examination

Most of the normal pelvic inlet is beyond the range of the examining fingers. The forepelvis can be reached by passing a finger behind the symphysis pubis, but the sacral promontory can only be reached in a contracted pelvis. The concavity or straightness of the sacrum and the convergence or separation of the side walls of the pelvis will give some idea of the size of the pelvic cavity. The bony outlet can be readily palpated, and the width of the subpubic arch and the ease with which the ischial spines can be felt should be noted. The distance between the ischial tuberosities can be measured.

The fetal head is the best pelvimeter. If it has already descended into the pelvis or can be made to do so on examination, there is only the pelvic outlet left for it to pass. Pure outlet contraction is very rare, and if the biparietal diameter enters the pelvic inlet with only two-fifths or less of the head palpable abdominally (*see* Fig. 4.19), with the lowest point of the head at the level of the ischial spines, disproportion is extremely unlikely.

If the head will not enter the pelvis disproportion may be suspected but in many cases, with strong uterine contractions and moulding of the head, vaginal delivery will occur. Pelvic examination is of limited value in predicting the outcome as so much depends on the moulding and the efficiency of the contractions.

Diagnosis of disproportion during labour

Failure of the head to descend and of the cervix to dilate during labour may indicate unsuspected disproportion. Excessive moulding and caput formation may occur, and in some cases the caput may be felt at the level of the ischial spines even though the maximum diameter of the fetal head has still not passed through the pelvic inlet.

In modern practice, if a partogram indicates that the cervix is not dilating normally, an oxytocin drip with good analgesia is likely to be tried, unless examination shows clear evidence of disproportion. If augmentation of the uterine action does not lead to progress in cervical dilatation disproportion is probable and vaginal delivery should not be attempted.

Radiological or CAT diagnosis of disproportion

X-rays are now used much less often to investigate suspected disproportion than they used to be. In Britain it is rare for a woman to have a pelvis which is too small to allow the passage of a normal sized fetus, provided that there are good uterine contractions. X-rays have the same inability as clinical examination to predict uterine efficiency in labour. Present practice is only to X-ray the pelvis when disproportion is suspected but confirmation is needed for management, and then only one lateral radiograph is taken in late pregnancy with the woman standing. This view allows measurement of the anteroposterior diameters of the pelvis but not of the transverse diameters (*see* Fig. 5.29). The latter could be measured by taking further views of the pelvic inlet and outlet but these are not now thought to be justified in view of the limited information gained and the additional irradiation of the fetus. Apart from the anteroposterior diameters, the erect lateral radiograph also shows the curve of the sacrum, the width of the sacrosciatic notches, the angle of inclination of the pelvic brim and the degree of engagement of the fetal head.

X-rays are now rarely used during labour for the diagnosis of disproportion, as they add little to the clinical assessment of progress and nothing to management.

The management of disproportion

When cephalopelvic disproportion is suspected as a result of antenatal examination a decision has to be taken whether or not to allow the woman to go into labour in the hope of vaginal delivery, or to deliver her before the onset of labour by elective caesarean section.

Indications for elective caesarean section

The indications are:

- disproportion so severe that vaginal delivery is unlikely
- lesser degrees of disproportion when there are complicating factors such as breech presentation, previous caesarean section, or a previous difficult vaginal delivery causing perinatal death or morbidity
- women with serious medical complications such as, for example, diabetes, or severe hypertension, where delivery well before term is required.

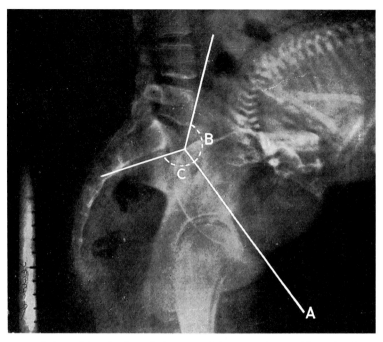

Fig. 5.29 Lateral X-ray of the pelvis, the fetus is in an occipitoposterior position. A, plane of the inlet; B, angle between the front of the 5th lumbar vertebra and the pelvic Brim; C, angle of inclination of the sacrum

Trial of labour

Except for the cases just mentioned, women with disproportion should be allowed a trial of labour. By this is meant that the labour is carefully watched for evidence of continuing progress as shown by dilatation of the cervix and descent of the head. The assessment of progress is made as in normal labour by repeated abdominal and vaginal examinations, and the results are recorded on a partogram.

In the majority of cases of trial of labour good uterine contractions will bring about steady progress to vaginal delivery. In the multiparous the head may remain high until the cervix is fully dilated and then descend rapidly through the pelvis during the second stage.

Augmentation of labour in cases of suspected disproportion

Good uterine contractions are essential if mild disproportion is to be overcome. In primigravidae

delay in labour caused by poor uterine action may be treated by the cautious use of an intravenous oxytocin infusion. Great care must be taken to avoid overstimulation leading to excessively strong contractions which may harm the fetus, or even rupture the uterus. Continuous monitoring of the fetal heart with a scalp electrode, as well as continuous recording of the uterine contractions are essential whenever labour is being augmented, but are particularly important in cases of possible disproportion.

Deciding when a trial of labour has failed

Although no hard and fast rules can be laid down, failure to progress in a trial of labour after a period of 4 hours or more of good contractions suggests that the disproportion is too great to be overcome, and that caesarean section should be performed. The trial may also have to be terminated at any time if fetal distress develops.

Uterine rupture is a rare event in primigravidae, but less rare in multigravidae. It usually only occurs after some hours of strong contractions. The development of a Bandl's ring, pain and tenderness over the lower uterine segment, or acute fetal distress are all indications that labour has been allowed to go on for too long and that rupture is impending; immediate caesarean section is required.

The management of trial of labour for suspected disproportion should be under the control of an experienced obstetrician in a fully equipped unit, as considerable skill and judgment may be needed to decide how long the labour may safely continue.

Symphysiotomy

Division of the ligaments holding the pubic bones together at the symphysis through a small suprapubic incision produces an increase in the available circumference of the pelvic ring and, by widening the subpubic arch, an increase in the available anteroposterior diameter at the outlet. The operation is hardly ever performed in the UK, but has been recommended in developing countries in association with a vacuum extraction where some women are reluctant to agree to caesarean section, and may not return to hospital for subsequent deliveries.

Shoulder dystocia

Shoulder dystocia is when the shoulders remain impacted in the pelvis after delivery of the fetal head. It is a serious, but fortunately rare, complication which usually occurs with babies weighing well over 4 kg. Perinatal mortality and morbidity in fetal macrosomia are not inconsiderable. The warning signs are unusually slow crowning of the fetal head, difficulty in delivering the baby's face by extension and slow restitution of the occiput to a lateral position. Once the condition has occurred it is important to get the woman into either the lithotomy or the left lateral position and to do an episiotomy if one has not been done already.

Examination should reveal whether or not the anterior shoulder is arrested above the pelvic brim. If it is not, it may be possible to deliver it from the pelvic cavity by a combination of suprapubic pressure and moderate lateral flexion and traction on the baby's head. If the anterior shoulder is still above the pelvic brim it will be the posterior shoulder that has descended into the sacral bay.

Under these circumstances lateral flexion and traction on the baby's head will only result in tearing of the brachial plexus and an Erb's palsy, which can be a serious permanent disability for the infant. The obstetrician will therefore have to consider delivering the posterior arm and shoulder. To do this the operator's fingers will have to be insinuated into the very restricted space available in front of the baby's chest in order to flex the posterior arm at the elbow and then bring it down. If it is possible to rotate the posterior shoulder a few degrees so that it lies in the oblique diameter of the pelvis, bringing the posterior arm down can be made a little more easy. Once the posterior arm has been brought down it will be much more straightforward to reach the anterior axilla for traction and delivery. As a last resort, and only if the manoeuvres already described have failed and the fetus has died, an experienced obstetrician would consider division of the clavicles or even symphysiotomy as a method of delivering the baby.

If a woman has had shoulder dystocia she and her attendants will be anxious to avoid this complication in future pregnancies. To this end, she should have ultrasound estimates of fetal weight at term and a caesarean section if there is evidence of a baby that is as large, or larger than the one which previously caused the problem.

OBSTRUCTED LABOUR

Labour is said to be obstructed when there is no progress in spite of strong uterine contractions. This may be shown by failure of the cervix to dilate or failure of the presenting part to descend through the birth canal. It is a most dangerous condition if it is untreated, and can be fatal to both mother and fetus.

Causes of obstructed labour

Obstructed labour may arise from maternal or fetal conditions, or both. The following list of the causes is long; the causes marked with an asterisk are rare.

Maternal conditions

    contraction or deformity of the bony pelvis
    pelvic tumours
        uterine fibroids
        ovarian tumours

*tumours of rectum, bladder or pelvic bones
*pelvic kidney
abnormalities of the uterus or vagina
    *stenosis of the cervix or vagina
    *obstruction by one horn of a double uterus
    *contraction ring of uterus.

Fetal conditions

large fetus
malposition or malpresentation
    persistent occipitoposterior or transverse
        position
    *mentoposterior position
    brow presentation
    breech presentation
    *shoulder presentation (rare in the UK
        but common in countries with inadequate
        antenatal care)
    *compound presentation
    *locked twins
congenital abnormalities of the fetus
    *hydrocephalus
    *fetal ascites or abdominal tumours
    *hydrops fetalis
    *conjoined twins.

Most of these abnormalities can and should be
detected during pregnancy so that early treatment
is possible, or a plan of action can be made before
labour. These conditions and their management are
fully described in the appropriate chapters; all that
is given here is a description of the effects of
obstructed labour if it is left untreated.

Symptoms and signs of obstructed labour

The importance of the early detection of possible
obstruction in labour is obvious, for if labour is
allowed to progress to the point of absolute
obstruction the death of the fetus is almost certain
and the life of the mother is endangered. In a
primigravida complete obstruction leads within 2
or 3 days to a state of uterine exhaustion or
secondary hypotonia; any relief which this gives to
the mother and fetus is only temporary. In a
multigravida obstruction becomes established
much sooner and progressive thinning of the lower
segment may lead to uterine rupture in a few
hours.

Probably the earliest sign of impending obstruction is deterioration in the patient's general condi-

tion. She looks tired and anxious and behaves as
though she is beginning to lose her ability and will
to cooperate. Between the contractions she seems
unable to relax and her anxiety increases.

The presenting part is often above or at the level
of the pelvic brim. The membranes rupture early in
labour because the presenting part is badly applied
to the lower segment. The amniotic fluid drains
away and there is retraction of the placental site,
which causes reduction in the maternal blood flow
to the placenta, severe fetal distress (shown by
marked slowing of the fetal heart and thick meconium in the fluid) and eventual fetal death from
hypoxia.

The woman's pulse rate and temperature rise.
The quantity of urine secreted diminishes and it is
concentrated and deeply coloured. Ketone bodies
are present in the urine and can also be smelt in the
woman's breath.

The possibility of obstructed labour should be
suspected when labour fails to progress. In the first
stage dilatation of the cervix should be progressive,
although sometimes it is not rapid even in normal
cases. A partogram will give early warning that
progress has ceased. Descent of the presenting part
should be continuous, especially in the second
stage. Any failure in the progress of labour calls for
careful abdominal and vaginal examination to
exclude any possible cause of obstruction, particularly in the case of previously undiagnosed disproportion or malpresentation.

If for some reason the diagnosis of obstruction is
missed for a time, the dangerous condition of over-
retraction of the uterus (generalized tonic retraction) may occur. In the course of normal labour
some retraction of the upper segment persists after
each contraction and the upper segment becomes
slightly shorter and thicker, while the lower segment becomes stretched and thinner. If the fetus is
unable to descend because of obstruction the total
length of the uterine cavity must remain constant,
so that as uterine contractions continue progressive retraction causes abnormal stretching and
thinning of the lower segment. The line of junction
of the upper and lower segments becomes very
evident and is known as the retraction ring of
Bandl (*see* Fig. 5.30). It can be seen or felt on
abdominal examination. Eventually rupture of the
lower segment occurs.

In advanced obstructed labour the uterus is
found, on abdominal examination, to be moulded
to the shape of the fetus. It feels hard all the time

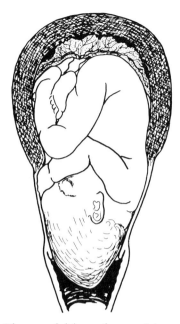

Fig. 5.30 Obstructed labour from pelvic contraction. The extreme retraction of the upper segment and the extreme stretching of the lower segment has formed a Bandl's ring. The head is greatly moulded, with a large caput succedaneum

When tonic retraction is present the fetus is certainly dead and the aim of treatment is to deliver the mother immediately by the safest possible method. Intrauterine manipulations are very liable to cause rupture of the abnormally thin lower segment. Internal version is particularly dangerous. In some of the cases it is possible to deliver the fetus vaginally after a destructive operation, but (except for perforation of a hydrocephalic head) these procedures are lengthy and difficult and caesarean section is usually less hazardous. Antibiotics, blood transfusion and modern anaesthesia have combined to reduce the risks of section in these cases.

In all cases of prolonged labour, especially if operative delivery is required, there is a high risk of puerperal sepsis and appropriate antibiotics should be given.

RUPTURE OF THE UTERUS

Prolonged obstruction in a primigravida often leads to temporary cessation of uterine activity from exhaustion (secondary hypotonia), but in a multigravida it is more likely to lead to uterine rupture particularly if there is a previous caesarean section scar. In obstructed labour this usually occurs obliquely at the junction of the upper and lower segments, but occasionally the uterus splits vertically at the side near the point of entry of the uterine vessels. The peritoneum may or may not be involved. Bleeding may occur into the peritoneal cavity or may track downwards between the bladder and upper vagina. Uterine contraction may expel the fetus and placenta through the laceration into the peritoneal cavity.

FISTULA FORMATION

In obstructed labour the fetal presenting part becomes impacted in the maternal pelvis and the result is ischaemia of the maternal soft tissues that lie between the presenting part and the walls of the maternal pelvis. If the obstruction is neglected this ischaemia can lead to tissue necrosis and the formation of vesicovaginal or rectovaginal fistulae when necrotic slough separates several days after delivery. Obstetric fistulae should not occur in countries where women have easy access to well organized hospital-based maternity services.

and does not relax. It is tender to palpation and Bandl's ring may be evident. Fetal parts are not easily felt and the fetal heart sounds are absent. The presenting part is fixed at the level of obstruction.

On vaginal examination the vagina is found to be oedematous and feels dry. The oedematous cervix is only loosely applied to the presenting part. If the head is presenting there will be a large caput succedaneum and extreme moulding of the skull. The presenting part is tightly fixed, and even under anaesthesia it cannot be pushed upwards without the danger of uterine rupture. If there is a shoulder presentation the oedematous arm of the fetus will have prolapsed, with the hand projecting from the vulva.

Treatment

Excessive retraction of the uterus should never be allowed to develop. The cause of the obstruction should have been discovered during pregnancy or in early labour, and treatment should have been applied.

PRETERM LABOUR AND PREMATURE RUPTURE MEMBRANES

INTRODUCTION

Preterm birth is hazardous for the fetus for it has not yet developed physiologically and is therefore not sufficiently adapted for extrauterine life. Five per cent of births occur preterm and these contribute over 50 per cent of the neonatal morbidity and mortality. By definition any delivery of less than 37 completed weeks is preterm; however improvements in neonatal care mean that at above 30 weeks' gestation most neonates survive. Even at 24 weeks, there is a survival potential (Table 5.2). The best results are achieved when the fetus is delivered atraumatically in a well oxygenated condition and when it is provided with expert neonatal intensive care from the moment of delivery. This can only be achieved if the obstetrician and paediatrician collaborate closely.

AETIOLOGY AND DIAGNOSIS

Preterm problems may present as either preterm labour or premature preterm rupture of membranes. In the majority of cases there is an under-

Table 5.2 Survival by gestational age at delivery

Gestation (wks)	% Survival	% Increase in survival if pregnancy could be prolonged by 2 weeks
22	0	45
23	42	10
24	45	30
25	52	5
26	75	10
27	57	30
28	85	8
29	85	10
30	93	4
31	97	1
32	97	0
33	98	0
34	95	0
35	98	0
36	98	0
37	98	0

lying cause, such as chorioamnioitis, placental abruption, urinary tract infection, multiple pregnancy, polyhydramnios or cervical incompetence. Intrauterine infection is the most important cause of problems, but infection may be subclinical and only identified if the placenta and membranes are examined histologically after delivery.

When labour is established, or when there is a large leak of fluid from the vagina, there is no diagnostic difficulty, but where there is doubt a logical approach is essential. The diagnosis of labour is the same whatever the gestation and requires the presence of both uterine activity and cervical dilation. Uterine activity alone is normal in the mid-trimester and may be painful, particularly in those experiencing significant personal or psycho-social stress. An increase in vaginal secretions during pregnancy, or a small involuntary leak of urine, may raise the suspicion of membrane rupture and this requires investigation. Clarification of the diagnosis rests on careful examination by an experienced obstetrician and will require vaginal examination to diagnose preterm labour with a speculum examination if there is a suspicion of membrane rupture. It may be possible to see fluid draining from the cervix; alternatively fluid may cause a rise of vaginal pH which can be inprecisely identified by using a nitrazine swab. Because of the possibility of introducing infection, a finger should not be passed through the cervix once the membranes are ruptured. However, gentle digital examination itself will not provoke labour and may be needed to assess the cervix when labour is suspected.

PRINCIPLES OF MANAGEMENT

The major determinants of outcome are the gestation at delivery, the condition of the baby at birth and the quality of neonatal care. It can be seen from the survival figures in Table 5.2 that at between 24 and 28 weeks' gestation an increase in gestation by as little as 2 weeks can lead to a dramatic increase in the likely survival. Hence the obstetrician will want to keep the baby in the uterus, unless the intrauterine environment is more hostile than a neonatal intensive care unit. The availability of neonatal intensive care is critical; preparations must be made as if all women with threatened preterm labour or premature preterm rupture of membranes will deliver on the day of admission. This requires liaison with the neonatal paediatricians about the availability of intensive care facilities and, in some cases, an *in utero* transfer of the fetus, transferring the mother from the receiving hospital to a second hospital, where neonatal facilities are available. This is preferable to delivering the baby and subsequently transferring it in an incubator, which is associated with a lower rate of survival.

A woman with preterm problems must be supervised by an experienced obstetrician working to an established protocol so that all members of the medical and nursing staff know what is expected of them. It is important to take a positive view of the survival of the fetus even at very low gestation, in order to give the fetus the optimum care and because gestation is sometimes underestimated and the prospects then believed to be worse than they actually are. The options open to the obstetrician are limited because preterm labour and premature preterm rupture of membranes are so often associated with uncorrectable, underlying pathology. Nevertheless, in the absence of fetal compromise, an attempt should be made to delay delivery. When delivery seems to be inevitable it should be conducted as atraumatically as possible, by an experienced obstetrician or a midwife trained in the management of preterm birth. The main principle is to avoid significant fetal handling in fetuses below 32 weeks' gestation. Fetal handling is associated with significant bruising which can lead to anaemia and disseminated intravascular coagulation in the neonate. Because of the need to avoid handling, and because we have no instruments suitable for assisted delivery, manoeuvres such as difficult breech delivery, internal version or forceps, which would be performed at term, are replaced by caesarean section in the very preterm group.

Between 32 and 37 weeks, the fetus is more robust and a delivery may be more readily achieved when complications occur. As always the obstetrician must balance the maternal risk of caesarean section against the potential fetal benefits. Caesarean section is, therefore, usually restricted to those fetuses of 26 weeks and above, that is, those who have a substantial chance of survival. Those presenting by the breech have the greatest benefit from caesarean section before 32 weeks' gestation.

PREDICTING AND PREVENTING PRETERM LABOUR

Attempts to predict preterm labour have so far been unsuccessful and even among the highest risk group, those who have laboured preterm previously, most will not suffer this complication again. Once labour is established, the success of attempts to arrest the progress will depend on the underlying cause. It is inappropriate to delay delivery in the face of major abruption or established infection. Tocolysis is rarely successful when labour is associated with ruptured membranes or when the cervix is dilated more than 4 cm. However, delivery can usually be delayed for up to 24 hours and this may be important to allow *in utero* transfer or to give sufficient time for steroid treatment. The administration of two doses of dexmethasone to the mother will help mature the fetal lungs and will reduce the risk of hyaline membrane disease. Because an increase in gestation is associated with such a significant increase in survival in fetuses below 33 weeks, it is always worth attempting to halt labour and to keep the baby in the uterus, provided that there is no evidence of fetal compromise which necessitates immediate delivery.

When there is clear evidence from the history, or from examination in between pregnancies, that the woman has cervical incompetence, preterm labour may be successfully prevented by supporting the cervix with an encircling suture. This procedure is known as cervical cerclage. Occasionally, a woman will present in mid-trimester with a widely dilated cervix, but no evidence of labour. If there is no evidence of infection a rescue cerclage can be performed and this will be successful in 50 per cent of cases.

TOCOLYSIS

The complex intracellular mechanisms which lead to smooth muscle contraction can be inhibited by three main groups of drugs acting at different parts of the actino-myosin pathway. Beta-adrenergic agonist drugs such as ritodrine, salbutamol and terbutaline are the most widely used and the most successful agents, but cyclo-oxygenase inhibitors, such as indomethacin, which prevent labour by inhibiting prostaglandin production and inhibiting calcium channel blockers, such as nifedepine, are also effective. Combined statistical analysis of the available controlled trials, shows that tocolytics can prolong pregnancy. Whether this procedure will improve fetal survival will depend not only on the presence of other associated conditions but at what gestation they are used, since only at certain gestations will prolongation of pregnancy increase survival significantly. Beta-adrenergic agonists and cyclo-oxygenase inhibitors are equally effective; the former are preferred mainly because the anti-prostaglandin drugs are associated with premature closure of the ductus arteriosus and constriction in other vascular beds, such as the cerebral vascular circulation and the kidney, with potential ischaemic complications. The place of calcium channel blockers in clinical practice has yet to be established.

The beta-adrenergic agonist drugs, although effective, are associated with significant maternal hazards and their use should therefore be limited to 24 to 48 hours. The principle complications are cardiovascular; these are maternal tachycardia, tachyarrhythmias and hypotension. Prolonged tachycardia may be associated with cardiac ischaemia or pulmonary oedema, the latter particularly when the tachycardia is associated with circulation overload. In addition, prolonged usage is associated with biochemical disturbance, notably hyperglycaemia and hypokalaemia. The beta-agonist is administered intravenously, starting at a dose which rapidly inhibits uterine contractions and produces a moderate tachycardia. The dose may be increased provided that the maternal pulse does not exceed 120 beats per minute. Treatment should be maintained at the minimum level which controls uterine contractions consistently. Once contractions have ceased for a period of 12 hours, the drug can be discontinued. There is no evidence that continued treatment, for example, by the oral route, conveys any advantage. In fact any continued treatment is accompanied by continuing biochemical disturbance, side effects and diminishing effectiveness, due to tachyphylaxis.

FETAL ASSESSMENT

In preterm labour or following membrane rupture, the fetus must be assessed very carefully. Ultrasound is essential to assist in establishing presentation, fluid volume, biophysical behaviour as an index of hypoxia, and for example, if there is bleeding, the presence or absence of placenta praevia. Even if caesarean section is not contemplated, fetal monitoring by cardiotocograph (CTG) is essential during labour, so that both obstetrician and paediatrician are cognizant of whether the fetus has been significantly hypoxic prior to delivery. Assessment of the CTG trace is slightly more difficult in the preterm infant. This is because tachycardia is common and produces a secondary loss of variability which does not reflect hypoxia. Nevertheless, severely abnormal fetal heart patterns do indicate a hypoxic state and indicate a need for early delivery. Causes of fetal hypoxia in posterm labour are shown in Table 5.3.

TECHNIQUES OF DELIVERY

Between 24 and 32 weeks special precautions must be taken to avoid trauma to the fetus during delivery. If the membranes are left intact through the first stage of labour, they protect the fetus both from bruising and from hypoxia caused by cord compression. It is often possible to deliver the fetus in an intact protective bag of membranes and the sac can be broken as the fetus comes down the vagina immediately prior to delivery. An episiotomy would be unhelpful with such a small baby and is therefore an unnecessary burden on the mother. The delivery must be conducted by someone with experience of preterm delivery, which is quite different to delivery at term. Delay or difficulty in the second stage at this gestation

Table 5.3   Causes of fetal distress in labour

Maternal
Poor placental perfusion
Hypovolaemia – haemorrhage, dehydration
Hypotension – shock, drugs, epidural
– supine hypotension/
Myometrial hypertonus – prolonged labour
– excess oxytocin
Fetal
Cord compression – oligohydramnios
– entanglement
– prolapse
Pre-existing hypoxia or growth retardation
Infection
Cardiac
Placental
Bleeding or separation (abruption)

should be treated by caesarean section, as manoeuvres such as internal version and forceps delivery are excessively traumatic to the fetus.

If caesarean section is required, the technique must be modified to provide an atraumatic delivery. The lower segment is often poorly formed, so that the uterus appears waisted. In this case a low vertical mid-line incision must be made in the anterior wall of the uterus. Around 14 cm of uterine incision are required for a fetus of this size and it is easy to see whether this space is available in the lower segment or whether a mid-line incision should be used. The available length of transverse uterine incision in the lower segment can be increased by making a U-shaped incision in the lower segment, extending up in a smile on both sides. This requires quite a well-formed lower segment however, and if there is any doubt a midline incision should be used. Having opened the uterus, pressure on the fundus will deliver the amniotic sac containing the fetus through the uterine incision, closely followed by the placenta. The sac is then opened, the cord clamped and the fetus wrapped in a warm towel to avoid heat loss, before passing it to the paediatrician.

Between 32 and 37 weeks the fetus is larger and more robust. Delivery with ruptured membranes is less traumatic and the perineum may sometimes obstruct the delivery so that episiotomy is required. There was a fashion for prophylatic forceps delivery in the 1970s, but it is now clear that this confers no benefit to the fetus.

FETAL DISTRESS DURING LABOUR

There are many causes of fetal distress in labour (Table 5.3) but uterine contractions are the most important factor. During each contraction the maternal blood flow through the placenta is impeded; the venous outflow is obstructed before the arterial supply. The fetus is usually unaffected by this, but if uterine contractions are prolonged or if there has been previous impairment of placental function then fetal hypoxia, hypercapnia and respiratory acidosis occur, with a fall in pH. In severe hypoxia the fetal tissues meet their energy requirements by anaerobic glycolysis, burning up glycogen but producing lactic acid in the process, and this causes further depression of the pH by metabolic acidosis. Although the fetus will withstand and recover from degrees of hypoxia and acidosis which would be fatal in an adult, if the pH falls below about 6.9 death or brain damage is almost inevitable.

The fetus also shows clinical signs of distress with the changes in hydrogen ion concentration. Distress is an ill-defined term, and may not invariably be due to hypoxia, although this is the commonest cause. The first effect of hypoxia is a rise in fetal heart rate from sympathetic action. Fetal tachycardia in cases of hypoxia is of varying degree, but a rate persisting at well above 160 beats per minute has considerable significance. Fetal tachycardia may also occur if the mother has high fever or is dehydrated, but in all other cases this sign must be regarded as evidence of fetal distress.

If the hypoxia persists and is of more severe degree, the fetal heart rate will slow after each uterine contraction, and there will also be persistent bradycardia. Any case with slowing below 120 beats per minute should be carefully assessed, and urgent assessment is necessary if the rate falls below 100, or if there is cardiac irregularity.

Initial slowing of the fetal heart rate is probably a vagal effect, and vagal activity may also lead to contraction of the bowel and the passage of meconium into the fluid. However, it should be noted that premature fetuses, even if fatally hypoxic, seldom pass meconium.

Unfortunately the correlation between many of these clinical signs of fetal distress and real hypoxia is not very strong. Even the most serious combination of these signs, bradycardia with the passage of meconium, is found to be associated with a significantly hypoxic fetus in only 25 per cent of cases, so that if action were taken on these signs alone it might be unnecessary in three-quarters of the cases.

CONTINUOUS MONITORING OF THE FETAL HEART RATE

Ordinary auscultation of the fetal heart rate has

several limitations. It has to be performed inter-mittently and the intervals between observations may vary. If the heart is being auscultated for 60 seconds every 15 minutes, there are 14 out of 15 minutes when changes may not be observed.

Further, the main stress that affects the fetal heart rate is a uterine contraction, yet it is at this time that many doctors and midwives fail to hear the heart. Unless auscultation takes place very soon after the contraction any change due to the stress will be missed. In consequence efforts have been made during the last 20 years to record the fetal heart rate continuously, even during contrac-tions. This can be achieved by several methods:

Phonocardiography

A microphone is held on the mother's abdomen by a wide elastic belt, and picks up the fetal heart sounds. This works well enough when the woman is still, but as labour progresses and the woman moves about the microphone picks up too much background noise so that this method is now seldom used during labour.

Pulsed ultrasound

With pulsed ultrasound the fetal heart rate can be recorded using the Doppler effect when a pulse source and receiving head is placed on the mother's abdomen. An ultrasonic pulse is passed into the fetus; those waves that hit moving fluid or objects are returned at a different frequency from those reflected back from static objects, and the pulsa-tions of the fetal heart can be recognized. The instrument is less affected by movements and external sounds than the phonocardiograph.

Recording electrical activity

The best method currently available is to record the electrical activity of the fetal heart. When the cervix is more than 2 cm dilated and the mem-branes have ruptured a scalp electrode can be fixed to the fetal scalp; this reduces difficulties arising from signals from the electrical activity of the maternal heart. With a ratemeter a continuous estimate of the fetal heart rate is recorded on a trace on which the uterine contractions are also recorded with a tocometer.

A disadvantage of this method is that the elec-trical connection between the electrode and the recording machine restricts the mobility of the patient. This can be overcome by telemetry appa-ratus which transmits radio signals to the record-ing machine. The purpose of the investigation, and its possible contribution to the safety of the baby, should always be carefully explained to the moth-er.

Conventionally a range of 120 to 160 beats per minute is taken as the normal fetal heart rate during labour, but with better recording methods it is found that in most cases the rate is within a range of 130 to 150 beats per minute (*see* Fig. 5.31). Both baseline tachycardia (*see* Fig. 5.32) and base-line bradycardia (*see* Fig. 5.33) have a significant association with chronic fetal hypoxia. In most instances the heart first beats faster and then slows down if the stress continues.

Continuous records of the fetal heart rate nor-mally show continuous minor variations, with a range of about 10 beats per minute. This baseline variability is the response of the heart to a variety of factors, including vagal and sympathetic stimuli, catecholamines and oxygen tension. Loss of base-line variability (*see* Fig. 5.34) implies that the car-diac reflexes are impaired, either from the effect of

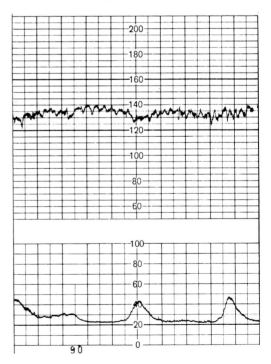

Fig. 5.31 Normal trace. The upper record is of the fetal heart rate; the lower record shows uterine activity

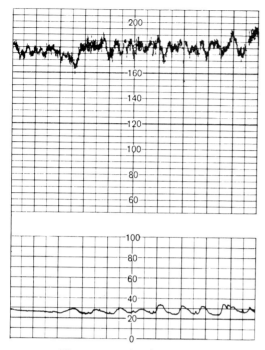

Fig. 5.32 Fetal tachycardia

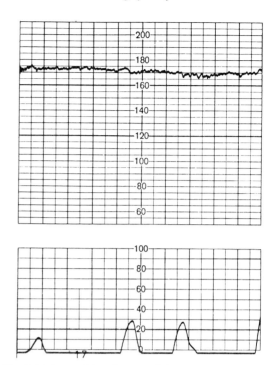

Fig. 5.34 Loss of baseline variability in fetal heart rate

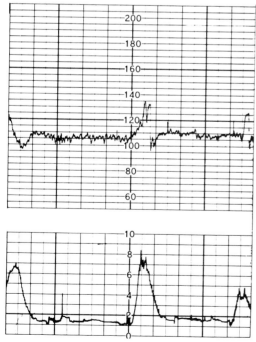

Fig. 5.33 Fetal bradycardia

hypoxia or of drugs such as diazepam (Valium). It may be serious, but it is also a physiological phenomenon that occurs cyclically.

Stress to the fetus during labour occurs during a uterine contraction, and the response of the fetal heart rate to contractions is shown on the rate-meter trace (*see* Fig. 5.35). With each contraction the rate often slows, but it returns to normal soon after removal of the stress. These early decelerations in the heart rate start within 30 seconds of the onset of the contraction and return rapidly to the baseline rate. They are not of serious significance as a rule and indicate that while the fetus is undergoing some stress the cardiac control mechanisms are responding normally.

Figure 5.36 illustrates a more serious condition. Here the fetal heart takes some time to respond to the stress of the uterine contraction, with the onset of late deceleration more than 30 seconds after the beginning of the uterine contraction, and a slow return to the baseline rate. Such late decelerations are commonly caused by hypoxia. They must be considered to be serious and their exact significance must be checked by taking a sample of fetal

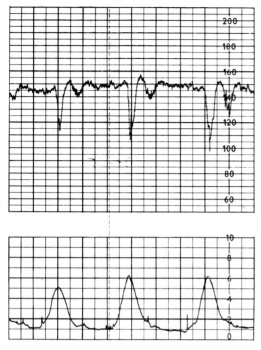

Fig. 5.35 Fetal heart rate; early decelerations

blood from the fetal scalp to determine its pH if the cervix is more than 2 cm dilated.

In practice variations in the fetal heart rate may be complex and difficult to interpret. Figure 5.37 shows variable decelerations with no consistent relationship to uterine contractions. These are sometimes caused by compression of the umbilical cord between the uterus and the fetal body, or because it is looped round some part of the fetus. Provided that they do not persist for more than a few minutes they may have little significance, but persistence for more than 15 minutes would call for an urgent fetal scalp blood sample.

Overstimulation of the uterine muscle with oxytocin will cause prolonged uterine contractions and severe changes in the fetal heart rate (*see* Fig. 5.38).

If the woman lies on her back compression of the inferior vena cava by the uterus may impede the venous return, depress the cardiac output and reduce placental blood flow. The effect of asking the woman to lie on her side should be observed (*see* Fig. 5.39); if the abnormal changes in the fetal heart rate still persist myometrial overactivity may well be the cause.

A rare cause of a slow fetal heart rate is congeni-

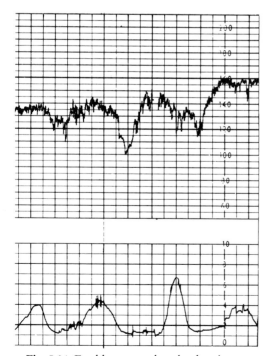

Fig. 5.36 Fetal heart rate; late decelerations

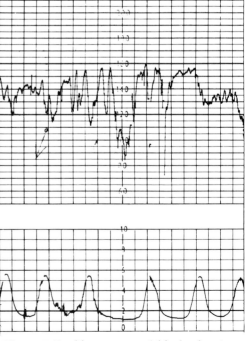

Fig. 5.37 Fetal heart rate; variable decelerations

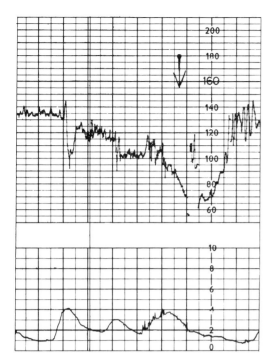

Fig. 5.38 Trace of fetal heart rate showing the effect of overstimulation of the uterus with oxytocin. The drip was turned off at the point marked with the arrow

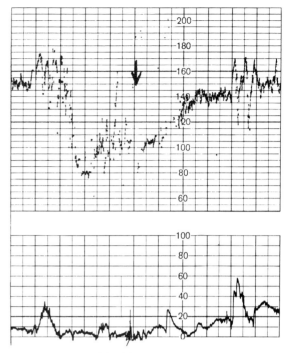

Fig. 5.39 Trace of fetal heart rate showing the effect of maternal supine hypotension. At the point marked with the arrow the mother was turned onto her side. As her blood pressure returned to normal the fetal bradycardia ceased

tal heart block. It carries a bad prognosis as it may be associated with other congenital cardiac abnormalities or maternal disseminated lupus erythematosis. The condition may be recognized by noting that the rate has always been slow, even in pregnancy, and by observing that the slow rate does not vary during labour.

An increase in the fetal heart rate can occur with intrauterine infection or with maternal pyrexia from any cause.

Precise continuous recording of the fetal heart rate in relation to uterine contractions is a great advance in monitoring the fetus during labour, but it should not be assumed that such traces are absolute indications for intervention. Any suggestion that the fetus is hypoxic should be checked by determining the pH of the fetal blood before caesarean section is undertaken. The most serious pattern of heart rate changes, namely fetal bradycardia with loss of baseline variability and late decelerations, is associated with significant fetal hypoxia in about 65 per cent of cases. Even among these there will therefore be about 35 per cent of

cases in which fetal blood sampling will show that immediate intervention is not necessary.

FETAL SCALP BLOOD SAMPLING

During labour, after the membranes have ruptured spontaneously or have been ruptured artificially, a sample of fetal blood can be obtained for evaluation of its acid–base status. It is possible to measure the partial pressure of oxygen, but this fluctuates very quickly, and the pH gives a better indication of the metabolic state of the fetus. The blood is obtained by passing an endoscope through the cervix and then using a very small blade held in a suitable handle to make a small incision in the fetal scalp. A drop of blood is drawn into a long fine tube. With 0.2 ml of blood the pH can be determined immediately with a micro-Astrup machine (*see* Figs 5.40 and 5.41).

Fetal blood sampling during labour is useful for excluding hypoxia and acidaemia in cases with clinical suspicion of fetal distress, so that unneces-

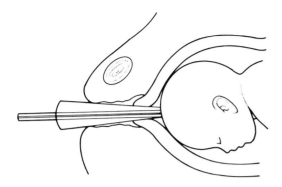

Fig. 5.40 A tapered endoscope is passed through the cervix to expose the fetal scalp. The scalp (Fig. 5.43) produces a blend of blood which is drawn off in a prelapanized tube

sary caesarean section may be avoided. If the pH is above 7.25 there is no hypoxia at the time of the observation, although the test has no predictive value. In doubtful cases a progressive fall in pH below 7.2 in successive samples during the first stage of labour is highly significant, but in the second stage a pH of 7.15 may be accepted as the lower limit of normal. Sampling is seldom performed during the second stage, because by this time it is usually simpler to deliver the fetus suspected of being hypoxic with forceps.

The state of the fetus is best monitored during labour by a combination of continuous recording of the fetal heart rate with blood sampling when necessary to confirm any suspicion of hypoxia. Obviously such monitoring is only possible in well-equipped and well-staffed units. In such units about a third of the women are usually judged to need continuous monitoring. At present continuous fetal heart rate monitoring is widely used in the UK but fetal blood sampling is less often

Fig. 5.41 Enlarged photograph of blade. The small rectangular blade is 2 mm long and the wide shoulder prevents deeper penetration of the scalp

performed. These methods do not cause an increased work load for the staff, because they may reduce the caesarean rate for unproven fetal distress. In some respects they make surveillance of the women easier, but they must never be allowed to interfere with the close personal relationship between mothers in labour and the staff. If women who are judged to have high risk pregnancies are transferred to units with good facilities for monitoring better management of fetal hypoxia will result, with a reduction in both perinatal mortality and morbidity.

TRAUMATIC LESIONS

RUPTURE OF THE UTERUS

Rupture of the uterus is a most serious condition which usually occurs during labour, although it can occasionally happen during the later weeks of pregnancy.

Causes

During pregnancy

During pregnancy the only common cause of rupture of the uterus is a weak scar after previous operations on the uterus. The higher the scar is placed on the uterus the greater the risk. The most dangerous scar is that of classical caesarean section; this is more dangerous than a hysterotomy scar. Rupture of a lower segment caesarean scar is uncommon during pregnancy. Rupture after myomectomy, tubal reimplantation or excision of a uterine septum, or following perforation of the uterus with a curette or cannula, is rare. Rupture of the uterus during pregnancy may follow a direct blow on the abdomen, and a perforating or gunshot wound may injure the uterus and the enclosed fetus.

During labour

During labour rupture may be caused by:

* obstructed labour – the rupture may be spontaneous or follow manipulations carried out for the relief of the obstruction
* intrauterine manipulations, such as internal version or manual removal of an adherent placenta
* forcible dilatation of the cervix. A cervical tear in delivery may extend up into the body of the uterus
* the injudicious use of oxytocic drugs
* a weak scar in the uterus after caesarean section, or in rare instances other uterine surgery or perforation of the uterus with a curette or cannula
* in women who have had numerous pregnancies uterine rupture occasionally occurs without any evident preceding abnormality.

Pathology

Ruptures of the uterus are divided into:

incomplete or extraperitoneal
complete or intraperitoneal

depending on whether the peritoneal coat is torn through or not.

Obstructed labour

In obstructed labour rupture of the uterus generally takes place in the overstretched and thinned lower segment, to which it may be limited, but sometimes it spreads upwards or downwards. The life of the mother is threatened by shock and intraperitoneal bleeding. In cases of obstructed labour the fetus may be dead before the rupture occurs, but in any case it will perish if complete rupture occurs.

Rupture of a scar

The traditional teaching that rupture of scar in the uterus is more likely to occur during labour rather than during the latter weeks of pregnancy is being questioned. In the UK a weak caesarean scar is now the commonest cause of rupture of the uterus. Overdistention of the uterus, by twin pregnancy for example, will increase the risk. Healing of a uterine scar may be imperfect if sepsis occurred in the puerperium, or if the edges of the incision had been inaccurately sutured. If the placenta is implanted over the scar in a subsequent pregnancy the risk of rupture is increased. The scar may give way if caesarean section is repeated several times and the current operating incision is made through the scar tissue left by previous operations. A lower segment scar may be just as likely to rupture as an upper segment one in either pregnancy or labour.

A caesarean scar in the lower segment may stretch gradually, so that the uterine wall in the region of the scar is only represented by attenuated and avascular fibrous tissue. When the weak area finally dehisces during labour there is sometimes relatively little intraperitoneal bleeding. The membranes may bulge through the rent, and will eventually give way, when the fetus or placenta may pass through it.

Symptoms and signs

Rupture through a uterine scar

In cases of rupture through a uterine scar during pregnancy the history of the previous operation will be available, and the scar in the skin will be seen, although a low transverse incision may be hidden by pubic hair. The abdominal incision of a classical (upper segment) caesarean section is made one-third above and two-thirds below the umbilicus. A longitudinal incision below the umbilicus signifies a lower segment section, as does the one commonly performed across the abdomen an inch or two above the symphysis.

Rupture during pregnancy may be so gradual that the symptoms may be very slight with a silent rupture. There is abdominal pain (which may be wrongly attributed to the onset of labour) but at first there is little change in the general condition. At this stage diagnosis may be difficult and it may be necessary to observe for a time before a conclusion is reached. If the rupture becomes complete and part of the uterine contents are extruded into the peritoneal cavity more severe pain and shock occur.

Rupture of a scar often occurs during labour, and the scar gives way more suddenly than during pregnancy, so the symptoms are more dramatic, with severe pain, shock, fetal distress and some bleeding from the vagina. Unless the contents of the uterus pass into the peritoneal cavity uterine contractions may continue. The possibility of rupture of the scar should always be considered if a woman who has had a caesarean section suddenly complains during labour of severe pain which is not synchronous with the uterine contractions. The accident does not only occur after a long and difficult labour, and for that reason the woman's general condition is better, and the risk of infection less, than in cases of rupture due to obstructed labour.

Spontaneous rupture during obstructed labour

The labour will have been prolonged, or there will have been violent uterine action almost without intermission between the pains, so that the woman may be exhausted before the rupture occurs. There may be signs of disproportion or of a malpresentation such as a transverse lie, although these signs may have been overlooked before the accident. There may be fetal distress. At the moment of rupture the woman complains of a sharp, tearing pain in the lower abdomen. After the rupture there is shock, with pallor, sweating and constant lower abdominal pain. The pulse becomes thready and rapid and the blood pressure falls. With an incomplete tear the signs of shock may not be so severe.

Some degree of vaginal haemorrhage is commonly present. On abdominal examination there is marked tenderness over the site of the rupture. The presenting part may not be reached *per vaginam* unless it is impacted in the pelvis. If the fetus is completely extruded into the peritoneal cavity uterine contractions may cease, but in other cases they often continue. With complete extrusion the fetus may be felt all too easily in the abdominal cavity with the retracted uterus beside it.

Extended cervical lacerations

In some respects these injuries are usually produced with the forceps at a difficult delivery, especially if the cervix is not completely dilated, but seldom extend far enough to open the peritoneal cavity. Brisk external haemorrhage may occur, or a large haematoma may form in the broad ligament. Vaginal bleeding in the third stage of labour with the uterus empty and firmly retracted should always suggest the possibility of this type of injury, which can be confirmed by visual examination of the cervix. For this effective retractors, a good assistant and a good light will be required.

Rupture caused by oxytocic drugs

Rupture of the uterus can follow the administration of oxytocin before the delivery of the child, particularly when obstruction prevents rapid delivery. The risk is much greater in multiparae. The danger is less if the oxytocin is given as a dilute intravenous drip given in an increasing fashion. The uterine contractions are carefully observed and controlled.

Rupture caused by direct injury to the abdomen

In the rare cases of rupture from this cause severe shock and abdominal pain, together with the history of the accident, will suggest the possibility of visceral injury. Precise diagnosis may be impossible without laparotomy, and in case of doubt this would be justified.

Prognosis

Rupture of the uterus in cases of obstructed labour has a high mortality because the accident usually occurs in cases of prolonged labour with manipulation, sometimes with inefficient obstetric aid, and often without proper aseptic precautions. The mortality after rupture of a caesarean scar is much less, as the accident does not usually follow prolonged labour, and the woman is usually confined in hospital where the rupture is more quickly detected.

In England and Wales the incidence of maternal death from uterine rupture has fallen from 6.0 per million maternities in the triennium 1976–78 to 1.9 in 1985–87 (six cases). In two of the six women, uterine rupture occurred spontaneously during normal labour. Four were caused by cervical lacerations, one due to oxytocic overstimulation and three to instrumental extraction (one forceps; two Ventouse) through a scarred or incompletely dilated cervix. There were no cases of caesarean scar rupture among these reported deaths.

The prognosis for the fetus varies with the degree of interference with placental circulation in instances of scar rupture which may be minimal in silent rupture or dehiscence to total, the overall perinatal morbidity rate being 50 per thousand total births.

Dangerous intrauterine manipulations are now rarely performed in the UK and obstructed labour is less often seen owing to timely caesarean section.

Treatment

Prevention

A woman who has had a caesarean section, hysterotomy or any operation on the uterine wall must be delivered in a hospital where all obstetric facilities are available. However, the fact that she has already had one caesarean section does not mean that a subsequent pregnancy must also be treated in this way. If the first operation was not performed for disproportion but for some non-recurrent condition, such as placenta praevia, vaginal delivery should be allowed, with the proviso that if labour does not progress smoothly section will be repeated. If the woman had an upper segment operation, or if gross uterine infection occurred, elective section may be advisable. Ultrasound scanning will show if the placenta overlies the uterine scar, when the risk of rupture is increased.

Treatment after rupture has occurred

Before operation the woman's general condition must be improved as much as possible by giving morphine, blood transfusion Haemacel and intravenous glucose solution until blood is ready.

Immediate laparotomy is required. In cases of scar dehiscence it is often possible to excise the edges of the rent and resuture the uterus. The bladder is sometimes adherent to the scar and may also be torn, in which case its wall must be freed and sutured in two layers and a suprapubic indwelling catheter inserted.

Some women with a uterine rupture during obstructed labour are best treated by hysterectomy, as efficient suturing of bruised and ragged tissues may be impossible. If the tear is accessible so that both ends can be seen and got at for suturing and the edges not too ragged it may be sutured, but the risk of rupture in a subsequent pregnancy is so great that it is usually wise to prevent this by occluding the Fallopian tubes.

Wide-spectrum antibiotics are given and paralytic ileus is treated by giving only intravenous fluids and maintaining gastric aspiration until bowel sounds reappear.

ACUTE INVERSION OF THE UTERUS

In this condition the body of the uterus becomes either partially or completely turned inside out after delivery of the fetus (*see* Fig. 5.42). It is an important but rare cause of shock in the third stage of labour and can be fatal if it is untreated. Inversion may take place before or after the delivery of the placenta.

Cause

The usual cause is over enthusiasm of the attendant

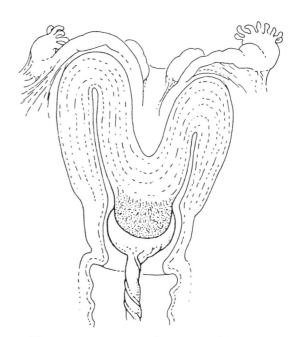

Fig. 5.42 Acute puerperal inversion of the uterus

in the third stage of labour by pulling on the umbilical cord or pressing on the fundus while the uterus is not contracting and the placenta has not separated. Very rarely it occurs spontaneously.

Symptoms and signs

The chief symptoms are those of shock, with some haemorrhage, and sometimes the appearance of the uterine fundus at the vulva. As a rule shock is severe and much greater than the blood loss warrants. Pain is of variable degree. Unexplained shock during the third stage of labour should always suggest the possibility of uterine rupture or that inversion has occurred and a vaginal examination should be made. In cases of inversion the body of the uterus will not be felt in its usual position and the round mass of the uterus will be felt protruding through the cervix. Inversion has a high mortality if it is undiagnosed or left untreated.

Treatment

Shock and possibly bleeding will continue until the uterus is replaced, hence it is desirable that it should be replaced at once. The inverted fundus soon becomes oedematous, making replacement more difficult.

As soon as the diagnosis is made the woman is anaesthetized and the uterus is replaced. After cleansing the vulva, vagina and inverted uterus, the placenta, if it is still attached to the uterus, is peeled off. The uterus is then squeezed in the hand and replaced, the part which became inverted last being replaced first, and the fundus last of all.

When replacement is complete further haemorrhage and recurrence of the inversion are prevented by intravenous injection of oxytocin. Treatment for shock is given concurrently.

Alternatively, replacement is possible by fluid pressure with warm saline delivered into the vagina through wide bore tubing from a container held at a height of about 60 cm. The vaginal exit is sealed by the operator's hand holding the labia major together. A surprisingly large amount of fluid is necessary.

LACERATION OF THE CERVIX

Minor lacerations of the cervix occur frequently but do not cause symptoms.

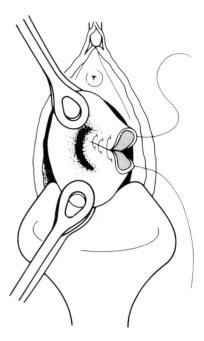

Fig. 5.43 The repair of a tear of the cervix by interrupted catgut sutures. The cervix is depicted outside the vulva for the sake of clarity

Extensive lacerations may be caused by precipitate labour, application of forceps with the cervix incompletely dilated, or rapid delivery of the head in a breech presentation. A scar in the cervix from previous injury may also tear. With a deep tear there is continuing haemorrhage during or after the third stage; this goes on even when the uterus is empty and retracted. The tear must be sutured (*see* Fig. 5.43). The woman is anaesthetized, a wide speculum is inserted, and the anterior and posterior lips of the cervix are held with sponge forceps and drawn well down, so that interrupted catgut or Dexon sutures can be inserted through the whole thickness of its wall. While waiting for arrangements to be made, bleeding may be temporarily controlled by applying sponge forceps to the edges of the tear.

LACERATION OF THE PERINEUM AND VAGINA

These lesions are of three degrees:

- the *first degree tear* involves only the skin and a minor part of the perineal body, and the related posterior wall of the vagina

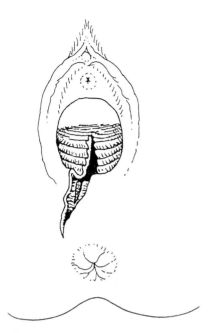

Fig. 5.44 Second degree perineal tear

- a *second degree tear* involves the perineal body up to (but not involving) the anal sphincter, with a corresponding tear in the vagina (*see* Fig. 5.44)
- a *third degree tear* includes the anal sphincter and usually extends for 2 cm or more up the anal canal. If a third degree tear is not repaired there will be incontinence of flatus and loose stool.

Extensive tears of the vaginal walls can occur without a tear in the perineum, and they should always be carefully inspected after delivery. Minor lacerations can also occur on the antero-lateral vaginal wall.

Treatment of first and second degree tears

It is important to repair all perineal lacerations immediately, to prevent any infection of the raw surface. The vaginal epithelium is sutured from the apex of the tear, which must be clearly identified, down to the introitus with either a continuous suture or interrupted sutures of fine catgut or polygycol. The latter is probably preferable for all perineal work for it is associated with fewer short- and long-term side effects. The perineal body is repaired with stitches of the same material. The skin edges are brought together without tension with fine catgut or polygycol sutures. The repair should be done carefully and accurately. Unless general anaesthesia or epidural anaesthesia is already being used, local anaesthesia with 1 per cent lignocaine is employed.

Treatment of third degree tears

The operation should be done by an experienced obstetrician in a properly equipped theatre with general or epidural anaesthesia. The tissues heal very well if the operation is done carefully soon after delivery, but if this is not done, or if the tissues fail to heal, the woman will have to undergo a more difficult operation later, or she will suffer from rectal incontinence.

Dexon sutures are used throughout. The anal mucosa is first repaired with fine stitches, tying the knots inside the bowel lumen. The ends of the anal sphincter are found and carefully brought together with interrupted sutures. The other tissues are repaired as described above for second degree tears (*see* Fig. 5.45).

After-care

Each day the perineum is washed with soap and water, and then carefully dried. Women with extensive tears sometimes get retention of urine, for which catheterization is necessary. If the bowel has not acted by the 4th day a glycerine suppository may be used.

Rarely the perineal wound becomes infected; sufficient stitches are removed to permit drainage, and antibiotics are given. Salt water bathing, preferably in a bidet, is continued until the wound is covered with granulation tissue, when secondary suture can be performed if necessary.

FISTULAE

Fistulae may occur as a result of pressure by the presenting part in prolonged labour, or by direct injury during operative procedures such as forceps or caesarean section. In obstructed labour prolonged pressure between the head and the pubic bone may cause local ischaemia and the subsequent necrosis of the anterior vaginal wall and base of the bladder leading to a *vesicovaginal fistula*. With very extensive necrosis the rectum may also be involved. A much more common cause of a *recto-*

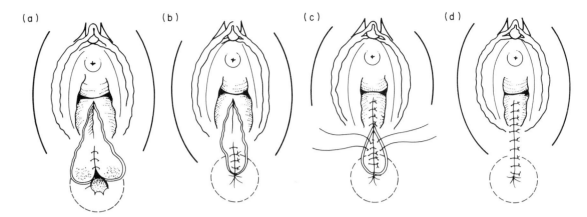

Fig. 5.45 Repair of complete perineal tear. (a) Suture of the rectal mucosa; 2, suture of the ends of the anal sphincter; 3, repair of the perineal body; 4, suture of the perineal skin

vaginal fistula is a third degree tear of the perineum, in which the lower part has healed but a defect remains higher up.

These varieties of fistulae are now very uncommon in the UK, but in countries with inadequate obstetric services they are still a frequent cause of distressing disability. Depending on the type of fistula the woman has urinary or faecal incontinence, and sometimes both. If the fistula is caused by operative injury the symptoms appear immediately, but if it is the result of pressure necrosis the symptoms do not appear until about the 8th day, when the sloughs separate. On examination an opening into the bladder or rectum is found.

When the wound granulates a small rectal fistula may heal, but this is unlikely with a vesicovaginal fistula. In the case of a fistula caused by direct injury immediate repair is performed, but in those caused by ischaemic necrosis repair is not attempted for 2 to 3 months, when the effects of trauma and superadded infection have subsided. Details of the surgical treatment are given in *Gynaecology by Ten Teachers*.

HAEMATOMA OF THE VULVA

This may be caused by rupture of a vulval varix, but more often it occurs after perineal repair when a vessel is in spasm at the time of repair, relaxes and bleeds later. It can occasionally occur after normal labour with an apparently intact perineum.

The haematoma appears fairly suddenly as a very tender purple swelling on one side of the vulva; it may become 10 cm or more in diameter.

Sometimes the blood tracks upwards to form a swelling at the side of the vagina. The pain is severe and there is sometimes shock. If a woman complains of severe perineal pain after delivery the perineum should always be inspected before giving her analgesics.

If the tension in the swelling is great or if the swelling is increasing in size it should be incised and the clot turned out. If the torn vessel can be found it should be ligated, but this is unlikely. A drain is left in the cavity and a firm dressing is applied.

BROAD LIGAMENT HAEMATOMA

In this uncommon accident a deep vessel has been torn at the time of delivery, goes into spasm and then relaxes later. A haematoma forms above the pelvic diaphragm and spreads into the base of the broad ligament. It may also be caused by extraperitoneal rupture of the uterus. The bleeding most frequently occurs during or soon after labour or caesarean section and the haematoma is discovered a few hours or often days later, because of pain and deterioration in the woman's general condition. The haematoma may be large enough to be palpable on abdominal examination, and it will displace the uterus upwards and to one side. The woman is often anaemic, and there may be slight fever.

A broad ligament haematoma usually undergoes gradual absorption, but will take several weeks if it is large. Infection is rare, but may occur and lead to abscess formation. Most cases are treated conservatively. Blood transfusion may be required, and antibiotics are given. If infection occurs and an abscess forms, this is opened wherever it points.

MATERNAL NERVE INJURIES DURING LABOUR

Foot-drop from paralysis of the dorsiflexor muscles of the leg can follow delivery.

In a few cases this is the result of pressure on the lateral popliteal nerve near the neck of the fibula by a leg support used to hold the woman in the lithotomy position. If the legs are placed outside the supports this cannot happen. Proper rubber padding and poles should be employed.

In the majority of cases it is a different type of injury involving the 4th and 5th lumbar nerve roots. This may be the result of sudden prolapse of an intervertebral disk or of pressure on the lumbosacral cord by the presenting part near the brim of the pelvis. The lesion is usually unilateral, and it often follows difficult labour, especially forceps delivery. Apart from the foot-drop there is an area of sensory loss on the dorsum of the foot and lateral aspect of the ankle. The prognosis is good, although recovery may take several months. During that time a toe-spring is attached to lift the foot during walking, and regular physiotherapy is given.

Rarely, prolonged sensory loss in the leg, without foot-drop, follows epidural anaesthesia.

POSTPARTUM HAEMORRHAGE

Postpartum haemorrhage (PPH) is excessive bleeding from the genital tract after the birth of the child. It is conventionally defined as a loss of more than 500 ml. The haemorrhage may be immediate (or *primary*) or if it occurs more than 24 hours after delivery until 6 weeks it is described as *secondary*.

In England and Wales maternal deaths from PPH have been reduced by two-thirds (18 to six) from 1970 to 1987. In the last 3 years of this period PPH accounted for over 60 per cent of all maternal deaths from haemorrhage because of reduction in the other causes of death from haemorrhage. From 1970 to 1987 the total number of deaths from haemorrhage (PPH, placental abruption and placenta praevia) fell from 30 to 10 in England and Wales.

PRIMARY POSTPARTUM HAEMORRHAGE

There are two sources of primary postpartum haemorrhage, the placental site and lacerations of the genital tract. The incidence is reported to be between 1 and 2 per cent of deliveries, although there is reason to suspect that the true rate is much higher than this.

PRIMARY HAEMORRHAGE FROM THE PLACENTAL SITE

Some blood always escapes as the placenta becomes detached from its site, but usually less than 200 ml. Further loss is normally prevented by the retraction of the uterine muscle fibres which surround the vessels in the wall of the uterus and compress them until intravascular thrombosis occurs. Although a loss of more than 500 ml is arbitrarily defined as a postpartum haemorrhage, any loss which appears excessive must be treated at once. Even a small loss may be dangerous in an anaemic patient.

Causes

Ineffective uterine contraction and retraction

Weak contraction of the uterus in the third stage of labour may fail to separate the placenta completely, so that it remains in the upper segment of the uterus and prevents effective retraction of the placental site. In other cases, though the placenta has been completely separated and expelled, severe haemorrhage can occur if the uterus fails to maintain its retraction.

If the uterus is not completely empty retraction will be ineffective. Although it is true that a very strong contraction, for example after the intravenous injection of ergometrine, may temporarily control bleeding if the placenta is still in the uterus, in cases in which the contractions are less strong even a retained placental cotyledon or succenturiate lobe or blood clot in the uterus will interfere with retraction.

Ineffective uterine action may occur:

- after a long labour caused by weak or incoordinate uterine action. It will also occur in cases of long labour caused by mechanical difficulty if uterine exhaustion occurs
- if prolonged or deep anaesthesia has been administered
- if the woman is a multipara with an atonic uterus
- if the uterus has been overdistended with polyhydramnios. With twin pregnancy it is probably the increased area of the placental site rather than uterine overdistension which accounts for postpartum haemorrhage
- where there has been antepartum haemorrhage. In placenta praevia the lower segment may not retract well enough to control bleeding from the placental site. In cases of placental abruption the damaged uterus may fail to contract.

Mismanagement of the third stage

After a normal delivery, if an oxytocic drug has not been injected at the end of the second stage, the uterus remains quiescent for a few minutes. The placenta is still completely attached, and no bleeding occurs. But if the uterus is manipulated during this interval the placenta may be partly separated, and bleeding will begin and must continue until uterine contractions complete the separation and allow proper retraction to follow. Injudicious attempts to expel the placenta before complete separation has occurred are a common cause of postpartum haemorrhage, and may even cause uterine inversion.

Abnormally adherent placenta

Sometimes part of the placenta is abnormally adherent. In rare instances most of the chorionic villi penetrate through the decidua (*placenta accreta*) or into the myometrium (*placenta increta*). Should they invade through the myometrium to the peritoneal surface of the uterus, they constitute a *placenta percreta*. In cases of placenta praevia the placenta may have a wider area of attachment than normal, and the lower uterine segment may fail to retract strongly enough to control bleeding effectively.

Disseminated intravascular coagulation

These are causes of slow but persistent and dangerous haemorrhage. DIC is especially associated with concealed placental abruption, but may also occur in cases of amniotic embolism and after a dead fetus has been retained in the uterus for some weeks. They are dealt with fully in Chapter 3.

Clinical events

The escape of blood is usually obvious and the question of major importance is whether it is coming from the placental site or from a laceration. In rare instances severe bleeding occurs into the cavity of an atonic uterus, with only some of the blood appearing externally. This should be suspected if the woman becomes shocked, the fundus of the uterus appears to be abnormally high in the abdomen or the uterus feels larger and softer than normal.

If haemorrhage continues the blood pressure falls, the pulse rate rises and, in severe cases, pallor and air-hunger occur.

Owing to the increase in blood volume during pregnancy, and the increase in total red cell mass, a previously healthy parturient woman stands haemorrhage comparatively well, provided that the loss is not extremely rapid. However, a woman who is already exsanguinated by an antepartum haemorrhage may die if even a relatively small amount of blood is lost after delivery, and in a similar way a comparatively insignificant secondary postpartum haemorrhage may have serious effects on a woman who has lost heavily at the time of delivery.

Postpartum necrosis of the anterior lobe of the pituitary gland is a rare sequel in cases of postpartum haemorrhage in which the blood pressure has remained at a low level for some hours.

Prevention

The prevention of postpartum haemorrhage is very important. Anaemia must be corrected during pregnancy because an anaemic woman tolerates haemorrhage badly. Any woman with a previous history of a postpartum haemorrhage, women with twins and grand multiparae should always be admitted to hospital for delivery.

Prolonged labour which might lead to uterine exhaustion can often be prevented by the use of an intravenous infusion of Syntocinon during labour;

if such an infusion is given it should not be stopped until the third stage is safely completed.

The correct management of the third stage of labour is the most important factor in avoiding postpartum haemorrhage. Intravenous or intramuscular Syntocinon or Syntometrine should be given immediately after delivery of the anterior shoulder of the fetus.

Treatment

Two principles govern the treatment of postpartum haemorrhage:

the bleeding must be arrested
the blood volume must be restored.

Bleeding from the placental site will stop when the uterus is empty and retracted. Treatment will differ according to whether the placenta is still in the uterus or has been delivered.

Treatment if the placenta has already been delivered

The first step is to place a hand on the abdomen, with the woman lying on her back, to ascertain whether the uterus is soft or hard. If the uterus is soft and relaxed a contraction is stimulated by rubbing the uterus gently with the abdominal hand, placing the thumb in front and the fingers behind the fundus. When the uterus contracts and bleeding is controlled any clots are expelled by fundal pressure, and an intravenous injection of Syntometrine 0.5 mg is given to maintain the contraction. The placenta and membranes are examined carefully to make sure that a cotyledon or succenturiate lobe is not retained; any retained placental tissue must be removed manually under anaesthesia.

In exceptional cases the uterus is hard and firmly contracted but bleeding continues. This is likely to be from a laceration of the cervix or vagina.

Treatment if the placenta is undelivered

The first step is again to ascertain whether the uterus is soft or hard. Unless the uterus is already firm the fundus is gently rubbed as described above to stimulate a contraction. The next step is to determine whether the placenta has separated. There are two clinical possibilities, as follows.

If the placenta has separated

If the placenta has separated and then been expelled from the upper into the lower uterine segment the uterus, when it contracts, will be felt as a firm rounded mass about 10 cm in diameter, at about the level of the umbilicus. It can be moved from side to side. The umbilical cord will have lengthened as the placenta separated and the lower part of the placenta can be felt per vaginam. If these signs of separation are present the placenta is delivered by Brandt–Andrews method. If the bleeding does not then stop an intravenous injection of Syntometrine 0.5 mg is given. (Even if the woman has already received 0.5 mg Syntometrine at the time of delivery this dose can be repeated safely once.) If the uterus does not contract well in spite of the Syntometrine *bimanual compression* is immediately performed. The right hand is inserted into the vagina and formed into a fist, which is placed in the anterior fornix above the cervix (*see* Fig. 5.46). The left hand is placed on the abdomen and pressed downwards onto the posterior wall of the uterus so that it is compressed between the two hands. This is an effective but temporary method of controlling uterine bleeding, although it is uncomfortable for the woman and tiring for the obstetrician. Firm pressure must be maintained until the uterus is felt to contract.

While this is in progress the placenta should be

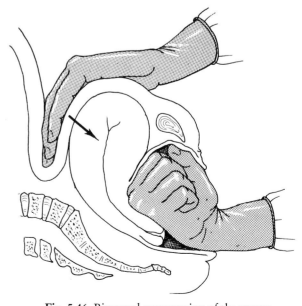

Fig. 5.46 Bimanual compression of the uterus

examined by an assistant to see that it is complete. If part of it is missing that will have to be removed digitally under general anaesthesia.

If the placenta has not separated

If bleeding is taking place and clinical examination, which may include vaginal examination, indicates that the placenta has not separated and that it remains in the upper uterine segment, then *manual removal of the placenta* under general anaesthesia is performed immediately. In these cases the uterus is still enlarged and soft, well below the umbilicus and, being splinted by the pelvic brim, it is relatively immobile. On vaginal examination the cord is felt passing up into the uterus but the placenta cannot be reached. Repeated or violent attempts to express the placenta by squeezing the uterus or pressing on it are unlikely to succeed and often produce shock.

If manual removal is to be performed it is best to withhold any further injection of oxytocic until after the removal of the placenta. When the anaesthetic has been induced a catheter may be passed to empty the bladder. For manual removal the left hand is placed on the abdominal wall to locate and steady the fundus of the uterus, and then the right hand is passed into the uterus to follow the cord to the placenta. The edge of the placenta is identified, and is gradually separated from the uterine wall with the fingers, while the external hand serves as a guide and reduces the risk of tearing the uterus (*see* Fig. 5.47). Only when the placenta is completely free should any attempt be made to remove it. The uterus is then explored carefully to ensure that no pieces of placental tissue are left behind; the beginner must realize that the site of attachment of the placenta is normally uneven and rough. The placenta is examined immediately to make sure that it is complete, which may be a difficult task as the placenta is now in pieces.

In the past manual removal of the placenta was a hazardous procedure, often performed after some delay on a shocked patient, with imperfect anaesthesia and a grave risk of infection. Today, if the operation is performed without delay, with good anaesthetic facilities, proper resuscitation, blood transfusion and antibiotic cover, the risk is slight. However, it is not an easy operation, and the beginner often diagnoses morbid adhesion of the placenta when none exists.

If bleeding still continues after removal of the

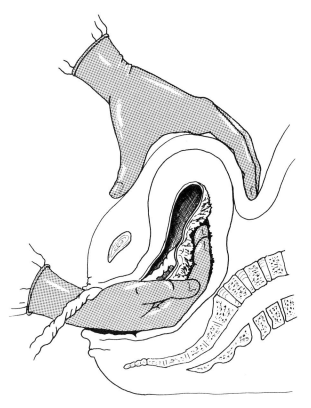

Fig. 5.47 Manual removal of the placenta

placenta bimanual compression may be necessary. In very exceptional cases, if bleeding continues in spite of all efforts (and is not caused by a clotting defect) tying of the internal iliac arteries or even hysterectomy may be considered as a last desperate resort.

Circulatory collapse caused by haemorrhage

In all cases of postpartum haemorrhage there is the danger of circulatory collapse, and resuscitation must be started as soon as possible, preferably *before* hypotension or tachycardia appear. Immediate blood transfusion is essential to restore the blood volume, and an infusion of plasma or saline may be started while the transfusion is being arranged. In an emergency, if the blood group is not known, rhesus-negative blood of the woman's ABO group (if it is known) may occasionally have to be used, but in every case a direct agglutination test of the donor's corpuscles and the woman's serum is essential. Dextran may interfere with the coagulation mechanism and should only be given after a sample of blood has been collected for cross-matching. It should not be used if DIV is suspected.

The woman is kept quiet and warm, with the foot of the bed raised. Her pulse, peripheral blood pressure and central venous pressure are monitored to assess transfusion requirements.

Problems of haemorrhage in a GP unit

If the woman is not delivered in a unit with full facilities for blood transfusion and emergency anaesthesia the dangers are increased, and the help of an emergency obstetric unit is essential. This flying squad, based on a maternity hospital, consists of an obstetrician, an anaesthetist and a midwife, and carries proper equipment for resuscitation and often for anaesthesia. A woman who has had a postpartum haemorrhage should not be moved until she has had an intravenous infusion set up, the placenta has been removed if possible and the bleeding has been controlled. There is much higher mortality if the woman is moved to hospital with the placenta still undelivered and after inadequate transfusion.

Haemorrhage caused by clotting disorders

If bleeding persists in spite of all other treatment described, then disseminated intravascular coagulopathy should be suspected. In an emergency the simple observation of failure of a sample of blood to clot in a test-tube may be sufficient to suggest the diagnosis.

PRIMARY POSTPARTUM HAEMORRHAGE FROM LACERATIONS

Primary postpartum haemorrhage can occur from lacerations of any part of the birth canal during labour, the most common sites of bleeding being either the cervix or the vaginal wall at the apex of an episiotomy wound or a tear. If bleeding continues after the placenta has been delivered and the uterus is firmly retracted, the vagina, cervix and lower uterine segment must be examined. This may be difficult until the woman is anaesthetized and placed in the lithotomy position. Proper retractors and instruments are needed to suture a high cervical tear. Profuse haemorrhage from a cervical tear involving a branch of the uterine artery can be temporarily controlled by clamping the highest part of the tear with sponge forceps until the woman can be taken to the operating theatre.

Bleeding from tears of the lower vagina, perineum or vulva should be controlled by pressure until the tear is sutured under local anaesthesia.

SECONDARY POSTPARTUM HAEMORRHAGE

This occurs more than 24 hours after delivery most often starting between the 5th and 10th days, although it can occur up to 6 weeks after delivery. It is usually caused by retention of a piece of placenta, and it is frequently complicated by intrauterine infection, with pyrexia. Ultrasound examination will show whether there is retained placental tissue. Secondary postpartum haemorrhage may also be caused by separation of an infected slough which has formed in a cervical or vaginal tear, or in a lower segment caesarean wound. A rare cause is infection and sloughing of subendometrial fibroids.

Treatment

Under general anaesthesia the uterine cavity is explored with the finger or with sponge forceps to discover and remove any placental tissue; the suction curette may also be used. The cervix often remains open when there is something retained in the uterus. If there is infection a uterine swab should be taken and the appropriate antibiotic is chosen according to the result of the cultures.

SHOCK IN OBSTETRICS

Obstetric shock does not differ from surgical shock; it results from the depression of many functions, in which reduction of effective circulation volume and blood pressure are of basic importance. The consequent inadequate perfusion of all the tissues leads to oxygen depletion and the accumulation of metabolites.

Most cases of shock in obstetrics are associated with severe haemorrhage, but other factors may be added; for example prolonged labour may be associated with electrolyte imbalance, general anaesthesia and trauma during operative delivery. Placental abruption can cause severe shock, and in the third stage of labour postpartum haemorrhage may occur. Maternal exhaustion, especially if combined with infection, will increase the effect of other factors.

Less frequent causes of shock include rupture of the uterus, acute inversion of the uterus, amniotic embolism, pulmonary embolism and adrenal haemorrhage. Bacteraemic shock is an additional form of shock.

Prolonged postpartum shock and hypotension may cause ischaemic necrosis of the anterior lobe of the pituitary gland.

Treatment

Unless prompt resuscitation is undertaken death may occur rapidly. It is essential to restore the blood volume as quickly as possible by intravenous infusion. If blood is not immediately available transfusion with saline or reconstituted plasma should be started while it is being obtained. The management of profuse blood loss after delivery (more than 2 litres) is an emergency for which every obstetric team should be trained. Clear lines of guidance should be issued in labour ward protocols and tested at intervals. A typical protocol is shown in Table 5.4.

Constant monitoring of the pulse rate, arterial blood pressure and central venous pressure is required to regulate the circulatory balance. Oxygen is sometimes required, and a small dose of morphine if there is severe pain. Attempts to raise the blood pressure by administration of vasoconstrictor drugs are usually ill-advised. If the limbs are pale and cold vasoconstriction is already present, and further vasoconstriction may only decrease the venous return still further. In many of these cases there is a risk of infection, and broad-spectrum antibiotics may be given intravenously.

POSTPARTUM PITUITARY NECROSIS

Severe postpartum collapse due to haemorrhage may be followed by ischaemic necrosis of the anterior lobe of the pituitary gland. Thrombosis occurs in the vessels which supply the anterior lobe, and necrosis of the whole lobe occurs, except for a thin rim of tissue which may survive at the surface of the lobe. Death may occur soon after delivery, but if the woman survives she will show the clinical picture of Simmonds' disease (sometimes also known as Sheehan's syndrome).

All the endocrine functions of the anterior lobe

Table 5.4 Typical management protocol for massive haemorrhage

Administration
- Contact the duty Obstetric Registrar (if not present) and the duty Anaesthetic Registrar.
- Contact the blood bank via the switchboard. Send at least 20 ml of blood for further cross-matching. Ask for at least 6 units of cross-matched blood.
- Inform the duty Haematologist of the clinical situation, the rate of blood loss, and of the clotting problems.
- One midwife should be assigned to record keeping and should record the following:
  (a) pulse
  (b) blood pressure
  (c) maternal heart rate, preferably via ECG
  (d) central venous pressure
  (e) urine output
  (f) amount and type of fluids the woman has received
  (g) dose and type of drugs the woman has received.
- Inform the duty Obstetric Consultant.
- Prepare for theatre as appropriate.

Clinical
- Put in two large bore (at least 16 gauge) intravenous cannulae.
- Give the following fluids:
  (a) up to 2 litres of Hartmann's solution
  (b) Haematocel or Hespan, up to 1.5 litres
  (c) uncrossed-matched blood, Rh negative, the woman's group
  (d) cross-matched blood as soon as available
  (e) give group 0 Rh negative blood only as a last resort.
- Insert a central venous line to monitor fluid replacement.
- Stop the bleeding.
  If the bleeding is from the placental bed of an intact uterus:
  (a) give 0.5 mg of ergometrine intravenously and set up an infusion of 40 units of syntocinon in 1 l of Hartmann's to run at 40 drops/min
  (b) commence bimanual compression of the uterus.
  (c) give carboprost i.m.
  If bleeding is from damage to the genital tract, repair it promptly under appropriate anaesthesia.
- In order, the following surgical procedures should be considered:
  (a) direct intramuscular injection of 0.5 mg of prostaglandin E2 into the exposed uterus
  (b) open the broad ligament and ligate the uterine artery on both sides
  (c) ligation of internal iliac arteries
  (d) hysterectomy.

of the pituitary gland are disturbed. There will be failure of lactation due to lack of prolactin. Because of lack of thyrotrophic hormone the woman becomes lethargic, abnormally sensitive to cold and usually gains weight. Her basal metabolic rate falls, and her glucose tolerance is increased. Because of lack of corticotrophic hormone she will also have asthenia, a low blood pressure and will respond poorly to infection. Lack of gonadotrophic hormones will lead to genital atrophy, with superinvolution of the uterus, amenorrhoea, and atrophy of the breasts. Hair will not regrow in the pubic region if it was removed for an abdominal delivery.

Less severe cases may only have part of the complex clinical picture, and in very rare instances a further pregnancy has followed, with regeneration of the pituitary gland.

Prompt treatment of collapse due to postpartum haemorrhage should prevent this disastrous complication. Once the necrosis has occurred substitution therapy may maintain the woman in fair health. Thyroxine will be required for the hypothyroidism; for the failure of suprarenal function suprarenal cortical hormones are given. There is no useful purpose in giving gonadotrophic hormones, but testosterone has been found to supplement the action of cortical hormones.

FURTHER READING

Garrey M., Govan A., Hodge C. and Callander R. (1992) *Obstetrics Illustrated*. Churchill Livingstone, Edinburgh.

Ingemarsson I., Ingemarsson E. and Spender J. (1993) *Fetal Heart Rate Monitoring – a Procedural Guide*. Oxford University Press, Oxford.

6

NORMAL PUERPERIUM

The puerperium is the time following labour during which the pelvic organs return to their pre-pregnant condition. By convention the puerperium is said to last for 6 weeks, although it may take much longer for some of the organs to return completely to normal.

The management of the early puerperium consists of keeping careful watch upon the physiological processes during that time, and in being prepared to intervene if they should show signs of becoming pathological. Attention is given to the general mental and physical welfare of the mother and baby. Some special points of management are set out in the following sections.

Prevention of infection

Every precaution should be taken to prevent the implantation of exogenous pathogenic organisms into the birth canal during labour and the puerperium. The vulva and perineum should be kept as clean and as dry as possible. The woman may leave her bed to visit the lavatory within hours of birth and may have a shower or bath the following day after a normal delivery. There is no need to swab the vulva or to pour antiseptic solution over it during delivery or in the puerperium. The vulva is simply kept covered with a dry sterile pad; this is changed whenever it is soaked or soiled.

Relatives, nurses, students or doctors who have septic foci due to streptococcal or staphylococcal lesions, or who may carry infection from recent contact with septic conditions must be excluded from labour and lying-in wards. Women who develop signs indicative of sepsis should be separated from the normal cases until it can be established that any infection organism is not pathogenic to others.

In order to reduce the risk of infection spreading, modern maternity units are designed with single rooms or small wards of four to six beds. Ventilation should be by a system which introduces fresh air rather than one which draws in air from other parts of the hospital. Many organisms, in particular *Staphylococcus aureus* and the haemolytic streptococcus, are found in dust and blankets in hospitals. Single rooms must be available for suspect cases. The wide range of antibiotics now available should not justify lower standards in the prevention of infection.

Time of getting up

After the physical and mental strain of pregnancy and labour a woman needs a period of rest from hard work and mental worry. The time spent in hospital varies according to the type of delivery and also the home conditions to which she will be returning. The mean stay in the UK is now 3 days, but providing steps are taken for adequate supervision by a midwife and doctor there is much to be said in favour of allowing the woman to return home before this time.

If the labour has been normal and there has been no gross injury to the pelvic floor or other complications, the woman is allowed out of bed the day of delivery. This early ambulation must not be made an excuse for shortening the total time of rest and time for breastfeeding to be comfortably established. Moderate exercise encourages recovery in the tone of the pelvic floor, the circulation is improved in the legs and the incidence of venous thrombosis is reduced.

After operative deliveries it may be necessary to keep the mother in bed for a further day before active mobilization.

Temperature and pulse

The temperature may briefly rise to 37.9°C in the first 24 hours, but afterwards should fall to normal and remain so. Provided that the woman feels and looks well the temperature need only be taken morning and evening.

It is strongly emphasized that when the temperature remains raised for more than a few hours, especially if there is a corresponding rise in the pulse rate, this should be regarded as being due to infection arising in the genital tract until the contrary is proved. All cases of fever during the puerperium should be carefully investigated (*see* Chapter 7).

For the first few hours after a normal delivery the pulse rate is likely to be raised, but it should return to normal by the 2nd day. A rise in the pulse rate must be regarded as seriously as a rise in the temperature. It may indicate severe anaemia, venous thrombosis or infection of the birth canal, urinary tract or breast.

Onset of lactation

During pregnancy there is considerable hypertrophy of the glandular tissue of the breasts, but secretion of milk does not start until after the birth of the child. Many women who intend to breastfeed put the baby to the breast within minutes of delivery. The consequent release of oxytocin may assist in keeping the uterus well contracted. With the activation of prolactin the breasts become more active, and there is increased vascularity. Breastfeeding usually becomes established on the 4th or 5th day. The release of milk is due to a monosynaptic reflex release of oxytocin, the let-down reflex.

Involution of the uterus

Immediately after delivery the fundus of the uterus lies about 4 cm below the umbilicus. The height of the fundus diminishes daily and it cannot normally be felt above the pubis after the 10th day. Any estimate of the height of the fundus should always be made when the bladder is empty. The uterus rapidly decreases in size for the 1st week, then decreases more slowly, being completely involuted in about 8 weeks. At the end of labour the uterus weighs 1000 g, by the end of the 1st week about 500 g, and by the end of the puerperium about 70 g. Involution is accomplished by autolysis of the muscle fibres, their protoplasm being broken down by enzymes, liquefied and removed in the bloodstream. The end products are excreted in the urine. Delay in involution occurs in the presence of uterine infection, retention of placental products or fibroids in the uterine wall, but in the absence of other signs of abnormality delay in shrinkage of the uterus is of no significance.

Retention of urine

A few women have difficulty in passing urine for the first day or two after delivery. Retention is liable to occur after a difficult labour which causes bruising or lacerations in the vulva, after epidural anaesthesia or when many perineal stitches had to be inserted. It is less likely with women who are ambulant.

If the bladder is atonic, residual urine accumulates and may become infected. If retention occurs or residual urine in some quantity is suspected a catheter must be passed with careful aseptic precautions. If catheterization is required on a second occasion an indwelling catheter should be left for 2–3 days, and prophylactic antibiotics given. Bladder overdistension that is overlooked may lead to detrusor instability.

Incontinence of urine

True incontinence, now a rare event in the UK, results from a vesicovaginal fistula, due either to a tear involving the bladder from instrumental delivery, or to pressure on the soft tissues from long labour and the formation of a slough involving the bladder. In the latter case incontinence does not appear until the slough breaks down some days after delivery.

Stress incontinence, or the leaking of urine on coughing, laughing or sneezing, is not uncommon in late pregnancy and may worsen after delivery. If it occurs soon after labour it may be only temporary, but if it persists surgical treatment is required (*see Gynaecology by Ten Teachers*). The operation is not advised until several months have elapsed because in many cases improvement occurs with the help of active pelvic floor exercises and return of tone in the pelvic floor musculature.

Cystitis and pyelonephritis

Urinary tract infection with high fever may arise in the 1st week of the puerperium. Symptoms such as frequency and discomfort on micturition are often absent in puerperal infections, but tenderness over the kidneys may be found, more commonly on the right than the left. The infection is usually caused by coliform organisms, and may be an exacerbation of a preceding chronic infection of the urinary tract, or it may follow catheterization. It is difficult to obtain a satisfactory midstream specimen for bacteriological examination at this time, but one may be obtained by catheter or suprapubic aspiration. Treatment is by encouraging the woman to drink adequate quantities of fluid and by giving appropriate antibiotics (e.g. amoxicillin).

Constipation

This is common in the puerperium and is due to a combination of factors. The woman's food intake has been interrupted, there may be dehydration during labour, the abdominal muscles are lax and perineal lacerations make defecation painful. Constipation may be prevented by giving bran or bulk-forming drugs such as methylcellulose, ispaghula or sterculia from the 1st day, but if this does not act, laxatives, suppositories or an enema may be required.

The lochia

The lochial discharge comes from the placental site. For the first 3 or 4 days the lochia are red in colour. As the site begins to heal the discharge decreases in amount and its colour changes to pink and finally becomes serous. Although it is said that the lochia disappear by the 10th day the average time before they become colourless is, in fact, usually 3 or 4 weeks. Lochia which remain red and excessive in amount indicate delayed involution, which may be associated with retention of a piece of placental tissue within the uterus, or with infection. If placental tissue is retained the uterus may be enlarged and the internal os of the cervix remains open. The retained products can be shown by ultrasound examination. Curettage is likely to be required, especially if there is an increase in red loss or the passage of clots.

Offensive lochia may indicate infection of the uterus, although the organisms may only be saprophytes. Virulent infection with haemolytic streptococci is not accompanied by an offensive smell.

Sleep and avoidance of anxiety

It is important to see that the mother not only gets a good night's rest, as free as possible from disturbance by the baby, but also that she has a rest in the afternoon. Pain from perineal stitches or engorged breasts is a common cause of sleeplessness. If she is excitable and sleeping badly hypnotics such as temazepam should be given for the first few nights. After that sleeplessness, unless it is habitual, should arouse anxiety as it may be the first sign of the onset of puerperal psychosis.

Diet

The day after a normal delivery the woman should be given a normal diet. During lactation she will need an adequate but not excessive intake of fluid. Animal fats, fruit and vegetables will supply necessary vitamins, and protein in the diet should be increased. Milk is a very good source of nutrition and the woman should be encouraged to drink up to half a litre a day.

Perineal stitches

The perineum should be washed after the early days only with soap and water and dried. A dry sterile vulval pad is then applied, this is changed frequently. Unabsorbable stitches are removed on the 5th day.

Care of the breasts

During the first 2 or 3 days after delivery the breasts secrete colostrum only, but it is important that the baby is put to the breast in order to promote bonding between the mother and baby, to stimulate the secretion of milk, and to teach the baby to suck.

If the mother cannot or does not wish to breastfeed, lactation is suppressed and she is helped to establish bottle feeding. Although oestrogens are effective in suppressing lactation their use has been given up because of the increased risk of thrombosis and embolism. Instead the secretion of prolactin is inhibited with bromocriptine in doses of 2.5 mg twice daily for 14 days. It has the disadvantage of causing nausea. Adequate support for the breasts and the administration of analgesics for the discomfort of engorgement is often all that is required.

Postnatal exercises

In many hospitals women are given breathing exercises, exercises for the abdominal and pelvic muscles and exercises for the legs to reduce the risk of thrombosis, but for most women early ambulation has reduced the need for this. It is probably better to concentrate the use of physiotherapy on those who have some special need for it.

Postnatal examination

This should be carried out at the end of the 6th week. Apart from discussing any problems or anxieties which the woman may present, and following up any complication of pregnancy or labour, the doctor enquires about her general health, whether the lochia have ceased, about bladder function, especially to exclude stress incontinence, and about any infant feeding problems. Tactful inquiry should be made to determine if sexual intercourse has occurred and if there were problems. Many breast-feeding women suffer from lack of vaginal lubrication and simple remedies, such as K-Y jelly, may prevent long-term psychosexual problems.

The abdomen is examined and the state of the musculature noted. Pelvic examination is performed to check that any lacerations have healed normally, that there is no prolapse of the vaginal walls and that the uterus has involuted normally. The uterus is not infrequently found to be retroverted. In most cases this causes no symptoms, or the uterus is known to have been in this position before pregnancy, and no treatment is required.

A speculum is passed and if a cervical smear was not taken in the antenatal period this is done. A cervical erosion is a common finding. Many of these regress spontaneously before the 12th week and the woman should be seen again then. Only if the erosion persists and causes discharge which is sufficient to trouble the woman should it be treated (*see Gynaecology by Ten Teachers*).

Women not infrequently complain of backache at a postnatal visit. Most cases of lumbar backache are due to poor posture (persistence of the lordosis of pregnancy) or fatigue, and spontaneous recovery often occurs. Retroversion or other pelvic lesions will not cause backache at this level. Persistent backache, or any other disability discovered at the postnatal examination, may call for further investigation or treatment.

Sometimes conditions such as hypertension or urinary tract infection call for prolonged follow-up.

Family planning advice

Most women are anxious to space their pregnancies if not to limit them. Practical contraceptive advice should be made easily available at the time of the postnatal visit. The choice of method lies with the individual couple but it is necessary to discuss with the woman the pros and cons of the various methods. She may be prepared for family planning by discussion during pregnancy, or while she is in the lying-in ward, of the methods available. If it is impossible for her to attend for advice in the postnatal period an intrauterine device may sometimes be inserted in the puerperium, or a prescription and instructions given for the use of the contraceptive pill. For a woman who has completed her family, and if she and her partner are certain that they will not change their minds later, sterilization may be performed in the early puerperium, but it is often better performed 6–12 weeks later, when the woman has had time to reflect whether she really wants it done, and the Fallopian tubes have returned to their normal size. Clips and bands put on hypertrophic tubes do not always occlude the lumen fully, leading to a failure of the sterilization process. For this reason, very few women have laparoscopy sterilization in the weeks immediately after delivery but wait 10–12 weeks.

Contraceptive pills containing oestrogen should not be given to women who are breastfeeding because they may inhibit lactation. Progestogen-only pills do not inhibit lactation and are useful.

Details of contraceptive methods are given in *Gynaecology by Ten Teachers*.

THE BABY

At the moment of delivery the infant is launched into an independent existence and must start breathing, or die. The circulation has to adjust quickly to the needs of pulmonary gas exchange. A normal body temperature must also be achieved rapidly. Shortly afterwards the infant must establish an appropriate nutritional intake and digestive functions. At birth profound physiological changes must occur in a very short space of time.

THE RESPIRATORY SYSTEM AT BIRTH

The fetus *in utero* demonstrates movements of breathing in a phasic manner. These movements last for varying lengths of time and are interrupted by long periods of rest. There is a net outflow of amniotic fluid from the alveolar spaces through the bronchial tree into the upper airway. During fetal breathing this fluid is drawn in and out of the nasopharynx in volumes approximating to one-tenth of that seen during the normal breathing of air after birth. These respiratory movements, although small, are an important rehearsal for what is to follow later.

During labour and immediately after birth the lungs are cleared of fetal lung fluid. Approximately one-third is squeezed out through the mouth and nose and the remainder is absorbed directly across the alveolar walls into the pulmonary lymphatic and blood capillary systems. Meconium residues in the fluid may also be inhaled during this period and can cause inflammatory changes within the lungs.

The moderate hypoxia encountered by the fetus during labour, with a slight fall in arterial oxygen tension and a rise in arterial carbon dioxide tension, heightens the responsiveness of the newborn infant's respiratory centres to a variety of stimuli. With more severe hypoxia these centres become depressed, leading to apnoea. This may be followed by gasping as a means of initiating respiration.

The thorax is squeezed as the body passes through the birth canal, and fluid is expelled from the chest, being replaced by air during the elastic recoil after completion of delivery of the trunk. Once the baby is born the body extends from the fetal position, the spine straightens and the change in shape of the chest and descent of the diaphragm facilitate respiration.

The fall in the environmental temperature after delivery is an added stimulus to breathing, particularly if cool air impinges on the sensitive area around the mouth and nose. A combination of these factors is thought to be responsible for the establishment of full respiration. Any anaesthetic or sedative administered to the mother is shared by the fetus and may depress the respiratory centres.

In normal circumstances the first breath draws air into the bronchial tree, unless there is obstruction from inhaled amniotic fluid or meconium. A negative intrapleural pressure is created and with successive respiratory movements the thoracic cage expands and the lung volume increases. The difference between this negative pressure and the atmospheric pressure of the air in the bronchial tree causes the fetal lungs to expand by entry of air into the alveoli. The unexpanded alveoli are filled by a thin layer of fluid and the surface tension between the walls must be overcome before full expansion can be achieved.

The maturing lung is able to produce *surfactant* from cells (type II pneumocytes) in the alveolar wall. The main constituent of surfactant is a phospholipid, lecithin, which reduces the surface tension within the alveoli and so permits easy and equal expansion throughout the lungs. If the production of surfactant is diminished by prematurity or hypoxia, or both, there is incomplete expansion of the lungs which may result in respiratory distress syndrome (hyaline membrane disease). Normal alveolar expansion is at first rapid, but several hours elapse before complete expansion of the lungs occurs. Lung expansion takes longer after caesarean section than vaginal delivery. The time to full lung expansion is longer following caesarean section when it is performed electively than when done during labour. There is a cessation of the production of lung fluid during labour.

Although the first breath is usually taken immediately after birth, the onset of breathing is sometimes delayed. The newborn infant appears to be less susceptible to the ill effects of prolonged hypoxia than the adult, but nevertheless severe hypoxia lasting for only a few minutes can cause bradycardia and falling blood pressure leading to impaired cerebral circulation, and can result in cerebral palsy or developmental delay.

Changes in the fetal circulation at birth

The fetal circulation has already been described in Chapter 1; *see also* Fig. 6.1a. Remarkable changes occur soon after birth. The infant's first breaths expand the lungs and open up the pulmonary vascular bed. This produces a sharp fall in the pulmonary vascular resistance and a considerably increased flow of blood to the lungs, where it is oxygenated. Immediately afterwards there is contraction of the ductus arteriosus to about half its original diameter, and complete contraction of the umbilical arteries. It is thought that the ductus arteriosus and umbilical arteries have a natural tendency to contract but are prevented from doing so by prostaglandins which are circulating at high concentrations in the fetus. It is suggested that there are enzyme systems in the lungs which break down the prostaglandins once the pulmonary blood flow is established after birth. The fall in the level of circulating prostaglandins, along with the rise in the arterial oxygen saturation, allows these vessels to contract. It is possible that the use of prostaglandin inhibitors to suppress premature labour might lead to premature closure of the ductus in the fetus.

The systemic arterial pressure rises as the low resistance of the placenta is replaced by the high resistance of the peripheral circulation in the baby when the cord is clamped. With the concomitant fall in the pulmonary artery pressure the pressure gradient across the ductus is reduced. The fall in pulmonary vascular resistance produces a fall in right sided pressures, and the increase in systemic resistance a rise in left sided pressures. These changes, together with the increased pulmonary venous return, lead to a rise in left atrial pressure

over and above that in the right atrium and consequent functional closure of the foramen ovale. (Fig. 6.1b)

Ultimately, but not for several days or weeks, the ductus becomes obliterated by endarteritis, and the final transition from fetal to adult circulation is complete. The umbilical cord dries and separates, leaving a raw area which heals by granulation and is then covered by epithelium. The umbilical vein is now a redundant vessel and closes by aseptic thrombosis. The ductus venosus atrophies and disappears, while the umbilical vein remains as the *ligamentum teres*. The umbilical arteries show retrograde closure as far as the hypogastric arteries and persist as sclerosed remnants.

Changes in the gastrointestinal tract

Although the placenta carries out nutritional and excretory functions for the fetus, the gastrointestinal tract is active during intrauterine life. During the later months of pregnancy amniotic fluid containing desquamated epithelial cells is continually being swallowed. Mucus and debrided intestinal epithelial cells are added to the gut content and, with the swallowed fluid, are digested by intestinal enzymes to form meconium, which is coloured greenish-black by bile pigments. By late pregnancy the meconium has transversed the gut as far as the rectum, and if strong intestinal contractions occur during labour, meconium may be voided into the amniotic fluid. If premature respiration is induced by hypoxia, meconium may be inhaled into the bronchial tree, and this can cause respiratory difficulties from meconium aspiration after delivery.

With the first inspiration air also enters the stomach and rapidly traverses the gut, reaching the ascending colon within 3 hours.

Hydrochloric acid is present in the stomach in relatively high concentration on the 1st day of life, but in only moderate amounts during early infancy. Amylases from saliva and from the pancreas are at a low level during the early weeks of life, and this explains the difficulty in digesting certain carbohydrates which is shown by some infants. The trypsin and lipase of pancreatic secretions are present in sufficiently high concentration, even during intrauterine life.

Regulation of body temperature

The fetal temperature is determined by the surrounding maternal environment and the fetal heat regulating centre which is thought to be relatively dormant. After birth, unless special precautions are taken, the infant's temperature can fall by 1 to 3°C and external heat must be supplied. The temperature regulating centre is active immediately after birth and causes an increase in the rate of metabolism and oxygen consumption in order to maintain a constant body temperature in the face of environmental variation. Brown adipose tissue has an important role in maintenance of body temperature in the newborn, and deficiency of this tissue in preterm and small-for-dates babies contributes to their thermoregulatory difficulties.

Defences against infection

Placental transmission of maternal IgG antibodies affords the infant non-specific, passive immunity against certain diseases. This immunity wanes during the first few months of life, and the infant becomes at risk from common infections until an acquired immunity is developed. Immune protection from the mother may be specific for diseases against which she has developed an active immunity, such as diphtheria, tetanus, measles, mumps and poliomyelitis. The infant's own IgG levels are very low at birth but increase steadily during the first 2 to 3 months of life. Serum IgA levels are also low, but the infant should be receiving IgA from the mother's breast; it is in high concentration in colostrum and continues to be secreted in mature milk. This immunoglobulin plays an important part in protecting the baby against infections acquired through the alimentary tract, and possibly also against the development of allergic reactions.

THE IMMEDIATE CARE OF THE HEALTHY INFANT

The infant may be limp when delivered, with only moderate muscle tone, and there may be purplish-blue cyanosis of general distribution.

Infants usually clear their airway when they start breathing shortly after birth. If there is a delay, the nasopharynx should be cleared of fluid with a sterile mechanical sucker regulated to avoid a negative pressure of more than 30 cm of water. The procedure must be carried out with the utmost gentleness to avoid damage to the mucous membranes or the production of laryngeal spasm.

Once respiration begins, the colour of the skin rapidly changes to pink. Cyanosis of the extremities may persist for a time, but this is not of pathological significance.

Care of the umbilical cord

After clamping, the cord is kept dry and clean until it shrivels and comes away leaving a small area of granulation tissue, which heals over after a few days.

General care of the baby

After ensuring that respiration is well established and the Apgar score has been recorded, the infant is inspected rapidly to note whether there are any obvious abnormalities before being wrapped in a warm towel and handed back to the mother to be cuddled and put to the breast to suckle the colostrum.

A recent fashion is to place the naked baby on the mother's abdomen immediately after delivery so that skin to skin contact will promote emotional bonding. The baby should be under the cover of a towel to avoid heat loss.

After this initial period of close contact, the baby is wrapped firmly in warm towels and laid supine in a cot alongside the mother with head turned to one side and facing her. Unless there is need for special care, it is important that the baby should not be removed from close proximity to the mother. Even if she falls asleep from exhaustion it is reassuring for her to find her baby beside her when she wakes.

During this initial period it is important to check the baby's axillary temperature. The rectal temperature should only be taken if the axillary temperature is low.

Identification

In hospitals the infant must be identified so that confusion between babies cannot arise. A plastic wrist band with the mother's name marked in indelible ink is commonly used with a copy on the baby's ankle.

Examination of the newborn infant

At the earliest convenient time after the birth a systematic physical examination should be made.

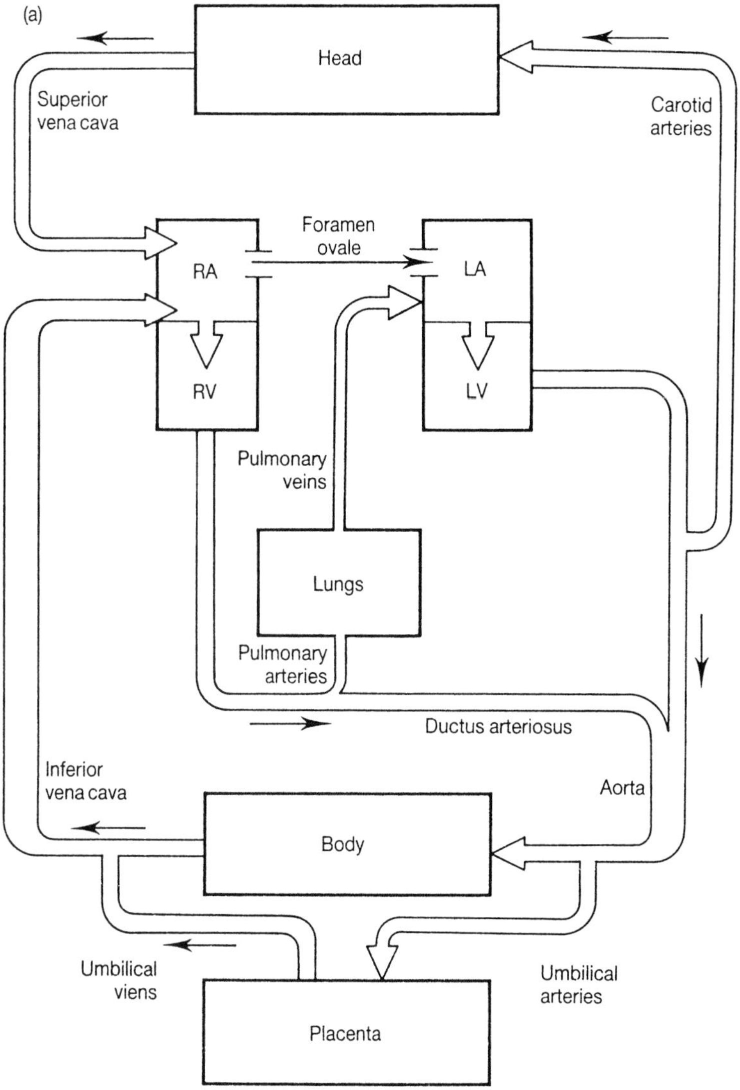

Fig. 6.1 (a) The fetal circulation. The right and left sides of the heart are in communication through the foramen ovale while lungs are bypassed by the ductus arteriosus

External examination is of great importance and attention should be paid to any minor congenital defects such as skin naevi, preauricular skin tags or supernumerary digits. Defects such as cleft palate, spina bifida or club feet may escape notice if of minor degree, and especial attention should be given to possible omphalocoele, penile abnormalities, and imperforate anus.

The skull should be palpated and the state of the sutures noted. The circumference of the head should be recorded. Widening of the sutures and an increase in the circumference of the skull may be found in cases of subdural haematoma or hydrocephalus. Areas of softening or an encephalocoele may be found.

Examination of the heart should be performed, but the significance of any murmur may only become apparent after observation for several

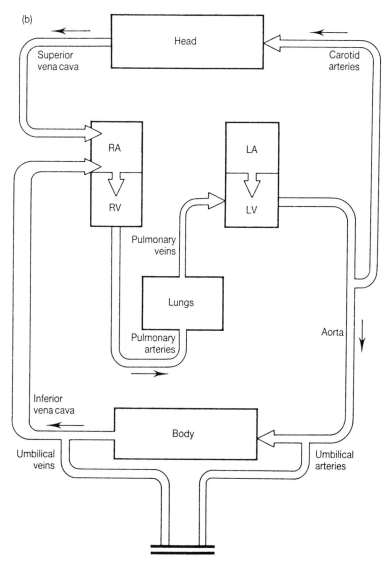

Fig. 6.1 (b) Changes in circulation at birth. The placenta is removed. The foramen ovale closes while the ductus arteriosus goes into spasm. Blood now passes from the right heart to the left heart only via the lungs

days. Palpation of the femoral pulses should be carried out to exclude coarctation of the aorta.

On examination of the abdomen the liver edge can usually be felt just below the costal margin, and the tip of the spleen is often palpable. In a well-relaxed abdomen the lower poles of the kidneys may be easily felt, but this is not of pathological significance.

The baby should be handled and the limbs moved to ensure that no fracture or paralysis is present. The hips should be examined for congenital dislocation. The range of movement of the ankles must be tested to detect minor degrees of talipes.

Normal reflexes

Certain reflexes are normally only present in the newborn infant.

The Moro reflex is a startle reflex elicited by banging the table on which the infant is lying, by letting the head drop backwards by 1–2 cm or by flexing the head on the trunk. Abduction and extension of the arms is followed by adduction of the arms like an embrace, and the legs are extended. This reflex is usually symmetrical and indicates normal neuromuscular coordination. Variations from the normal response are seen when there is a nerve palsy, a fractured clavicle or long bone, or an intracranial lesion. It is not necessary or desirable formally to test the Moro reflex as it upsets the baby and parents. The same information can be obtained by observing the baby's limb movements.

If the infant is touched on the area near the mouth the head will turn and the infant will try to suck the touching object. This rooting reflex is followed by the sucking reflex which seems to be designed to enable the baby to draw the nipple well back into the mouth so that the gums can gently massage the milk sinuses just below the surface of the areola and stimulate milk ejection.

A grasp reflex is seen when the palm of the hand or the sole of the foot is touched and is normal in the newborn infant. A walking reflex is also usually seen in the normal infant.

NORMAL PROGRESS OF THE FULL TERM INFANT

During the 1st week of life the baby continues to adjust to the extrauterine environment.

Traumatic lesions such as a caput succedaneum a chignon after vacuum extraction disappear rapidly. The baby begins to assume an infantile position instead of that of intrauterine folding. The feet are less dorsiflexed and the hands less firmly clenched. The chest becomes less flattened and the head is moved freely. The cry is definite and begins to assume characteristics for hunger and pain.

WEIGHT GAIN

There is loss of weight for the first 2 or 3 days of up to 10 per cent of the birth weight. After feeding is established the weight increases and will usually return to birth weight by the 7th to the 10th day. If this is not achieved it may indicate inadequate feeding, or some condition which is preventing normal progress, but some breastfeeding mothers take longer than others to establish lactation, and their babies may take 2 to 3 weeks to regain their birth weight without any harmful effects.

Temperature

At birth the infant is thermolabile, and the temperature may fall below 36°C, but it is unwise to allow the rectal temperature to fall below this. The infant responds quickly to the application of external heat, and in a few hours temperature regulation will be stabilized in the full term infant. The temperature shows a normal daily variation from 36 to 37.2°C.

Normal stools

Meconium is normally passed within a few hours of birth, but may be delayed for 12 or even 24 hours. Although this is not necessarily pathological it demands observation, as it may indicate intestinal obstruction. The first meconium stool is often preceded by a plug of whitish mucus, and if this is large the appearance of meconium can be delayed for 24 hours and the abdomen may show some distension.

In the next 2 or 3 days meconium is passed several times a day. Normal meconium is sticky, odourless and greenish-black in colour. After a few days changing stools are passed. These are non-homogeneous, thin and yellowish-brown in colour. Undigested milk elements can often be seen in the stool.

Towards the end of the 1st week milk stools are passed. The stool of the breast-fed infant is smooth, pasty, and mustard or golden-brown in colour. The motion of the artificially fed infant is paler in colour, alkaline, pasty and non-homogeneous.

The breast-fed baby may have a bowel action only once in the day, or up to six or seven, in the 24 hours and yet be normal. When the stools are frequent they may be green and this should not lead to a hasty diagnosis of gastroenteritis in a baby who is otherwise well. It is usually due to a temporary excess of lactation when the milk *comes in* and it is important to leave the baby to continue to feed normally and maintain the stimulus to lactation. Later, the maternal diet may affect the baby's bowel habit as many aperient substances are secreted in the milk.

Micturition

The baby usually passes urine shortly after delivery, but micturition may not occur until 24 hours have elapsed, especially if the infant has voided urine just before or during delivery. During the first day or two the napkin may be stained pink from the deposit of urates.

Breast engorgement

The breasts of infants of either sex may swell and become engorged during the 1st week of life. Milk may be secreted and a female may bleed from the uterus. These effects are due to transfer of maternal oestrogen and require no treatment. On no account should the engorged breasts be handled or squeezed as they readily become infected to produce neonatal mastitis or breast abscess.

Skin

In Caucasian infants the skin soon changes from the red colour seen after birth to a paler hue. The skin may become dry and scaly as it dries out after the long intrauterine immersion. Fissures and cracks may appear in the folds of the ankles and wrists.

INFANT FEEDING

The rate of growth during infancy is greater than at any other stage of life after birth. The weight of a normal baby doubles in 5 months and trebles in 12 months. To achieve this, the intake of food must be relatively great and of a quality that allows easy digestion. During the early months it must contain, in liquid form, a balanced mixture of protein, fat, carbohydrate, minerals and vitamins.

DAILY FLUID REQUIREMENTS

Infants need water in relatively greater volume than at any other period of life. The ability to concentrate urine is less well developed in infants than adults, and to remove the waste products of metabolism the urinary output must amount to nearly half the total fluid intake. The losses of fluid from the skin, lungs and faeces are proportionately greater than in adult life. After the 1st week of life infants need approximately 150 ml of fluid per kg of body weight during 24 hours. Infants who are growth retarded require an even larger volume of fluid ranging between 150 and 200 ml/kg body weight per 24 hours.

ENERGY REQUIREMENTS

The diet should provide 110 kcal/kg in 24 hours. This calls for 70 kcal per 100 ml of feed. Growth retarded infants require more calories per kilogram – up to 150 kcal per kg in 24 hours.

PROTEIN, FAT AND CARBOHYDRATE

Sufficient of these are supplied in a diet of breast milk or infant milk formula

MINERALS

At birth, babies have only limited reserves of iron. During the period of marked growth in early infancy there is a corresponding increase in the total number of red cells. The reserve stores of iron may become exhausted in forming haemoglobin and unless adequate supplies of iron are available in the diet the infant will develop hypochromic anaemia. Breast milk usually supplies enough iron during the first 4 months of life particularly as the lactoferrin content ensures rapid and efficient absorption. Modern infant milk formulae contain additional iron but unmodified cow's milk does not. The bioavailability of iron is much less in cow's milk formulae than human milk. In the premature infant who has depleted iron stores at birth and an even more dramatic increase in size than term infants it is recommended to supplement the diet with additional iron.

VITAMINS

Clinical vitamin deficiency is rare in developed countries. The optimal intake of vitamins has not been determined. The recommended intake of vitamin D is 400 IU per day (10 µg).

Vitamin K deficiency is commoner in breastfed infants and in most units vitamin K is given as prophylaxis against haemorrhagic disease of the newborn either as a single intramuscular injection or as several oral doses. Vitamin supplements are recommended for preterm infants and those who have a poor nutritional intake.

BREAST OR BOTTLE FEEDING?

When artificial feeds became widely available the popularity of breastfeeding declined markedly. Artificial feeding became the new and fashionable way in the developed world. More recently following the rigorous promotion of breastfeeding the proportion of infants receiving breast milk during the first 6 months of life has increased markedly. The advantages and disadvantages of breastfeeding are listed in Table 6.1.

Breast milk is clearly the ideal milk for human infants. In addition it helps secure the intense psychological attachment of a mother for her baby and provides intimate, close contact for the baby with his or her mother.

However not all mothers wish to breastfeed. They may be embarrassed, find the thought distasteful or may simply prefer the convenience of bottle-feeding. Some women are unable to establish or provide adequate lactation. Breast milk is not only the most appropriate milk for human babies, but also has other advantages. It protects against infection, especially enteric infections. Several mechanisms are involved. These include maternal secretory IgA and lysozymes, which either kill organisms or prevent their adherence to the intestinal wall; lactoferrin which ensures rapid absorption of iron so that it is not available for the essential needs of replicating organisms; and encouragement of the growth of *Lactobacillus bifidus* to the exclusion of other pathogenic organisms which flourish in the more alkaline intestinal environment provided by the ingestion of cow's milk. The reduction of enteric and other infections in breastfed infants is very important in developing countries, where enteric infections remain an important cause of mortality and long term morbidity. In developed countries, a reduction in the number of enteric and other infections in infancy is also seen.

Breastfeeding provides some reduction in the severity of eczema in atopic babies. The secretory IgA in breastmilk is thought to be important in providing protection from allergic reactions to cow's milk protein.

There are other potential benefits of breastfeeding. There has also been considerable interest recently in the suggestion that the incidence of hypertension and vascular insufficiency in adult life is lower if breastfed in infancy. There is also some evidence to suggest that breastfeeding reduces the risk of breast cancer in later life.

From a global perspective the promotion of breastfeeding is of crucial importance. In developing countries breastfeeding dramatically improves survival during infancy from its protective effect against enteric and other infections. Breastfeeding is also important in helping to increase the time interval between pregnancies. Although breastfeeding cannot be relied upon as a contraceptive, the practice of such feeding has an important effect in reducing population growth in developing countries.

LACTATION

Anatomy and physiology

The breast is composed of about 20 segments arranged radially from the nipple, the glandular tissue being mainly peripheral (*see* Fig. 6.2). The branching duct system from each segment unites to form a single duct which opens on the nipple. Immediately before opening on the nipple each duct has a dilatation called a lactiferous sinus. This lies immediately below the areola of the nipple and is a thin-walled part of the duct which can be dilated by milk to a calibre of 0.5 to 1 cm. The ducts and alveoli of the glandular tissue are surrounded by myoepithelial contractile cells, and there are smooth muscle cells under the areola. The pectoral fascia ensheathes the whole breast and sends laminae between the lobes and alveoli. There are many fat cells and a rich supply of blood vessels.

During pregnancy the enlargement of the breast is mainly due to hyperplasia of the glandular tissue from the 12th week onwards. It results from increasing concentrations of placental lactogen,

Table 6.1   Advantages of breast feeding (adapted from Booth, 1992)

Anti-infective properties of breastmilk

Humoral

Secretory IgA	90 per cent of immunoglobulin in human milk. Mucosal protection
Bifidus factor	Promotes growth of *Lactobacillus bifidus*. Resulting low pH may inhibit growth of gastrointestinal pathogens
Lysozyme	Bacteriolytic enzyme
Lactoferrin	Iron binding protein. Inhibits growth of *E. coli*
Interferon	Antiviral agent

Cellular

Macrophages	Phagocytic
Lymphocytes	T-cells may transfer delayed hypersensitivity responses to infant

Nutritional properties of breast milk

Protein quality

Hypoallergic	More easily digested curd May reduce subsequent atopic disease

Lipid quality

Breast milk lipase	Improved digestibility and fat absorption Enhanced lipolysis

Minerals/electrolytes

2:1 calcium:phosphorus ratio Low renal solute load	Prevents hypocalcaemic tetany and improves calcium absorption
Iron content	Bioavailable (40–50 per cent absorption)

Infant-mother bonding is promoted

Contraceptive effects

Protection against breast cancer

Table 6.2   Disadvantages of breast feeding (adapted from Booth, 1992)

Transmission of drugs	Antithyroid drugs, cathartics, antimetabolites
Nutrient inadequacies	Prolonged breast feeding without introduction of appropriate solids may lead to iron or vitamin D deficiency
Breast milk jaundice	Mild, self-limiting, unconjugated hyperbilirubinaemia, and not a contraindication
Transmission of infection	HIV identified in breast milk
Vitamin K deficiency	Preventable by routine administration of vitamin K at birth

prolactin and chorionic gonadotrophin. By the second trimester placental lactogen begins to stimulate secretion of colostrom. Milk production is inhibited before delivery by prolactin inhibitory hormone. The withdrawal of inhibitory hormones at birth, together with increasing prolactin secretion from the anterior pituitary leads to milk production. Prolactin release is stimulated by sucking at the breast as in oxytocin release from the posterior pituitary. Oxytocin stimulates the contraction of myoepithelial cells in the breast and results in the ejection of milk into the terminal ducts, the let down reflex. The milk is emptied by the baby feeding on the breast.

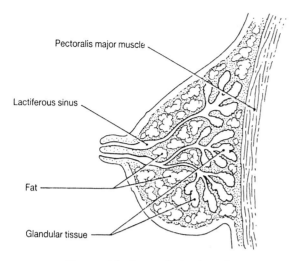

Pectoralis major muscle

Lactiferous sinus

Fat

Glandular tissue

Fig. 6.2 The breast during lactation

When the baby takes the breast the nipple is drawn into the arch of the hard palate and held there by the tongue to be exposed to suction, which delivers the milk to the back of the mouth to be swallowed. The lips form a seal on the nipple and areola and the gums champ on the areola to compress the lactiferous sinuses and propel their contents into the infant's mouth.

The active secretion by the breast is due to the contractile myoepithelial network of cells which invest the alveoli and ducts. Stimulation of the nerve endings in the nipple initiates impulses which reach the posterior lobe of the pituitary gland and provoke a pulsatile release of oxytocin into the bloodstream. Release of oxytocin may begin in response to the baby crying even before stimulation of the nipple. The oxytocin causes contraction of the myoepithelial cells and propulsion of the milk along the ducts.

After delivery colostrum is secreted for about 48 hours and then the breasts become engorged and milk is secreted.

Secretion of milk

This is the transformation of amino acids, glucose, lipids and minerals present in the blood plasma into caseinogen, lactalbumin, lactose and milk fats which are secreted into the alveoli by the activity of the alveolar epithelial cells. Only a portion of the milk yielded at a feed is preformed; the major portion is secreted during feeding.

Prolactin is mainly responsible for the secretion by the alveolar cells. This pituitary hormone is secreted by cells in the anterior lobe of the pituitary glands, the lactotrophs. Oestrogens cause

After delivery colostrum is secreted for about 48 hours and then the breasts become engorged and milk is secreted.

Excretion of milk

The baby is often said to suck milk. Sucking is a relatively unimportant part of the complex process, and the word wrongly suggests that suckling is entirely dependent on an activity of the baby. When lactation is well established milk will continue to spurt or flow from the nipple if a sucking infant is removed from the breast, indicating active secretion by the breast, separate from any activity of the infant.

Composition of colostrum and breast milk

For the first 2 days colostrum is secreted and on the 3rd and 4th days the secretion changes to normal breast milk. Colostrum is a yellow fluid containing large fat globules, the colostrum corpuscles, and it has a high mineral, moderate protein and relatively low sugar content. Colostrum has a high content of antibodies, especially secretory IgA, which play an important part in protection against infection. Colostrum may help to clear the small intestine if it becomes contaminated by infected material swallowed during the birth process. Colostrum is said to possess laxative qualities, but no laxative constituent has ever been demonstrated.

When the secretion changes from colostrum to milk its colour changes to bluish-white.

Breast milk protein contains three fractions:

caseinogen
lactalbumin
lactoglobulin.

Table 6.3 Composition of colostrum and breast milk

	Protein (%)	Fat (%)	Carbohydrate (%)
Colostrum	2.25	3.15	4.00
Breast milk	1.25	3.50	7.25

The latter is present only in small amounts and the proportion of lactalbumin to caseinogen is 2 to 1. The calorie value of breast milk is 70 kcal per 100 ml.

BREASTFEEDING

The essence of the initial management of breastfeeding is a gradual increase in the time at the breast and the amount which the infant takes.

Interval after birth when first feeding is permitted

It is to be hoped that whenever possible the mother will be encouraged to handle her baby soon after birth and that she can put the infant to the breast for a short time. This will enable the baby to obtain the valuable colostrum while starting the stimulus to lactation.

Frequency of feeding

In the past maternity hospitals have tended to a rather strict regimentation of baby care. Because most babies ultimately settle down to an approximate 4 hour interval between feeds, it became customary to establish babies on a 4-hourly regimen as soon as possible. This policy was not always successful, and seldom so without recourse to a night supplement. Where there has been complete acceptance of a policy of rooming in with freedom for the mother to feed her baby on demand it has been found that there is a wide range in the frequency with which healthy babies feed during the early stages of lactation. In one investigation this varied from a minimum of six feeds per day up to a maximum of 24, with a mean of 11.

It may take 2 or 3 weeks before the baby settles down to 3- to 4-hourly feeds, and a further few weeks before the night feed is willingly foregone.

Oxytocin release occurs not only in response to stimulation of the nipple during suckling, but also to other factors such as the cry of the baby. However, it is inhibited by maternal anxiety, including anxiety engendered by fear of being unable to feed the baby, so that encouragement and some simple explanation of these automatic responses may be helpful.

The completely flexible demand-feeding schedule has not been associated with an increased incidence of cracked nipples, possibly because the baby is only actively sucking when the milk is flowing freely. On the contrary, establishment of lactation is easier and engorgement of the breasts is rarely seen.

Time at the breast

Strict adherence to times is not necessary as babies tend to regulate their own time of feeding. During the 1st week the frequency may be increased but the time at the breast is short. Short frequent feeds of the more dilute milk which is yielded in the early part of a feed will provide the baby with a reasonable quantity of water, while the breasts are receiving the maximum stimulus to lactate. The time at each breast increases gradually as the frequency of feeds declines.

Amount taken from each breast

It is best to allow the baby to regulate the supply of milk according to his or her own needs and the mother's milk yield. The time taken for a full feed may be anything between 3 minutes each side to 15 minutes or longer each side.

Position during feeding

The infant is usually nursed by the mother propped up in bed or sitting in a low armchair. The infant is held in the crook of the arm on the same side as the breast which is to be given, with the weight of the baby supported on the forearm. The head should be allowed to extend beyond the bend of the elbow as most infants feed better with the neck extended. With the baby in this position the mother should bend forward slightly and allow the nipple to fall into the baby's mouth. If the breast is large the redundant tissue can be restrained with the fingers of the other hand and the nipple allowed to protrude between the extended fingers.

DIFFICULTIES IN BREASTFEEDING

Difficulties may arise from causes in the mother or the infant.

Causes in the infant

Reluctance to feed is an early sign of illness in babies. Infants who are tachypnoeic or are infected or ill from other causes are often reluctant to feed. Any baby who is not feeding should be examined to check for an underlying cause. If a baby is not feeding a blood glucose level and weight should be checked. Some babies if they are unable to establish feeding in the first few days become dehydrated and their jaundice increases. Rehydration is required with milk via a nasogastric tube or by intravenous fluids while breastfeeding is established.

When the baby is very small there may be disproportion between the size of the nipple and the baby's mouth, but this will be remedied as the baby grows. Cleft lip and cleft palate cause less difficulty in feeding than might be expected. Micrognathos can cause difficulty in feeding. Difficulty will also arise if there is obstruction to nose breathing.

Causes in the mother

Anatomical

When the nipples are poorly developed or retracted the infant may have difficulty drawing them into the correct position in the mouth for satisfactory feeding. In some cases nipple shells are used.

Engorgement

On the 3rd or 4th day after delivery the breasts may become engorged and painful so that milk cannot be taken and the mother is not able to tolerate the infant at the breast. This condition can develop very rapidly, and it should be recognized before it is fully established, when it may be relieved by allowing the baby to suckle frequently. The breasts may be emptied by manual expression or by breast pump. Cracked nipples and breast infections are considered in Chapter 7

Deficient lactation

If, in spite of the assistance of the midwifery and medical staff, lactation cannot be established successfully, an infant formula will need to be given. A complementary feed is one given to augment feed from the breast. A supplementary feed is one which replaces the breastfeed. Many midwives and health professionals believe that mixing breast and bottlefeeds is inadvisable as infants find bottlefeeding easier and will therefore stop taking milk from

the breast. In practice many mothers use both forms of feeding very successfully. Test weighing, where a baby is weighed both before and after a breastfeed, has been found to be very inaccurate as a guide to milk intake. All babies should be weighed regularly (weekly is reasonable) in the first few weeks of life to ensure adequate weight gain. Breastfed babies who are not receiving enough milk may appear content but simply fail to gain weight appropriately.

CONTRAINDICATIONS TO BREASTFEEDING

The virus causing HIV has been identified in the breast milk of some HIV positive mothers and there is evidence to suggest that HIV may be transmitted postnatally via breastmilk. There has been considerable controversy over whether HIV positive mothers should avoid breastfeeding. In the UK avoidance has been recommended but this will not be applicable to countries where there is no satisfactory alternative.

Secretion of drugs in breast milk

No drug should be given to a lactating mother unless there is some definite clinical indication for its use, because some of the drug, or its degradation products, is likely to be secreted in the milk. However, most drugs are only secreted in small amounts, so that breastfeeding need not be discontinued except in the case of a few drugs.

Cytotoxic drugs and radioactive iodine are absolute contraindications. If a thyrotoxic mother is taking carbimazole it is necessary to monitor the baby's thyroid function and discontinue breastfeeding if it is impaired.

Warfarin, senna, phenobarbitone, phenytoin, digoxin and steroids pass into the milk in harmless amounts.

Antibiotics are excreted in extremely small amounts, but there is a theoretical possibility of sensitization of the infant.

ARTIFICIAL FEEDING

Infants who are not breastfed will need to be fed on an infant formula. The milk should meet the infant's nutritional requirements, be well digested and not result in short- or long-term morbidity.

Cow's milk based formulae

Infant formulae have been modified extensively to make them resemble human milk as closely as possible. One of the problems of this is that the composition of breast milk varies considerably between mothers and even from the same mother between feeds and according to postnatal age. In addition although the milk has been extensively modified it is still based on cow's milk and our knowledge of the essential nutrients found in breast milk is limited.

Cow's milk not only contains more protein than human milk, but the protein in it is predominantly curd protein (casein), whereas breast milk protein consists of about equal parts of casein and whey protein (mainly lactalbumin). The fat content of breast milk is approximately equal to that of cow's milk, but the fat is of a different quality; it is more easily absorbed and it contains more polyunsaturated fatty acids, which are probably essential for man. All modern infant formula milk preparations are now low solute with an osmolality similar to that found in human milk.

In making a low solute preparation some manufacturers reduce the milk to a demineralized whey, containing whey protein and a low content of minerals, and then add a small amount of skim milk to provide some casein, further lactose and some minerals in their naturally occurring forms, while other minerals such as iron, copper, zinc and manganese may also be added. The fat which is added consists of a mixture of animal and vegetable fats, the composition of which approximates to that of breast milk.

Alternatively, the relative concentration of protein and minerals can be reduced by simply adding carbohydrate.

Infant formulae are now available in prepared bottles. These preparations are gaining in popularity but most parents use dried milk powder in the UK. The safety of infant formulae is dependent on the correct reconstitution of feeds with accurate use of the scoops provided and strict attention to sterility.

Soy-based formulae

Soy-based formulae have become increasingly popular in the belief that their use may lead to less atopic disease although convincing evidence of this is still lacking. Although they will support normal growth there is reduced bioavailability of some essential amino acids and of some vitamins, minerals and trace elements, particularly zinc. About 20–30 per cent of infants with cow's milk protein intolerance who are fed on a soy milk will subsequently develop soy intolerance. Soy-based formulae are best reserved for infants with clinical evidence of cow's milk protein intolerance.

Unmodified cow's milk

Doorstep milk should not be given to infants in the first 6 months and preferably year of life. Unmodified cow's milk contains too much protein and electrolytes. This may lead to hyperosmolar dehydration, hypocalcaemic fits from phosphate overload and bowel obstruction from casein curd. In addition there is inadequate iron and vitamins. Semi-skimmed or skimmed milk should not be used as they have low energy and vitamin A content. If doorstep milk is introduced at 6 months of age infants will require vitamins and iron supplements unless they are receiving a good diet of mixed solids. Alternatively a 'follow-on' formula can be used which contains more protein and sodium than infant formulae and is fortified with iron and vitamins (*see* Table 6.4).

General principles of artificial feeding

After the 7th day of life infants take an average of 150 ml of fluid/kg of body weight daily. However there is considerable variation in intake between babies. The volume of milk provided by a lactating mother is at first quite small and gradually increa-

Table 6.4 The composition of human milk, cow's milk and infant cows milk formula (per 100 ml)

	Human milk	Cow's milk	Modified cow's milk
Energy (kcal)	70	67	65–69
Protein (g)	1.3	3.5	1.5–1.9
Carbohydrate (g)	7.0	4.9	7.0–8.6
Casein:whey ratio	40:60	63:37	40:60–63:37
Fat (g)	4.2	3.6	2.6–3.8
Sodium (mmol)	0.65	2.3	0.65–1.1
Calcium (mmol)	0.88	3.0	0.88–2.1
Phosphorus (mmol)	0.48	3.2	0.9–1.8
Iron (μmol)	1.36	0.9	8–12.5

ses. Similarly infants fed on artificial milk will generally take about 60 ml/kg of milk on the 1st day and thereafter increase the total volume by 30 ml/kg each day until up to 150 ml/kg. A fully constituted artificial feed provides 65–70 kcal/100 ml (270–290 kJ/100 ml).

Although there is no need to keep rigidly to a strict time schedule for bottle feeding, and a certain amount of flexibility is appropriate for most babies, the same frequency of feeds is not required as in the early days when lactation is being established. Babies tend to settle down to a 3 or 4 hourly rhythm, and some flexibility enables each infant to show his or her own preference.

Advantages of artificial feeding

Artificial feeding is popular as it allows mothers more freedom as their presence is not required to feed the baby. This is particularly useful when mothers return to work when their babies are still young. Fathers and other family members can participate with feeding. In addition the volume of milk taken can be readily seen. Babies can also be fed easily in public places.

Disadvantages of artificial feeding

Modern artificial formulae are highly modified preparations of cow's milk trying to mimic breast-milk (*see* Table 6.4). As new information becomes available they will have to be modified to provide essential nutrients or remove potentially harmful ones. Breast milk does not suffer from any of these problems as it is designed for human babies. In developing countries because of the increased risk from infections in infancy it has been estimated that one-third of all deaths in young children can be attributed to the absence of breastfeeding. Breastfeeding is therefore a key part of the World Health Organization strategy to improve the survival of infants worldwide.

FURTHER READING

Hall D. and Stephenson T. in *Turnbull's Obstetrics* (1995) Ed. Chamberlain G. (Section 8; Normal Puerperium). Churchill Livingstone, Edinburgh.

Illingworth R.W. (1993) *The Normal Child*. Churchill Livingstone, London.

7

ABNORMAL PUERPERIUM

The causes of pyrexia following delivery are:

- genital tract infection
- urinary tract infection
- breast infection
- thrombophlebitis
- wound infection
- respiratory tract infection
- intercurrent febrile illness.

GENITAL TRACT INFECTION

Childbed fever was the scourge of the first lying-in hospitals; the mortality from it became so great that a woman was held to be fortunate if she survived her stay in hospital. It was not realized until the middle of the nineteenth century that the cause of epidemic puerperal infection was lack of cleanliness on the part of the attendants, who carried the infection from one woman to another.

Louis Pasteur discovered that surgical infection was caused by micro-organisms. Following this antiseptics (and later asepsis) were introduced into obstetric practice and quickly diminished the risks of infection. However, infection still remained the most important cause of maternal mortality. In 1935 the sulphonamide group of drugs produced dramatic results in the treatment of the most lethal type of infection – that due to the haemolytic streptococcus. Penicillin and other antibiotic drugs

have further reduced the dangers of infection, and the great majority of cases occurring today are of the mild, localized type.

Infection now accounts for less than 5 per cent of all direct deaths in the UK but up to a third of deaths in the world. Although the incidence and severity of birth canal infection has fallen dramatically in recent years in the UK, there is danger in complacency. New strains of organisms, resistant to the commonly used antibiotic drugs, appear from time to time and cause outbreaks of serious infection. It is only by strict attention to asepsis and the maintenance of a high standard of obstetric practice that they can be controlled.

Aetiology

When the placenta separates from the uterine wall it leaves a raw area on the uterine wall. Further down the birth canal other wounds may also follow: the cervix, the fourchette and sometimes the perineum. These wounds may become infected, and the symptoms and physical signs resulting in genital sepsis.

Bacteriology

Organisms which may cause infection of the birth canal during labour include:

Endogenous
   coliform organisms
   enterococci (*Streptococcus faecalis*)
   anaerobic streptococci
   gonococci *Chlamydia*
   less commonly, streptococci groups
     B, C, D and G
   other anaerobic bacteria (*Bacteroides* spp)
   Clostridium perfringens
Exogenous
   haemolytic streptococcus, group A
   Staphylococcus aureus

Endogenous organisms

The endogenous organisms are present in the patient's body before the onset of labour, often in the lower intestinal tract, on the perineum and in the vagina. Up to a quarter of women harbour one of these organisms in the vagina, but they seldom cause more than transient and localized infection.

In exceptional cases a localized infection due to anaerobic streptococci may spread into the pelvic veins and cause lung abscesses from septic emboli, or cause thrombosis of the femoral vein.

A clostridial infection may cause collapse, jaundice and haemoglobinuria. These rare infections may be rapidly fatal. They occur when there has been much bruising and tissue damage or when there is retention of a macerated fetus.

Exogenous organisms

The exogenous organisms come from other infected women or from attendants who either have an infection or are carriers of infection. The organisms may pass to the birth canal of a woman in labour on the hands of birth attendants, by droplet infection or, in the case of the haemolytic streptococcus, group A, from infected dust.

Women and attendants suffering from colds and sore throats are liable to harbour haemolytic streptococci or staphylococci in their noses and throats, and a small proportion of healthy individuals are also found to be carrying these organisms. In both types of carrier the organisms can be cultured from the hands.

β-Haemolytic streptococci can be responsible for serious maternal infections; it caused severe epidemics in the past. Mild localized infections occur with group B organisms and uncommonly with other groups. Some commensal organisms which cause little harm to the mother are easily transmitted to the newborn infant such as haemolytic streptococci of group B. The virtual elimination of the haemolytic streptococcus as the cause of serious infection is the result of both a diminution in its virulence and its sensitivity to penicillin. The main danger is now from *Staphylococcus aureus*, which may be penicillin-resistant.

There may be carriers of resistant strains of staphylococci among hospital staff, in dust and in the bedding. Staphylococci may colonize the umbilical stump of the newborn child and the organisms may be spread from this site to other infants in the nursery.

Pathology

After the organisms have entered the tissues subsequent events depend on:

- the virulence of the organisms
- the resistance of the patient
- the amount of trauma and resultant dead tissue
- the speed with which effective antibiotic treatment is begun
- the effective blood supply to the infected area allowing antibiotics to get to the tissues.

Three degrees of severity of infection may be seen:

Mild infection

The infection remains localized to perineal, vaginal or cervical lacerations, to the birth canal or at the placental site.

Moderate infection

Spread of infection occurs from the vagina or cervix into the pelvic cellular tissue and may cause pelvic cellulitis. Infection may also spread from the uterine cavity to involve the Fallopian tubes and pelvic peritoneum, giving rise to acute salpingitis and pelvic peritonitis.

Severe infection

When the organisms are particularly virulent, as in the case of the haemolytic streptococcus group A, the infection may involve the general peritoneal cavity to cause peritonitis, or into the bloodstream to produce septicaemia. The pleura, pericardium or joints may also be infected. When there is an overwhelming infection the woman rapidly becomes acutely ill from the effect of toxins produced by the organisms, although the local inflammatory response at the site of entry of the organisms in the birth canal may be minimal, and a perineal laceration, for example, may look quite clean.

Symptoms and signs

The earliest sign of puerperal infection is fever. There may be a slight and transitory rise of temperature associated with the activity of labour and again on the third postpartum day when the milk secretion really starts causing breast engorgement. Thereafter the puerperium should be apyrexial.

The fever may appear within 24 hours of delivery, and only exceptionally does it appear later. The rise of temperature may be abrupt and is occasionally accompanied by a rigor, or it may be step-like, taking several days to reach its maximum. The pulse rate is raised and the woman feels hot, with headache and backache. Spread to the pelvic peritoneum is shown by lower abdominal pain and tenderness on examination of the uterus and adnexa. Pelvic cellulitis causes persistent pyrexia and a mass to one or both sides of the vagina and uterus which may take several weeks to resolve.

Women with general peritonitis are severely ill with a rapid, thready pulse; abdominal pain and distension; vomiting and diarrhoea. There is generalized abdominal tenderness and there are diminished bowel sounds. The fever is usually persistently high, but in the very worst cases, and terminally, it may be slight.

In septicaemia rigors are common with continuous high fever. The women are very ill and there may be no localizing signs.

Diagnosis

Pyrexia following labour or miscarriage and persisting for more than 24 hours must always be assumed to be due to infection of the birth canal until this has been excluded. In each case a general clinical examination should be made. Involution of the uterus may be delayed and it is often tender on abdominal examination. There may be offensive lochia. The perineum should be examined to see if any lacerations or an episiotomy are infected. If infected they will not be healing normally and there will be swelling with surrounding redness and a purulent exudate. A high vaginal swab should be taken and a midstream specimen of urine collected for examination and culture.

Judgement is required in interpreting the bacteriological findings because organisms such as coliforms and non-haemolytic streptococci are commonly found in the vagina in the absence of clinical infection, but a profuse growth of streptococci must be investigated and the organisms must be typed. When coliforms and anaerobic streptococci are found the lochia is often purulent and foul-smelling, but virulent haemolytic streptococcus group A infections may spread to the general peritoneal cavity or cause septicaemia with relatively few local abnormal physical signs, and with lochia which is not offensive. If the severity of the illness or the type of fever leads to suspicion of septicaemia, blood is taken for culture, preferably when the temperature is at its highest or during a rigor. Several blood cultures are sometimes needed to establish a diagnosis.

In salpingitis the Fallopian tubes and ovaries are swollen and inflamed with adherent omentum and bowel, but the tenderness of these structures in the rectovaginal pouch makes them difficult to distinguish on bimanual examination. In most cases of pelvic cellulitis a vaginal examination will show induration of the parametrium extending to the lateral pelvic side wall. If it is unilateral it may push the uterus to the other side of the pelvis. With extensive cellulitis the whole pelvis feels solid, and the induration may even be palpable in the lower abdomen.

Prevention

During labour vaginal examinations should only be done with sterile gloves and adequate aseptic precautions; trauma at delivery should be reduced to a minimum.

Exogenous infection

The risk of exogenous infection is diminished by the usual aseptic technique with instruments and towels, and the use of sterile gloves and gown. Masks are not being worn in every labour room now. The major source of genital tract infection is flora from the woman's own skin. This is best coped with by careful perineal cleaning before delivery. The next most common source is the skin of the attendants; pre-delivery hand washing and glove wearing prevent this. Naso-oral transmission is now an uncommon source and so the use of masks has been stopped in many centres. This renders the process of childbirth more human.

Infected women must be isolated. Any member of the medical or nursing staff with any infection such as sore throat, boil or paronychia is excluded from the labour rooms. If there is more than a sporadic case of puerperal infection bacterial swabs are taken from everyone working in the unit so that the source of infection can be traced.

Endogenous infection

The prevention of endogenous infection is far more difficult. It is impossible to sterilize either the vagina or vulva completely. In normal circumstances the vagina does not contain organisms of high virulence, but the vulva and perianal region are covered with intestinal organisms such as *E. coli* and *S. faecalis*. These organisms are potentially pathogenic and it is impossible to pass anything, even a sterilized instrument, into the vagina without also conveying organisms from the introitus. To minimize this risk the vulva and perianal region are swabbed with an efficient antiseptic such as chlorhexidine before making a vaginal examination. When any vaginal procedure is necessary during labour the same technique should be employed as in an operating theatre; after swabbing the vulva with antiseptic the surrounding skin, especially the perianal region, is covered with sterile towels.

Although the indiscriminate use of antibiotics for prophylaxis is to be condemned because it encourages the production of resistant organisms, such treatment may be considered after premature rupture of the membranes or amniotomy if labour has not started within 24 hours. Antibiotics may also be given in a case of long labour, especially if this is terminated by caesarean section. The antibiotic chosen should be safe for mother and fetus; ampicillin, or a cephalosporin such as cephalexin, are often used.

Treatment

If the woman is not already in hospital she must be

transferred there, so that all facilities for nursing, diagnosis and treatment are available. She must be made comfortable by good nursing, with analgesics if she is in pain and sedatives to ensure rest. The fluid intake must be adequate. Haemoglobin estimations should be done every 2 or 3 days because anaemia is often associated with severe sepsis, and transfusions of fresh blood may be required. Antibiotic treatment will be chosen according to the infecting organism and its sensitivities. If the woman is not very ill treatment can wait for the results of the bacteriological report, but in many cases treatment is started immediately with co-trimoxazole (Septrin) or with a broad-spectrum antibiotic such as ampicillin or cephalexin and is subsequently modified if necessary.

If there is an infected perineal wound the stitches should be removed.

Operative treatment has little place and is usually confined to exploration of the uterine cavity to remove retained pieces of placenta. It is sometimes necessary to drain a pelvic abscess by an incision in either the posterior vaginal fornix or rectum. Pelvic cellulitis may sometimes result in an abscess which points above the outer end of the inguinal ligament and needs drainage there.

URINARY TRACT INFECTION

Infection of the urinary tract is the commonest cause of puerperal pyrexia, and is generally caused by *Escherichia coli*. The infection is almost always introduced by catheterization, which is frequently necessary during labour and sometimes in the puerperium. Alternatively, the bruising of the tissues during delivery may be sufficient to give rise to recurrence of a pre-existing chronic and symptomless infection.

Diagnosis

This is made by examination of a midstream specimen of urine. Part of the specimen is sent to the laboratory for bacterial culture and sensitivity tests, but immediate examination of a drop of urine under the microscope will often settle the diagnosis; a large number of pus cells in the field makes the diagnosis of a urinary infection most likely. A bacterial colony count of more than 100 000 organisms/ml signifies infection. Symptoms such as dysuria and frequency are often equivocal, since they may occur after delivery when there is no infection, and may be absent in the presence of infection.

Treatment

When there is a history of recent urinary infection a prophylactic course of ampicillin or co-trimoxazole may be given during labour and the first 4 days of the puerperium.

In the established case the fluid intake must be at least 3 litres in the 24 hours. A full course of ampicillin or co-trimoxazole is given.

BREAST INFECTION

Acute mastitis or a breast abscess may cause puerperal pyrexia.

THROMBOPHLEBITIS

The most common time for pyrexia from this cause is from the 4th to the 10th day after delivery. The fever i.s slight except in cases of extensive iliofemoral thrombosis and those cases caused by pelvic infection.

The aetiology and pathology of thrombophlebitis during pregnancy and the puerperium are described later in the chapter. Apart from the usual case of deep venous thrombosis, which starts in the veins of the calf and which is caused by stasis rather than bacterial infection, it is possible that in some cases of puerperal sepsis the initial thrombosis is in the deep veins of the pelvis. Anaerobic

streptococci may occasionally invade pelvic veins, proliferate in blood clot, and then give rise to septic emboli which pass to the lungs or other sites with episodes of high fever.

WOUND INFECTION

About 1 in 10 caesarean section wounds becomes infected and many obstetricians now recommend prophylactic antibiotics be given at all caesarian sections. If a wound does become infected the signs of infection are local redness, tenderness and, on occasions, the discharge of sero-sanguineous or purulent fluid. Swabs should be taken and appropriate antibiotic therapy instituted. Infected wounds sometimes break down and major wound dehiscence (burst abdomen) may occur particularly with vertical subumbilical incisions; a burst abdomen is exceptionally rare with transverse suprapubic (Pfannenstiel) incisions.

RESPIRATORY TRACT INFECTION

When an anaesthetic has been given during labour, postanaesthetic chest complications such as basal collapse or bronchopneumonia may occur.

INTERCURRENT FEBRILE ILLNESS

Tonsillitis, influenza, or any of the acute specific fevers may occur in the puerperium, as well as surgical conditions such as appendicitis. A general clinical examination should always be made to exclude these.

BREAST DISORDERS IN THE PUERPERIUM

ENGORGEMENT

On the 2nd or 3rd day after delivery the breasts become engorged and secretion of milk begins. If the baby does not empty them sufficiently they rapidly become overdistended. The breasts are enlarged and covered with distended veins; the skin over them may be slightly congested. They are very tender and feel hard and knotty. Nodules of enlarged breast tissue may be palpable in the axilla. The pain may prevent the woman from sleeping.

Treatment

In the early stages of engorgement the baby may be able to take enough milk to relieve the congestion, but once the condition is fully established the congestion and pressure on the ducts prevent the flow of milk and the infant cannot empty the breasts. A little milk should be manually expressed before putting the baby to the breast as this will help to promote an easy flow of milk. Sometimes the mother cannot tolerate this, and an electric breast pump, which produces rhythmical negative pressure in a soft rubber breast cup, may give relief. Analgesics may be required and the breasts should be effectively supported. Occasionally, breast engorgement can be relieved by one or two doses of bromocriptine 2.5 mg. Used in this way, there is no risk of suppressing lactation.

CRACKED NIPPLES

The nipple may become sore and painful from two conditions. One is the loss of the epithelium covering a considerable area of the nipple, with formation of a raw area which is very tender. The other is a small deep fissure situated at either the tip or the base of the nipple, which is also very painful. The two conditions may exist simultaneously, and are referred to as cracked nipples.

After delivery a flat nipple, or one that is not kept aseptic and dry, tends to become sore. If there is not sufficient milk in the breast a hungry baby will suck too vigorously and its gums cause abrasions of the epithelium. Another cause is leaving the baby too long at the breast, but with demand feeding this does not occur so often. The nipples may also become sore if the mother does not depress the breast away from the baby's nostrils when it is suckled, because if the baby cannot breathe through its nose it has to let go of the nipple repeatedly and then take it up again. Thrush (moniliasis) is another occasional cause of soreness of the nipple.

Cracked nipples cause tenderness and pain during suckling. There is also a risk of a breast abscess forming, as the ducts are not emptied because pain and engorgement is unrelieved, and perhaps because the raw area allows access for infecting organisms.

The baby may draw blood from a fissure into its stomach with the milk. If the blood is regurgitated it may give rise to a false diagnosis of gastric haemorrhage.

Treatment

A cracked nipple will heal spontaneously if the trauma that produced it ceases. The first essential of treatment is therefore rest, and the baby must not be put to the breast on that side until the crack has healed. While the crack is healing the breast is emptied by manual expression or by an electric breast pump. When breastfeeding is recommended the baby should only be put to the breast for a few minutes at first, otherwise the nipple will crack again. If the lesion is recognized at an early stage it may heal within 48 hours, but once the crack has become extensive or indurated healing will be far more difficult, and in many cases breastfeeding cannot be re-established.

Various local applications have been recommended. It is important that any such application does not stick and drag away any newly formed epithelium when it is removed, and it must not be harmful to the baby or need to be cleaned off before a feed. Flavine in liquid paraffin is one of the most suitable and harmless antiseptics, which will not adhere.

ACUTE PUERPERAL MASTITIS

The infecting organism is almost always *Staphylococcus aureus*. All that has been said about the prevalence and problems of infection with this organism in puerperal sepsis applies equally to breast infections. In addition the baby is a frequent source of infection to its mother's breasts. It has long been known that breast infections are liable to occur in association with skin infections of the baby, and that such infections spread rapidly in a nursery unless isolation and careful aseptic techniques are practised. The umbilical cord is a common site of infection, and must be inspected daily.

Bacteriological investigations carried out in epidemics of breast infections with *S. aureus* have shown that if one baby in a nursery has the organism in its nose or mouth almost all the other babies can become similarly affected within 2 or 3 days. These infected babies, who may be thriving and not ill, are a dangerous source of infection to their mothers.

Acute mastitis may arise at any time in the puerperium. The onset is rapid, with pain in the breast and fever which may rise as high as 40.5°C within a few hours. In both types the clinical picture is the same, the infection being limited at first to one lobe. A wedge-shaped area of cutaneous hyperaemia is seen, and the affected lobe is tense and tender; there may be general malaise. Unless early treatment is successful the condition will progress to a breast abscess.

Treatment

Prophylaxis consists of scrupulous attention to

aseptic technique. Any mother or baby with an infection must be isolated. Ideally mother and baby should always be kept together and the mother should, as far as possible, attend to her baby's needs herself, thus reducing to a minimum the risk of cross-infection from handling by other persons.

As soon as mastitis occurs breastfeeding on the affected side is suspended and the breast is emptied by gentle expression or with an electric breast pump. It is firmly supported over a large pad of cotton wool. A sample of milk is sent to the laboratory for culture and sensitivity tests; it is almost always possible to grow the infecting organism from the milk. If antibiotic treatment is going to prevent an abscess forming it must be started at once. Most of the infecting organisms in hospital are likely to be penicillin-resistant, but the prevalent strain may be known so that an appropriate antibiotic can be given. If the likely strain is not known treatment with flucloxacillin (Floxapen) 250 mg orally 6-hourly may be given until laboratory reports are available.

BREAST ABSCESS

A breast abscess follows acute mastitis. A segment of the breast becomes painful and tender, with oedema and usually redness of the overlying skin. The temperature is raised and the axillary glands become tender and enlarged. The abscess may form near the surface or in the substance of the breast. If it is neglected a deep abscess may burrow in several directions and lead to almost total disorganization of the breast.

Treatment

Breastfeeding and proper treatment of the abscess are incompatible. The baby must be taken from the breast and alternative feeding must be arranged. Lactation is suppressed with bromocriptine 2.5 mg twice daily for 14 days. As soon as an abscess forms it should be drained under general anaesthesia. It is wrong to wait for fluctuation; brawny oedema of the skin is sufficient for diagnosis.

The incision can radiate from the nipple to avoid cutting the ducts. This is usually done in the lower two quadrants but a circum-areola approach leaves a better aesthetic result in the upper two quadrants. The ducts penetrate deeply under the nipple and it is unlikely that any incision will damage them or

affect breastfeeding in the future. Since these abscesses have loculi running in different directions, and not infrequently have superficial and deep portions connected by a narrow tract, the incision should be adequate and a finger inserted into the abscess to break down any septa or loculi. A drainage tube is inserted, and with large abscesses it is sometimes necessary to make a counter-incision to obtain dependent drainage. Antibiotic treatment is given.

INHIBITION OF LACTATION

Although oestrogens will inhibit lactation they are no longer used because of the risk of venous thrombosis. Bromocriptine 2.5 mg twice daily for 14 days effectively inhibits lactation, but it is expensive, has many side effects and there may be rebound lactation after the drug is discontinued. In most cases all that is required is to support the breasts firmly night and day and to await the cessation of activity. Limitation of fluid intake, but not to a degree which causes thirst, may help and it may be necessary to prescribe some analgesic agents.

If lactation is already established weaning is achieved by omitting successive feeds, and only if there is an urgent need to suppress lactation, as in a case of breast abscess, should bromocriptine be used.

GALACTOCOELE

A galactocoele is a retention cyst of one of the larger mammary ducts. Its content is chiefly milk. A local fluctuating swelling is felt, and the skin over it may be reddened, but there is not severe pain and no constitutional disturbance. A galactocoele that is small and deeply situated may be mistaken for a carcinoma in the breast. The cyst should be excised or aspirated with cytological examination of the fluid obtained.

CARCINOMA OF THE LACTATING BREAST

When carcinoma occurs in the lactating breast it develops with great rapidity and spreads widely. The affected breast is larger than the other, the nipple is flattened or retracted and may be fixed, and the skin over the tumour is oedematous – *peau d'orange*. There may be redness if the growth is close to the surface.

Immediate recognition is of the utmost importance. The rapidity with which the enlargement appears and the reddening of the skin suggest mastitis, but the lack of pain and the fact that the tumour is scarcely tender to palpation should give a clue to the correct diagnosis. This must be confirmed by biopsy, and the treatment will then often be by irradiation. Results are poor.

DISORDERS OF HAEMOSTASIS IN PREGNANCY

During pregnancy there are profound physiological changes in the blood and dramatic changes in the coagulation factors and haemostatic mechanisms; these protect the pregnant woman from haemorrhage, particularly at delivery, and they tip the balance of haemostasis towards clotting. However, along with the beneficial effects are a number of unwanted effects; deep vein thrombosis and pulmonary embolism are found more commonly in women who are pregnant than in non-pregnant women. Haemorrhage remains a risk and, together with thromboembolic disease, deep vein thrombosis and pulmonary embolism, is a significant cause of the rare event of maternal death.

Some of the nomenclature of coagulation is confusing, especially as there are so many opposing reactions with a myriad of factors, stimulators and inhibitors. In order to understand these coagulation and related mechanisms a brief account of how haemostasis is achieved outside pregnancy is given. This is followed by a description of the changes that occur in pregnancy and the coagulation disorders specific to pregnancy.

HAEMOSTASIS

Haemostasis is dependent on the following factors:

the platelet–vessel wall interaction
the coagulation system
the anticoagulation system
the fibrinolytic system.

These factors interact in a fine and balanced way. They form an intricate interlacing system that interdigitates between the coagulation, fibrinolytic, kinin and complement cascades. If the balance is upset, haemorrhage or thrombosis may result.

THE PLATELET–VESSEL WALL INTERACTION

Platelets can initiate thrombosis and can also provide a surface on which the interaction of coagulation factors may take place, by providing the necessary phospholipid.

When the vessel wall is damaged vasoconstriction occurs and any endothelial defect is bridged by the adhesion of platelets to the underlying collagen fibres. Platelets aggregate and, together with fibrin, form a haemostatic plug.

The platelets which adhere to collagen undergo a change in shape and release the contents of their dense body cytoplasmic granules. These contain adenosine diphosphate (ADP), adenosine triphosphate (ATP) and serotonin, which promote further aggregation.

Phospholipid in the platelet membrane is hydrolysed to arachidonic acid. This is converted to a labile cyclic endoperoxide called thromboxane A_2 with a half-life of only 30 seconds. It is a potent releaser of the dense granules and hence is a powerful platelet aggregator; it is also a vasoconstrictor.

Phospholipid in the endothelium is also hydrolysed but a different labile endoperoxide is produced – prostacyclin. This is a powerful inhibitor of platelet adhesion and aggregation and is a vasodilator.

The balance between thromboxane and prostacyclin determines whether there is local adhesion of platelets to endothelium or not. In addition, the platelets have a negative surface charge, as does the endothelium, and the resultant electrostatic repulsion also contributes to the antithrombogenic properties of normal blood vessels.

Platelets and the coagulation cascade are interrelated. Platelet aggregation releases platelet factor 3, which in turn stimulates both the intrinsic and extrinsic coagulation pathways. Platelet factor 3 also promotes irreversible platelet aggregation and clot retraction, which reduces vessel occlusion by means of the contractile platelet protein, thrombasthenin.

PLATELET AND VESSEL WALL INTERACTION

Platelet function and the interaction with the vessel wall are measured by the bleeding time. There are a number of commercially available disposable devices which produce a cut in the skin of standard depth and length. Venostasis is maintained at a pressure of 40 mmHg by applying a sphygmomanometer above the elbow while cuts are made in the volar surface of the forearm. The cuts are carefully blotted with filter paper every 15 seconds and the length of time the cut bleeds is measured. This is a reflection of platelet numbers, platelet adhesion to the subendothelium and platelet aggregation. The normal bleeding time is less than 10 minutes.

Platelet aggregation is assessed in an aggregometer. Change in the transmission of light through a suspension of platelets is measured in response to various aggregating agents such as ADP, adrenaline, ristocetin and collagen. As the platelets aggregate the light transmission increases.

Platelet turnover is usually assessed by radioisotope labelling techniques but this is not applicable in obstetric practice. An estimate of turnover can be made from platelet size; young platelets are larger. Platelet size can be measured automatically on some sophisticated cell counters but can also be assessed on a well-spread blood film. The megakaryocytes on a bone marrow aspirate may be increased if there is increased turnover.

THE COAGULATION SYSTEM

All the coagulation factors are normally present in the blood in an inert form. When there is a break in vascular integrity they are activated in the vicinity of the break. Fine control mechanisms ensure that the activation is contained within the immediate area of the wound and that the circulating blood remains fluid.

There are two pathways leading to the activation of the final common pathway. These comprise the intrinsic pathway, involving factors found in the blood and the extrinsic pathway, involving factors found in damaged tissue. They both lead to the activation of factor X. These two pathways inter-

act with each other and to separate them is artificial but for simplicity they can be described separately (*see* Fig. 7.1).

Intrinsic system

The intrinsic system begins with contact activation. Factor XII, prekallikrein and factor XI are activated with the participation of the cofactor, high molecular weight kininogen. It is of interest that the contact factors are also components of the fibrinolytic and kinin systems, illustrating the close interrelationships between these various physiological systems.

The activated contact factors are specific proteases and they lead to the activation of factors IX, VIII and then X. The activation of factor X probably involves the formation of a complex at the cell surface comprising activated IX, VIII, Ca^{2+} and phospholipid.

Extrinsic system

The extrinsic system bypasses the contact phase. Direct activation of factor X is achieved by a complex formed from tissue factor, phospholipid from the cell surface, Ca^{2+} and factor VII.

The activated factor X formed by either pathway then forms another complex, the so-called prothrombinase complex which consists of factor X, Ca^{2+} and factor V. This results in activation of prothrombin to thrombin, which in turn converts soluble fibrinogen to insoluble fibrin, the skeleton of a firm clot. The fibrin clot is then further strengthened by the cross-linking effect of factor XIII (*see* Fig. 7.1).

The coagulation cascade

This is an amplification system. The concentration of factor XII in the plasma is 20 mg/l and activation of that tiny amount of contact factor can result in the conversion of fibrinogen with a concentration of 2 to 4 g/l to fibrin.

Coagulation tests

The platelet count

This is measured electronically by modern automatic cell counters. The natural tendency of platelets to aggregate may produce falsely low counts.

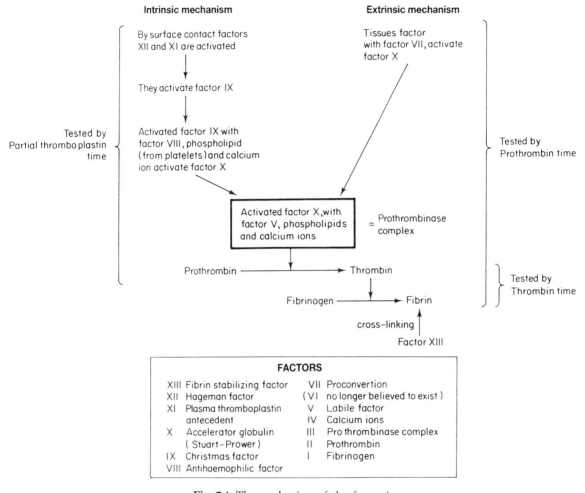

Fig. 7.1 The mechanism of clot formation

The interaction between the platelet and the vessel wall is best measured by the bleeding time.

The intrinsic coagulation pathway

This is measured by the activated partial thromboplastin time (APTT). Unfortunately the terminology is confusing and the test is also known as the partial thromboplastin time with kaolin (PTTK) and the kaolin cephalin clotting time (KCCT). The normal range of the APTT is 35 to 43 seconds. It is prolonged by deficiencies in factors XII, XI, X, IX, VIII, V or prothrombin and also by the presence of coagulation inhibitors such as the lupus anticoagulant. It can detect haemophilia A and B and some cases of von Willebrand's disease.

The intrinsic coagulation pathway is also affected by the presence of heparin and may be used to monitor heparin therapy. However, it may be more reliable to use the protamine sulphate neutralization test to measure therapeutic levels of heparin and the inhibition of activated factor X to measure the very small levels used in low dose heparin prophylaxis.

The extrinsic coagulation pathway

This is reflected by measurement of the prothrombin time. The title of the test is a misnomer because apart from measuring prothrombin it also measures factors V, VII and X. The normal prothrombin time is 10 to 14 seconds. The pathway may be prolonged in liver disease and by treatment with the coumarin anticoagulants such as warfarin. The results of the test are often expressed as a ratio with an international standard – the International Normalized Ratio (INR).

The thrombin time

This measures the end of the coagulation cascade with the conversion of fibringen to fibrin. Both the length of the test and the appearance of the clot give important information. The normal range for the thrombin time is determined in each individual laboratory, but is usually about 10 seconds. Abnormalities may be due to a reduction in fibrinogen but also to the presence of interfering substances. For example, the test is very sensitive to the presence of heparin and to the products of fibrin and fibrinogen degradation (FDP). The fibrinogen level may also be measured directly.

THE ANTICOAGULATION SYSTEM

A variety of mechanisms exist to restrict activation of the coagulation factors to the site of vessel damage and prevent further spread. The activation of thrombin occurs most readily where the platelet phospholipids are available, i.e. where aggregation is occurring. There are also a number of naturally occurring anticoagulants. Antithrombin III (ATIII) not only inactivates thrombin but also the activated factors IX, X, V and VIII. It is also the cofactor for heparin. Protein C is another naturally occurring anticoagulant. It inactivates activated factors V and VIII and promotes fibrinolysis. Protein S is a cofactor for protein C.

Families with deficiencies in these factors may have a hypercoagulable state with a particular predisposition to venous thrombosis. It is interesting to note that, like factors II, VII, IX and X, proteins C and S are dependent on vitamin K for the last stage of their synthesis. They are thus similarly reduced by coumarin anticoagulant therapy and there is the possibility of increasing the hypercoagulable state in susceptible individuals in the early stages of anticoagulation. Skin necrosis has been described in protein C deficiency treated with warfarin due to the development of thromboses in the small vessels of the skin.

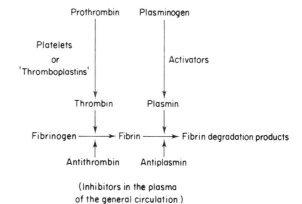

Fig. 7.2 The mechanism of fibrinolysis

FIBRINOLYSIS

During fibrinolysis fibrin clots are digested, rendered soluble and vessel patency is restored. A summary of fibrinolysis is given in Fig. 7.2.

Plasmin is the effector enzyme of the fibrinolytic system but it has wide substrate specificity. There are, therefore, inhibitors to the activation of the precursor (plasminogen) and to plasmin itself, e.g. α_2-antiplasmin.

At the time of clot formation plasminogen becomes bound to fibrin where it is relatively protected from its inhibitors. Plasminogen activators diffuse from the tissues into the clot and activate plasmin, which digests fibrin to form fibrin degradation products. The uterus is particularly rich in such activators. When there is systemic overactivity of the fibrinolytic system a bleeding state may be produced. The products of fibrinolysis may themselves act as anticoagulants, interfering with the polymerization of fibrin and prolonging the thrombin time.

Various fibrinolytic activators are used therapeutically to lyse clots. The best known are streptokinase and urokinase. However there are some newer preparations which are expensive and not fully assessed but which may come to have wider applications. These include single chain urokinase, and acylated plasminogen activator (APSAC). Recombinant tissue plasminogen activator Rt-PA and streptokinase have been used extensively in the treatment of acute myocardial infarction.

COAGULATION AND FIBRINOLYTIC SYSTEMS IN PREGNANCY AND LABOUR

In pregnancy the balance of haemostasis swings towards a hypercoagulable state with increases in the coagulation factors, particularly fibrinogen, factor VIII and von Willebrand factor antigen. Allowing for the changes in plasma volume the amount of circulating fibrinogen virtually doubles by the end of pregnancy. By contrast, the fibrinolytic system is suppressed, with an increase in tissue plasminogen activator inhibitor. The source of the reduced fibrinolytic potential is thought to be the placenta because immediately after delivery of the placenta the fibrinolytic system returns to normal. The overall effect is to protect the mother from haemorrhage during labour and delivery. However, the reverse side of the coin is that there appears to be an increased susceptibility to thromboembolic disease and to disseminated intravascular coagulation. Apart from protein S levels, which fall in pregnancy, levels of the natural anticoagulation factors, protein C and ATIII, remain approximately the same.

The number of platelets in a normal pregnancy probably does not change, although there may be a slight decrease in platelet lifespan in the last trimester. However, in pre-eclampsia there is certainly an increase in platelet turnover.

Despite the dramatic changes in the coagulation and fibrinolytic factors, probably the most important factor in preventing torrential bleeding at the time of placental separation is effective myometrial contraction. This rapidly cuts down the flow of blood to the placental site. It is facilitated by changes that occur in the structure of the spiral arteries towards the end of pregnancy. Without effective myometrial contraction all the coagulation changes that take place in pregnancy would be to no avail.

The placenta is rich in tissue factor. This activates the extrinsic coagulation cascade and results in the deposition of up to 10 per cent of the total body fibrinogen on the placental bed when the placenta has separated. At the time of delivery of the placenta the tests of coagulation and fibrinolysis show activation. Levels of fibrinogen and plasminogen fall and platelet turnover increases. After a short early puerperal increase the levels return to normal by the 4th postpartum week.

COAGULATION DISORDERS SPECIFIC TO PREGNANCY AND LABOUR

HAEMORRHAGE

Haemorrhage is an important cause of mortality in obstetric practice. Between 1985 and 1987, there were 10 maternal deaths due to haemorrhage and a further 36 deaths in early pregnancy including ectopic pregnancies and abortions, many of which were largely due to blood loss. Together this represents 17 per cent of maternal deaths.

DISSEMINATED INTRAVASCULAR COAGULATION (DIC)

DIC occurs outside of pregnancy, but a number of specific obstetric disorders predispose the pregnant woman to the development of this serious bleeding diathesis.

The changes in the coagulation and fibrinolytic systems in pregnancy result in a tendency towards hypercoagulability but paradoxically can result in severe bleeding. Conditions associated with the systemic activation of both systems result in the deposition of multiple fibrin–platelet thrombi in the microcirculation. This results in the consumption of clotting factors and in clinical bleeding; such activation is known as disseminated intravascular coagulation (DIC).

DIC is always a secondary phenomenon, it is not a disease entity in its own right. Its treatment depends upon the alleviation of the underlying primary condition and replacement of depleted factors. Obstetric DIC is usually associated with intrauterine pathology and when the uterus is emptied the DIC fades away. It is particularly associated with placental abruption, retention of a dead fetus *in utero*, amniotic fluid embolism, intrauterine infection due to septic abortion, eclampsia or prolonged shock. The release of tissue thromboplastins or vascular endothelial damage by septicaemia may result in a chronic compensated DIC with an increased turnover of platelets and coagulation factors. Alternatively, there may be a rapid depletion of factors and platelets resulting in catastrophic bleeding with oozing from venepuncture sites, surgical incisions and vaginal blood loss.

In the full blown DIC syndrome the platelet and fibrinogen levels fall and all the coagulation screening tests are prolonged. The blood film shows a microangiopathic picture because red blood cells are fragmented when they are squeezed through intravascular fibrin strands. Fibrin degradation products (FDPs) are formed and these interfere with the polymerization of fibrin. The thrombin time is very sensitive to the presence of FDPs and is a quick, easy test to use in an emergency to monitor the progression of DIC. It also assays the fibrinogen level. It may be useful to monitor the platelets, although the assay of FDPs can often wait until after the cross-matching of blood and the preparation of blood products, which takes priority. The presence of FDPs cannot be taken as diagnostic of DIC because they may be raised in other circumstances such as the normal fibrinolysis of a large internal bleed.

In women with massive bleeding, as in a failure of myometrial contraction, the coagulation factors and platelets may be depleted without there being DIC because the blood used in replacement does not contain all the coagulation factors.

Every obstetric unit should have a written code for the management of major obstetric haemorrhage. In obstetric cases with DIC the precipitant is usually easily recognized. Treatment depends on removal of the underlying cause, correction of hypovolaemia and treatment of the haemostatic disorder.

Treatment of DIC and massive haemorrhage

More than 1 peripheral line may be required to achieve an adequate transfusion rate to maintain volume replacement. A central venous line can monitor volume replacement therapy but its insertion may need to be delayed while the coagulation defects are corrected. Transfusion of fresh frozen plasma (FFP) and cryoprecipitate may alleviate clinical bleeding that is associated with prolonged coagulation tests. Depending on the size of the FFP unit, it may take half an hour to thaw, and volume replacement would be needed during that time including crystalloids or plasma substitutes. It was previously the usual practice to replace missing coagulation factors in a preset formula using a unit of FFP for every 4 to 6 units of transfused red cells. However, current guidelines stipulate that FFP should never be used in this way because it is a blood component capable of transmitting viral infections. The use of FFP should be limited to treating clinical bleeding associated with coagulation factor deficiencies that have been demonstrat-

ed by laboratory coagulation tests. In cases where the clinical evidence of a hypocoagulable state is overwhelming, the coagulation tests may be done retrospectively but they should still be used to monitor the period of treatment. Both FFP and cryoprecipitate are good sources of fibrinogen. FFP contains all the coagulation factors and the naturally occurring anticoagulants including antithrombin III. There may be some advantage in using cryoprecipitate as a concentrated source of fibrinogen and factor VIII.

When hypovolemia is corrected, delivery can be expedited although there may be continuing severe coagulation defects. However, because DIC is a secondary phenomenon, full coagulation correction may not be possible until after delivery. In this situation oxytocics and amniotomy may be used, but caesarean section may carry the risk of operative bleeding. Uncrossmatched blood may be required if haemorrhage is very fast with the first choice being ABO and RhD compatible blood. Almost all antenatal women have their blood groups determined and an antibody screen for atypical red cell antibody performed. If the woman is known to have no atypical antibodies, uncrossmatched, group specific blood is very unlikely to be unsuitable. Crossmatched blood can commonly be available within 20 min with modern laboratory techniques.

Heparin and antifibrinolytics are almost never used in the treatment of DIC. Certainly the use of heparin should not be contemplated unless the uterus is empty and the myometrium well contracted.

Placental abruption

The amount of bleeding from the vagina cannot be used to predict the severity of the haemostatic defect because the bleed may be occult.

Prolonged retention of a dead fetus

If a dead fetus is retained for more than 5 weeks a haemostatic defect due to DIC may result. The defect should be corrected before careful induction of delivery. Although in the past this may have been an important cause of DIC, modern obstetric practice allows for the early recognition of a dead fetus and the retention of a dead fetus is now a rare cause of this condition.

Endotoxic shock

When endotoxic shock follows peripartum uterine infection the mainstay of treatment is to eliminate the cause and the underlying infection and to give supportive treatment for the coagulation defect.

Amniotic fluid embolism

This is a rare but devastating event that may occur during or just after delivery. The amniotic fluid gains access to the circulation and results in extreme shock, obstruction to pulmonary blood flow, massive depletion of coagulation factors and exsanguinating vaginal blood loss. Treatment revolves around cardiopulmonary resuscitation, supportive transfusion therapy and immediate delivery.

Pre-eclampsia

In pre-eclampsia there is an increase in platelet turnover, deposition of fibrin in the microcircula-tion, increased levels of FDPs and microangio-pathic haemolysis. This is a chronic DIC state which increases in severity as the pre-eclampsia worsens. The differential diagnosis of the throm-bocytopaenia rests between autoimmune throm-bocytopaenia, thrombotic thrombocytopaenia and DIC secondary to pre-eclampsia.

COAGULATION DISORDERS COINCIDENT WITH PREGNANCY

There are a number of bleeding diatheses which occur in women and which may complicate preg-nancy. The inherited disorders of haemophilia A and B are sex-linked and female carriers may have suboptimal levels of factors VIII or IX. Von Will-ebrand's disease is an autosomal dominant condi-tion which may be associated with a low factor VIII. In all these conditions it may be necessary to identify affected individuals because they may require replacement therapy or special obstetric management and also because they may need genetic counselling and antenatal diagnosis.

THROMBOCYTOPAENIA IN PREGNANCY

Platelets play an integral part in haemostatis, inter-relating with the coagulation cascade and the vessel wall. The commonest abnormality in platelet pathophysiology is a reduction in number, to below 150×10^9/l – thrombocytopaenia. This may be diagnosed incidentally during pregnancy but adverse effects with bleeding in mother or fetus are uncommon. The management of mother and fetus varies with the cause of thrombocytopaenia.

In normal pregnancy, the platelet count tends to reduce inside the normal range; this is exaggerated in women developing pre-eclampsia. Thrombocy-topaenia may be seen in the HELLP syndrome (the triad of haemolysis, elevated liver enzymes and low platelets), which is the extreme end of the spectrum of pre-eclampsia, particularly directed at the liver. Thrombocytopaenia may also occur in pregnancy as a consequence of severe megaloblastic anaemia due to folic acid deficiency. It then co-incides with low white cell counts and a raised red cell mean cell volume (MCV). Disseminated intravascular coagulation causes thrombocytopae-nia by increased consumption of platelets and is described in another part of this section.

INCIDENTAL THROMBOCYTOPAENIA OF PREGNANCY

Incidental thrombocytopaenia discovered by chance on routine screening of well pregnant women has no adverse effects on mother or infant and may be the mild end of the spectrum of AITP (auto-immune thrombocytopaenia). The rate of neonatal thrombocytopaenia in infants born to healthy women without a history of AITP or pre-eclampsia but with incidental thrombocytopaenia is the same as for normal mothers without throm-bocytopaenia. These infants may be mildly throm-bocytopaenic but they are unlikely to have clinical bleeding. The maternal platelet counts usually return to normal by 7 days postpartum. Therefore healthy women with incidentally discovered thrombocytopaenia and no previous history of auto-immune thrombocytopaenia before pregnancy require no treatment and their mode of delivery should be determined by obstetric indications only.

AUTO-IMMUNE THROMBOCYTOPAENIA

Autoimmune thrombocytopaenia (AITP) is relatively uncommon in women of childbearing age (1 or 2:10 000 pregnancies). The mechanism of immune destruction of platelets has been shown to be due to auto-antibodies directed against platelet surface antigens. This has special relevance in pregnancy because the placenta has receptors for the constant fragment (Fc) of the IgG immunoglobulin molecule facilitating its active transport across the placenta to the fetal circulation. Immunoglobulin passage becomes more marked as pregnancy progresses and may result in fetal thrombocytopaenia.

Clinical presentation of AITP

Overt cases may present with skin bruising, and platelet counts between 30 and 80 × 10⁹/l. However, it is rare to see severe bleeding associated with very low platelet counts. The correlation between maternal and neonatal platelet counts is poor. Rarely, infants have bleeding complications which may be serious. The clinical fear is that intracranial haemorrhage will occur in thrombocytopaenic infants delivered vaginally. It has been assumed that delivery by caesarean section is less traumatic to the fetus than vaginal delivery and while that premise could be debated, distinguishing incidental from pathological maternal thrombocytopaenia will avoid many unnecessary caesarean sections.

Diagnosis of AITP

AITP is a diagnosis of exclusion with peripheral thrombocytopaenia but normal or even increased megakaryocytes in the bone marrow and the absence of other diseases. Its diagnosis requires the exclusion of SLE, lupus anticoagulant and anticardiolipin antibody as these may co-exist with thrombocytopaenia.

Many tests have been devised to predict fetal thrombocytopaenia. However platelet antibody on the platelet membrane and free in the plasma, demonstrated by tests analogous to the direct and indirect Coomb's tests on red cells, have been slow to enter the repertoire of routine haematological laboratories. This is because they are fraught with technological difficulties such as the intrinsic reactivity of platelets and the presence of some platelet associated immunoglobulin in normal individuals. While some cases of AITP have normal or increased amounts of immunoglobulin some women have no demonstrable platelet associated IgG.

The free circulating maternal antiplatelet antibody test may also be a good discriminator in some hands. The absence of free antibody at the time of delivery is associated with a very low risk of neonatal bleeding. However, not all centres have reliable tests for circulating antiplatelet antibodies. Currently available serological tests cannot be used to determine management.

Therapy of AITP

Women who enter pregnancy and are known to have had AITP may already be on therapy and so may have normal platelet counts. Rarely, women will present for the first time in pregnancy. If the platelet count falls below haemostatic levels (about 50×10⁹/l) especially as term approaches, treatment will be necessary although no safe platelet count has ever been determined. It is sensible to avoid trauma and the use of anti-platelet drugs. The mainstay of treatment in pregnancy is prednisolone commencing with a dose of 1–1.5 mg/kg, and gradually reducing. Corticosteroids alter immune function and improve capillary fragility. However, because of well-known adverse effects, they are generally only used in the short term or at very low dosages. They may raise the platelet count sufficiently to enable the use of epidural anaesthesia and operative delivery.

Splenectomy, the mainstay of treatment in the non-pregnant state, is not generally recommended during pregnancy although occasionally it may be required for resistant cases. While treatment with steroids or a splenectomy may result in clinical remission, a normal maternal platelet count does not reduce the known risk of thrombocytopaenia to the fetus. In other cases, splenectomy may be indicated after the puerperium.

Corticosteroid resistant cases may be treated with high dose intravenous immunoglobulin. The mechanism of action of this treatment is unknown but may in part be related to Fc (the constant fraction of IgG) receptor blockade of the reticuloendothelial system or to feed back suppression of IgG synthesis. This expensive therapy may produce increases in the platelet count more quickly than with corticosteroids and the effect may last 3 weeks. When used 3–4 weeks before delivery, the IgG will cross the placenta and may raise the fetal platelet count.

Fetal diagnosis and mode of delivery

Cases of AITP diagnosed prior to the index pregnancy may benefit from fetal blood sampling at about 36 weeks' gestation to assess the fetal platelet count. This carries very little risk to the fetus. However, fetal blood sampling is not universally available. Where the fetus has a normal platelet count and there are no obstetric contraindications, vaginal delivery is indicated. Moderate to severe fetal thrombocytopaenia is currently considered an indication for caesarean section. The exact level of thrombocytopaenia that determines this is constantly changing in the downward direction. However fetal thrombocytopaenia due to the auto-immune condition AITP should not be confused with fetal thrombocytopaenia due to alloimmunization of the mother by the fetus, which is a far more serious disease for the fetus. In the latter condition there is a functional platelet defect as well as a thrombocytopaenia, but the maternal platelet count is normal.

Cases with other obstetric indications for caesarean section should not be investigated by fetal blood sampling because this would not further contribute to management since an operative delivery is planned anyway.

In centres with no access to fetal blood sampling all cases of maternal AITP diagnosed prior to the index pregnancy should probably be delivered by caesarean section although local expertise and preference will determine this in the individual case. This may present a risk of maternal bleeding at the incision and from soft tissue injury in the presence of thrombocytopaenia. Platelets from random donors, although rarely used, could provide some haemostatic assistance in resistant cases but would be subject to the same consumption as the woman's own platelets.

Epidural anaesthesia can probably be safely administered with platelet counts above $80 \times 10^9/l$, in the presence of a normal bleeding time and a coagulation screen within the normal range.

Neonatal thrombocytopaenia

Neonatal thrombocytopaenia due to maternal AITP may not reach its nadir until 24 to 72 hours post delivery. It may be some weeks before the platelet count recovers fully. Very low neonatal platelets may require therapy with platelet transfusions, corticosteroids and intravenous IgG.

THROMBOCYTOPAENIA AND HUMAN IMMUNODEFICIENCY VIRUS

Thrombocytopaenia is a well recognized complication of HIV infection following the use of therapeutic drugs and severe infection. However, women with HIV may also have thrombocytopaenia otherwise indistinguishable from AITP. This may be due to immune platelet destruction directed at molecular cross-reaction between the human immunodeficiency virus and platelet glycoproteins. This may explain why AIDS-free women may develop HIV associated AITP. It has also been suggested that disturbances in the B cell subset, CD5, in HIV infected women causes immunological changes correlating with the platelet count. While most HIV positive patients have so far been young men, it is possible that this infection will become commoner in young pregnant women although the degree of heterosexual spread of HIV is uncertain. Certainly, young pregnant women in a high risk group for HIV and with thrombocytopaenia should be considered for HIV testing.

THROMBOCYTOPAENIC PURPURA AND HAEMOLYTIC URAEMIC SYNDROME (HUS)

These conditions are both due to the presence of platelet thrombi in the microcirculation which cause ischaemic dysfunction and microangiopathic haemolysis. In HUS, the brunt of the disease process is taken by the kidney. It has rarely been associated with childbirth particularly in the puerperium. It has also been seen during pregnancy and reported in association with ectopic pregnancy. It has been postulated that endothelial damage is mediated through neutrophil adhesion in association with infection and leads to the formation of platelet thrombi.

In thrombocytopaenic purpura (TTP), the focus shifts to multisystem disease, often with neurological involvement and fever. It has been associated with pregnancy and the postpartum period and with the platelet anti-aggregating agent, ticlopidine. The underlying aetiology of TTP in pregnancy remains unknown and these various abnormalities may only be epiphenomena.

The clinical pentad of fever, normal coagulation tests with low platelets, haemolytic anaemia, neurological disorders and renal dysfunction are virtually pathognomonic. The platelet count may range from 5 to $100 \times 10^9/l$ (TTP). The clinical picture is severe with a high maternal mortality.

It has been suggested that plasma supplies a factor lacking in women with TTP that stimulates the release of prostacyclin. Steroids alone may be effective in mild cases. Severe cases may benefit from intensive plasma exchange but where that is difficult, intensive plasma infusion is indicated. Unresponsive cases may benefit from cryosupernatant infusions. The use of anti-platelet drugs is not helpful. In order to prevent recurrence, plasma infusion should be gradually reduced but continued until all objective signs have been reversed. Platelet infusions are contraindicated. There is no evidence that delivery alters the course of the disease although it does simplify maternal management. Many women go on to further normal pregnancies.

THROMBOEMBOLIC DISEASE IN PREGNANCY

Deep vein thrombosis (DVT) may be clinically inapparent, its first clinical sign may be an embolic complication. There must, therefore, be a high index of suspicion tempered by effective objective diagnosis because treatment carries risks for both the mother and fetus. Also, clinical diagnosis of DVT is notoriously unreliable and is wrong half the time. Recent reductions in deaths from pulmonary embolism (PE) are probably due to improved prophylaxis and the reduced use of oestrogens for suppressing lactation.

The incidence of thromboembolic disease in pregnancy lies between 0.3 and 12 per 1000 deliveries. Pulmonary embolus is the commonest cause of death associated with pregnancy. In the period 1985–87, in the UK , there were 30 deaths due to pulmonary embolus out of a total of 265 representing 11 per cent of maternal deaths. Less than half of the cases of fatal thromboembolism and almost one-quarter were before 17 weeks of pregnancy.

The risk of recurrence of thromboembolism in women who have had a previous episode has been estimated at 12 per cent.

While pregnancy itself is a risk factor for venous thromboembolism the risk of pulmonary embolism is increased in women of greater parity and age and in those with a previous history. The usual risk factors of obesity and any postoperative state continue to apply in pregnancy.

There is an increased risk of thromboembolism in women with congenital deficiencies of antithrombin III and protein C and with some sickle cell anaemia syndromes. The lupus anticoagulant is an acquired condition which predisposes to thrombosis. This lengthens *in vitro* tests of coagulation while causing thrombosis *in vivo*. It also causes fetal loss.

It is very important to investigate and identify thromboembolic diseases and to come to the correct diagnosis, because therapeutic and prophylactic anticoagulation in the index pregnancy and in future pregnancies carries important hazards. DVT should be confirmed by ultrasound; this can show the clot and lack of venous expansion of the affected vein during the Valsalva manoeuvre. Previously, venography was used to confirm the diagnosis with radiation limited to the midthigh and with good pelvic screening. The benefits of ultrasound over venography are avoidance of exposure to radiation and that it can be repeated as often as necessary to monitor progress.

In a woman who is not shocked and who has pleuritic pain but no changes on chest X-ray, the diagnosis of pulmonary embolism can be made by lung scan. Simple perfusion scans in a woman with a normal chest X-ray may show a deficit. Where there is an abnormal chest X-ray a ventilation-perfusion scan will discriminate between pneumonia which will show a matched deficit, and pulmonary embolism, where there will only be reduced perfusion. The radiation dose from such a scan is less than that given in a standard chest X-ray.

Treatment of thromboembolism

Massive pulmonary embolism may cause sudden death. If the woman survives long enough surgical embolectomy may be indicated. Fibrinolytic therapy is rarely used in pregnant women but may be indicated in women in the postpartum period.

Less severe cases require immediate anticoagulation with therapeutic heparin, starting with 40 000 units per day by intravenous infusion. This limits the progression of the clot and reduces mortality. Heparin does not cross the placenta. It is continued for about a week and is monitored by the APTT or the protamine sulphate test. Thereafter low-dose subcutaneous heparin is given until 6 weeks into the puerperium. The dose is about 10 000 units twice daily, reducing to 8000 units after delivery, and it is usually self-administered by the woman. It has no effect on the coagulation factors or on coagulation tests and it serves only to prevent the spontaneous conversion of factor X to the active form. Activation of factor X is the important link between the intrinsic and extrinsic pathways. The level of heparin achieved by this low-dose regime can only be measured by inhibition of activation of factor X.

The major short-term side effect of therapeutic heparin is maternal bleeding. With therapeutic levels of heparin surgical delivery is contraindicated but there is no such restriction in the use of prophylactic heparin levels.

After the acute phase, the anticoagulant, warfarin, is used in some centres. It is monitored by the prothrombin time. Bleeding is the principle side effect in the mother and this is particularly important as delivery approaches. As the effects of heparin can be reversed more quickly than those of warfarin, heparin often replaces warfarin from 36 weeks. Heparin virtually disappears from the circulation in 6 hours whereas it takes 3 days for the effects of warfarin to dissipate. However, in an emergency warfarin can be reversed with an infusion of FFP. It is not good policy to reverse the warfarin effect with vitamin K because this is said to produce a rebound effect in the mother and it may be very difficult to reinstitute warfarin therapy later. In an emergency heparin can be reversed with protamine sulphate, although in practice withholding the drug usually suffices. It must be remembered though, that protamine sulphate is an anticoagulant in its own right. Whichever drug is used it is continued until the 6th week postpartum. Women on heparin may choose to change to warfarin in the puerperium.

Warfarin carries an additional risk in pregnancy because it crosses the placenta and can cause bleeding and congenital abnormalities in the fetus in a dose-dependent fashion. It should therefore be avoided in the first trimester during the period of organogenesis. When it is used it should be well controlled and the dose should be maintained at the lower end of the therapeutic range. Its use cannot be avoided in women with replacement heart valves because heparin cannot prevent thrombosis effectively enough in this situation.

Neither warfarin nor heparin are excreted in the breast milk and are therefore safe to use during lactation.

Prophylaxis of thromboembolism

Heparin is sometimes given to pregnant women with a history of thromboembolism. Bone demineralization is associated with heparin treatment; the effects may be cumulative over successive pregnancies. It is likely that the risks of recurrence of thromboembolism may not be as great as was previously suspected, so that there is no need for prophylaxis in most women with a previous solitary event and no other risk factors. In view of the risk of bone loss, heparin may be given from delivery for six weeks in women who have had a single episode of thrombosis. Changing to warfarin after a week further reduces the risk of bone loss.

FURTHER READING

Samuals P. *et al.* (1990) Estimation of the risk of thrombocytopenia in the offspring of pregnant women with presumed immune thrombocytopenic purpura. *New England Journal of Medicine*, **323** (4); 229–35.

Turnbull's Obstetrics (1995) Ed. Chamberlain G. Section 6. Puerperium. Churchill Livingstone, Edinburgh.

8

OBSTETRIC PROCEDURES

EXTERNAL VERSION

External version involves manipulation of the fetus through the abdominal wall to change its position, usually from a breech presentation to a cephalic presentation and, more occasionally, from a transverse lie (shoulder presentation) to a cephalic presentation. If the lie is turned from transverse to a breech this is known as podalic version.

The use of external version in breech presentation has varied with obstetric fashion. Although being out of favour for some years, it is once again receiving wide usage. Around 3 per cent of fetuses present by the breech at term, but as many as 50 per cent assume this position at some time between 30 and 34 weeks. External version at this stage of pregnancy is very easy, but reversion to a breech is relatively common. Controlled studies have shown that version at this stage of pregnancy does not reduce the final incidence of breech presentation at term.

External version at 36–37 weeks of pregnancy will, if successful, lead to a consistent cephalic presentation. Because the fetus is larger and there is a lower amniotic fluid volume at 37 weeks, version must be assisted by uterine relaxation, using a tocolytic agent, such as ritodrine, given intravenously. Short-term administration of tocolysis is quite safe and enables version to be achieved in around 50 per cent of cases. The benefit of such a manoeuvre is to reduce the overall need for caesarean section in breech presentation for otherwise, a high proportion of babies will be delivered by caesarean section; this can be as the obstetrician's choice or because of difficulties arising mechani-

cally during labour. External version is associated with a theoretical risk of cord entanglement or of feto-maternal haemorrhage, leading to isoimmunization. Cord entanglement can lead to acute fetal distress and therefore version should only be performed where immediate facilities for caesarean section are available and when the mother has been appropriately prepared. All rhesus negative women should receive anti-D gamma globulin to prevent the risk of isoimmunization.

INDUCTION OF LABOUR

Indications

Induction of labour is indicated in some cases of:

> pregnancy induced hypertension
> prolonged pregnancy
> intrauterine growth retardation
> antepartum haemorrhage
> unstable lie
> diabetes mellitus
> haemolytic disease
> fetal abnormality
> fetal death.

It is important that induction of labour should only be carried out for sound obstetric or medical reasons, although a few women will have exceptionally important domestic problems for which they request induction. The hazards of induction are few, but there will be occasional cases where unexpected fetal distress occurs during induced labour which would not have occurred in natural labour, or where attempts at induction fail and a

Table 8.1   Modified Bishop score

Score	0	1	2	3
Dilatation of cervix (cm)	0	1 or 2	3 or 4	5 or more
Consistency of cervix	Firm	Medium	Soft	–
Length of cervical canal (cm)	>2	2–1	1–0.5	<0.5
Position of cervix	Posterior	Central	Anterior	
Station of presenting part (cm above ischial spines)	3	2	1 or 0	Below

caesarean section, which would not otherwise have been required, has to be performed.

Where induction is essential for severe fetal or maternal problems, it should be attempted regardless of the ripeness of the cervix after attempts to soften it with local prostaglandins. In this situation if the induction fails, caesarean section is reasonable whereas if an induction is successful, caesarean section has been satisfactorily avoided. Where the indication for induction is less absolute, then the cervix must be carefully assessed beforehand by vaginal examination. This is because the success of induction will depend in part on the extent of pre-labour changes in the cervix. The important features are the length of the cervix and its dilatation with the degree of softness, position and station of the presenting part giving additional secondary information. In research studies these features have been quantified using the modified score devised by Bishop (Table 8.1).

DEVELOPMENT OF INDUCTION METHODS

In the 1970s an effective and standard method of induction was developed which involved low amniotomy and simultaneous administration of oxytocin. The dose of oxytocin was increased in increments until optimal uterine activity was achieved. It was found that around 80 to 85 per cent of women could be successfully delivered by this approach, although often prolonged periods of time were required and high doses of oxytocin may have to be used. In the 1980s the advent of vaginal prostaglandin preparations enabled obstetricians to mimic the natural process leading to the onset of labour. This has significantly improved success rates. Ninety five per cent of women can be induced successfully at the first attempt using prostaglandins, with fewer fetal and maternal complications; thus the neonatal hyperbilirubinaemia

seen after high dose oxytocin administration is avoided, fewer women have a prolonged stay on the labour ward or require a caesarean section for failed induction. A further advantage from the maternal point of view is that the use of prostaglandins allows mobility during the latent and early part of the active phase of labour. This is not possible when the woman has to be connected to an infusion pump for the administration of oxytocin.

MANAGEMENT OF INDUCED LABOUR

Women who are about to undergo induction of labour should have careful counselling so that they are thoroughly aware of the reason for the procedure and the likely course of events. It is particularly important that they do not believe that delivery is guaranteed on the day that induction commences as it may be necessary to administer prostaglandins on more than one consecutive day to initiate the labour process. There must be a clear protocol for the use of oxytocic drugs such as prostaglandins or oxytocin and all staff must be aware of the need to monitor the fetus carefully during uterine stimulation, because of the risk of hypertonus and fetal hypoxia. As in the management of normal labour, it is helpful to record findings consecutively on a partogram. Latent phase abnormalities are difficult to interpret but once the active phase of labour is reached the indications for the use of syntocinon are as in spontaneous labour (Chapter 5).

DRUGS USED IN THE INDUCTION OF LABOUR

Prostaglandins

Although there are various prostaglandins available, the one used for the induction of labour with a live fetus is prostaglandin E_2 (dinoprostone). This may be administered vaginally in a gel containing a dose of either 1 mg or 2 mg of dinoprostone or as a tablet containing 3 or 6 mg of dinoprostone. The pharmacokinetics are more consistent and reliable with the vaginal gel which has recently become the preferred method of administration. Prostaglandins can be administered orally, sublingually or intravenously, but these routes are unreliable and have been superceded by the vaginal approach. Intravenous administration is particularly problematic as prostaglandins are metabolized very quickly by a single passage through the lungs, therefore, high doses have to be given leading to gastrointestinal side effects. Other prostaglandins such as 15 methyl analogues or gemeprost have a more powerful oxytocic effect and their use is usually reserved for situations when the fetus is dead as the risk of hypotonus is much greater.

Oxytocin

Oxytocin is administered by continuous intravenous infusion preferably using an infusion pump. Because of the antidiuretic effect of oxytocin, care must be taken not to administer large volumes of intravenous fluid concomitantly as this may lead to water intoxication, a major disturbance of electrolyte affecting both mother and fetus. This can be avoided by using concentrated solutions of oxytocin in electrolyte containing solutions such as sodium chloride injected by a pump into an intravenous line carrying fluid at a constant and safe rate. The dose of oxytocin administered is calculated in milli-units (mU) per minute and it is important to record the dose in this nomenclature because of the many different concentrations of solutions used in different hospitals making nomenclature such as drops per minute or ml per hour widely variable in terms of the actual dose administered. The starting dose of oxytocin is usually 2 mU per minute and will be increased either arithmetically or geometrically at regular intervals until optimum uterine activity is achieved. It takes between 35 and 40 minutes for the effect of each increment to stabilize and the dose should not be increased more frequently than every 30 minutes. Up to 32 mU per minute may be required in the early stages of labour to stimulate uterine activity. Once labour is established much lower doses, between 7 and 10 mU, will usually be sufficient to maintain good uterine activity. Reduc-

ing to these lower maintenance doses reduces the risk of uterine hypertonus.

Amniotomy

When amniotomy and simultaneous oxytocin infusion was the standard method of induction of labour, all women had an amniotomy early. The advent of prostaglandins has meant that early amniotomy is not required and indeed is best avoided after recent administration of prostaglandins because of the surge of endogenous prostaglandin which it produces, risking hypertonus. In certain high risk cases amniotomy may be beneficial to allow early internal fetal monitoring. Controlled studies have shown no significant benefit overall of early amniotomy and the membranes can safely be left intact until close to delivery unless progress is abnormal. There is a theoretical risk that oxytocin administration in the face of intact membranes might lead to amniotic fluid embolus. Therefore when oxytocin administration is commenced after labour has been induced with prostaglandins, amniotomy is usually performed.

In some cases when induction of labour is planned the cervix is found to be widely dilated. If the Bishop's score is more than 10 it may be possible to induce labour simply by amniotomy. This is successful in multiparous women, though this is less likely to be successful in primiparae. This approach is particularly favoured in grand multiparae because of an anxiety that oxytocic drugs might cause hypotonic uterine action.

Management of uterine hyperstimulation

When hypotonic uterine contractions occur there may be an acute fetal bradycardia matching the persistent excessive uterine contraction. This acute bradycardia is not dangerous to the normal fetus, but may be of severe consequence to a compromised fetus. If oxytocin is being administered, it should be switched off immediately and unless the uterine contraction rapidly resolves, tocolysis should be induced by administering a bolus dose of ritodrine or salbutamol intravenously. Where there is acute fetal bradycardia, preparation should be made for an immediate caesarean section, but usually this will not be required as the problem will spontaneously resolve.

Uterine hyperstimulation is a particular problem where there is an obstructed labour.

EPISIOTOMY

An incision in the perineum enlarges the introitus. It is used to aid delivery when the perineum is obstructing progress or when more room is required, such as for an instrumental delivery with forceps or vacuum extractor, during assisted breech delivery or when there is shoulder dystocia. The routine use of episiotomy in primagravid women is one of the obstetric interventions which led to a conflict between consumer groups and the medical profession in the 1970s. It is now recognized that routine episiotomy confers no benefit and that this operation should only be employed for specific indications. Around 30 per cent of primagravidae will require an episiotomy, at a normal delivery either because of significant delay in progress, fetal hypoxia in the second stage of labour or when there is a very large baby and a substantial tear would otherwise result. An episiotomy is not so commonly needed in multiparous women, unless a previous delivery has been associated with a substantial and severe perineal tear. Although there is no doubt that an episiotomy may be easier to repair than a bad perineal tear, in the majority of cases, care at delivery can limit the extent of tearing so that a small second degree tear of a lesser extent than a planned episiotomy occurs. The women then have less perineal trauma to recover from.

TECHNIQUE OF INCISION AND REPAIR

The need for episiotomy is first explained to the woman, who should be aware of the nature of the procedure from her antenatal classes. The perineum must be infiltrated with local anaesthetic and then the fingers of the left hand introduced into the vagina to protect the fetus and stretch the perineum. The incision is made in the perineum with a pair of scissors, to the extent which is required. There are three main types of incision made: midline, right medial lateral or J-shaped. Figure 8.1 shows the J-shaped episiotomy which commences with a mid-line incision and then curves to avoid the anus. In the majority of women a mid-line episiotomy will suffice and this will be easier to repair and more comfortable in the puerperium than a right mediolateral approach. Where the fetus is large or the peritoneum short the J-shaped episiotomy will allow a greater incision to be made or a right mediolateral approach can be used which

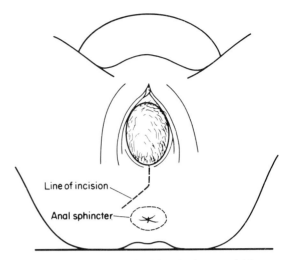

Fig. 8.1 Episiotomy. An incision is shown which starts as a median episiotomy and has been extended lateraly

commences at the same point at the fourchette but extends in a straight line at 8 o'clock to completely avoid the anus.

The repair of an episiotomy requires skill and experience and should only be performed by a trained obstetrician or midwife. Adequate analgesia must be provided by the use of local infiltration, although analgesia may be available if an epidural block has been used for pain relief during labour. In the latter event, infiltration with local anaesthesia is still a benefit as this has been shown to prevent too tight a suturing of the perineum and to reduce pain in the puerperium. Lignocaine 0.5 per cent is used; this dilute solution is equally effective as are higher concentrations but allows a volume of 40 ml to be used for the dose limit of 200 mg. This allows wide and adequate infiltration. Pain in the puerperium depends on the degree of tightness used in the repair, the fineness of the suture material and the way in which the skin is closed. The best results are achieved with Vieryl suture material of 2/0 or 3/0, sewing the perineum fairly loose to allow for the inevitable postoperative oedema and closing the skin with a subcuticular suture. Care must be taken to avoid the rectum and the minimum amount of suture material compatible with careful apposition of the tissues should be used.

There is no question that an episiotomy is painful in the puerperium and that regular analgesia must be provided. Recently it has become common practice to administer a non-steroidal anti-inflammatory analgesic such as diclofenac at the time of the procedure in the form of a suppository. This will provide analgesia in the 12 hours after delivery and can be supplemented later by orally administered drugs. The non-steroidal anti-inflammatory drugs, such as diclofenac have a greater effect than, conventional analgesics such as paracetamol and should be provided as routine, rather than on-demand analgesia, for at least the first 72 hours after delivery.

The majority of episiotomies heal extremely well and cause no problems in the future. A small minority of women will have minor problems due to fibrous tissue or granulation tissue in the wound or the vagina will heal badly causing discomfort or dyspareunia. In these cases, minor corrective surgery will be required 3–4 months after delivery and should always be able to resolve the problem.

FORCEPS DELIVERY

The original obstetric forceps were invented by one of the two elder Chamberlen brothers, members of a Huguenot family that settled in England in about the year 1600. The invention was kept secret in the family for over 100 years.

The Chamberlen forceps consisted of two blades, curved to fit the fetal head, with short handles, the two halves being strapped together where they met because they had no lock. In 1744 William Smellie devised the double-slotted English lock which allowed easy application and gave additional strength. The pelvic curve was added by Levret of Paris in 1747.

A new type of forceps was developed by Christian Kjelland in 1915 for the purpose not only of traction, but also of rotation of the head. Lastly, light, short curved forceps were introduced by Wrigley in 1935 for the particular purpose of outlet application.

The forceps are used to apply traction to the head of the fetus in a pelvis of adequate size, never to attempt to overcome disproportion. With Kjelland's forceps the head may also be rotated into a more favourable position before traction is applied. The forceps are designed to fit the head, not the breech, but forceps may be applied to the aftercoming head in breech delivery.

The presentations suitable for the forceps operation are vertex, face with the chin anterior and the aftercoming head of the breech. The vertex may have the occiput anterior, tranverse or directly posterior, but with the latter two, certain additional factors, referred to later, must be taken into account.

Indications for the use of forceps

Forceps are used in the second stage of labour for

- delay
- maternal distress
- fetal distress
- prevention of extra maternal effort (e.g. in heart disease)
- some cases of preterm delivery
- delivery of the aftercoming head of a breech presentation.

Delay in the second stage of labour

In primigravidae the active part of the second stage of labour should not last for much over 1 hour, and in multiparae it is often much shorter, usually being less than 30 minutes in the absence of an epidural anaesthetic. A second stage is considered to be delayed if these limits are exceeded, but this is only a general guide and if progress is not being made assistance should be given sooner.

Conditions likely to cause delay in the second stage of labour are:

- inadequate uterine contractions and poor voluntary effort
- resistant pelvic floor and perineum
- large fetus
- persistent occipitoposterior or deep transverse arrest of the head
- other malpresentations such as face presentation or brow presentation
- contraction of the pelvic outlet.

A common cause of delay in the second stage of labour is the resistance of the pelvic floor, but this should be treated by episiotomy rather than by the application of forceps.

Epidural anaesthesia is used in about 20 per cent of labours delivering in the UK; sometimes this is accompanied by decrease in the tone of the pelvic floor so that the fetal head is less directed into its rotation and then the second stage is longer. In consequence, if a woman has an epidural running, it is good practice to allow a longer time in the second stage, counting the temporal limits from when active pushing starts.

Maternal distress

Maternal distress in the second stage of labour usually occurs because there has been a long and poorly managed first stage.

Strictly, maternal distress shows in a rise in pulse rate without a significant rise in temperature and is always due to dehydration.

Mental distress occurs in women who have had a long tedious first stage and are unable to cooperate in aiding expulsion during the painful second stage. This sometimes means that analgesics have been too sparingly used.

Fetal distress

The following conditions may cause fetal distress during the second stage of labour:

- prolapse of the cord, or a tight loop or knot in it
- placental bed under perfusion associated with hypertension, antepartum haemorrhage or postmaturity
- prolonged or difficult labour.

Anything that interferes with fetal oxygenation will cause fetal distress. With gross placental lesions such as separation or infarction there may be severe fetal distress, but diminished uteroplacental blood flow may occur to a lesser degree in cases of postmaturity and maternal hypertension. Each uterine contraction causes decreased inflow in the vessels passing through the myometrium; in long labour, or in cases in which a great deal of retraction of the upper segment of the uterus occurs, fetal hypoxia may eventually arise.

Prolapse of the cord is a serious threat to the fetus. If the cervix is fully dilated and the fetus is alive expeditious forceps delivery is required otherwise immediate Caesarean section is required. Less acute cord compression may arise from tightening of a loop of cord round the neck or a knot in the cord.

When there are signs of fetal distress in the second stage of labour delivery should be effected at once. While preparations for delivery are being made oxygen is given to the mother in the hope of improving the supply to the fetus.

Prevention of maternal effort

In some conditions, maternal effort should be limited. Mitral stenosis, respiratory disease and severe hypertension are three examples where the second stage should be shortened to about 15 minutes. If the fetal head is not descending well with contractions, a prophylactic forceps delivery should take place.

Preterm delivery

When the fetus is small and the head is liable to injury during passage through the vagina a careful forceps delivery with episiotomy may reduce the risk although often an episiotomy will usually suffice.

Conditions necessary for the application of forceps

To apply forceps safely and successfully certain conditions must be fulfilled:

The cervix must be fully dilated

If the cervix is not fully dilated there will be difficulty in applying the forceps without including and tearing the cervix. More traction will be required to overcome the resistance of the cervix. The attempt to deliver may fail because the factor which prevented cervical dilatation may be uncorrected.

The presentation must be suitable

The forceps can only be used to extract the head of the fetus when it presents as a vertex or as a face with the chin anterior (leaving aside the use of forceps for the aftercoming head of a breech). A brow or a mentoposterior face presentation must be corrected before forceps can be used for extraction. The forceps must be applied accurately to the sides of the fetal head, otherwise dangerous compression of the head will occur causing intracranial stress and haemorrhage, and probably the attempt to deliver will fail.

The head must be engaged

If the head is not engaged in the second stage this may be because of cephalopelvic disproportion or because the head is extended.

The pelvic outlet must be of adequate size

Even with the head engaged the pelvic outlet must be of adequate size. If the subpubic angle is narrowed there must be room in the posterior part of the pelvic cavity.

The membranes should be ruptured

If the forewaters are still present the membranes should be ruptured and only if delivery does not follow would forceps be used.

The bladder should be empty

A catheter is passed to empty the bladder before forceps delivery, not only to avoid any risk of injuring it but also to remove its inhibiting influence on retraction of the uterus after delivery, and thus to reduce the risk of postpartum haemorrhage.

The uterus should be contracting

Contractions may be stimulated with an oxytocin infusion and then delivery completed with forceps. Intravenous Syntocinon should be given as soon as the fetal shoulders are born.

Types of obstetric forceps

Forceps in use today are shown in Fig. 8.2:

- short curved forceps
- long curved forceps
- Kjelland's forceps.

Each of these instruments consists of two halves meeting at the lock. Each half has a blade which is joined to the handle by a shank. The blades have two curves; the cephalic curve, in which the blade is curved to fit the head and a curve on the edge, the pelvic curve, to correspond with the curved axis of the pelvis. The length of the handles and shanks varies with the different forceps and the type of handle differs too. Kjelland's forceps have special characteristics.

Short curved forceps

This instrument, designed by Wrigley, is for use in the low or outlet forceps operation. The shank is short (2.5 cm) and the whole instrument is light.

Long curved forceps

The total length is 35 cm, with shanks of 6.5 cm,

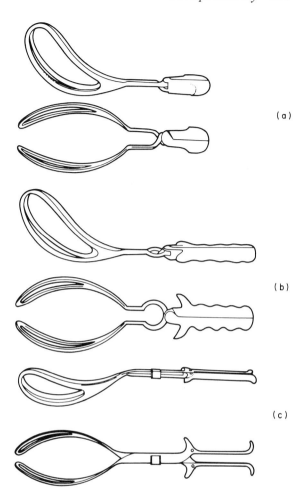

Fig. 8.2 (a) Short curved (Wrigley's) forceps; (b) long curved forceps; (c) Kjelland's forceps

and the instrument is relatively heavily built. It can be used for delivery of a head from the pelvic cavity.

Kjelland's forceps

The advantage of this instrument is that it allows accurate cephalic application, no matter what the position or level of the head in the pelvis. If the occiput is posterior or lateral the application is made and then the head can be rotated with the forceps to bring the occiput anterior. The forceps have a lock which allows one blade to slide on the other in the long axis, and thus the blades can lie at different levels when applied to a head which is lying transversely and is tilted. The other special characteristic is the pelvic curve. This is initially in a backward direction and then it quickly begins its forward sweep, but the tips of the blades never quite reach the plane of the shanks and handles. It is this feature which makes rotation with the forceps safe without risk of injury to vagina, bladder or rectum.

Preliminary steps before the application of forceps

Some form of analgesia will be required. This may be pudendal block, caudal, epidural, or general anaesthesia. Pudendal or epidural block should be used whenever possible in preference to general anaesthesia. A general anaesthetic should only be given by someone experienced in dealing with the complications which may arise. Unless it is reasonably certain that the woman's stomach is empty a stomach tube must be passed before the anaesthetic is begun. It is essential that a suction apparatus is available and that the head of the obstetric bed can be lowered without delay in case of unexpected gastric regurgitation. Prolonged general anaesthesia may increase the risk of postpartum haemorrhage and depress the respiratory centre of the newborn infant.

The woman is placed in the lithotomy position with a wedge under one buttock in order to take the pressure of the uterus off the vena cava. After swabbing the vulva and perineum with an antiseptic solution sterile towels are arranged and the bladder is emptied with a catheter.

A careful preliminary vaginal examination is made. It must first be established that the cervix is indeed fully dilated and then the level and position of the head is determined. If the sutures and fontanelles are obscured by caput formation it may be necessary to insert the whole hand to feel for an ear; the pinna will be directed toward the occiput. Episiotomy may be needed for this. Finally the size of the pelvic outlet is assessed. This is not always easy, but the points to note are the prominence of the ischial spines, the width of the subpubic arch and the length of the sagittal diameter from the lower border of the symphysis pubis to the sacrococcygeal joint.

If all these points are satisfactory and the occiput is near to the midline in front the forceps can be applied, but if the occiput is lying obliquely or transversely manual rotation will be needed first, unless Kjelland's forceps are being used, when the head may be rotated with the forceps.

If the occiput is posterior, unless the head is so low that the perineum is already being stretched, it is almost always best to rotate the occiput to the front before the head is delivered. The disadvantage of not doing so is the increased risk of intracranial stress and of extensive perineal laceration that goes with delivery of the head in the occipitoposterior position.

If the head has not descended to the pelvic floor then a more serious cause for delay is possible. A flattening of the sacrum might be felt, or the head may be seriously deflexed or extended. If there is doubt about the size of the pelvis with the head arrested at this level caesarean section is the safest course.

Application of obstetric forceps

The left blade of the forceps is applied first. The fingers of the right hand are passed into the vagina. The left blade of the forceps is held by its handle between the fingers and thumb of the left hand and is passed between the fetal head and the palmar surfaces of the fingers of the right hand (*see* Fig. 8.3). The handle is held well over the mother's abdomen, and inclined to the mother's right side so that it is almost parallel with her right inguinal ligament. As the blade passes up into the birth

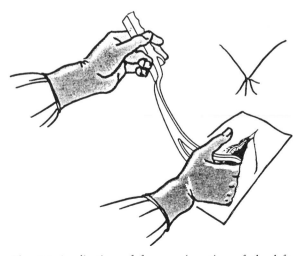

Fig. 8.3 Application of forceps; insertion of the left blade

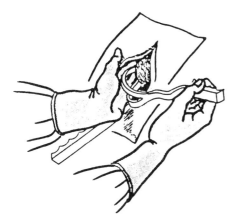

Fig. 8.4 Application of forceps; insertion of the right blade

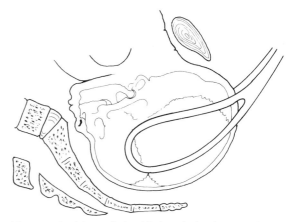

Fig. 8.6 Position of the blade of the forceps during extraction of a vertex presentation. The blades of the forceps lie in the submentovertical diameter and grasp the biparietal diameter

canal the handle is carried backwards and towards the midline, thus following the direction of both the pelvic and cephalic curves of the instrument. After ascertaining that the blade lies next to the head and is in the correct position the fingers of the right hand are withdrawn.

The fingers of the left hand are now introduced along the right side of the pelvis, and the right blade of the forceps is held and passed in a similar manner to the left. Its external visible portion will thus lie above and across the handle of the left blade (*see* Fig. 8.4). The shanks are now pressed backward against the perineum and the handles should lock and come to lie in a horizontal position (*see* Fig. 8.5).

If there is difficulty in locking the forceps it is probable that the position of the head has not been diagnosed correctly, and that the instrument has been applied to a head in the oblique or transverse diameter of the pelvis. If the forceps will not lock,

or if the handles will not close together, the blades must be removed and the position of the fetal head re-examined.

The ordinary curved forceps must always be applied correctly to the fetal head, with the parietal eminences lying within the fenestrations of the blades and the sagittal suture lying midway between the blades (*see* Fig. 8.6). These forceps must also be correctly placed in the pelvis. It does not matter if the head is slightly oblique in the pelvis, with the occiput pointing less than 45° away from the midline. With such a slight degree of obliquity the head and forceps will turn into the midline as traction is applied (*see* Fig. 8.7), but if the head is more oblique than this its position *must* be corrected by manual rotation or straight forceps before the ordinary type of forceps are applied to it.

Application of forceps to a face presentation

If the head is delayed in the pelvic cavity and there is a mentoanterior position the forceps may be used to assist delivery (*see* Fig. 8.8). If there is a mentoposterior position delivery with the forceps is only possible after the chin has been rotated forwards. This is possible manually or with Kjelland's forceps, but it is safer to treat this uncommon condition by caesarean section for the skills of the obstetrician may not include the manoeuvres these days.

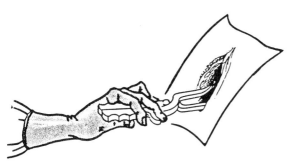

Fig. 8.5 Forceps articulated before traction

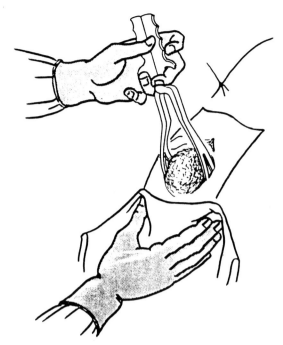

Fig. 8.7 Delivery with forceps

Application of forceps to the aftercoming head

The forceps are applied to the aftercoming head in breech delivery as described in the section on breech delivery.

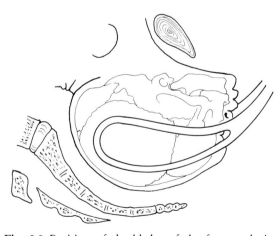

Fig. 8.8 Position of the blades of the forceps during extraction of the head in the mentoanterior position. The blades of the forceps lie in the occipitomental diameter and grasp the head in the bizygomatic diameter

Extraction with the forceps

The operator sits facing the perineum. Traction is applied to the handles of the forceps in a backwards and downwards direction in the axis of the birth canal. Traction is intermittent, timed with the uterine contractions, and each pull should only last for a few seconds. As the head descends the handles are gradually raised to about 45° above the horizontal, when an episiotomy is performed through the stretched perineum (unless this has already been done). The head is then gently guided through the vulva until crowning takes place with the handles of the forceps in vertical position. The forceps blades are then removed before delivery of the face by manual extension of the head.

Excessive force should never be used. Slow intermittent traction reduces the risk of intracranial injury.

The use of Kjelland's forceps (*see* Fig. 8.9)

Kjelland's forceps should only be used by those who have been trained in their use. In inexperienced hands this is a dangerous instrument, and may be the cause of severe vaginal and bladder lacerations if it is used incorrectly. The popularity of Kjelland's forceps has waxed and waned; at present manual rotation of the head is less preferred than rotation with these instruments.

Before applying the forceps the exact position of the head in the pelvis must be determined. To avoid error the operator is advised to hold the articulated instrument before the vulva in the position to be taken up when it is applied to the head. The blade which is to be inserted first is selected, and the other blade is put aside. If the head lies anteroposteriorly, or only slightly obliquely, the application of the forceps is not difficult. The forceps are always applied so that, after any rotation, the pelvic curve will be positioned correctly. For example, if the occiput lies posteriorly and it is intended to rotate it with the forceps, the forceps must be applied with the pelvic curve in the reverse direction so that *after* rotation the curve will be correctly orientated to the birth canal.

Unlike ordinary forceps this instrument can be applied to a head lying transversely in the pelvis; there are special methods of application for this. In both methods the anterior blade is applied first. In the wandering method the anterior blade is guided

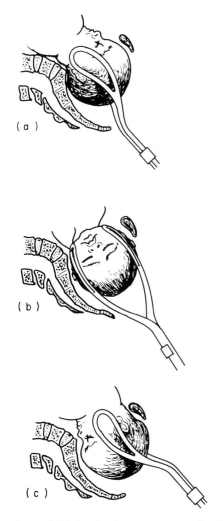

Fig. 8.9 Use of Kjelland's forceps. (a) Application to head in occipitoposterior position before rotation; (b) application to head in occipitolateral position before rotation; (c) position of forceps after rotation

into the lateral side of the pelvis beside the head, with the cephalic surface of the blade properly facing the head. It is then slid gently round the pelvis, keeping as close as possible to the head and passing over the forehead until it comes to rest fitting the anterior parietal eminence. The direct method of application is only used when the head is low down and the fit is not too tight. The tip of the anterior blade, with the handle as far back as possible, is applied to the side of the head anteriorly and then slipped over the anterior parietal eminence behind the symphysis pubis, with the

cephalic surface of the blade kept as close as possible to the head while it is in transit. The posterior blade is then inserted behind the head over the posterior parietal eminence.

Force is never used for rotation of the head, and it should be done with the fingers of one hand only. Rotation is first tried at the level of arrest, but if this is not immediately easy the head may be pushed up in the pelvis slightly and rotation tried again. Sometimes rotation will be easier after traction has brought the head down to a roomier level of the pelvis.

Traction is made in the line of the handles and is exerted with only two fingers hooked over the proximal shoulders of the handles. Rotation and traction are never done together, although spontaneous rotation commonly occurs during traction and should not be impeded. The handles should never be compressed together, as this will have a crushing action on the head.

The problems of forceps delivery

Some of the dangers of forceps delivery are due to the circumstances calling for the operation rather than to the operation itself. For example, delivery with forceps may be called for at the end of a long tedious first stage of labour because of maternal or fetal distress.

Some of the dangers of forceps delivery are:

Mother
dangers of general anaesthesia
lacerations of cervix, vagina or perineum
postpartum haemorrhage from uterine atony or
   lacerations
puerperal genital infection.

Infant
intracranial haemorrhage
facial palsy
cephalhaematomata.

Spastic diplegia in a child is sometimes attributed to instrumental delivery. It is improbable that intracranial haemorrhage will have this effect and if the haemorrhage is not fatal complete recovery is likely. On the other hand, prolonged cerebral hypoxia will also leave permanent damage; forceps may have been used in such a case but could hardly have caused the hypoxia.

Bad results from forceps delivery usually result from bad forceps delivery. The dangers will be minimal when proper regard is given to the indications for the operation; when the exact position of the head is determined by a preliminary vaginal examination and when any necessary correction of the position of the fetus is made before traction.

Failure to deliver with forceps

Fortunately failure of forceps is now rare in Britain, for the dangers to both mother and baby are serious and are directly proportional to the lack of skill and the force employed during the unsuccessful attempts. Criticism is not so much against attempting to apply forceps or attempting traction, but against *persisting* in such endeavours in the face of obvious difficulties.

The causes of failure to deliver with the forceps are mostly commonplace; either the cervix is not fully dilated or the head is in a malposition, usually with the occiput posterior. Fortunately, attempts to apply forceps with the head not yet engaged are not made nowadays, for the dangers are even greater. A rare cause of failure to deliver with forceps is a contracted pelvic outlet, which is often only diagnosed late in labour.

Severe laceration of the cervix may be caused by attempts to apply forceps with the cervix incompletely dilated, either because the blades are applied outside the cervix or because the head is forcibly dragged through it. Sometimes an extensive tear of the upper vagina is caused by attempts to rotate the head with the forceps. If the obstruction is at the brim the uterus itself may be perforated by a blade of the forceps. A complete tear of the perineum into the anal canal comes from a combination of too strong traction and misapplied force. The effect on the woman of these obstetric insults, even if she has avoided a badly given or prolonged general anaesthetic, is to cause shock from trauma or haemorrhage, or both. Sepsis is likely to follow.

The fetus may suffer intracranial haemorrhage and it may be dead before delivery is completed.

Any case of failed delivery with forceps must be immediately reassessed. Not infrequently a more experienced operator is able to effect an easy delivery with forceps after correcting a malposition. In other cases caesarean section has to be performed.

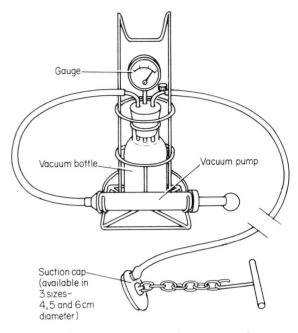

Fig. 8.10 The vacuum extractor (by courtesy of Down's Surgical PLC). The cups, which are of various sizes, are attached to a rubber tube which is connected to the reservoir, pressure gauge and hand vacuum pump. Traction is applied with chain and handle

VACUUM EXTRACTOR (VENTOUSE)

The idea of extracting the fetal head by means of a vacuum cup applied to the scalp has been considered since Younge in 1706 tried a glass suction cup; an instrument was designed by Simpson in 1849, although little use was made of it. The modern vacuum extractor was introduced by Malmström in 1954. Opinions differ about the value of this method of assisting delivery. In some clinics, particularly on the European continent, the vacuum extractor is preferred to the forceps, but it is chosen much less often in Britain and America.

Malmström's extractor (*see* Fig. 8.10) consists of a metal suction cup attached by a chain to a metal handle. The cavity of the cup is connected by rubber tubing to a reservoir with a pressure gauge and a hand vacuum pump. The suction cups are made in three sizes, and the largest possible cup is used, according to the dilatation of the cervix. The modern Ventouse extractor has silastic cups which can in some instances provide better suction, and may inflict less trauma to the fetal scalp.

Method of use

Local infiltration of the perineum with 1 per cent lignocaine is usually the only anaesthetic needed, a great virtue to the method.

The cup is introduced sideways into the vagina by pressing it backwards against the perineum. It is then guided into place on the scalp as near as possible to the posterior fontanelle, taking care that neither the cervix nor any part of the vaginal wall comes between the cup and the scalp. While the obstetrician holds the cup in the correct position an assistant uses either a hand pump or an operator controlled electrical suction pump to create a vacuum, gradually increasing the negative pressure by $0.2\,\text{kg/cm}^2$ at 1 minute intervals, until $0.8\,\text{kg/cm}^2$ is attained. Failure to maintain the vacuum indicates that either the cup is incorrectly applied or that the apparatus is faulty. The negative pressure causes an artificial caput succedaneum or chignon to be formed within the cup and when the vacuum reaches $0.8\,\text{kg/cm}^2$ the cup is completely filled with scalp, ensuring maximum adhesion, and traction can be started.

Traction on the handle is made as nearly vertically to the cup as possible, because an oblique direction of traction will tend to pull it off. Traction is made intermittently with the uterine contractions, the direction of pull changing as the head descends through the birth canal. Unless there is obvious descent during three or four contractions the use of the Ventouse should be reconsidered.

After delivery the vacuum is reduced as slowly as it was created as this tends to diminish the risk of damage to the scalp. Immediately after removal of the cup the baby has an unsightly chignon where the cup was applied, but this rapidly diminishes and within a few hours only a faint ring can be seen.

Indications

These are similar to those for the use of obstetric forceps in the second stage of labour. If the occiput is not anterior the extractor may still be used. It is applied as near to the vertex as possible, and with traction forward rotation of the occiput often occurs. However, many obstetricians would prefer to use Kielland's forceps or even to perform manual rotation in such a case. The vacuum extractor can be used by skilled operators in the final stage of labour for fetal distress in a multiparous mother. The cup can be manouvered through the cervix at 7.8 cm dilation and after careful checking, full dilation can be achieved with a very few minutes traction. This is much quicker than a caesarian section.

Contraindications

The vacuum extractor cannot be applied to the breech or face.

The operation takes too long for urgent cases of fetal distress and so forceps delivery is preferred.

There is some doubt as to its safety when used on preterm babies, when there is possibly a greater risk of intracranial haemorrhage than with forceps.

Necrosis of the scalp, cephalhaematoma, subaponeurotic haematoma and intracranial haemorrhage have all been reported after vacuum extraction, but in a number of these cases the cup had been injudiciously applied for long periods and strong traction had been applied to overcome some degree of disproportion.

INTERNAL VERSION

In internal version a hand is passed through the cervix, which must be fully dilated, or nearly so, into the uterus to turn the fetus. It is always podalic version, and one or both legs are brought down as part of the operation. The woman must be anaesthetized or have an effective epidural block. It is a difficult and hazardous procedure for both mother and child.

Before caesarean section and forceps delivery were introduced internal version was the only way out of many obstetric difficulties. Today the commonest indication is for the correction of a transverse lie of a second twin after delivery of the first twin and when external version has failed.

Internal version is extremely dangerous when labour has been in progress for some time and the lower segment is stretched and thin because of the risk of uterine rupture.

Figures 8.11 and 8.12 show the method of performing internal version.

CAESAREAN SECTION

Caesarean section is the operation by which a potentially viable fetus is delivered through an

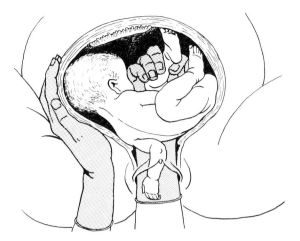

Fig. 8.11 Internal podalic version; stage 1, shoulder presentation. The operator's right hand has been introduced to grasp the leg of the fetus and bring it down. The left hand, working outside the sterile sheet, is coaxing the head out of the iliac fossa

incision in the abdominal wall and the uterus. A similar procedure undertaken to terminate a pregnancy of a non-viable fetus is referred to as hysterotomy.

Indications

A list of indications for caesarean section can easily be devised, but in many cases there is more than

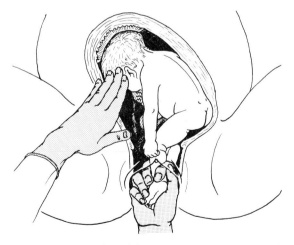

Fig. 8.12 Internal podalic version; stage 2, shoulder presentation. The leg is brought down causing the back to rotate forwards. The head has almost reached the fundus

one indication. As the dangers of the operation has diminished there is a tendency for it to be performed more often for fetal indications such as distress during labour, placenta praevia, severe hypertension, diabetes mellitus, haemolytic disease and prolapse of the cord.

A few indications are absolute in the sense that delivery by any other method would be extremely dangerous, for example gross disproportion or placenta praevia. However, in most cases the indications are relative and caesarean section is carried out when it is thought that the balance of maternal and fetal dangers will be reduced by section rather than vaginal delivery. The indications include:

Faults in the birth canal

Cephalopelvic disproportion. Gross obstruction to delivery from pelvic contraction will obviously justify caesarean section; it is in the borderline cases that judgement is required.

A trial of labour often has to be terminated because of fetal distress, or because of abnormal uterine action. It is not always certain what is causing delay in labour, when a woman has been in labour for some hours and has received cautious oxytocic augmentation of uterine action yet cervical dilatation is not progressing, but caesarean section should not be delayed.

If a trial of labour in a previous pregnancy is judged to have failed because of disproportion, delivery should be by elective section shortly before term.

In obstructed labour with a dead fetus, now fortunately very rare in Britain, caesarean section is safer than craniotomy in most doctors' hands.

Pelvic tumour. Impaction of an ovarian cyst or a fibroid in the pelvis are rare indications for section.

Cervical or vaginal stenosis are also rare indications for section.

With a *double uterus* obstruction during labour may occur because the unimpregnated horn lies below the presenting part, because the narrower part of the uterus fails to dilate or because of a vaginal septum.

Section is advised after a previous operative repair of a fistula or successful treatment of stress incontinence, but after most repair operations vaginal delivery with episiotomy is preferable to section.

Fetal malpresentations

Caesarean section is the best treatment for a brow presentation with the head at the level of the pelvic brim.

Impacted mentoposterior face presentation is also an indication for section.

Shoulder presentation. Section is performed for a shoulder presentation if the fetus is alive in preference to internal version.

Some cases of *locked twins* are treated with caesarean section.

Section is now performed in at least 60 per cent of cases of *breech presentation*. Section would be advised if the fetus was in breech presentation when the mother goes into labour from 26 to 36 weeks if the mother is over 35 years of age, has been infertile or has an adverse obstetric history, has a small or android pelvis or a large fetus or indeed has any additional complication. Operation should also be performed during labour if the breech does not descend onto the pelvic floor after a few hours of regular contractions.

In exceedingly rare cases of *conjoined twins* a classical caesarian section may be required.

Abnormal uterine action

In many cases of abnormal uterine action there may also be some mechanical difficulty and precise diagnosis of the cause of delay may be uncertain, but if a woman has been in labour for more than 12 hours, the membranes are ruptured and there is no progress with an oxytocic infusion, section should be considered.

Antepartum haemorrhage

Caesarean section is the best treatment for both mother and fetus in most cases of *placenta praevia*.

Placental abruption. Section is seldom performed in severe cases of placental abruption because the fetus is usually dead. In unusual cases in which the fetus is alive but bleeding is continuing caesarean section is justified near term. Placental separation, even if it is slight, is a major threat to fetal survival, and during

labour section may become necessary for fetal distress.

Other maternal indications

For cases of cardiac or respiratory disease, and indeed most intercurrent medical disorders, assisted vaginal delivery is preferable to caesarean section unless there is some obstetric impediment to easy delivery. Section should never be performed merely to permit sterilization; this is better done some months later by laparoscopy

Fulminating pregnancy induced hypertension. This may have to be treated by section if the chances of speedy vaginal delivery are small. The woman may not go into labour after induction, or the severity of the hypertension may demand delivery before the 34th week in a primigravida, when induction is uncertain. In other cases of hypotension or renal disease section is performed in the fetal interest.

Fetal indications

If *placental bed malperfusion* is suspected because of inadequate fetal growth, maternal hypertension or proteinuria, many are treated by induction of labour before term, but if labour does not follow, if the fetus is very small, or if fetal distress occurs during the first stage of labour section is the best method of delivery.

Fetal distress may occur during labour in cases of postmaturity or hypertension, in cases of long labour or abdominal uterine action, or without explanation. Whatever the cause, if there is clinical evidence of fetal distress during the first stage of labour, confirmed by monitoring of the fetal heart rate and by scalp blood sampling, caesarean section is urgently performed.

Prolapse of the cord when the cervix is not sufficiently dilated to allow immediate vaginal delivery is treated by emergency section.

A *bad obstetric history*, when the woman has had several stillbirths or neonatal deaths, may be a sufficient indication for caesarean section. In such cases the timing of sections is very important.

In cases of *diabetes mellitus*, when blood sugar levels have been difficult to control, complications are likely to occur. These include fetal macrosomia, polyhydramnios and pre-eclampsia. Such women are frequently delivered before term, usually by caesarean section.

Caesarean section should not be performed routinely on *older primigravidae*, but if there is an additional factor such as minor disproportion, a breech presentation, hypertension or even postmaturity, section may be indicated.

Gross prematurity is becoming a more frequent indication for caesarean section, in cases of breech presentation. Between weeks 26 and 32 the fetus is particularly prone to intracranial haemorrhage secondary to acute hypoxia during labour; this may be avoided by caesarean section.

Repeated caesarean section

If a woman has already had a caesarean section there is a risk of rupture of the scar in any subsequent labour. If the indication for the section persists, for example a contracted pelvis, operation should obviously be repeated. If, however, the indication was a non-recurrent one, for example placenta praevia in a previous pregnancy, then the risk of a second elective operation outweighs the risk of rupture of a lower segment scar at vaginal delivery. When a woman has had more than one previous section the indication for repetition of the operation may be stronger.

The upper segment scar is much less secure than the lower segment scar, and after an operation of the former type (now rarely performed) it is wise to repeat the section.

Technique of caesarean section

In the past the classical procedure was a vertical incision in the upper uterine segment. This incision was attended by free bleeding, could not be covered with a free layer of loose peritoneum and gave an insecure scar which might give way during a subsequent pregnancy or labour. It is seldom performed today. The standard procedure is the transperitoneal lower segment operation with a transverse incision in the uterus.

With any operation the best results are obtained when there is ample time to prepare the patient, but the need for caesarean section may only become evident during the course of labour. If section is elective it should be performed a few days before the expected date of labour, but if there is doubt about the maturity it may be wiser to wait until labour begins.

Anaesthesia for caesarean section

Caesarean section often has to be performed at inconvenient times, but the services of a skilled anaesthetist are essential. Deaths from avoidable complications of anaesthesia still make a tragic contribution to maternal mortality statistics. In emergency cases the woman may be ill-prepared for operation and it may be necessary to pass a gastric tube and empty the stomach before anaesthesia. Fifty milligrams of ranitidine given intravenously together with 30 ml of a 0.3 M solution of sodium citrate minimize the risks of acid reflux.

Morphine and allied drugs which may depress the respiratory centre of the newborn infant should not be used for premedication. The woman should not be under the influence of the anaesthetic for longer than is absolutely necessary before the operation begins and all preparations for the operation must be completed before inducing anaesthesia.

The anaesthetist must be able to insert a cuffed endotracheal tube. A very common technique is to induce sleep with a small dose of intravenous thiopentone, to obtain relaxation with an injection of scoline and, after insertion of a cuffed endotracheal tube, to maintain anaesthesia with nitrous oxide and oxygen. Recently propofol (Diprivan) has tended to replace thiopentone.

An alternative method is to use a lumbar epidural block (*see* Chapter 4, The relief of pain in labour).

Regional infiltration of the abdominal wall and parietal peritoneum can be employed, but this method is time-consuming and is not suitable for apprehensive women or emergency operations.

LOWER SEGMENT CAESAREAN SECTION

The bladder is emptied with a catheter before the operation is started. An intravenous glucose–saline drip is set up. A litre of cross-matched blood should be available (although it may not be used); in cases of placenta praevia, when blood loss may be heavy, this is essential.

The operating table is tilted laterally by 15° to the left. This prevents the uterus from compressing

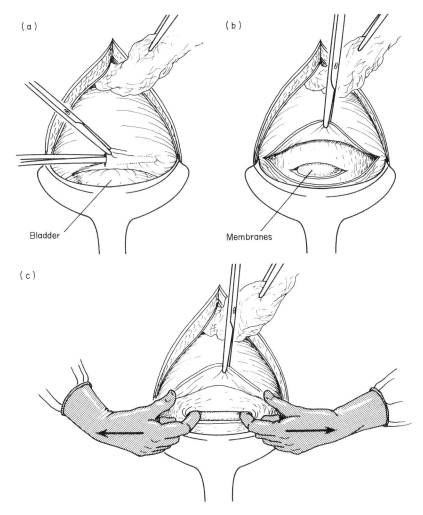

Fig. 8.13 Lower segment caesarean section. (a) The loose peritoneum over the lower uterine segment is lifted up before it is incised; (b) the lower segment is exposed and a short incision made through it down to the fetal membranes; (c) the incision in the lower segment can be enlarged by finger traction with very little bleeding

the inferior vena cava and impeding the venous return.

The peritoneal cavity is opened by a low transverse or vertical subumbilical incision. A wide Doyen's retractor is inserted into the lower end of the wound. The peritoneum of the uterovesical pouch is divided transversely for about 10 cm and the bladder is pushed gently down off the lower segment. A transverse incision about 2 cm long is made in the midline of the lower segment and deepened until the membranes bulge (*see* Fig. 8.13);

the amniotic sac should be kept intact at this stage if possible. The two index fingers are slipped into the incision and by exerting finger traction the incision can be extended to about 10 cm in length.

The membranes are ruptured and the head is delivered by slipping a hand beside it or by applying first one blade of Wrigley's forceps and then the other. The head is delivered gently with the forceps. As the head is delivered the anaesthetist injects 10 units of Syntocinon intravenously.

The shoulders are carefully delivered, easing

them out of the wound to avoid dangerous lateral splitting of the uterine wound.

The fetus, now delivered, is held head downwards at about the same level as the placenta while the mouth and pharynx are cleared of fluid with a soft catheter attached to a suction apparatus. The cord is clamped and divided and the infant is handed over to the care of a paediatrician.

The placenta soon separates and is delivered through the wound. The uterine incision is sutured with two layers of catgut or polyglycolic acid. It is important to secure complete haemostasis and to remove any blood, amniotic fluid and vernix from the peritoneal cavity. The abdominal wound is closed in the ordinary way.

UPPER SEGMENT CAESAREAN SECTION

Although this operation is often still described as classical it has been virtually replaced by the lower segment operation, except possibly for a case of transverse lie of the fetus, or if a fibroid prevents access to the lower segment. A paramedian incision is made. The uterus is incised vertically in the midline of the anterior wall of the upper segment and the membranes are ruptured. The fetus is grasped by the feet and delivered in a manner similar to breech extraction. The incision in the uterus is closed in two layers, and the abdomen is closed in the ordinary way.

STERILIZATION

Sterilization may be performed at the time of caesarean section, but section should never be done just to sterilize the patient.

Sterilization is sometimes advised if a woman has had many previous sections, when the risk of further section may exceed that of normal delivery, and it may be recommended for various medical reasons.

The irreversible nature of the operation must be explained to both the woman and her partner; although the partner's consent is not necessary if there are medical reasons for the operation, it is wise to obtain it. It is also important, because of the risk of litigation, to point out that there is a very small percentage of failures, no matter what method is used.

Because it is not possible to guarantee that any baby will survive and thrive, unless there are very strong medical reasons to advise against further

pregnancy it may be wise to postpone sterilization until the baby is 4 to 6 months old and when sterilization can be performed by a laparoscopic method.

For sterilization at the time of caesarean section part of the tube is drawn up into a loop and a ligature is tied tightly round the base of the loop, which is then excised. Within a few days the catgut gives way and the two sealed ends of the tube separate (Pomeroy method).

CAESAREAN HYSTERECTOMY

Hysterectomy at the time of caesarean section is rarely indicated, although it carries little additional risk except on account of the condition which led to it. The technique differs little from that of an ordinary hysterectomy. The tissues are more vascular, but the tissue planes separate easily. It is occasionally indicated in cases of:

- rupture of the uterus if there is gross damage to the uterine blood supply or evidence of severe infection
- fibroids. When a woman has to be delivered by section because of multiple or large fibroids, hysterectomy may be considered, but it is usually easier to perform the hysterectomy 3 months later
- carcinoma of the cervix. Cases that have reached late pregnancy may be treated by caesarean section followed by Wertheim's operation.

POSTOPERATIVE CARE AFTER CAESAREAN SECTION

A woman who has had a caesarean section is looked after in the same way as any woman who has had a major abdominal operation.

RISKS TO THE MOTHER

It is estimated that the maternal mortality after caesarean section is four times greater than after vaginal delivery. The risk to the mother is affected by the indication for the operation, her health before and during labour, whether there have been previous attempts at delivery, the length of labour and the skill of the surgeon and anaesthetist. The risks in lower segment caesarean section performed as a planned procedure are low, but if emergency cases are considered the risk is higher.

Immediate risks

We have already emphasized the risks of anaesthesia in an unprepared woman with an inexperienced anaesthetist.

Blood transfusion must always be readily available for women undergoing section, particularly in cases of placenta praevia. When emergency section is performed anaemia and shock must first be treated by blood transfusion.

Serious infection is uncommon in planned operations, but still may occur if section is performed after labour has been in progress for some time or any other attempt at delivery has been made. Such problems could often have been prevented if prophylactic antibodies had been used. Severe postoperative ileus is now rare.

Pulmonary embolism is more likely after caesarean section than after normal delivery. Together with anaesthetic complications, it now accounts for most of the fatalities. Pulmonary embolism is more likely in obese and anaemic women. Early ambulation and anticoagulant treatment will play an important part in reducing the danger.

Remote risks

Rupture of a caesarean scar is relatively rare after lower segment section, and then usually occurs during labour. An upper segment scar is much more likely to rupture, and may do so during pregnancy as well as during labour. Any woman who has had a caesarean operation should be under the care of a hospital obstetric unit for every subsequent delivery.

Intestinal obstruction from adhesions is rare since the upper segment operation has been largely abandoned.

RISKS TO THE FETUS

Caesarean section has an inherent small risk for the fetus, although the conditions for which the operation is performed may explain much of the increased perinatal mortality. Anaesthetics cross the placental barrier and may depress the respiration of the newborn; respiratory problems may also occur with premature babies or babies born to diabetic mothers.

Occasionally, when section is planned near term, the duration of gestation is wrongly estimated. The routine use of ultrasound in early pregnancy should eliminate this risk.

It is possible to cause intracranial damage by delivering the head without proper care when it has to be brought up from the pelvis, or even by delivering it through too small a uterine incision.

RISING INCIDENCE OF CAESAREAN SECTION

The incidence of caesarean section has been rising in the UK to around 13 per cent of deliveries partly due to the threat of litigation following the birth *per vaginam* of brain-damaged babies and partly to the tendency to deliver preterm and very small babies abdominally. It has been suggested that the increasing use of fetal monitoring in labour has been the cause of unnecessary operations which might have been avoided had the pH of fetal scalp blood been measured.

FURTHER READING

Planché W., Morrison J. and O'Sullivan M. (1992) *Surgical Obstetrics*. W.B. Saunders, Philadelphia.

9

PROBLEMS OF THE NEWBORN

RESUSCITATION

Skilled resuscitation of the newborn should be available wherever babies are born. Resuscitation of the newborn has been divided into basic and advanced resuscitation.

Basic resuscitation comprises:

artificial ventilation via face mask
external cardiac compression
giving naloxone if necessary.

Advanced resuscitation comprises:

tracheal intubation
drug administration
management of meconium aspiration
care of the very low birth weight infant.

The need for resuscitation of newborn infants may arise unexpectedly. Basic resuscitation needs to be initiated immediately. All health professions actively involved in the care of mothers and babies in the maternity unit should be proficient in basic resuscitation. A person skilled in advanced resuscitation should be available at all times and summoned immediately if required.

The need for advanced resuscitation may be anticipated when there are conditions which predispose to asphyxia and a person proficient in advanced resuscitation should be present (Table 9.1).

Table 9.1 Indications for advanced resuscitator to attend deliveries

DURING LABOUR AND DELIVERY
Fetal distress including low fetal pH
Thick meconium staining of the amniotic fluid
Caesarean section
Forceps or Ventouse delivery (except uncomplicated lift out)
Abnormal presentation
Prolapsed cord
Antepartum haemorrhage

MATERNAL CONDITIONS
Diabetes
Fever
Heavy sedation
Concomitant illness
Severe hypertension and/or proteinuria

FETAL CONDITIONS
Multiple pregnancy
Preterm labour (less than 37 weeks)
Post-term labour (more than 42 weeks)
Intrauterine growth retardation
Isoimmunization
Abnormal baby

OBSTETRICIAN OR MIDWIFE EXPRESSING CONCERN

PATHOPHYSIOLOGY

Normally the lungs are filled with fluid at birth. The first few breaths need to fill the lungs with air driving fluid out of the alveoli into the circulation. In babies who are subjected to continuous, prolonged asphyxia there is initially a period of gasping which is followed after 1–2 minutes by the onset of primary apnoea (*see* Fig. 9.1). At this stage

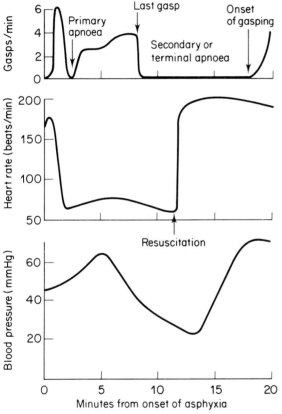

Fig. 9.1 Changes in Rhesus monkeys during asphyxia and on resuscitation by positive-pressure ventilation. Brain damage was assessed by histological examination some weeks later.

the infant is blue but the blood pressure is maintained although the heart rate is falling. This is called blue asphyxia or *asphyxia livida*. A further period of gasping which lasts 8–10 minutes occurs following which secondary or terminal apnoea ensues. The blood pressure is now extremely low and the heart rate is slow and deteriorating. Peripheral perfusion is poor and the baby lies limp. At this stage the infant is white and shocked which is described as *asphyxia pallida* or white asphyxia. In terminal apnoea breathing will not begin unless positive pressure ventilation is provided. The condition of infants shortly after birth is recorded using the Apgar score at 1, 5 and 10 minutes (Table 9.2).

The events described above are based on observations in animals subjected to a sudden, severe and continuing asphyxia. This occurs only rarely in the human infant, such as following a cord prolapse, but more often the fetus is subjected to intermittent asphyxia over a prolonged period. It should not be assumed that all babies requiring resuscitation have experienced birth asphyxia. Other causes of infants failing to establish respiration at birth include:

- trauma
- analgesia and anaesthetic agents given to the mother during labour or at delivery
- caesarean section, when there is often delay in clearance of lung liquid
- meconium aspiration
- the very low birth weight infant
- problems within the infant, e.g. diaphragmatic hernia.

BASIC RESUSCITATION

An outline of the management of basic resuscitation is shown in Fig. 9.2. The baby who, within the first minute of life, establishes regular breathing, has a good heart rate and is well perfused requires

Table 9.2 Apgar score

Sign	0	1	2
Heart rate	Absent	< bpm 100	> bpm 100 ↓
Respiratory effort	Absent	Slow, irregular	Good, crying
Muscle tone	Limp	Some limb flexion	Active
Response to stimulus (nasal catheter)	Nil	Grimace	Vigorous cry
Colour	Blue, pale	Body pink, limbs blue	Pink

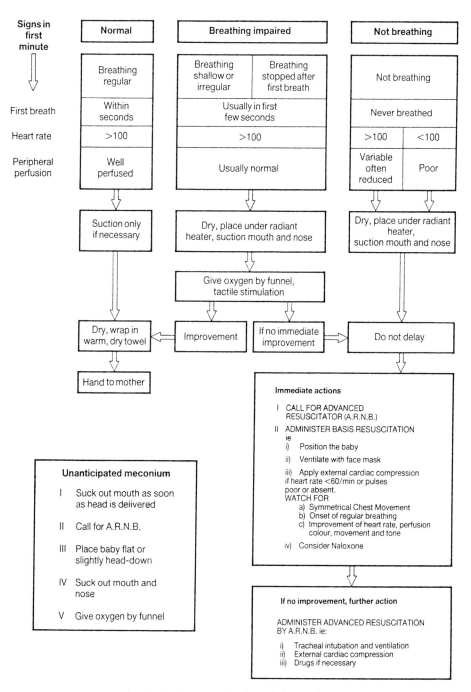

Signs in first minute →	**Normal**	**Breathing impaired**		**Not breathing**	
	Breathing regular	Breathing shallow or irregular	Breathing stopped after first breath	Not breathing	
First breath	Within seconds	Usually in first few seconds		Never breathed	
Heart rate	>100	>100		>100	<100
Peripheral perfusion	Well perfused	Usually normal		Variable often reduced	Poor

Normal: Suction only if necessary → Dry, wrap in warm, dry towel → Hand to mother

Breathing impaired: Dry, place under radiant heater, suction mouth and nose → Give oxygen by funnel, tactile stimulation → Improvement / If no immediate improvement

Not breathing: Dry, place under radiant heater, suction mouth and nose → Do not delay

Immediate actions

I   CALL FOR ADVANCED RESUSCITATOR (A.R.N.B.)

II   ADMINISTER BASIS RESUSCITATION
ie
i)   Position the baby
ii)   Ventilate with face mask
iii)   Apply external cardiac compression if heart rate <60/min or pulses poor or absent.
WATCH FOR
a) Symmetrical Chest Movement
b) Onset of regular breathing
c) Improvement of heart rate, perfusion colour, movement and tone
iv)   Consider Naloxone

If no improvement, further action

ADMINISTER ADVANCED RESUSCITATION BY A.R.N.B. ie:

i)   Tracheal intubation and ventilation
ii)   External cardiac compression
iii)   Drugs if necessary

Unanticipated meconium

I   Suck out mouth as soon as head is delivered

II   Call for A.R.N.B.

III   Place baby flat or slightly head-down

IV   Suck out mouth and nose

V   Give oxygen by funnel

Fig. 9.2 Basic rescuscitation of the newborn

no resuscitation at all. The baby can simply be wrapped in a warm dry towel and handed to his or her mother.

Babies whose breathing is impaired with either shallow or irregular respiration with a good heart rate and whose peripheral perfusion is good should be placed under a radiant heater and the mouth and nose cleared of mucus and blood. Oxygen is given to the baby's face together with tactile stimulation, such as firm drying of the infant with a towel, if needed. In most cases, regular respiration is rapidly established.

For the baby who does not breathe at birth, after placing the baby under the radiant heater and clearing the airways, basic resuscitation is initiated. The baby is correctly positioned with the neck slightly extended and ventilation initiated with a face mask. The face mask needs to be applied to cover the nose and mouth to ensure an adequate seal. Intermittent positive pressure ventilation is best obtained using the artificial ventilation circuit on the resuscitation trolley. A starting pressure of up to 30 cm of water may be needed for the first few breaths. The initial breath should consist of an inspiratory time of 1–2 seconds to adequately expand the lungs. Thereafter a pressure of 10–15 cm of water is usually sufficient at a rate of 30–40 breaths per minute with shorter inspiratory times. The chest is carefully observed to ensure that the artificial ventilation is achieving adequate and symmetrical movement of the chest wall.

While this is initiated, a paediatrician or obstetrician capable of advanced resuscitation is called.

If the baby is bradycardic (heart rate less than 60 beats per minute or the pulse is absent or difficult to palpate), external cardiac compression is applied to the chest. This is best done by grasping the chest with both hands, placing the thumbs over the junction of the middle and lower third of the sternum and applying pressure using both thumbs.

Naloxone is given if the mother has received a narcotic analgesic within 4 hours of delivery and the baby fails to breathe regularly when artificial ventilation is stopped.

ADVANCED RESUSCITATION

If mask ventilation fails, the baby should be intubated and ventilated promptly. Intubation should be performed by someone proficient in advanced resuscitation of the newborn. This is done under direct vision using a laryngoscope and an appropriate size of tracheal tube. After tracheal intubation, one needs to check that there is satisfactory

excursion of the chest wall. If not the tracheal tube may not be in the trachea. If the movement of the chest is asymmetric it may be because the tracheal tube is too far down and into the right main bronchus or there may be a pneumothorax, diaphragmatic hernia, pulmonary hypoplasia or pleural effusion. During artificial ventilation cardiac compression is continued if indicated.

If the baby still does not recover drugs may be administered. Adrenaline (1 in 10 000, 0.1 ml/kg) can be given via the endotracheal tube or intravenously into the umbilical vein. Sodium bicarbonate (4.2 per cent, 5 ml/g) is often recommended, but its use is controversial as it may further lower the intracellular pH. After flushing the catheter with 0.9% saline, calcium gluconate (10 per cent solution 1 ml/kg) is given for persistent bradycardin or absent pulse. If the infant remains pale in spite of the application of the measures given above and if the history is suggestive of fetal blood loss such as placental abruption, a transfusion of Group O negative blood should be given. If respiratory support is still needed the baby should be transferred to the neonatal unit while continuing to provide artificial ventilation.

Throughout all resuscitation it is most important to ensure that the baby does not become hypothermic. The baby needs to be rapidly dried immediately after birth, kept covered and under a radiant warmer. Resuscitation needs to be performed quickly and skilfully.

MECONIUM ASPIRATION

If thick meconium is present on the skin or in the pharynx, the vocal cords must be checked under direct vision and the oropharynx aspirated. If meconium is present in the trachea a tracheal tube is inserted and the meconium aspirated. If the tube becomes blocked it may be necessary to reintubate the baby. If the baby becomes bradycardic positive pressure ventilation will need to be started. It is important that this procedure is performed in skilful hands as prolonged attempts at intubation may make the situation worse by inducing bradycardia by stimulating vagal tone.

LOW BIRTHWEIGHT INFANTS

Low birth weight is defined as a birth weight less than 2500 g. About 7 per cent of babies in England and Wales are of low birth weight. Infants who are born before 37 weeks of gestation are called preterm and those whose weight is below the 10th centile for gestation are usually called light for dates. Infants weighing less than 1500 g at birth are called very low birth weight (VLBW) infants and those weighing less than 1000 g at birth are often described as extremely low birth weight (ELBW) infants.

The most reliable way to assess an infant's gestation is from the mother's dates or else from fetal size at an early ultrasound scan. After birth the infant's gestation can be assessed by using charts which rely on a range of physical characteristics which change with maturity such as skin thickness, the formation of auricular cartilage and genital development as well as a range of neurological characteristics which also change with gestation (Dubowitz score). The accuracy of the assessment is ± 10 days and it is of less value in infants of below 28 weeks' gestation. The presence of fused eyelids suggests that the baby is extremely premature and is associated with a high mortality.

An infant's gestation is a better guide than birth weight to the likelihood of problems in the neonatal period and of long-term morbidity and mortality. Infants who are light for dates may be growth retarded or normal. Most are normal small babies whose weight simply falls below the 10th centile based on a population distribution. Those who are malnourished are thin with little subcutaneous fat. When malnutrition occurs in the latter part of pregnancy the infant's weight is disproportionately low compared with the head size as the growth of the brain is spared at the expense of the rest of the body. In this case the depletion of glycogen stores in the liver and heart and of fat stores makes them more susceptible to intrauterine hypoxia and intrauterine death and postnatally to hypoglycaemia and hypothermia. When the intrauterine malnutrition has commenced in the early part of pregnancy head size will also be reduced.

NEONATAL CARE

There has been a dramatic reduction in the neonatal mortality of low birth weight infants over the last 20 years (*see* Fig. 9.3). This has been from improvements in both obstetric and neonatal care. About 1–2 per cent of all infants require intensive care. Intensive care has been categorized as high and medium dependency intensive care (Table 9.3).

Because of the specialized nature of this work, requiring skilled medical and nursing staff and expensive monitoring equipment, intensive care

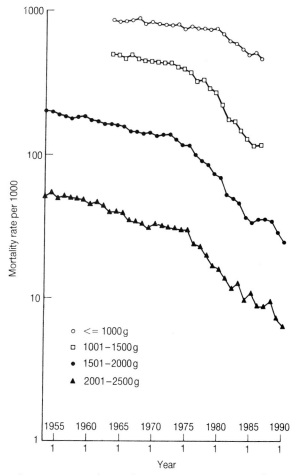

Fig. 9.3 Neonatal mortality rate (birthweight specific) in England and Wales

Table 9.3 Clinical categories of neonatal care (British Association of Perinatal Medicine)

MAXIMUM INTENSIVE CARE
Babies:

- receiving assisted ventilation (including intermittent positive airway pressure, intermittent mandatory ventilation and constant positive airway pressure and in the first 24 hours after its withdrawal)
- of less than 27 weeks' gestation for the first 48 hours after birth
- with a birth weight of less than 1000 g for the first 48 hours after birth
- who require major emergency surgery for the pre-operative period and post-operatively for 48 hours
- on the day of death
- being transported by a team including medical and nursing staff
- who are receiving peritoneal dialysis
- who require exchange transfusions complicated by other disease processes
- with severe respiratory disease in the first 48 hours of life requiring an FiO_2 of >0.6
- with recurrent apnoea needing frequent intervention
- with significant requirements for circulatory support

HIGH DEPENDENCY INTENSIVE CARE
Babies:

- requiring total parental nutrition
- who are having convulsions
- being transported by a trained skilled neonatal nurse alone
- with arterial line or chest drain
- with respiratory disease in the first 48 hours of life requiring an FiO_2 of 0.4–0.6
- with recurrent apnoea
- who require an exchange transfusion alone
- who are more than 48 hours post-operative and require complex nursing procedures
- with tracheostomy for first 2 weeks

has to be centralized at regional units. About 10 per cent of all babies will require special care in the newborn period. Special care may be provided in the Special Care Baby Unit or, particularly for the more mature infant, on the postnatal wards in a Transitional Care Nursery. This has the great advantage of not separating mother and baby. Infants who are acutely ill or where problems are anticipated will need to be admitted to the baby unit. Most infants under 34 weeks' gestation and weighing less than 1800 g at birth are best admitted

to the baby unit. Many of these babies will need to be in hospital for several weeks.

Although only 7 per cent of infants in England Wales are low birth weight, 66 per cent of neonatal deaths in 1988 were in this group of infants. Very low birth weight infants are the most vulnerable and require expert perinatal care to minimize morbidity and mortality. They are easily bruised and traumatized at delivery, whether this is vaginal or by caesarean section. They often require help to establish breathing and efficient resuscitation must be provided. Following resuscitation very low birth weight infants should be transferred to the baby unit in a transport incubator. Many will require artificial ventilation during transfer. On arrival the baby's condition needs to be fully stabilized. Every baby unit should be able to provide this whether or not the infant will remain there for long-term intensive care.

VERY LOW BIRTH WEIGHT INFANTS

For very low birth weight infants two experienced attendants need to be present at delivery. It is particularly important for the baby to be kept dry, warm and covered during resuscitation. Active resuscitation is initiated early unless the baby is vigorous, breathing normally and pink. A high percentage of very low birth weight infants need tracheal intubation and artificial ventilation at birth.

STABILIZING THE VERY LOW BIRTH WEIGHT INFANT

On arrival on the baby unit the infant is quickly weighed, placed in an incubator and briefly examined to assess the baby's clinical condition and identify any congenital abnormalities.

ARTIFICIAL VENTILATION

Many infants of less than 30 weeks' gestation will initially require ventilatory support whether or not they have respiratory distress syndrome. Artificial ventilation or additional ambient oxygen may be required because of lung liquid or simply because of immature lungs and weakness of the muscles of respiration combined with a highly compliant chest wall. Adequate ventilation and oxygenation can prevent the development of hypoxia, lung atelectasis, secondary surfactant deficiency and circulatory failure.

CIRCULATION

The infant's circulation is assessed from the heart rate and skin perfusion and blood pressure. A very helpful guide is the toe-core temperature differential which should not exceed 2°C. Circulatory support is often required with colloid and inotropic drugs.

MONITORING

The infant's heart rate, respiratory rate and temperature are monitored from the time of arrival on the baby unit. The infant's oxygenation needs to be closely monitored to avoid hypoxaemia which may result in ischaemic damage and hyperoxaemia which may produce retinopathy of prematurity (retrolental fibroplasia). The PaO_2 is kept between 7 and 12 kPa (50–90 mmHg). This monitoring can be achieved using transcutaneous electrodes, oxygen saturation monitors or directly from blood gas analysis. Repeated arterial blood gases will be required for babies requiring assisted ventilation. This is best achieved by inserting an arterial catheter, usually into the lower aorta through an umbilical artery, or by cannulation of a peripheral artery, usually the radial artery. The arterial line also allows direct, continuous measurement of the blood pressure while blood samples can be taken from it without disturbing the baby.

A chest X-ray is taken to assess the infant's lungs and the position of the endotracheal tube and umbilical artery catheter.

TEMPERATURE CONTROL

The baby must not be allowed to become hypothermic. Special attention needs to be paid to this not only during resuscitation but during transfer to the baby unit or else these small babies readily become cold.

FLUID BALANCE AND METABOLIC MONITORING

The fluid requirements of very low birth weight infants are highly variable during the first few days of life. Usually 40–60 ml/kg of fluid is sufficient on the 1st day excluding colloid support. However

this needs to be reassessed several times a day. The fluid intake is altered according to the infant's urine output, blood pressure, acid-base balance and clinical assessment of the peripheral circulation. In addition hypoglycaemia is prevented by establishing an intravenous infusion containing dextrose. Monitoring for hypocalcaemia is also undertaken.

ANTIBIOTIC THERAPY

Very low birth weight babies requiring ventilatory support are usually given antibiotics as group B streptococcal infection cannot be reliably distinguished from respiratory distress syndrome both clinically and radiologically.

MINIMAL HANDLING

The infant's circulation and oxygenation are adversely affected by handling and painful procedures. The number of procedures should be kept to a minimum and performed as rapidly and efficiently as possible. Analgesia should be given before painful procedures and usually is in the form of opiates (morphine, dimorphine) administered as a continuous infusion.

PARENTS

Although the doctors and nurses are kept very busy during this period of stabilizing very low birth weight infants, time must be found for the parents to allow them to see their baby and for a full explanation. If the mother has had a caesarean section or is unable to come to the neonatal unit, the neonatal staff will need to go to see her on the labour or postnatal ward. Instant photographs of the baby and booklets about the unit which describe the management of very low birth weight infants are of considerable help to them.

SPECIFIC PROBLEMS

There are a number of problems which are specific to very low birth weight infants (Table 9.4).

TEMPERATURE CONTROL

Low birth weight infants become hypothermic very readily. They have a large surface area relative

Table 9.4 Problems of very low birth weight infants within the neonatal unit

---

Temperature control
Respiratory problems
- Respiratory distress syndrome
- Pulmonary air leaks (pulmonary interstitial emphysema, pneumothorax)
- Recurrent apnoea
- Bronchopulmonary dysplasia
Nutrition
Osteopenia of prematurity (rickets)
Intraventricular haemorrhage and ischaemic brain injury
Patent ductus arteriosus
Infection
Anaemia of prematurity
Necrotizing enterocolitis
Retinopathy of prematurity
Growth

---

to their weight which permits rapid heat loss and they have relatively little subcutaneous fat for insulation. They also have relatively little of the adipose tissue (brown fat) present in term infants which they can metabolize rapidly to produce heat. Preterm infants cannot shiver and are often nursed naked to allow proper observation. It has long been recognized that keeping infants warm to avoid hypothermia improves their survival.

An infant's oxygen consumption is increased when the environment is either too cold or too hot. There is a neutral temperature range when the infant's oxygen consumption is minimized and this is the most suitable temperature to nurse them. This neutral temperature is highest in preterm infants and is further increased in the first few days of life when their skin is thin and poorly keratinized or if they are nursed naked.

Hypothermia occurs when a baby is not able to produce sufficient heat in a cool environment. However a normal body temperature does not necessarily imply that the ambient temperature is satisfactory as it may be achieved by increasing heat production. This increased energy consumption will be achieved at the expense of growth.

INCUBATORS AND RADIANT WARMERS

Very low birth weight infants are nursed in closed incubators or under radiant warmers to provide a sufficiently warm environment. They should be clothed as far as is possible while still ensuring constant observation and ready access. Infants nursed naked in a closed incubator lose heat by radiation and evaporation. Radiant heat loss can be reduced by keeping the intensive care room in the neonatal unit very warm (26–28°C), by using incubators with double walls and by covering the baby with a plastic heat shield or with an insulating fabric such as bubble plastic sheeting. Evaporative heat loss can be minimized by humidification. The disadvantage of humidification is that it increases the risk of colonization with *Pseudomonas* which flourishes in a damp environment. The main disadvantage of closed incubators is the lack of accessibility for staff and parents. Practical procedures, especially those done in an emergency, are much more awkward to perform when babies are nursed inside closed incubators.

Sick infants can also be nursed under radiant warmers. The output of the heat is adjusted according to the infant's skin temperature. This has the advantage of allowing ready access to the baby. They are also more effective in keeping very immature babies warm. Their main disadvantages are that the evaporative heat loss is considerably higher because of increased evaporative water loss from the skin which makes the control of fluid balance much more difficult. In addition heat loss by convection is increased from draught. This can be reduced by covering the infant with an insulating fabric.

As soon as babies are well enough they should be clothed. Clothing markedly reduces heat loss. In infants, the head is a large part of the surface area and bonnets will increase thermal insulation. For infants below 2 kg the room temperature needs to be about 26°C, for those over 2 kg it should be about 24°C to achieve a neutral thermal environment.

RESPIRATORY PROBLEMS

RESPIRATORY DISTRESS SYNDROME (HYALINE MEMBRANE DISEASE)

This is almost entirely confined to preterm infants; infants of diabetic mothers and those with rhesus isoimmunization are also at increased risk. It occurs in about 9 per 1000 live births overall.

Respiratory distress syndrome (RDS) occurs when there is deficiency of surfactant because the production of type II pneumocytes in the lungs is reduced. After about 36 weeks' gestation surfactant is usually produced in adequate quantities. After 32 weeks' gestation it is usually possible to manage respiratory distress syndrome successfully without long-term sequelae provided full intensive care facilities are available. This is important when deciding to deliver babies before term. In the past it was helpful to have information regarding surfactant production by measuring the lecithin content of the amniotic fluid. Since absolute measurements will be dependent on the degree of dilution by amniotic fluid this was commonly expressed as a ratio of lecithin to sphingomyelin (L/S). When the L/S ration is 2:1 or greater, RDS is unlikely but a ratio of under 1.5 means that there is a high risk of hyaline membrane disease, of the order of 75 per cent. With modern intensive care widely available the L/S ratio is now rarely measured.

Administration of corticosteroids to the mother will increase the production of surfactant if the steroids are given at least 24 hours preceding delivery. Betamethasone or dexamethasone are usually used. More recently thyroid releasing hormone (TRH) given antenatally also promotes surfactant production and is combined with corticosteroids. In many preterm deliveries there is an insufficient time interval between giving the corticosteroids and delivery for the drug to be effective.

Respiratory distress syndrome (RDS) occurs in about 60% of infants born before 30 weeks gestation, but in only 35% after a full course of antenatal corticosteroids.

Respiratory distress develops within 4 hours of birth in babies with RDS. The clinical course is of gradual deterioration for the first 2–3 days with worsening of hypoxia, hypercapnia and acidosis. Recovery is often heralded by a reduction in the additional oxygen required and an increase in the urinary output and of the infant's muscle activity. The chest X-ray is characteristic with a ground glass appearance throughout the lung fields and an air bronchogram where the air in the bronchi contrasts with the surrounding opaque lung fields and heart. When the disease is severe the lungs appear as a white out when they cannot be differentiated from the heart. In fatal cases autopsy reveals almost airless lungs and a hyaline membrane-lining the terminal air passages (which takes up eosinophilic stains).

All affected babies should be stabilized and if necessary transferred at an early stage to a centre with facilities for neonatal intensive care. Management of babies with RDS requires expert medical and nursing care and comprehensive monitoring. Added oxygen should be given as required and ventilatory support started early if the disease is severe. Sufficient peak pressure is given to expand the lungs and positive end expiratory pressure (PEEP) to maintain airway patency on expiration. Careful attention must be paid to maintain a satisfactory blood pressure and circulation. Analgesia is given to babies requiring mechanical ventilation. Muscle paralysis is used for babies with severe disease or if the ventilator and the infants breathing cannot be synchronized. Antibiotics are given as group B streptococcal infection may mimic RDS.

The major recent advance in the management of RDS has been the development of surfactant therapy which is given directly into the lungs via the tracheal tube. The surfactant may be synthetic, porcine or bovine. Human surfactant, harvested from amniotic fluid is not readily available. Surfactant therapy has been studied extensively in numerous multicentre trials and has been shown to reduce mortality by about 40 per cent, with no increase in morbidity.

Most of the serious complications of preterm infants relate to the complications of RDS. This includes pulmonary interstitial emphysema where air tracks into the lung interstitium dissecting along the tissue plains; this shows up on the chest X-ray. Pneumothorax and bronchopulmonary dysplasia are important respiratory complications. There is an increased incidence of patent ductus arteriosus in infants with RDS. Intraventricular haemorrhage and ischaemic injury to the brain are the most serious neurological complications and are commoner in babies with RDS. Problems in establishing feeding often requiring parenteral nutrition, retinopathy of prematurity and infection are all commoner in babies requiring prolonged ventilatory support for RDS.

PNEUMOTHORAX

A pneumothorax may occur spontaneously or more commonly as a complication of mechanical ventilation. It may cause increasing respiratory distress or sudden hypoxaemia with an increased oxygen requirement. On examination reduced movement of the chest wall may be noted accompanied by reduced breath sounds on the affected side with a shift of the mediastinum. It can be confirmed most quickly by transillumination of the chest wall with a cold light source or on the chest X-ray. Most pneumothoraces will require direct drainage, often with suction. A pneumothorax is a particularly serious complication in preterm infants as it causes marked swings in oxygen and carbon dioxide tensions and in blood pressure. These result in marked fluctuations in cerebral perfusion which are associated with an increased risk of intraventricular haemorrhage.

APNOEA OF PREMATURITY

Some VLBW infants develop recurrent apnoeic attacks in the first few weeks of life. In most, it is related to their prematurity. They may, however, be an early sign of infection, and this always needs to be carefully considered and excluded. Other causes include anaemia, a patent ductus arteriosus and aspiration of milk feeds. Any underlying cause needs to be treated. Giving a theophylline preparation will often help reduce the frequency and severity of apnoeic attacks. Breathing will usually be re-established by tactile stimulation. Additional ambient oxygen or ventilatory support is sometimes required.

BRONCHOPULMONARY DYSPLASIA

When infants who required artificial ventilation still require supplemental oxygen at 4 weeks (28 days of age) this is called bronchopulmonary dysplasia (BPD). The number of babies with BPD has increased with the improved survival of very low birth weight infants. In this condition there are also chronic changes on the chest X-ray. These include generalized hazy shadowing from areas of collapse and fibrosis alternating with areas of overinflation and emphysema. Prolonged treatment with oxygen and mechanical ventilation is often required and may be necessary for a period of weeks or months. Courses of corticosteroids may result in a dramatic reduction in oxygen requirement. It allows some infants to be weaned off the ventilator. Once off the ventilator additional oxygen may continue to be required. It may be delivered via a nasal cannula or prong attached to a circuit delivering a low flow of oxygen. When additional oxygen is required for some weeks or months this can be provided in the infant's home if there is adequate equipment and support for the parents. Bronchoconstriction may occasionally require the use of bronchodilators. Some infants may benefit from diuretic therapy.

Many of these infants fail to thrive and are short. Vomiting from gastro-oesophageal reflux is not uncommon and may be associated with apnoeic attacks. They are at increased risk of developing problems from recurrent respiratory infections, particularly during the first 2 years of life, and recurrent episodes of wheezing due to bronchial hyperreactivity. Some infants die from respiratory infection or cor pulmonale. The majority show gradually improving lung function. Adequate nutrition is essential for adequate lung growth and repair. Although they have persistent abnormalities on detailed respiratory function testing, they develop sufficient respiratory reserve to be able to undertake all the normal activities of life. The long-term effects of these chronic changes in adult life are not yet known.

NUTRITION

Most preterm infants will need assistance with feeding. Those of less than 34 weeks' gestation are usually unable to suck sufficiently well to feed on their own and will require nasogastric tube feeding. Infants less than 30 weeks' gestation, or preterm infants who require ventilatory support, are often unable to tolerate enteral feeds initially and will require parenteral nutrition.

BREASTFEEDING

It is possible to establish breastfeeding in most preterm infants even those requiring intensive care. The milk will initially need to be expressed from the breast, and it may be more difficult to establish breastfeeding when milk expression is required for a long period of time. Mothers need to be well motivated but given good support and encouragement by the staff it can often be successfully achieved.

Preterm infants fed on breast milk appear to absorb it more readily than formula feeds and to be at a lower risk of developing necrotizing enterocolitis. A study of preterm infants who were breastfed were found to have a higher intelligence quotient in childhood. Breast milk has a lower energy content than formula feeds and infants grow slower on it. Preterm infants often become hyponatraemic because of their high urine sodium loss and the low sodium content of breast milk. Sodium supplements are often required and calcium and phosphate supplementation may also be needed if infants are purely breastfed.

Banked breast milk, where milk is pooled from a number of donors, has a lower concentration of protein, minerals and fat than breast milk expressed from the infant's mother. VLBW babies grow much more slowly on banked breast milk than maternal breast milk. Banked breast milk must be heat treated because of the recent concern about the possible spread of human immunodeficiency virus (HIV) in breast milk. In the UK most neonatal units have therefore stopped running milk banks.

FORMULA FEEDS

Special preterm infant formulae have been developed. They contain a higher concentration of protein and minerals and additional sodium, phosphate and calcium to take into account the preterm infant's increased requirements.

NASOGASTRIC FEEDING

Most infants of less than 30 weeks' gestation, or those with respiratory distress or who are ill, are not given enteral feeds as food absorption is poor under these circumstances and there is a risk of regurgitation and aspiration. In addition respiration can be compromised by gastric distension. An intravenous infusion of glucose is usually given while the infants are stabilized. Thereafter enteral feeding is introduced. In preterm infants the milk is usually given in frequent small boluses via the nasogastric tube to reduce the risk of problems from abdominal distension and of aspiration. The feeds are sometimes given continuously via a syringe pump but this is less physiological than bolus feeding. Another technique to reduce the risk of regurgitation and aspiration is to use a transpyloric tube to bypass the stomach. These silastic feeding tubes are more difficult to position correctly than nasogastric tubes.

PARENTERAL NUTRITION

Parenteral nutrition is often required to supplement or replace enteral feeding in infants receiving intensive care. It is used most often for VLBW infants on artificial ventilation for RDS when there is a delay in establishing full enteral feeds. It is also required in infants with necrotizing enterocolitis or following bowel surgery. Parenteral nutrition can be given for short periods via a cannula in a peripheral vein but the infusion site needs to be observed very carefully as extravasation may result in tissue necrosis and scarring. Long-term parenteral nutrition is best given via a central catheter placed in the right atrium. Parenteral nutrition should only be undertaken in a fully equipped intensive care unit. The infants need to be meticulously monitored for complications, the commonest and most serious of which is infection. Other complications include hyperglycaemia, fluid and electrolyte imbalance, metabolic acidosis and cholestatic jaundice.

OSTEOPENIA OF PREMATURITY (RICKETS)

Preterm infants are prone to osteopenia of prematurity. Their bones may become poorly mineralized and in severe cases they develop rickets with the metaphyses becoming splayed and cupped. Fractures of the ribs and other sites may occur. Infants at most risk are those receiving prolonged parenteral nutrition when it is difficult to give adequate phosphorus parenterally. Infants receiving breast milk alone are also at increased risk. All very low birth weight infants should have their plasma phosphate and calcium levels monitored regularly and an adequate plasma level maintained. Additional phosphate and, if necessary, calcium are given as required. Vitamin supplements, which include additional vitamin D, are usually given to VLBW infants when they are a few weeks old.

INTRAVENTRICULAR HAEMORRHAGE AND ISCHAEMIC BRAIN INJURY

Periventricular haemorrhage and cerebral ischaemia are now a major cause of morbidity and mortality in low birth weight infants. Periventricular haemorrhage occurs in the rich network of fragile capillary vessels which lie alongside each lateral ventricle. The haemorrhage begins under the lining of the ventricle in the germinal matrix and then may burst into the ventricle or the parenchyma. The main complications result from this spread: intraventricular blood may lead to ventricular dilation and hydrocephalus while parenchymal haemorrhage destroys the brain substance. The relationship between haemorrhage and ischaemic lesions is complex and incompletely understood. Ischaemic lesions may, after some weeks, become cystic and filled with cerebrospinal fluid. The cystic lesions may be localized or generalized when it is called periventricular leucomalacia. Intraventricular haemorrhages may be the primary lesion but may also result from bleeding into an ischaemic area, a venous infarct.

Pathogenesis

The more preterm the baby the higher the incidence of periventricular haemorrhage. About half of all VLBW babies have evidence of periventricular haemorrhage on ultrasound. Most of the lesions occur within 48 hours of birth. Infants at most risk are those who have experienced asphyxia or who have RDS. Episodes of hypotension and acidosis also increase the risk of haemorrhage. Any episode that causes an abrupt change in cerebral blood flow may lead to a haemorrhage from rupture of the fragile capillaries in the germinal layer or to ischaemic damage. The changes in cerebral blood flow which occur with a pneumothorax or blockage or dislodgement of an endotracheal tube place an infant at increased risk of periventricular haemorrhage.

Diagnosis

Periventricular haemorrhages can be readily identified on intracranial ultrasound scans performed through the anterior fontanelle. They are usually graded as:

Grade I: small haemorrhage confined to the germinal matrix
Grade II: intraventricular haemorrhage
Grade III: intraventricular haemorrhage with ventricular dilation
Grade IV: parenchymal haemorrhage.

Ventricular dilatation

Ventricular dilatation can be readily identified and its progress followed on ultrasound scanning. If the ventricular dilatation progresses it may be necessary to remove cerebrospinal fluid using a drain inserted into the ventricle or with the ventriculoperitoneal shunt. Although cerebrospinal fluid may be removed by repeated lumbar puncture or ventricular taps, a prospective randomized control trial was unable to show that this improved the long-term prognosis.

Ischaemic lesions

Ischaemic lesions may be evident as an echodense flare within the parenchyma. This may resolve or progress to single or multiple cystic lesions. Cystic lesions become evident at several weeks of age. Haemorrhagic lesions may also become cystic.

Prognosis

Intracranial ultrasound has been used as a prognostic aid. Infants with normal or germinal layer haemorrhages (Grade I) are at low risk for neurodevelopmental disorders. Those with large parenchymal haemorrhages, ventricular dilatation requiring surgical intervention, or those with widespread periventricular leucomalacia are at markedly increased risk. Although the ultrasound appearance allows one to place the child in a risk category it does not allow accurate prediction for an individual child.

Prevention

It is thought that periventricular haemorrhages and cerebral ischaemia result from fluctuations in cerebral blood flow. To prevent these lesions these fluctuations in cerebral perfusion should be avoided. It is therefore particularly important to prevent asphyxial insults intrapartum, at delivery or postnatally. During the first few days of life, when most haemorrhages occur, meticulous attention needs to be paid to the infant's ventilation and circulation to keep them as normal as possible.

PATENT DUCTUS ARTERIOSUS

The patent ductus arteriosus is much larger and its musculature less well developed in preterm than term infants. Delay in closure is common in preterm infants, particularly when they require artificial ventilation because of RDS. About 20 per cent of infants with a birth weight of less than 1.75 kg develop haemodynamically significant shunting across the patent ductus. The more preterm the infant the higher the incidence. At birth the pulmonary vascular resistance is high but falls over the first few days of life. When the resistance in the pulmonary circulation becomes lower than in the systemic circulation, shunting across a patent ductus is from the aorta to the pulmonary artery (left to right). The incidence of left to right shunting is increased by circulatory overload from a high fluid intake during the first few days of life.

Clinical significance

Significant left to right shunting causes pulmonary oedema and venous congestion. In infants requiring artificial ventilation there is an increase in ventilatory requirements and difficulty in weaning infants from the ventilator. It is associated with an increased incidence of bronchopulmonary dysplasia. Infants not requiring artificial ventilation develop respiratory distress or apnoeic attacks or may be asymptomatic. With severe left to right ductal shunting there is retrograde diastolic flow of blood. It has been suggested that this may reduce blood flow to vital organs and could result in ischaemic damage to the bowel wall causing necrotizing enterocolitis and to the brain causing cerebral ischaemia.

Diagnosis

The peripheral pulses become bounding on palpation. There may be a systolic murmur in the pulmonary area and left sternal edge and the cardiac impulse may become increasingly active. In infants on ventilators it is often difficult or impossible to hear murmurs. Other clinical signs are tachypnoea, tachycardia and hepatomegaly from cardiac failure.

The chest X-ray may show cardiomegaly and pulmonary plethora when ductal shunting is severe but these are often masked by the baby's lung disease. On echocardiography circulatory overload of the left side of the heart can be identified and using Doppler ultrasound shunting across the ductus itself can be confirmed.

Management

Infants who are symptomatic can be treated by fluid restriction, diuretics, indomethacin or surgery. Fluid restriction will reduce the infant's circulatory volume but is likely to compromise nutritional intake and growth. Infants on diuretics need to have their serum electrolytes monitored regularly as they readily become hyponatraemic. Indomethacin, a prostaglandin inhibitor, is reserved for infants with symptoms from cardiac failure despite fluid restriction and diuretic therapy. In many infants indomethacin closes the ductus, but is less effective in infants who are less than 28 weeks' gestation or more than 3 weeks old. Its main side effects are oliguria and gastrointestinal haemorrhage.

Surgical ligation of the ductus is sometimes required in infants with RDS in whom the ductus is adversely affecting their ventilation or is preventing the infant from being weaned off the ventilator and in whom medical therapy has failed. It is a relatively straightforward operation.

INFECTION

This is discussed later in this chapter.

ANAEMIA OF PREMATURITY

Blood transfusion is often required in infants receiving intensive care. It is given if the infant is anaemic at birth or to replace blood taken for tests. As the infant's blood volume is only 80–90 ml/kg, a significant proportion of an infant's blood volume is removed from repeated blood tests. Early anaemia of prematurity is due to deficient erythropoiesis combined with reduced red cell survival. It is preferable to maintain the haemoglobin of sick newborn infants above 12 g/dl.

As preterm infants grow, the haemoglobin levels fall to their lowest levels at 4–10 weeks of age. Blood transfusion is usually required if the haemoglobin concentration drops below 8 g/dl or if infants become symptomatic with poor feeding, poor weight gain, respiratory distress or apnoea. Their haemoglobin concentration needs to be monitored during this period. Supplemental iron is given from about 3 weeks of age, unless the infant has been recently transfused (to prevent iron deficiency anaemia from developing). The role of erythropoetin therapy to reduce the need for blood transfusions is being assessed.

NECROTIZING ENTEROCOLITIS

Necrotizing enterocolitis (NEC) is a severe illness usually seen in preterm infants at 2–4 weeks. It is an inflammatory condition of both large and small bowel characterized by the presence of gas in the bowel wall and sometimes in the portal tract which may lead to bowel perforation. The exact cause is not known but is probably due to a combination of ischaemia of the bowel wall and infection. Predisposing factors include hypotension, early feeding and umbilical catheterization. Most cases are sporadic but from time to time epidemic outbreaks occur in neonatal units.

The baby presents with poor feeding or vomiting and abdominal distension, and the passage of blood stained stool. The skin of the abdominal wall becomes tense and shiny and discoloured. The infant may rapidly become shocked. An abdominal radiograph shows distended loops of bowel and the bowel wall is oedematous and may contain intramural gas, a diagnostic feature. There may also be bowel perforation which can be detected on the X-ray or by transilluminating the abdomen.

Treatment is to stop oral feeds and commence nasogastric suction. The baby will need intravenous fluids. Broad-spectrum antibiotics to cover both aerobic and anaerobic organisms are given. Shock and acidosis are treated with plasma and blood. Surgery is required for some infants who deteriorate further or develop perforation of the bowel. NEC has a significant morbidity and mortality at all gestations. Once the acute illness has resolved, the baby's nutrition is provided parenterally. Some infants who survive the initial illness subsequently develop malabsorption or strictures of the bowel.

RETINOPATHY OF PREMATURITY

VLBW infants are prone to retinopathy of prematurity, a disorder of the retinal vasculature. In its mildest and commonest form there is early proliferation of the retinal vessels which regresses fully. Sometimes it progresses to fibrosis, distortion and scarring of the retina, which may result in retinal detachment, severe visual impairment or blindness.

A dramatic increase in blindness from retinopathy of prematurity was noted in the 1950s when preterm infants were nursed for long periods in high concentrations of oxygen. When oxygen therapy was restricted to the concentration required by the baby the incidence of retinopathy of prematurity fell markedly. However, in spite of the careful monitoring of oxygen tension that is now available, retinopathy of prematurity is seen in very low birth weight infants, particularly those who have severe lung disease and require prolonged oxygen therapy. It therefore appears that oxygen toxicity is not the only factor in its aetiology. All very low birth weight infants should have their eyes checked by an ophthalmologist when they reach the equivalent of 32–34 weeks' gestation. With severe disease cryotherapy to the retina is sometimes used. Squints and reduced visual acuity are more common in VLBW infants.

GROWTH

Although intrauterine growth rates are widely used as a yardstick for growth, this is often difficult or impossible to achieve. In practice one has to accept the fastest obtainable growth without causing metabolic upset. A slower growth rate is achieved in preterm infants fed on breast milk alone than with those on special preterm infant formulae.

PARENTS

A preterm delivery often leaves a mother and her partner psychologically unprepared for a baby at this early stage of pregnancy. Additional anxiety and stress comes from seeing their tiny fragile baby transferred to the neonatal intensive care unit. This is compounded by the complex, high technology of the unit and the uncertainty of the baby's short- and long-term outlook.

When possible, parents find it helpful to see the baby unit and meet some of the staff before delivery. After admission, parents are helped to get to know their baby by allowing them and their close family to visit whenever they want, by providing a friendly, caring atmosphere and by good communication with the staff at all times. Parents should also be encouraged to get to know their baby by touching and stroking, by spending time with their infant in the unit and by helping with their baby's care according to the baby's clinical condition and their own wishes.

DISCHARGE FROM HOSPITAL

Infants can be discharged from hospital when they are feeding well, their condition is stable and when the parents are able to care for them. Many mothers of babies who have received intensive care and have been in hospital for many weeks find it helpful to return to hospital for a few days prior to the baby's discharge, during which time they can assume responsibility for the care of their baby in preparation to going home. Once home, parents value continuing support and advice from the hospital and community.

All very low birth weight infants and infants with neonatal problems who may have long-term sequelae need to have their developmental progress closely monitored. Immunizations should be given as for other children at the same chronological age from birth. The parents need to be advised whether pertussis immunization is contraindicated or not. In the very preterm infant the first dose may already be given before they leave the baby unit. Very low birth weight infants are also given additional iron, folic acid and vitamin supplements. Parents are encouraged to attend their general practitioner or community health clinic to keep a close check on the baby's growth and development. The neonatal unit or community paediatrician will conduct more detailed developmental assessment periodically. All VLBW infants should have their hearing checked as they are at increased risk of sensorineural hearing loss. They should also have their vision checked for visual acuity and squints.

Readmission to hospital during the 1st year of life is increased approximately four-fold in very low birth weight infants. Infants with bronchopulmonary dysplasia are more susceptible to recurrent wheezing and chest infections. The development of an inguinal hernia is not uncommon in these infants.

NEURODEVELOPMENTAL OUTCOME

Survival and outcome are related to both gestational age and birth weight. Very low birth weight infants are at risk of a wide range of neurodevelopmental problems. These include visual disability, hearing loss, cerebral palsy and severe learning disability. A particularly helpful neurodevelopmental follow up study is the detailed, population based study performed on all babies of birth weight less than 1750 g born in Scotland in 1984. At $4\frac{1}{2}$ years of age 71 per cent had survived. They were shorter and thinner than normal children. Overall 16 per cent were disabled, of whom 8 per cent had cerebral palsy, 1 per cent were blind and 2 per cent were deaf using hearing aids. Although the mean IQ of this population was somewhat lower than children of normal birth weight, only 5 per cent had an IQ below 70, although a further 3 per cent could not be tested. However, some of the children have specific learning difficulties, in particular marked delay in their language development. Overall, they also performed less well on visual recognition, verbal comprehension, number and fine motor skills, than children born at term. Some had a poor attention span and they had more behavioural problems than their siblings. In general, the ones with the most problems were the extremely low birth weight babies (weighing less than 1000 g).

Overall, however, very low birth weight infants only account for a very small proportion of the total neurodevelopmental handicap in the population. In the UK less than 2 per cent of serious handicap can be attributed to prematurity.

THE ASPHYXIATED INFANT

There has been a dramatic reduction in the incidence of birth asphyxia in full-term infants in developed countries in recent years. However it still remains the most common cause of perinatally acquired brain damage in term infants. Its reported incidence varies according to its definition. A recent report from the UK showed an incidence of 6 per 1000 live births. Moderate and severe asphyxial encephalopathy occurred in 2 per 1000 births. One-quarter of the infants with post-asphyxial encephalopathy showed intrauterine growth retardation.

Unfortunately, the term birth asphyxia is used differently in different situations. CTG monitoring is widely used and when it remains normal, it serves as a sensitive predictor of well-being at birth for the infant. When abnormal, in general it is a poor predictor of severity of any asphyxial insult and therefore other indicators of the fetus's condition are sought. The commonest is to use fetal scalp blood samples for biochemical assessment of asphyxia looking for evidence of hypoxaemia, hypercarbia and metabolic acidosis. These do serve as a marker of intrapartum asphyxia when they are very severely deranged but in most clinical situations their predictive power is poor.

As severely asphyxiated infants experience secondary apnoea at birth, failure to establish respiration at birth or low Apgar scores have been widely adopted as criteria for birth asphyxia. However, as described in the section of resuscitation, there are many reasons why babies fail to establish respiration at birth. Most babies who require resuscitation with artificial ventilation at birth do not have biochemical evidence of asphyxia. Only about 1 in 5 babies with low Apgar scores at 1 and 5 minutes have a severe metabolic acidosis and only a similar proportion of acidotic babies have a low Apgar score. In clinical practice the most sensitive indicator of long-term problems is the development of clinical evidence of hypoxic ischaemic encephalopathy. Intrauterine, intrapartum and postnatal asphyxia may all contribute to the newborn's condition and can be difficult to separate out. Antenatal events, for example haemorrhage from placental separation are responsible for about 25 per cent of cases. Intrapartum complications, for example prolonged and difficult labours, instrumental deliveries, acute placental abruption or prolapse of the umbilical cord, are present in about 40 per cent. In a further 25 per cent there is both antepartum and intrapartum asphyxia. In the remaining 10 per cent of infants the asphyxial insults are mainly postnatal.

Hypoxic ischaemic encephalopathy is usually classified as mild, moderate or severe:

In mild encephalopathy there is no alteration in conscious level but the infants seem to be hyperalert. They spend time awake and restless, often with staring eyes. They are jittery and respond excessively to stimulation. Complete recovery usually occurs within 48 hours.

In moderate encephalopathy there are seizures and lethargy with reduction in spontaneous movements. The limbs are hypotonic, the tone in the legs being greater than in the arms. Tendon reflexes are exaggerated. Sucking is poor. Breathing is rapid from hyperventilation. The seizures tend to develop from 12 to 48 hours of life. Some improvement usually occurs by the end of the 1st week but complete recovery, if it occurs, may take several weeks.

Infants with severe encephalopathy are comatose with severe hypotonia and require respiratory support from birth. They are profoundly hypotonic with no spontaneous movements. Seizures are frequent and may be prolonged. The infants do not suck and tendon reflexes are absent. The pupils are poorly responsive or unresponsive to light. There may be accompanying myocardial ischaemia resulting in hypotension and poor cardiac output. Ischaemia to the kidneys may result in acute renal failure. If the infant survives, the level of consciousness may gradually improve but there may be persistent neurological abnormalities. Particularly important both for the infant's management and prognosis is the inability to suck and swallow because of disturbed coordination. Prolonged nasogastric tube feeding may be required. There may be reduced tone and weakness in the proximal muscles of the limbs together with hypertonicity of the neck muscles on extension.

Intracranial ultrasound may initially show a diffuse increase in echogenicity from ischaemia. Raised intracranial pressure and oedema may result in small compressed cerebral ventricles. The EEG may be abnormal or even isoelectric in the most severely affected infants. Meticulous attention is paid to maintain respiration, systemic blood pressure and peripheral perfusion. Artificial ventilation may be required. The plasma sodium, glucose and calcium need to be monitored closely. Seizures are treated, though their response to anticonvulsant therapy is often poor. Fluid intake is reduced to a minimum because of the cerebral oedema. Hyperosmolar infusions with mannitol, barbiturates and steroids have all been used to treat the cerebral oedema though data showing that any are beneficial are lacking. At several weeks of age evidence of ischaemic lesions may be seen on computerized tomography (CT) or nuclear magnetic resonance (NMR) scans. Further information may be gleaned from a number of techniques such as cerebral blood flow velocity studies using Doppler ultrasound, NMR spectroscopy, auditory and visual evoked responses and power spectral analysis of the electroencephalogram (EEG)·but their clinical use has not been determined.

PROGNOSIS

In general, acute hypoxic episodes do less damage than prolonged severe hypoxaemia. Overall, full-term infants who are able to feed normally and have no abnormal neurological signs by 10 days of age have an excellent long-term prognosis. Those who are still abnormal at this age, particularly those who have not re-established normal feeding have a variable prognosis. Infants who have sustained permanent brain damage from hypoxic ischaemic encephalopathy usually have motor deficits and develop cerebral palsy. This may be predominantly spasticity, choreoathetosis or ataxia. A variable degree of severe learning difficulties with or without seizures may co-exist.

BIRTH TRAUMA

Birth injuries vary from minor skin abrasions to severe internal haemorrhage. Their prevention depends on the art of obstetrics. It is a reflection of the improvement in perinatal care in recent years that serious birth trauma is now so uncommon.

BIRTH INJURIES TO THE HEAD

Caput succedaneum

This is caused by oedema of the subcutaneous layers of the scalp (*see* Fig. 9.4). It lies over the presenting part of the head as it passes through the birth canal. The swelling is maximal at birth and resolves within a few days. A localized caput or chignon may be produced at a Ventouse delivery.

Cephalhaematoma

This is a subperiosteal haematoma which most commonly lies over one of the parietal bones (*see* Fig. 9.5). Spontaneous absorption occurs, but this may take several weeks. No treatment is necessary.

Subaponeurotic haematoma

This very uncommon injury occurs when blood lies in the loose areolar layer between the aponeurosis and the periosteum. The haematoma is not limited to a single bone, and a large collection of blood may even extend as far as the cervical region. Slow absorption takes place.

It can result in anaemia and jaundice.

Skull fractures

These are now rarely seen. They used to occur in a

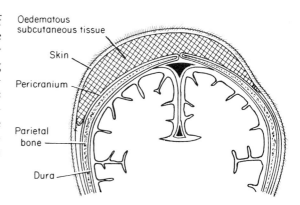

Fig. 9.4 Caput succedaneum. The swelling is oedema of the structures lying superficial to the pericranium. Note that it is not limited to one bone

parietal or frontal bone as a result of difficult forceps delivery. Active treatment is rarely indicated although it is best to seek neurosurgical advice about their management.

INTRACRANIAL INJURIES

Traumatic intracranial injuries have become rare. They may occur following difficult instrumental deliveries of infants with malposition or breech deliveries. Preterm infants are particularly vulnerable.

Periventricular haemorrhage

This is predominantly a problem of preterm

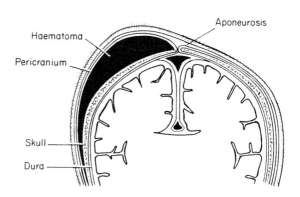

Fig. 9.5 Cephalhaematoma. The swelling is blood lying between the pericranium and the skull and is limited to the area of bone on which it started because the pericranium is attached to sutures around the bone

infants. It is rarely seen in term infants but may accompany trauma and perinatal hypoxia.

Subdural haemorrhage

This may occur following trauma during delivery. Although rarely seen in term infants, it is encountered in the preterm when there is increased compliance of the skull and a prolonged or precipitous delivery, particularly when the infant has an abnormal presentation. Subdural haemorrhage results from damage to the superficial veins where the great vein of Galen and the inferior sagittal sinus combine to form the straight sinus. It may also result from laceration of the tentorium and falx, particularly at their junction. After breech delivery in preterm infants there may be diastasis of the occipital bone leading to blood collecting in the posterior fossa.

The clinical presentation is similar to severe hypoxic-ischaemic encephalopathy. Intracranial ultrasound is not reliable to detect subdural haematomas and a CT scan should be performed if the infant is stable. Skull X-rays may show fractures or occipital diastasis on lateral views.

A major tear of the tentorium or falx is invariably lethal. The size and site of smaller subdural haemorrhage will determine the clinical course and their long-term prognosis.

Subarachnoid haemorrhage

In the newborn subarachnoid haemorrhage is of venous origin and is quite different to the large arterial haemorrhage seen in adults. There is bleeding from veins within the subarachnoid space which may follow mild trauma or asphyxia at birth. The condition may be asymptomatic but can present with irritability or seizures during the first few days. Later hydrocephalus may develop. The condition may be fatal. The CSF is uniformly blood stained. A CT scan is required for diagnosis.

Most of the infants require no treatment and their long-term prognosis is good. It is only the rare instance of massive haemorrhage that has a poor outlook.

FRACTURES OF THE LONG BONES

These are now uncommon. The clavicle is the bone most frequently broken from shoulder dystocia. Injuries are usually noticed when the infant fails to move his or her arm as freely as the other side because of pain. Sometimes a lump can be felt over the bone from callus formation; X-rays show the fracture and healing occurs without specific treatment.

Other bones are rarely fractured. Orthopaedic advice should be obtained for their management. Healing is usually rapid and complete.

PERIPHERAL NERVE INJURIES

Facial nerve

The facial nerve can be damaged by pressure from the tip of the forceps; palsy can also follow an apparently normal delivery.

On the affected side the infant is unable to shut the eye completely, the corner of the mouth drops and the nasolabial fold is less marked than on the other side. When the infant cries, the mouth is drawn to the normal side. In most cases recovery occurs spontaneously within a few days.

Brachial nerve palsy

This may occur from overstretching or tearing of the nerves by lateral traction on the neck, usually during a difficult delivery either from shoulder dystocia or a breech presentation. The clinical picture depends on the nerve roots involved. If the roots of C5 and C6 are damaged (Erb's palsy) the arm lies at the side of the trunk with internal rotation at the shoulder, the elbow is extended, there is pronation of the forearm and flexion at the wrist, the waiter's tip position of Edwardian restaurants. Rarely, the roots C7, C8 and T1 are involved when the wrist and finger flexors are affected resulting in a claw hand deformity with Klumpe's palsy.

Orthopaedic advice should be obtained. Usually, the nerves have been compressed or stretched and recovery occurs. If the nerve itself has been ruptured the injury may be permanent. In such cases surgical repair of the damaged nerve roots is being undertaken in some specialist centres.

RESPIRATORY DISTRESS

Respiratory distress in newborn infants is manifest by tachypnoea, a respiratory rate greater than 60 breaths/min, recession of the chest wall and expiratory grunting. The infant may be cyanosed. There are a number of important causes (Table 9.5).

Infants with respiratory distress need a chest X-ray which will usually identify the cause. It will also exclude important causes requiring immediate treatment, particularly a pneumothorax, pleural effusion or diaphragmatic hernia. All infants with respiratory distress should have continuous monitoring of their heart and respiratory rate and of their oxygenation by a transcutaneous oxygen or oxygen saturation monitor. Blood gases are measured as required. The blood pressure and peripheral perfusion also need to be monitored. Additional ambient oxygen, mechanical ventilation and circulatory support are given if necessary.

PULMONARY CAUSES

Common conditions

Respiratory distress syndrome and immature lungs

These are problems of preterm infants and are considered in that section.

Transient tachypnoea of the newborn

The resorption of lung liquid, mainly into the lymphatics, and the progressive expansion of the alveoli by respiratory movements is usually rapid, occurring within a few minutes of birth, but total expansion may not be attained for some hours. Delay in the resorption of lung liquid causes respiratory distress and is called transient tachypnoea of the newborn. It is by far the commonest cause of respiratory distress in term infants, and is seen especially following caesarean section. The chest X-ray shows hyperinflation of the lungs with fluid spreading from the mediastinum and in the horizontal fissure. Additional ambient oxygen may be required. Although the condition usually settles within a few hours, it is sometimes more severe and can take several days to resolve.

Uncommon conditions

Meconium aspiration

The passage of meconium before birth is seen in 8–20 per cent of deliveries. It is primarily related to the infant's gestation. It is rarely passed by preterm infants and becomes increasingly common the greater the gestational age, affecting 20–25 per cent of deliveries at 42 weeks. Infants who are asphyxiated may start gasping either *in utero* or immediately after birth and may inhale thick meconium into the large and small airways. Meconium is irritant to the lungs and results in both mechanical obstruction and a chemical pneumonitis. It may also cause infection.

Following meconium aspiration, infants develop respiratory distress. The lungs are over-inflated and accompanied by patches of collapse and consolidation. There is a high incidence of air leak leading to pneumothorax and pneumomediastinum. Ventilatory support is often required in

Table 9.5 Causes of respiratory distress in newborn infants

PULMONARY
Common
     Respiratory distress syndrome
     Immature lungs
     Transient tachypnoea of the newborn
Uncommon
     Meconium aspiration
     Pneumothorax
     Pneumonia
     Pulmonary haemorrhage
     Pleural effusions
     Pulmonary hypoplasia
     Diaphragmatic hernia
     Tracheo-oesophageal fistula
     Airways obstruction (e.g. choanal atresia)
     Persistent fetal circulation
       (persistent pulmonary hypertension of the newborn

OTHER CAUSES
     Congenital heart disease
     Intracranial birth trauma/asphyxia
     Severe anaemia

affected infants, who may develop secondary persistent fetal circulation (persistent pulmonary hypertension of the newborn) which may result in inadequate oxygenation.

Whenever possible the condition should be prevented. With improved obstetric care the number of affected infants has declined markedly. When there is thick meconium it should be aspirated from the upper airway immediately after the head is delivered. Meconium seen beyond the cords on laryngoscopy is removed by intubating the baby and aspirating the tracheal tube and the brochi beyond.

Pneumonia

Infants with pneumonia develop respiratory distress. Group B streptococcal infection has become an important pathogen in the newborn baby. The infection is acquired from the maternal genital tract, the current rate of carriage in the UK being about 15–20 per cent. Colonization of the throat or the body surfaces occurs in about 20 per cent of these infants but only 1–2 per cent of those who are colonized develop systemic disease. The presenting features can be identical to RDS. There is an increased risk of infection in infants of low birth weight and if there has been prolonged rupture of the membranes. A chest X-ray may show patchy consolidation or the diffuse granular appearance of RDS. Treatment consists of systemic antibiotics and full respiratory and circulatory support.

Infection from other organisms, for example, *Listeria monocytogenes* and Gram negative infections may present in a similar way.

Pneumothorax

A pneumothorax may occur spontaneously or more commonly as a complication of mechanical ventilation.

Diaphragmatic hernia

This occurs in about 1 in 4000 births. In most cases there is a left sided hernia through the foramen of Bochdalek. It usually presents with failure to respond to resuscitation but sometimes with increasing respiratory distress. The apex beat and heart sounds are displaced to the right side of the chest and there is poor air entry into the left chest. Vigorous resuscitation may cause a pneumothorax

of the normal lung thereby aggravating the situation. The diagnosis is confirmed by an X-ray of the chest and abdomen showing the displaced mediastinum and loops of bowel in the chest. Once the diagnosis is suspected, a large nasogastric tube is passed and suction applied to prevent distension of the intrathoracic bowel. Optimal respiratory support is essential.

In most infants with this condition the main problem is pulmonary hypoplasia due to compression of the herniated viscera preventing the development of the lung. The condition carries a high mortality because of accompanying pulmonary hypoplasia. In some centres extracorporeal membrane oxygenation (ECMO) has been used pre- and post-operatively to provide respiratory support during the first few days of life.

Persistent fetal circulation (persistent pulmonary hypertension of the newborn)

This condition is usually associated with meconium aspiration, RDS, asphyxia, or diaphragmatic hernia. It may, however, occur as a primary disorder. Cyanosis occurs soon after birth. Because of the high pulmonary vascular resistance there is right to left shunting within the lungs and at atrial and ductal levels. Murmurs and signs of cardiac failure are not usually present. A chest X-ray shows that the heart is of normal size. There may be pulmonary oligaemia but this is often difficult or impossible to appreciate in the newborn period. An urgent echocardiogram is required to establish that the child does not have congenital heart disease.

Most infants require mechanical ventilation and full intensive care. Circulatory support is often required. Tolazoline is often given but is not always beneficial for it is a general vasodilator and may be accompanied by systemic hypotension which may need correction with colloid support. Prostacyclin can also be beneficial. With meticulous attention to their ventilation and circulatory support most affected infants can be managed successfully. Recently extracorporeal membrane oxygenation, with the infant on cardiopulmonary bypass for several days, has been used for very severe cases.

OTHER CAUSES OF RESPIRATORY DISTRESS

Congenital heart disease

Newborn infants in heart failure usually present with respiratory distress. Examination of the infant may reveal abnormal heart sounds and/or heart murmurs. An enlarged liver from venous congestion is an important sign of heart failure in infancy. Crepitations of the lung bases are not usually a feature. As the commonest cause of heart failure in newborn infants is coarctation of the aorta it is important to palpate the femoral arteries. On the chest X-ray there may be an enlarged heart and pulmonary plethora although these are often diffi-cult to be certain of in newborn infants. The structural abnormality can be identified by echo-cardiography.

Intracranial birth trauma and birth asphyxia

Infants who have sustained intracranial birth trau-ma or birth asphyxia will often have respiratory distress from hyperventilation.

Severe anaemia

Severe anaemia in the newborn may present with respiratory distress from heart failure. The infant's haemoglobin level or haematocrit needs to be measured to identify this.

JAUNDICE IN THE NEWBORN INFANT

Jaundice is common in newborn babies as the breakdown of red blood cells results in the pro-duction of indirect, unconjugated bilirubin which is insoluble in water and is carried in the serum bound to albumin. This bilirubin is conjugated in the liver by the glucuronyl transferase system of enzymes to form water soluble bilirubin glucur-onide which is excreted in the bile.

As the haemoglobin concentration is much higher in fetal life than after birth, there is a marked breakdown of red blood cells in the first few days of life. In addition, the glucuronyl trans-ferase system in the liver is less well developed in the immediate newborn period. The concentration of unconjugated bilirubin often rises for a few days and causes physiological jaundice. However, there are also a number of conditions which may cause jaundice.

Jaundice in the newborn is important for two reasons. The first is its potential to damage the brain. The second is that it may be a manifestation of an underlying disorder which requires identi-fication and treatment. Unconjugated bilirubin may cross the blood–brain barrier and stain the brain, particularly the basal ganglia. This causes neuronal damage (kernicterus); in the early stages the infant may be irritable, reluctant to feed and has increased muscle tone manifest by hyper-extension of the back and neck. The long-term problems are sensorineural deafness, cerebral pal-sy, particularly with choreo-athetoid movements, and mental retardation. It may also be fatal. For-tunately with modern perinatal care kernicterus has become very rare. In clinical practice the management of jaundice can be most readily dif-ferentiated according to the baby's age when it presents (Table 9.6).

JAUNDICE PRESENTING AT LESS THAN 24 HOURS OF AGE

Haemolysis

Haemolytic jaundice is by far the most severe in the newborn period. Rhesus incompatibility can cause the most dramatic escalation of bilirubin levels in the first few hours of life. It is now mostly prevented or treated before birth. ABO incompati-bility remains a problem but the jaundice is much less severe, as is the jaundice from other blood group incompatibilities.

G6PD deficiency

This X-linked recessive disorder is a deficiency of

Table 9.6 Causes of jaundice in newborn infants according to age at presentation

---

LESS THAN 24 HOURS OF AGE

Haemolysis	Rhesus incompatibility
	ABO incompatibility
	Other blood group
	incompatibilities
	G6PD deficiency
	Congenital spherocytosis

Bruising
Polycythaemia
Intrauterine infection

JAUNDICE AT 24 HOURS TO 14 DAYS OF AGE
Physiological jaundice
Breast milk jaundice
Infection
Haemolytic jaundice
Dehydration

PROLONGED JAUNDICE, MORE THAN 14 DAYS
Breast milk jaundice
Congenital hypothyroidism
Galactosaemia
Neonatal hepatitis syndrome
Bile duct obstruction e.g. biliary atresia

Table 9.7 Drugs which may cause haemolysis in G6PD deficiency

---

Anti-malarials	Primaquine
Antibiotics	Some sulphonamides, eg
	co-trimoxazole
	nitrofurantoin
	nalidixiic acid
	ciprofloxacin
Others	dapsone, naphthalene (moth balls)

---

an enzyme which is essential for the survival of erythrocytes. It is not uncommon in Asiatic and Mediterranean races. A less severe form occurs in black Africans and North Americans. It can be diagnosed by measuring the enzyme level in the blood. Haemolysis in later life may be precipitated by a number of drugs (Table 9.7). The parents need to be informed about this.

Congenital spherocytosis

This autosomal dominant disorder is an uncommon cause of haemolytic anaemia. Spherocytes can be seen on the blood film. Splenectomy is often required.

Bruising and polycythaemia

Jaundice is likely to be exacerbated by the presence of bruising or from polycythaemia at birth.

Intrauterine infection

Intrauterine infection with rubella, cytomegalovirus, toxoplasmosis and syphilis may cause a conjugated jaundice. Infants with jaundice are likely to have other stigmata of infection.

JAUNDICE AT 2–14 DAYS

Physiological

The commonest cause of jaundice in the newborn period is physiological. This usually starts on the 2nd or 3rd day of life and lasts for only a few days. There is no constitutional upset although some infants appear to feed less well and are somewhat more lethargic.

Infection

Increasing jaundice may be an early sign of bacterial infection, particularly urinary tract infection. This should be considered in all newborn infants with jaundice at any age.

Haemolytic jaundice

Jaundice from haemolysis may sometimes only become evident at this age.

Breast milk jaundice

Jaundice becomes more severe in some infants who are breastfed. The evidence for this is that the jaundice declines if breastfeeding is omitted and returns on its re-establishment. It is not necessary to interfere with breastfeeding.

PROLONGED JAUNDICE (> 14 DAYS)

Breast milk jaundice

Breast milk jaundice is the commonest cause of prolonged jaundice. It gradually declines over several weeks. The jaundice is unconjugated.

Congenital hypothyroidism

Prolonged neonatal jaundice may be an early manifestation of hypothyroidism. The jaundice is unconjugated. All newborns are screened for this during the first week of life with a dried blood spot, the Guthrie spot. In babies with prolonged jaundice it should be ensured that the baby's thyroid function is normal.

Galactosaemia

This rare metabolic disorder may present with prolonged jaundice, vomiting and hepatomegaly. It is excluded by showing that there are no reducing substances present in the urine.

Infection

Bacterial infection may cause prolonged jaundice. It is best to exclude a urinary tract infection in these infants.

Neonatal hepatitis and biliary atresia

It is important to diagnose biliary atresia as early as possible as the prognosis from surgical treatment is better the earlier it is performed. The jaundice is conjugated. A useful clinical marker is that the stools become pale or white and the urine is dark in colour. Detailed investigation is required to differentiate biliary atresia from other causes of neonatal hepatitis and bile duct obstruction. Biliary atresia is treated surgically. Successful treatment by liver transplantation has also been reported.

MANAGEMENT OF JAUNDICE

Jaundice can be detected clinically by the yellow to green discoloration of the skin. As the level of jaundice increases the skin discoloration spreads from the face to the periphery of the limbs. Accurate assessment of moderate or severe jaundice requires measurement of the bilirubin level in the blood. Particular care needs to be taken in the clinical assessment of black babies or those from the Indian subcontinent when the conjuctiva of the eyes becomes an important site of assessment. Assessing the clinical importance of the jaundice, requires consideration of the baby's age and gestation as well as the absolute level of the bilirubin. Significant jaundice within the first 24 hours of life needs to be investigated and closely monitored. It is most likely to be haemolytic and the bilirubin may rise very rapidly. The rate of change of the bilirubin needs to be checked. This is readily done by plotting the levels on a graph.

Treatment is by phototherapy or exchange transfusion. Light converts bilirubin into harmless pigments. The baby is usually placed under a bright light. As the baby is nursed exposed, the infant is at risk of hypo- or hyperthermia. The fluid intake is increased because of the marked fluid loss through the skin which may be accentuated by loss in the gut. Infants under phototherapy may develop an erythematous rash. The eyes are covered because of the brightness of the lights. The eye patches may cause parental anxiety, particularly when they slip off. Although no long-term sequelae of phototherapy have been reported, it should not be used indiscriminately. More recently a light source in the form of a pliable blanket has been developed which can be incorporated into the baby's vest. Phototherapy reduces the maximum serum bilirubin reached and the number of exchange transfusions required.

There is no specific level of bilirubin which is known to be either safe or damaging. In term infants with haemolytic disease it was found that clinical evidence of kernicterus no longer occurred when the bilirubin level was kept below 360 μmol/l. This figure became widely adopted as an indication for exchange transfusion although many paediatricians would now accept bilirubin levels up to 420 μmol/l in a well term infant. Phototherapy and exchange transfusion will be initiated at lower levels in preterm infants, in infants who are ill or if a further rise in bilirubin is anticipated from haemolysis or bruising. Most units have adopted their own criteria for initiating phototherapy and exchange transfusion.

INFECTIONS

Infection is a common problem in the newborn period. The newborn infant, particularly the pre-term, has a poorly developed immune system resulting in increased susceptibility to infection. Preterm infants are born with very low levels of IgG immunoglobulins as most IgG is transferred to the infant in the third trimester.

The infant may have been exposed to infection or even have become infected before birth. There is an increased risk of infection if the mother develops chorioamnionitis, although only 1–6 per cent of infants born to mothers with clinical manifestations of this become infected. Prolonged rupture of the membranes is also associated with an increased risk of perinatal infection, with the risk increasing with duration of membrane rupture; again the actual rate of neonatal infection is low while chorioamnionitis itself may predispose to prolonged rupture of the membranes. Infections arise mainly from exposure to bacteria and viruses from the mother's genital tract when the newborn infant has no immunological protection against them. This means that the important bacterial pathogens in the newborn are *Group B streptococci*, gram negative organisms and *Listeria monocytogenes*.

Infants receiving intensive care develop a different range of infections as they undergo numerous invasive procedures. Their skin is often traumatized by monitoring equipment, and so loses its protective effect. Tracheal tubes and indwelling catheters are often left in place for a long time for monitoring, intravenous feeding, and intravenous infusions. Within the neonatal nursery many of the organisms causing infection are normally commensals or low grade pathogens. There is also the additional risk of cross-infection from nosocomial spread. Diligent handwashing by staff between handling babies is imperative and must be strictly adhered to in order to keep this to a minimum.

Clinical features

Infection in the newborn is often poorly localized and may spread rapidly. It is therefore important that diagnosis and treatment are initiated as quickly as possible; given the non-specific symptoms and signs at presentation this is not easy to achieve in practice (Table 9.8).

Table 9.8 Symptoms and signs of infection in the newborn infant

Respiratory distress
Fever or temperature instability
Poor feeding
Vomiting and diarrhoea
Apnoeic attacks
Irritability, drowsiness, lethargy
Jaundice
Abdominal distension
Hypoglycaemia or hyperglycaemia
Poor peripheral circulation
Shock

Infants may be severely ill at birth or more often present at several hours of age. This is early onset infection. In the newborn nursery, late onset infection, presenting after 48 hours of age, is most often a complication of monitoring and invasive lines. By far the commonest organism causing infection at this stage is coagulase negative staphylococcus (*Staphylococcus epidermidis*) although Gram negative organisms and *Group B streptococci* may also cause infection at this age. The initial presentation may be subtle with infants tolerating handling less well than previously, developing apnoeic attacks either for the first time or of increasing frequency or an increase in ventilatory support. There may be increasing jaundice, abdominal distension, deterioration on the chest X-ray, glucose intolerance, temperature instability or poor peripheral perfusion with an increase in the core to peripheral temperature gap. Infection may also be first suspected when a change is noted in the differential white cell count, with a fall in the platelet count or a rise in C-reactive protein or other markers of inflammation.

Investigation

In order to confirm or exclude infection and identify its location, a septic screen often needs to be performed. This may include a full blood count, including a white cell count and differential, a blood culture, urine and CSF for microscopy and culture. A chest X-ray will also be needed. C-reactive protein or other markers of inflammation can also be helpful to identify and follow an infective episode.

Treatment

In the newborn it is important to initiate treatment as quickly as possible, often on an empirical basis. The most appropriate antibiotics are selected, taking into account the mode of presentation and the baby's gestational and postnatal age. Infants presenting within the first 48 hours of life are likely to be infected from the mother's vaginal flora. Antibiotics suitable for *Group B streptococci*, Gram negative organisms and *Listeria monocytogenes* are used, e.g. a penicillin combined with an aminoglycoside or a third generation cephalosporin. If meningitis is suspected a third generation cephalosporin is usually favoured because of its CSF penetration.

Within the neonatal unit, for babies over 48 hours of age an antibiotic which covers *Staphylococcus epidermidis* is indicated, for example vancomycin. A second antibiotic (e.g. a third generation cephalosporin) is often added to cover a wider spectrum of bacteria. The importance of good supportive management is often underestimated. Infants need to be closely monitored and additional oxygen and ventilatory support provided as indicated. The baby's glucose and electrolytes need to be closely monitored. Circulatory support with colloid, sometimes in large volumes, and inotropic agents are titrated against the baby's clinical condition. Evidence of disseminated intravascular coagulation needs to be sought. Serious infection still has a significant morbidity and mortality in spite of the powerful antibiotics now available.

In the preterm infant there has been considerable interest recently in the use of immunoglobulin given by intravenous infusion to protect against infection. Infants of less than 32 weeks' gestation have a much lower serum IgG level at birth compared with term infants. A number of large multicentre studies have failed to consistently show a reduction in morbidity and mortality.

SOME SPECIFIC INFECTIONS

Urinary tract infection

Because of the non-specific way urinary tract infection presents in newborn infants it will only be identified if a satisfactory urine sample is taken and analysed whenever infection is suspected and no localized site identified. At this age, urine infections are often accompanied by septicaemia, particularly if there is urinary outflow obstruction, such as posterior urethral valves or other congenital abnormalities. Fortunately most, but not all, of these structural abnormalities can be identified at an antenatal ultrasound.

Unlike later life, when urinary tract infections are much commoner in females than males, this is not the case in the newborn. The commonest organisms are coliforms, *Streptococcus faecalis* and infrequently *Pseudomonas* or *Proteus*.

The main problem in indentifying urinary tract infections in neonates is in obtaining a satisfactory urine sample. Bag urines are often contaminated and their significance then becomes difficult to interpret. If antibiotics are going to be started a suprapubic aspiration will often be required to obtain a satisfactory sample. All infants with proven urinary tract infection should have a renal and urinary tract ultrasound to identify urinary tract outflow obstruction which may require urgent surgical treatment. A micturating cystogram will also be required at a later stage to identify vesico-ureteric reflux.

After treating the urine infection, infants are placed on a low dose prophylactic antibiotic until the results of the micturating cystogram are known.

Meningitis

Neonatal meningitis remains a serious condition with a high mortality and significant long-term morbidity among survivors. Initially there may be the non-specific features of ill health already described. In addition infants may develop seizures. A tense or bulging fontanelle and head retraction from neck stiffness are late signs of meningitis in the newborn. To diagnose meningitis CSF needs to be obtained by lumbar puncture. This should be performed whenever infants are thought to be infected without localizing features or if there is any clinical suspicion of meningitis. In the past the antibiotic of choice was chloramphenicol as it has good CSF penetration, but the drug may accumulate as it is metabolized by conjugation by glucuronide in the liver which is at a low level in the first few days of life. Overdosage may cause the grey baby syndrome with collapse and death unless the drug's blood level is closely monitored. For this reason chloramphenicol has largely been superseded by the third generation cephalosporins which also have good CSF penetration. Supportive measures and close monitoring are important in managing this condition.

Osteomyelitis

Osteomyelitis of the long bones or the maxilla may be a complication of infection by *Staphylococcus aureus* or occasionally by other organisms. In the newborn it is usually first noted when the infant stops moving the affected part and a swelling is found on careful inspection. Other non-specific signs of infection may or may not be present. The area over the affected bone is tender and may become red. An effusion or septic arthritis may develop in an adjacent joint. Full antibiotic treatment and orthopaedic advice is indicated.

Skin infection

The most common skin infection is the eruption of small pustules caused by *Staphylococcus aureus*. The umbilicus may be colonized by the organism from which the infection may spread. The infecting organism may also spread from the nose or skin of the parents, nurses or doctors or from other infected babies.

Bullous impetigo is caused by streptococcal or staphylococcal infection. Fluid-containing blisters are formed, the top of which is easily rubbed away to leave raw areas. The lesions may spread to other sites and it therefore needs to be treated vigorously with an appropriate systemic antibiotic.

Conjunctivitis

Conjunctivitis with a purulent discharge and swelling of the eyelids in the first few days of life needs to be taken seriously. Important causes are *Neisseria gonococcus* and *Chlamydia trachomatis*. An eye swab must be taken urgently for microscopy to identify gonococci on Gram stain as well as for culture and antibiotic sensitivity. *Chlamydia trachomatis* is identified rapidly using a monoclonal antibody test on the pus, which is placed on a special slide.

Gonococcal infection is treated with penicillin given parenterally and eye drops initially instilled hourly. For β-lactamase producing organisms a suitable cephalosporin, for example cefotaxime is used instead. Chlamydia infection is usually treated with oral erythromycin for 2 weeks and tetracycline topically to the eye. The mother and partner also need to be treated.

Sticky eyes are common in the first few days of life and generally settle on their own. The eyelids can be kept clean by wiping them with sterile water. When they are more troublesome a tropical antibiotic, for example neomycin, is often used.

HIV infection

Infants born to mothers who are positive for the human immunodeficiency virus have a positive antibody test at birth. About 15 per cent remain antibody positive at 18 months of age. Clinical manifestations of the disease usually develop after the neonatal period. Because of the risk of transmission of the virus breastfeeding is not recommended in the UK although in developing countries where adequate milk substitutes are not available breastfeeding may be preferable.

METABOLIC PROBLEMS

HYPOGLYCAEMIA

There is no agreed definition of hypoglycaemia in the newborn. In practice a level of 1.8–2 mmol/l has become widely used although most babies with levels below this remain asymptomatic. There is good evidence that prolonged, symptomatic hypoglycaemia can cause permanent neurological disability.

Hypoglycaemia is particularly likely to occur in

the first 24 hours of life in babies who are growth-retarded, but it can develop in any ill baby. In these infants the blood glucose is monitored 4–6 hourly or more frequently depending on previous blood glucose levels. All babies undergoing intensive care need to have their blood glucose monitored regularly. Infants of diabetic mothers and large-for-gestational age infants are also at increased risk of developing hypoglycaemia. The blood glucose should be measured in any newborn baby who develops a fever or temperature instability, refuses to feed, has apnoeic attacks, seizures, or develops abnormal neurological signs.

In the growth retarded infant there is depletion of the glycogen stores and the aim must be to replenish them. Asymptomatic hypoglycaemia has an excellent prognosis and symptoms can usually be prevented by early and frequent feeding. The blood glucose is usually monitored using a reagent strip at the bedside. If this suggests that the blood glucose is low a blood sample should be sent to the laboratory for confirmation as the reagent strips are not reliable at very low levels. However treatment should be initiated without delay and the plasma glucose level rechecked directly to ensure that the problem has resolved. If the baby is symptomatic, glucose is immediately given intravenously and the glucose level checked regularly to confirm that the hypoglycaemia has resolved. Intravenous glucose infusions need to be carefully monitored. If they stop suddenly because of extravasation into the tissues there is a risk of reactive hypoglycaemia.

Infants of diabetic mothers or large-for-gestational age infants have sufficient glycogen stores but hyperplasia of the islet cells in the pancreas secondary to poorly controlled maternal hyperglycaemia results in high blood insulin levels. Early milk feeding usually prevents hypoglycaemia. If the infant cannot be fed or becomes symptomatic, an intravenous infusion of 10–20 per cent dextrose is started immediately. Should this not correct the hypoglycaemia or if there is difficulty or delay in starting the infusion, intramuscular glucagon can be given. Glucocorticoids or intralipid are occasionally required for refractory hypoglycaemia.

HYPOCALCAEMIA

Newborn infants rarely develop symptoms unless the plasma calcium is lower than 1.7 mmol/l (7 mg/100 ml). Newborn infants are at risk of hypocalcaemia if they are preterm or severely ill during the first 48 hours of life or if they are infants of diabetic mothers. Hypocalcaemia can cause tetany, jitteriness and convulsions.

Hypocalcaemia can be prevented by giving intravenous calcium gluconate to ill infants not taking oral feeds. Infants with symptomatic hypocalcaemia are given calcium gluconate intravenously slowly under ECG control. The infusion site needs to be carefully observed as extravasation may cause severe tissue injury.

Inborn errors of metabolism

Many of these are not detected until after the neonatal period. Phenylketonuria and galactosaemia are rare but important disorders as early detection allows the introduction of appropriate dietary management which reduces long-term morbidity. They are detected by biochemical screening (the Guthrie test) performed on the 5th day of life on a blood sample. The blood is now also used to detect hypothyroidism which is much commoner than phenylketonuria and its prognosis is improved by prompt treatment. In some parts of the UK the blood samples are also used to screen for haemoglobinopathies and cystic fibrosis.

GASTROINTESTINAL PROBLEMS

VOMITING

Newborn babies often vomit. If the vomit is bile stained, contains blood or becomes persistent or is associated with other symptoms or signs, further investigation is required.

In the immediate newborn period babies sometimes vomit mucus which may be blood stained. It is often ascribed to swallowed blood causing gastric irritation. If the vomiting is marked, a gastric lavage is often performed to remove any blood from the stomach.

Important cases of vomiting include the following.

Intestinal obstruction

Vomiting which begins shortly after birth and is persistent may be from intestinal obstruction. It is usually bile-stained unless the obstruction is above the level of the ampulla of Vater where the pancreatic duct enters the intestinal tract. Bile-stained vomit should be assumed to be from intestinal obstruction until proven otherwise. In duodenal atresia the bile-stained vomiting is accompanied by delay in passage of meconium but abdominal distension is absent. Malrotation with volvulus is an important cause of bile-stained vomiting as a delay in diagnosis may result in extensive ischaemic damage to the bowel. When the obstructed bowel is more distal there is abdominal distension and peristalsis. Meconium may initially be passed. A plain abdominal X-ray shows dilated proximal gut with characteristic fluid levels with a paucity of gas distal to the obstruction. The stomach is kept empty with a nasogastric feeding tube which is aspirated and fluids are given intravenously. Large volumes of colloid may be required to compensate for pooling of fluid within the bowel. Referral needs to be made to a paediatric surgeon.

In necrotizing enterocolitis, which mainly affects preterm infants, vomiting is accompanied by abdominal distension and the passage of blood per rectum.

Pyloric stenosis is sometimes seen in babies who are still in the neonatal unit at 3–6 weeks of age.

Infection

Vomiting may be seen in babies with infection of the urinary tract, septicaemia, meningitis and of other sites. It is important to exclude urinary tract infection in all babies with persistent vomiting and also to consider other sites of infection (Table 9.8).

Uncommon causes of vomiting

Vomiting may be a symptom of raised intracranial pressure or cerebral irritation. It is also a feature of ileus associated with severe illness, for example RDS. It may also be seen in rare metabolic disorders, for example congenital adrenal hyperplasia.

DIARRHOEA

Diarrhoea is sometimes diagnosed when it is simply the normal variation in bowel habit of newborn infants. In breastfed babies the stools may be frequent and greenish in colour and may contain a moderate amount of mucus. No ill effects result and the stools subsequently become less frequent and normal in consistency.

Gastroenteritis

Diarrhoea caused by the dysentery group of organisms is uncommon in the neonatal period. Occasionally infections with pathogenic *E. coli* and other organisms of relatively low pathogenicity may occur. In newborn nurseries epidemics of diarrhoea may be caused by viral infection.

If an infant has diarrhoea, a stool specimen is sent for bacterial and viral culture. Infants with gastroenteritis must be isolated from other babies and nursed with full barrier precautions. Milk feeding is stopped until the diarrhoea improves. Dehydration is corrected with intravenous fluid if necessary.

HAEMATOLOGICAL PROBLEMS

RHESUS AND ABO INCOMPATIBILITY

In rhesus incompatibility the lifespan of the red cells of the fetus or newborn infant is shortened by the action of specific antibodies present on the red cells of the infant but not of the mother. These IgG antibodies are produced by the mother, and being small molecules, cross the placenta. Transfer is minimal before the 12th week of gestation, rising slowly to the 24th week and then increases rapidly until term. Antibodies causing haemolytic disease may be produced against all red cell antigens but most commonly it is against rhesus, A or B or Kell antigens. Because effective prophylaxis against haemolytic disease caused by the rhesus antigen is available the other antigens have become relatively more prominent.

The gene that makes most people rhesus positive is D. Cases of maternal immunization to one of the other antigens of the rhesus group occur only occasionally and are weak antigens.

Feto-maternal transfusion

Less than 0.1 ml of fetal blood is sufficient to cause the initial sensitization to rhesus D antigen but the maternal response varies greatly. It is uncommon for significant feto-maternal transfusion to occur during pregnancy, but it is relatively common during labour especially during the third stage. It is therefore uncommon for antibodies to be produced during a first pregnancy, but if the mother is sensitized she will show a rapid increase in antibody titre during any subsequent pregnancy if the fetus is rhesus positive. The presence of fetal cells in the maternal circulation can be demonstrated by a Kleihauer test where a film of maternal blood is mixed with a weak acid solution before staining. Fetal red cells contain haemoglobin which is more resistant to denaturation with acid than adult haemoglobin and this allows fetal red cells to be recognized and counted.

It might be expected that immunization of a mother of blood group O to an A or B factor inherited by the fetus from the father would be common. Naturally occurring antibodies to A or B antigens are IgM molecular size and cannot cross the placenta but smaller IgG antibodies to these antigens are acquired in some group O mothers. These can cross the placenta to destroy fetal red cells of group A or B. Thus haemolytic disease from ABO incompatibility may occur in a first pregnancy. Fortunately the disease is much less severe than rhesus incompatibility and relatively few babies will need treatment.

Pathophysiology

There is haemolysis of fetal red cells which are destroyed by the reticulo-endothelial system. In the most severe cases the fetus becomes severely anaemic and develops generalized oedema (hydrops fetalis) with pleural, pericardial and peritoneal effusions. Untreated, this may lead to intra-uterine death, usually after 28 weeks' gestation.

In less severe cases the infant is born alive with the clinical features of haemolytic disease. These are pallor, jaundice and enlargement of the liver and spleen. At birth the haemoglobin level in the cord blood is used as a guide to the severity of the disease. It is severe if the haemoglobin is less than 10 g/dl, of moderate severity if 11–13 g/dl and mild if greater than 14 g/dl. There may be an extremely rapid increase in the level of jaundice after birth. Before birth the excess bilirubin from haemolysis is excreted into the mother's circulation by the placenta, hence the cord bilirubin level at birth is unreliable as an index of severity.

Antenatal management

Every pregnant woman should have her blood group determined and red cell antibodies screened at her initial antenatal visit. If antibody is present its identity and concentration is determined. In practice, tests for rhesus, Kell and Duffy antigens are performed.

The concentration of maternal rhesus antibodies serves as a guide to the severity of the disease in the fetus and whether further investigation is required.

In the past the disease severity was monitored by measuring the bilirubin in amniotic fluid obtained by amniocentesis. Fetuses which were judged to be only mildly affected were left to proceed to term. Those of moderate severity were delivered before term and a few which were severely affected and at risk of early intrauterine death were treated by intrauterine blood transfusion given into the peritoneal cavity. Labour was induced as soon as the infant was considered to have a good chance of survival. At birth exchange transfusions, often repeated several times, were required to maintain the infant's haematocrit and avoid very high levels of bilirubin. Morbidity and mortality were high.

The antenatal management of rhesus incompatibility has been transformed by the development of cordocentesis and the ability to transfuse blood directly into the fetal vascular compartment. The disease severity can be measured directly from the haematocrit on the fetal blood sample. If indicated the fetus is given a blood transfusion at the same time. The development of oedema in the fetus is also monitored with serial antenatal ultrasound scans. Most affected infants are now born at or near term with a normal haematocrit. Relatively few even require an exchange transfusion.

Postnatal management

In infants affected by rhesus haemolytic disease the baby is anaemic at birth. The direct antiglobulin test (Coombs test) will demonstrate antibody on the red cells. The aim of treatment is to correct anaemia, thereby reducing the risk of death from heart failure and to prevent kernicterus from severe jaundice. The affected infant is placed under intensive phototherapy immediately after birth. The haemoglobin and bilirubin levels are monitored initially every few hours. If the baby is anaemic or the jaundice rapidly rises, an exchange transfusion is performed. In this procedure aliquots of blood are removed from the infant and replaced with donor blood. The procedure is continued until twice the infant's blood volume has been exchanged. Meticulous monitoring of the infant's clinical condition and electrolytes is required during the procedure. Traditionally, the exchange has been performed via a catheter placed in the umbilical vein as this was the simplest technique to cannulate a large blood vessel. When possible it is preferable to avoid cannulating the umbilical vein and to perform the exchange transfusion by withdrawing blood from an arterial line while infusing blood via a peripheral cannula.

Prevention of rhesus haemolytic disease

The formation of anti-D antibodies can usually be prevented by giving a rhesus negative mother anti-D immunoglobulin whenever feto-maternal transfusion is likely to occur:

- after delivery of a rhesus positive infant, whether live of stillborn
- after any abortion or termination
- after any external version
- after any amniocentesis or chorionic villus sampling.

There continues to be debate about the need to give anti-D to all rhesus negative women during pregnancy to prevent the development of rhesus antibodies to the small feto-maternal haemorrhages which occur throughout pregnancy.

HAEMORRHAGIC DISEASE OF THE NEWBORN

Newborn infants are deficient in vitamin K when compared with older children and adults. Bleeding without warning can occur in a small number of infants. Classicial haemorrhagic disease occurs during the 1st week of life. In most the bleeding is transient and slight but in a few it is life threatening including intracranial haemorrhage. Classical haemorrhagic disease can be prevented by giving vitamin K, 1 mg intramuscularly or 1–2 mg orally immediately after birth. Feeding with infant formula, which contains more vitamin K than breast milk, protects against classical haemorrhagic disease.

Late haemorrhagic disease is rare. It occurs between the 1st and 8th week of life. A significant proportion suffer from intracranial haemorrhage, half of whom are permanently disabled or die. Given intramuscularly on the day of birth, 1 mg of vitamin K will prevent this, but a single dose of oral vitamin K will not reliably do so. Intracranial haemorrhage occurs in infants who are wholly breastfed and who have not received prophylactic vitamin K; it seldom if ever occurs in formula-fed infants.

There is no clear consensus on the safest form of prophylaxis for newborn babies. As stated, vitamin K given intramuscularly will prevent both classical and late haemorrhagic disease. This is also advisable when mothers are taking anticonvulsants which antagonize vitamin K (particularly phenytoin). However, a recent study suggested that there might be a risk that vitamin K given intramuscularly is associated with the development of cancer in childhood, though this has been refuted by further studies. Several oral doses are required to provide complete protection.

POLYCYTHAEMIA

A venous haematocrit greater than 0.65 in newborn infants is often defined as polycythaemia. Blood viscosity rises exponentially at such high levels. Polycythaemia is caused by increased intrauterine erythropoeisis in growth retardation, maternal smoking, maternal diabetes and trisomy 21. It can also be secondary to red cell transfusion from placental transfusion from delayed cord clamping or materno-fetal or twin to twin transfusion. Hyperviscosity is associated with decreased flow both in the microcirculation and in the large veins and is associated with cerebral, renal or inferior vena caval thrombosis.

The infants often look plethoric but are usually asymptomatic. However clinical features ascribed to polycythemia include irritability, lethargy, respiratory distress, heart failure, convulsions, necrotizing enterocolitis and renal vein thrombosis. There is an increased risk of hypoglycaemia.

Several studies have shown an increased incidence of long-term neurological morbidity. The haematocrit at which treatment should be initiated has not been agreed. Some authorities recommend treatment if the venous haematocrit is greater than 0.65, while others recommend waiting until the haematocrit is 0.70 unless the baby becomes symptomatic. Treatment is with a dilutional exchange transfusion with fresh frozen plasma or albumin. It is best done using a peripheral arterial line to withdraw blood and intravenous fluids are given via a peripheral infusion.

CONGENITAL MALFORMATIONS

Congenital malformations are structural deformations present at birth. About 1.5 per cent of babies have a major defect so identified and many more, especially congenital heart disease, only become evident when the baby is older. The cause of many congenital malformations is unknown. Some congenital malformations are genetic whilst others may be due to the adverse effects on the fetus of drugs, radiation, or metabolic disturbances in the mother. Structural malformations may also result from a lack of space for the fetus to grow, for example oligohydramnios. However in most the cause is not known.

An increasing number of malformations are now recognized prenatally, from routine ultrasound scanning, or from prenatal testing of the fetus known to be at increased risk of specific disorders.

The parents will want full information about any congenital malformation.

CHROMOSOME DISORDERS

Down's syndrome, trisomy 21

This is by far the commonest chromosomal disorder, occurring in 1.4 per 1000 live births. It is due to 47 chromosomes, with three 21 chromosomes. The majority (95 per cent) are the result of nondysjunction during mitosis. The risk increases with maternal age from 1 in 350 at 35 years to 1 in 100 at 40 years and 1 in 30 at 45 years. Another 3–4 per cent are due to translocation and are not related to maternal age. If translocation affects the mother there is a 20 per cent risk of an affected child; if it affects the father the risk is about 6 per cent. One to two per cent are mosaic, with a normal cell line as well as a cell line with an extra 21 chromosome.

There is an intense interest in the antenatal diagnosis of Down's syndrome. Chromosome analysis can be performed on samples obtained by chorionic villus sampling or amniocentesis in women at increased risk because of maternal age or where either parent has a chromosomal translocation. This does not significantly reduce the number of affected infants born. Much attention is being focused on trying to identify structural markers by antenatal ultrasound. The flat shape of the occiput and the nuchal fat pad may allow early ultrasound identification of some infants. In addition biochemical screening of maternal blood samples has been introduced in some areas. The predictive value of this technique is improved by measuring two or more markers (low alpha-fetoprotein level combined with hCG and oestrogen levels). The efficacy of these methods has yet to be determined.

The characteristics of Down's syndrome include:

- eyes which slant upwards and outwards with epicanthic folds
- a round face, a small mouth with protruding tongue
- simple ears
- a flat occiput
- palpable third fontanelle
- hands which are short with incurving of the

little finger and single transverse palmar crease
- simian cleft between the great toe and next toe
- profound hypotonia
- increased incidence of duodenal atresia and congenital heart disease.

The diagnosis can sometimes be very difficult to make clinically. Whenever it is suspected a senior member of the paediatric staff should see the baby and parents.

The diagnosis is confirmed by chromosome analysis on a blood sample. If the chromosomes show non-dysjunction it is not necessary to check the parents' chromosomes. This is only required for a translocation. The parents will want to know not only about the medical problems associated with Down syndrome but also the long-term neuro-developmental outlook. Detailed written information produced by other parents and involved professionals is available. Some parents also wish to meet other affected families. Before discharge a plan needs to be made about which health professionals are going to be involved and the role of each of them clearly identified.

Trisomy 13–15 (Patau syndrome)

This condition occurs in about 1 in 12 000 births. The infants have multiple congenital abnormalities. These include growth retardation, microcephaly and abnormalities of the eye – microphthalmia, coloboma or cataracts. These abnormalities occur in about 80 per cent of cases. Other common abnormalities are localized scalp defects in the parietal–occipital area and cleft lip and plate, as well as congenital heart disease, genital anomalies and polydactyly. Most affected infants die in the neonatal period.

Trisomy 18 (Edward syndrome)

This occurs in approximately 1 in 8000 births. The infants are growth retarded. In addition they also have multiple congenital abnormalities which include abnormal facies with a narrow nasal bridge, short palpebral fissures and low set ears. There are characteristic abnormalities of the hands with overlapping, flexed fingers and rocker bottom feet. Most have congenital heart disease. Most die within the first few weeks of life.

MALFORMATIONS OF THE GASTROINTESTINAL TRACT

Cleft lip and palate

This occurs in about 1 per 1000 births. A careful check needs to be made for other abnormalities. Surgical repair of the lip is increasingly performed within the 1st week of life but this depends on the individual surgeon. Some feel that better results are obtained if surgery is delayed. The palate is usually repaired at several months of age.

A cleft palate may make feeding more difficult. Many affected infants can still be breastfed successfully. Milk may enter the nose and cause choking. Special teats and feeding devices are available. Orthodontic and speech therapy advice is required. The development of secretory otitis media, which is relatively common, needs to be monitored. There is a particularly active voluntary group of parents who have had affected children (Cleft Lip And Palate Association CLAPA) who often provide valuable support and advice for parents with an affected newborn infant.

Pierre–Robin syndrome

The Pierre–Robin syndrome is an association of micrognathia, glossoptosis and a midline cleft of the soft palate. As the tongue falls back there is obstruction to the upper airways.

There may be upper airways obstruction, cyanotic episodes and difficulty with feeding. Many of the infants fail to thrive during the first few months. If there is upper airways obstruction the infant may need to lie prone allowing the tongue and small mandible to fall forward. Persistent upper airways obstruction can be treated using a nasopharangeal tube which can be connected to a continuous positive airway pressure circuit. In the long term the mandible grows and these problems resolve. The cleft palate can then be repaired.

Oesophageal atresia and tracheo-oesophageal fistula

Oesophageal atresia with or without tracheo-oesophageal fistula occurs in approximately 1 in 3500 births. It usually occurs just above the level of the bifurcation of the trachea (in 85 per cent of cases) and is most commonly associated with a tracheo-oesophageal fistula in which the lower oesophagus joins the trachea just above the carina. Less than 10 per cent of cases have oesophageal atresia without a fistula. A tracheo-oesophageal fistula on its own also occurs but is very rare.

Hydramnios during pregnancy is a common feature and oesophageal atresia should be considered in all mothers with this complication. Presentation is with persistent salivation and drooling from the mouth after birth associated with choking and cyanotic episodes. If the diagnosis is not made at this stage and the infant is fed this will result in coughing, choking and vomiting of the milk. Some of the milk may be aspirated. Half of the babies have other congenital malformations, particularly of the skeletal, heart and genitourinary systems. Infants should be thoroughly examined both physically and radiologically before surgery is undertaken.

Any suspicion that an infant is unable to swallow on hydramnios during pregnancy should lead to prompt investigation by attempting to pass a moderately stiff feeding tube through the oesophagus into the stomach. A soft catheter may coil up in the segment above the atresia and delude the observer into believing that it has passed into the stomach. The aspiration of acid secretion and the emptying of the stomach of air provide proof that the oesophagus is patent. The feeding tube should be radio-opaque so that its position can be confirmed radiologically. Directly the diagnosis is established, the infant is transferred to a paediatric surgical unit with frequent or continuous suction of saliva from the mouth during the journey. Surgical treatment now offers a good chance of survival and satisfactory function in the longer term, provided multiple anomalies are absent.

CONGENITAL ATRESIA OF THE GUT

Duodenal atresia or stenosis

This condition occurs in about 1 in 6000 babies. It may be detected on ultrasound prenatally when a double bubble shadow is seen in the stomach and duodenum with empty gut beyond. In the immediate neonatal period it presents with vomiting, which is bile stained, and failure to pass meconium. A radiograph of the abdomen in the erect position shows a double bubble of air in the stomach and proximal duodenum but no gas beyond this level.

The lesion is associated with Down syndrome in up to one-third cases and as many as half the children with this condition have other congenital malformations. Treatment is surgical.

Jejunal and ileal atresia

This occurs in about 1 in 5000 infants. Single or multiple atretic segments of the small bowel result in vomiting, abdominal distension and failure to pass meconium after birth. An erect or lateral decubitus abdominal film will show gaseous distension of the bowel with fluid levels present above the site of the lesion. Treatment is surgical.

Meconium ileus

Almost all infants with meconium ileus have cystic fibrosis in which the pancreatic secretion into the bowel is abnormal. During intrauterine life the meconium is abnormally digested and becomes packed into the lower ileum forming a mass of putty-like consistency and appearance. This may be seen as abnormal bowel shadowing on ultrasound as early as 18–20 weeks' gestation. Immediately after birth the infant develops vomiting and abdominal distension and fails to pass meconium. X-ray examination shows gaseous distension and fluid levels in the small bowel with lack of gas in the distal bowel and a mottled opacity in the areas of the inspissated meconium. Surgery has largely been replaced by gastrograffin enemas in recent years. A diagnosis of cystic fibrosis needs to be sought using gene probes or a sweat test though it can be difficult to obtain sufficient sweat at this age.

Hirschsprung's disease (congenital megacolon)

This condition occurs in 1 in 5000 births with a male to female ratio of 4:1. The basic defect is an aganglionic segment of colon. This may affect a small or long segment of bowel.

Presentation is usually shortly after birth or in early infancy. There is constipation of varying degrees and lower intestinal obstruction with difficulty in passing meconium, abdominal distension and vomiting. Occasionally the condition may go unrecognized into early childhood. Rectal biopsy confirms the diagnosis. Treatment is to resect the aganglionic segment.

Imperforate anus

This occurs in about 1 in 5000 births. The diagnosis is made at birth. In most there is a rectal pouch and a fistula connecting the rectum with the perineum or the urogenital tract. Treatment is surgical.

Exomphalos and gastroschisis

In exomphalos there is protrusion of abdominal contents through the umbilical ring covered with a transparent sac. It may vary in size from a loop of bowel to containing other abdominal organs. In about 40 per cent there are other major anomalies.

In gastroschisis the bowel protrudes through a defect in the anterior abdominal wall. The umbilical cord itself is not involved. The exposed bowel is wrapped in clingfilm to avoid fluid loss. A large nasogastric tube is passed and aspirated on continuous suction. An intravenous infusion is established to prevent hypoglycaemia and maintain the intravascular fluid volume. Replacement of the intestine into the peritoneal cavity may be possible with primary closure of the abdomen; otherwise the intestine can be enclosed in a silastic sac sutured to the edges of the abdominal wall and the contents gradually returned into the abdomen.

CENTRAL NERVOUS SYSTEM

Neural tube defects

Neural tube defects result from failure of closure of the neural tube during early organogenesis at 21–28 days of gestation.

The commonest abnormality is failure of the neural plate to close distally causing spina bifida. As formation of the posterior arch of the vertebrae follows formation of the neural tube, it is accompanied by anomalies of the spine. Spina bifida is usually localized to the lumbosacral region. In the most severe form, myelomeningocoele, the spinal cord lies open. Neural tissue covered by meninges is visible at the surface. The posterior arch of one of more vertebrae is absent. In some cases of spina bifida, because the spinal cord is adherent to the skin when the trunk grows, the brain is pulled down into the foramen magnum causing obstruction and hydrocephalus, the Arnold–Chiari malformation. In a meningocoele only the meninges protrude through the body defect in contrast to the myelomeningocoele where the cord itself also herniates. In spina bifida occulta the skin and meninges are intact but the posterior part of the vertebra is missing and there may be a small tuft of overlying hair. In an encephalocoele there is herniation of brain tissue through a defect in the skull. It is usually occipital.

When failure of closure of the neural tube is at the cranial end, anencephaly occurs. This is associated with failure of ossification of the skull bones as they differentiate only under the influence of a normal neural tube. Pregnancy may progress to term and a normal weight infant is delivered. If the brain stem is intact death may take 1–3 days. Recently, there has been much discussion about the use of such infants for organ donation but they do not satisfy the criteria for brain death as the brain stem is still functional.

There has been considerable progress in the epidemiology, antenatal diagnosis and more recently prevention of neural tube defects. It has been known for a long time that there are large variations in the incidence of central nervous system (CNS) defects in different parts of the world and within the British Isles, where the incidence is highest in Ireland followed by Scotland, Wales and lowest in South East England. It is known that neural tube defects are less frequent in higher social classes. There is a risk of recurrence to siblings following a fetus or baby with neural tube defect of 5 per cent rising to 10 per cent if the mother has had two or more affected fetuses.

There has been a dramatic reduction in the number of babies born with neural tube defects from 3.2 per 1000 in England and Wales in 1957 to 0.4 per 1000 in 1987. This is from a combination of general decline in its incidence together with antenatal diagnosis and termination of pregnancy. Routine antenatal ultrasound examination allows the identification of the vast majority of fetuses with anencephaly and of many with spina bifida where a defect in the spine, a 'lemon shape' of the skull or accompanying hydrocephalus may be detected. In addition, in many districts, screening for an elevated alpha-fetoprotein in maternal blood is performed. If raised or where there is a strong family history, the alpha-fetoprotein in amniotic fluid is then checked.

Studies of mothers who have had one or more affected fetuses have shown a dramatic reduction in the risk of recurrence when given additional folate periconceptually during subsequent pregnancies. This has been extended to apply to all pregnant women, in whom periconceptual folate supplementation is now recommended.

Myelomeningocoele

About 70 per cent of spina bifida is myelomeningocoele. The potential short- and long-term problems for children with this lesion are:

- infection of the lesion and ventriculitis
- paralysis of the legs accompanied by loss of sensation
- hydrocephalus from the associated Arnold–Chiari malformation
- congenital dislocation of the hip, kyphosis or scoliosis of the spine
- absence of bladder and rectal tone which may be accompanied by malformations of the urinary tract, recurrent urinary tract infections and hypertension
- severe learning difficulties
- psychosocial problems in adolescence and adult life.

The management of these affected problems is complex and involves multidisciplinary teams. There are also difficult ethical dilemmas for both parents and professionals in deciding how best to treat and manage these major problems.

Hydrocephalus

This may be diagnosed antenatally on ultrasound scanning. With increasing intracranial pressure the head circumference increases and the sutures separate. Thereafter the fontanelle bulges and becomes tense. Postnatally it can readily be monitored by intracranial ultrasound. When surgery is indicated, ventriculo-peritoneal shunting is the treatment of choice.

CONGENITAL HEART DISEASE

The overall incidence of congenital heart disease is about 8 per 1000 live births. About 90 per cent comprise eight lesions – ventricular septal defect (VSD), atrial septal defect (ASD), pulmonary stenosis, aortic stenosis, coarctation of the aorta, patent ductus arteriosus, Tetralogy of Fallot and transposition of the great arteries.

Newborn infants with congenital heart disease present with cyanosis, heart failure or murmurs detected on routine examination.

Cyanosis

Cyanosis is best detected on examining the tongue. If the arterial oxygen tension in the baby's blood is measured it will be found to be low and does not rise above 15 kPa even after the baby is placed in 100 per cent oxygen for 10 minutes. Cyanotic congenital heart disease can then be confidently diagnosed in the absence of lung disease and persistent fetal circulation (persistent pulmonary hypertension of the newborn).

Cyanosis may be caused by abnormal mixing of systemic and pulmonary venous blood as with transposition of the great arteries. It is also caused by reduced pulmonary blood flow as with Tetralogy of Fallot, pulmonary atresia and tricuspid atresia. These infants with duct dependent pulmonary circulation become severely cyanosed when the duct closes. Maintenance of ductal patency is the key to early survival in these children.

Infants who are cyanosed from congenital heart disease should be given prostaglandin to maintain ductal patency. This reverses its natural tendency to constrict and so increases pulmonary blood flow. This allows time for the child to be transferred to a paediatric cardiac centre, a precise diagnosis established and surgery or other management initiated.

Heart failure

Infants in heart failure present with respiratory distress. An important sign of heart failure in newborn infants is an enlarged liver from raised venous pressure. In the first few days of life heart failure is usually from obstruction to the left side of the heart, particularly coarctation and interruption of the aorta and severe aortic valve stenosis. These children also depend on a patent ductus to maintain their circulation. Closure of the duct rapidly leads to severe acidosis, collapse and death.

Three heart defects which often present in the immediate newborn period are transposition of the great arteries, coarctation of the aorta and hypoplastic left heart syndrome.

Transposition of the great arteries

In transposition of the great arteries there are two parallel circulations, with the systemic venous return passing from the right atrium into the right ventricle then into the aorta and a separate circulation of pulmonary venous blood returning to the left atrium via the left ventricle and back into the pulmonary arteries. Mixing of blood between the two circulations must occur for this condition to be compatible with life. This is through associated abnormalities such as ventricular septal or atrial septal defects. These children usually present with severe cyanosis in the first day or two of life. An arterial oxygen tension of only 1–3 kPa is not unusual. A prostaglandin infusion is established to maintain the ductus arteriosus. If the child remains cyanosed at echocardiography a balloon atrial septostomy is performed. A catheter with an expandable balloon at its tip is passed through the umbilical femoral vein and through the right atrium and foramen ovale. The balloon is inflated within the left atrium and pulled through the atrial septum. This tears the atrial septum and allows mixing of the systemic and pulmonary venous blood within the atrium. Either a corrective operation or now more often an arterial switch procedure, which provides anatomic correction, can then be performed in the first few weeks of life.

Coarctation of the aorta

Coarctation usually affects the aorta distal to the left subclavian artery adjacent to the insertion of the ductus arteriosus. The presentation may be with heart failure or circulatory collapse or the infant may be asymptomatic apart from recognition of weak or absent femoral pulses detected on routine examination. Most will need surgery, though in some infants it may be possible to perform balloon dilatation without the need for surgery. Infants with interruption of the aortic arch present in a similar way. This is, however, a more serious condition as there is no connection between the aorta proximal and distal to the ductus arteriosus.

Hypoplastic left heart syndrome

In this condition there is underdevelopment of the entire left side of the heart. In addition the ascending aorta is small and almost always associated with coarctation of the aorta or interruption of the aortic arch. Ductal closure leads to acidosis and cardiovascular collapse. The diagnosis needs to be established by urgent echocardiography. The condition is generally inoperable. Recently neonatal

heart transplantation and multiple surgical operations have been performed. The use of these heroic procedures has been the subject of much debate.

UROGENITAL PROBLEMS

Bilateral renal agenesis (Potter's syndrome)

This occurs in about 1 in 3000 births. There is renal agenesis which is usually accompanied by characteristic facies with widely set eyes (ocular hypertelorism), low set ears, a flattened nose and micrognathia. Oligohydramnios or total absence of fluid is the rule. It is accompanied by lung hypoplasia and marked talipes equinovarus. Many are stillborn and the remainder die within the first 24 hours of life. Death is due to pulmonary hypoplasia.

Urinary tract abnormalities

Many of the important structural malformations of the urinary tract are now identified on antenatal ultrasound. They are caused by failure of normal embryogenesis. Union of the ureteric bud which forms the ureter, pelvis, calyces and collecting ducts with the nephrogenic mesenchyme is essential for nephrons to be formed. Failure of this union produces a non-functioning kidney filled with large fluid cysts – a multicystic kidney. Dysplastic kidneys are small and often contain cysts.

Abnormal caudal migration may result in a pelvic kidney or horseshoe kidney. The abnormal position may predispose to obstruction to urinary drainage. Premature division of the ureteric bud gives rise to a duplex system which often has abnormal drainage. Failure of fusion of the midline structures below the umbilicus results in exposed bladder mucosa – bladder extrophy. Absent abdominal musculature is associated with dilated ureters and a large bladder with a wrinkled abdominal skin – the prune belly syndrome.

Hydronephrosis is not infrequently identified on antenatal ultrasound. Obstruction to urine flow may occur at the pelvi-ureteric junction or at the vesico-ureteric junction. If obstruction occurs at the bladder neck the hydronephrosis will be bilateral. In a boy the most likely cause is an obstruction in the posterior urethra due to mucosal folds known as posterior urethral valves. The bladder wall becomes thickened and dilated. It may also be due to a neuropathic bladder as from spina bifida.

The perinatal management of these conditions is controversial. In the male fetus with posterior urethral valves which develops progressive bilateral hydronephrosis, intrauterine drainage of the bladder has been performed or the fetus has been delivered early. The place of these interventions is yet to be determined. After birth immediate surgical treatment is indicated. The management of urinary tract obstruction from other causes is less clear as their natural history is not yet known. Urinary tract anomalies predispose infants to urinary tract infection, which may result in renal damage if there is vesico-ureteric reflux or urinary outflow obstruction. Infants with significant urinary tract anomalies diagnosed antenatally are therefore placed on prophylactic antibiotics from birth while further investigations are performed. It is important that all such children are referred to paediatricians for investigation and follow up in order to identify significant outflow obstruction requiring surgery and for the prevention and early treatment of urinary tract infections. These infants usually require serial ultrasound scans of their urinary tract and radionucleide scans and micturating cystography to identify vesico-ureteric reflux and outflow obstruction. Bilateral hydronephrosis in a male requires urgent investigation to identify posterior urethral valves.

Hypospadias

In hypospadias the urethral meatus is ventral and proximal to the normal orifice. The most common site is at the coronal sulcus but the more serious grades have orifices along the shaft of the penis, scrotum or perineum. The penis may be curved from tethering by chordae. Treatment is surgical but may not be required for glandular hypospadias in the absence of chordae. Circumcision should not be performed in affected infants, as skin may be required for plastic repair later.

Undescended testes

The testes have not descended into the scrotum in 3 per cent of full-term infants and a much higher percentage of preterm infants. Most will descend during the first few months of life and by 1 year about 1 per cent of male infants have undescended testes. Most are unilateral. As few undescended testes (in contrast to retractile ones) descend spon-

taneously after 1 year of age, elective orchidopexy is generally performed in the 2nd or 3rd year of life in order to try to optimize future fertility.

Inguinal hernia

This is relatively uncommon in term infants but not uncommon in preterm infants particularly those with bronchopulmonary dysplasia. The treatment is surgical and should be performed promptly. In preterm infants with ongoing lung disease surgery may be delayed to enable the infant to tolerate the anaesthetic better. Urgent surgery is required if the hernia becomes acutely tender and cannot be reduced.

Circumcision

Circumcision in the newborn period is performed for religious or social reasons. There is some evidence that urinary tract infections are less common in circumcised males and it may be recommended in those with troublesome urinary tract infections. The foreskin should not be forcibly retracted during infancy.

LOWER LIMB MALFORMATIONS

Congenital dislocation of the hip

Congenital dislocation of the hip results from abnormal development of the acetabulum, the femoral head or the surrounding capsular tissues. It is more common if there is a family history, if the fetus has presented in breech position, and in female infants. Dislocated or dislocatable hips are identified in 1–2 per 100 infants. It is important that all infants have their hips checked for congenital dislocation during the first 24 hours of life. Abduction of the hip at birth may be restricted. More often the hip is dislocatable when the head of the femur can move out of and then be relocated back into the head of the acetabulum. It can be confirmed on ultrasound. At several months of age X-ray examination will allow identification of the position of the femoral head.

Treatment is to stabilize the hip in abduction and flexion. A harness is often used to achieve this. Many hips that are dislocatable at birth will become stable spontaneously. Some children will require surgery.

The screening test to detect congenital dislocation of the hip requires training and experience. It is important that the infant is not crying during the procedure. During the hip examination ligamentous clicks without movement of the femoral head into or out of the acetabulum may be detected but is not of long-term significance.

Talipes equinovarus (club foot)

This condition occurs in about one per 1000 births.

It is twice as common in males as females. It may develop *in utero* if there is lack of space, such as in oligohydramnios, or lack of fetal movement of the legs, for example from spina bifida. It may also be familial. Treatment is required if the foot cannot be moved into the plantigrade position. If this is not possible the baby should be promptly referred to an orthopaedic surgeon. Treatment may range from passive exercises to strapping or splinting. Surgery may subsequently be required.

FURTHER READING

Harvey D., Cooke R. and Levitt G. (1995) *The Baby Under 1000 grammes*. Wright, London.

Roberton N.R.C. (1992) *Textbook of Neonatology*. Churchill Livingstone, Edinburgh.

Royal College of Obstetricians and Gynaecologists. *Resuscitation of the Newborn* (1994) *Basic and Advanced Resuscitation*. London.

10

MEDICO-LEGAL PROBLEMS IN OBSTETRICS

The rapid increase in litigation against British doctors at the present time is a matter for concern. The combined specialities of obstetrics and gynaecology are among the foremost targets of such claims. Not only does this mean that large sums of money have to be paid in compensation but it also means that there may have to be changes in the style of practice. Defensive medicine, which results from these changes, implies unnecessary and expensive investigations and a hasty recourse to caesarean section.

So far as the financial aspects are concerned obstetric claims and awards now lead the way in the sheer magnitude of the sums of money involved. Settlements for the brain damaged infant, for example, are at present approaching £2–3 million. Until the 1st of January 1990 these sums of money had to be paid by the medical profession itself through its defence societies, which accounted for the high level of subscriptions to these societies. In the USA the individual doctor has personal insurance and any large claim can be recovered from the fees charged to patients. This is not possible in the UK in the context of a National Health Service but from the beginning of 1990 NHS claims arising from hospital work have been paid from Government funds, thus removing the threat of a huge decrease in the number of young doctors entering the speciality of obstetrics and gynaecology. Doctors in private practice and general practitioners need to subscribe to a defence system as in the past. Many doctors support the no-fault system used in New Zealand. By this method compensation is decided by a committee, and is appropriate to the disability or discomfiture incurred; doctors are not named. It is not a scheme that is favoured in the UK.

LITIGATION IN OBSTETRICS

Obstetrics is unusual in that the obstetrician is dealing with two people – the mother and the developing baby. This makes the task doubly demanding particularly in the present climate of a consumer orientated service. Parents have, quite rightly, a high expectation for the best outcome of pregnancy – a healthy, normal baby and a mother unscathed physically by pregnancy and delivery and fulfilled emotionally by the experience.

Advances in obstetrics have been considerable over the past 10 or 15 years and, as a result perinatal and maternal mortalities and morbidities have diminished. These results have been associated with techniques such as ultrasound scanning, cardiotocography and fetal blood sampling. Many mothers find these techniques unacceptably invasive in the sense that they feel that they interfere with the normal process of labour, the privilege of the mother and father to enjoy the experience of childbirth and their freedom of choice in planning the method of delivery. If obstetricians are not aware of these attitudes then hostility can arise and from this it is a short step to complaint and, of course, litigation.

We can therefore define the most important broad principles to avoid litigation and to maintain a good relationship between parents and obstetrician. Some of these principles apply equally to gynaecology, and indeed to all specialties. They are of equal importance.

Communication

Time and again failure of communication between doctor and patient is the root cause of litigation. If more time had been spent with the patient, not only giving explanations but ensuring that they were understood, and in answering questions, many a claim would never arise.

Time is often short in obstetrics, not only because of the demands which the specialty makes on frequently inadequate numbers of staff, but also because action – on behalf of a distressed baby for example – must be taken swiftly. But as doctors and midwives develop their skills in talking to women, speed need not be sacrificed in achieving understanding and cooperation.

Recording

Closely allied to communication is the need to make clear and concise notes written and signed at the time of, or just after, an event and containing not only the course of action decided upon but also what was said to the mother and her partner. Irritating and time consuming though these records may be they can ultimately avoid or modify potentially serious litigation.

The consent form should be signed whenever possible in obstetrics, as in any other specialty, although it may not be regarded by a Court of Law as evidence of *informed* consent. Unless action has to be taken so urgently that consent cannot be obtained, explanation should be given to the woman over and above the standard words used in hospital consent forms and that this has been done should be recorded in the woman's case notes. Again adequate records and full explanation made after the event will go a long way to avoid legal action. It goes without saying, however, that a doctor who takes urgent action to save life without proper consent is most likely to be supported in a Court of Law. But the court will not necessarily be supportive in a less than life and death situation where no evidence exists that information was given to the woman or a relative.

Observation

In many ways the part which the obstetrician plays throughout pregnancy, and even during labour, is that of observer. Pregnancy and delivery are essentially physiological processes and the midwife, who assists in a practical sense during labour, also has to play a part in the avoidance of unhappy litigation. The obstetrician, however, bears the greater responsibility. Antenatal care is concerned with observing the mother and the fetus closely for early warning signs such as hypertension, poor intrauterine growth and abnormal lie. Forearmed is forewarned and at the booking visit the doctor must take into account any points in the medical, obstetric or family history which may be of importance to the forthcoming pregnancy, recording the facts in the notes for future observers.

Most hospitals now run special high risk antenatal clinics for women with actual or potential problems, so that they may be seen by the more senior and experienced members of the obstetric team who are able to provide a high degree of vigilant observation and thus reduce the risk of litigation.

Referral

Junior obstetric staff should not take decisions without discussing the case with their seniors. Midwives too should share responsibility with medical personnel when their experience tells them that it is advisable.

PRE-PREGNANCY AND GENETIC COUNSELLING

Before pregnancy a couple may wish to seek advice. This may be because of their age, because of a family problem such as a Down's syndrome child, because of the medical condition of the mother (diabetes for example) or because of a previous difficult delivery. Many hospitals have been fortunate enough to be able to establish pre-pregnancy counselling clinics to which enquiring women can be sent by their family doctors. The clinics should be conducted by experienced doctors giving sound, correct advice because if the wrong advice is given and problems result litigation may well follow. For many straightforward conditions the non-specialized doctor can give advice but in more complex circumstances specialized advice must be obtained.

Hospitals may not have geneticists available but there is almost always a district centre to which women may be sent or where information may be obtained from an expert. The non-specialist should never be tempted to give advice on such matters unless absolutely sure of the facts.

Possible legal action may also arise following prenatal diagnostic tests themselves. Amniocentesis carries a 0.5–1 per cent risk of abortion; chorionic villus biopsy carries a higher risk. Ultrasound is safe but fetoscopy may be followed by miscarriage.

These procedures must be carried out by an expert, following comprehensive counselling on the risks involved. It should be explained that amniotic fluid may not be obtained, chromosome analysis may not be accurate or cell culture may fail.

Written consent must be obtained for all the invasive procedures with the implication that such consent is informed.

PREGNANCY

Antenatal care is, as we have seen, an exercise in diligent monitoring. There is a risk that repetition creates boredom which in turn may lead to carelessness.

Frequent appointments

The mother should be seen at frequent intervals. Shared care with the general practitioner and the midwife is a good method of decreasing the burden on antenatal clinics but the ultimate responsibility lies with the hospital under whose care the mother has booked. She should be seen by the obstetrician at important stages in early and late pregnancy.

Identify at risk cases

Special at risk cases must be identified and preferably seen in a special antenatal clinic by the more senior doctors available there.

Discussion of problems

Junior doctors must be encouraged to discuss all problems, however minor they may seem, with their seniors. Consultants must be involved with problems by telephone or personally.

Check records

At every visit the woman and her records should be scrutinized, dates should be checked, histories confirmed and notes made on the symptoms, if any, the findings of obstetric examination and the advice given.

If a problem arises later and a move is made towards litigation the woman's records will be scrutinized in detail by doctors, lawyers and the defendant's and plaintiff's experts alike. Inadequate antenatal care and poor records then become difficult to defend.

SPECIFIC PROBLEMS IN PREGNANCY

There are, of course, innumerable possible areas in antenatal care which could become subjects for litigation.

The small-for-dates baby

In this situation intrauterine prenatal anoxia may occur, with the risk of a brain-damaged baby after delivery. These babies can be identified using a variety of techniques from the simple clinical ones such as measurement of the fundal height to the more sophisticated techniques of serial ultrasound measurements. Once detected the mother ought to be admitted to hospital for intensive monitoring of her fetus, including cardiotocography and Doppler flow tests. The management requires experience and scrupulous records should be kept.

Medical conditions in association with pregnancy

Diabetes

This is perhaps the most commonly encountered medical condition and carries grave risks to the fetus and to the mother unless adequately managed. The best results are obtained in units where the expertise of an experienced obstetrician and a diabetic physician are combined, preferably at a joint clinic. Once again, careful record keeping is essential

Pre-existing hypertension

This is not common in obstetrics for these women since they represent a young population. Mutual sharing of the responsibility between obstetrician and physician is again desirable; in pregnancy induced hypertension the responsibility is usually entirely that of the obstetrician. It is useful in such cases to have a unit protocol for management which guides junior staff on action to be taken and is of value if litigation arises. However, it should not replace records of events and decisions written at the time.

Underlying good antenatal care are high levels of awareness, early and adequate treatment, appropriate referral for senior opinions, full explanations to women and accurate, signed, note keeping. These steps will not avoid complaints but they will go a long way to protect doctors and midwives from litigation.

LABOUR AND DELIVERY

Women have a high expectation of a safe delivery but also want a fulfilling and satisfying experience. Some may make special demands, that they should, for example, be allowed to adopt unusual positions for delivery, such as squatting, or sitting upright. Some insist on no mechanical monitoring by the cardiotocogram during labour; that the baby should be delivered in the dark; that the cord be left uncut while the baby is handed directly to the mother to be put immediately on the breast.

These and similar requests put the midwife and obstetrician in somewhat of a dilemma. If the requests of the mother are refused then complaints may arise. If they are agreed to and something goes wrong, for example, failure to resuscitate the baby adequately, litigation may follow.

In many maternity units now a compromise has been reached. A birth plan is either prepared by the woman or given by the clinic staff to the mother during the antenatal period in which she can mark off her particular requests and discuss them with the midwives and the doctors in advance of labour. Full discussion takes place with tact and understanding on the part of the advisers. Any difficulties or disagreements which arise should be noted. Once more the recurrent theme is communication and recording.

In the conduct of labour itself the most scrupulous vigilance is required on the part of the attendants. Traditionally, when problems arise the midwife will call the doctor on duty, who is often of quite junior status. It is imperative that a junior

doctor seeks help from a senior colleague (usually a registrar) where there is the slightest doubt over management and that the registrar, in turn, must never hesitate to discuss the situation with the consultant. It is advisable that the time when a senior is sought for advice is recorded in the notes and the outcome of discussions noted. Too often a defence will fail because of lack of such adequate recording.

Improvements in the organization of medical cover for the labour ward are being made in most hospitals. Sessions are now given to consultants purely for labour ward work and supervision. A junior staff team is assigned to labour ward duties, free from outpatient clinics or theatre sessions so that their availability to the labour ward is immediate and the midwifery staff know at once who to turn to when a problem arises.

Health Authorities and Trusts with increasing awareness of the potential for litigation in obstetrics should be appointing extra medical staff and ensuring that facilities are adequate.

Occasionally problems may arise which are not particularly associated with the process of labour. Retention of swabs and breakdown of the episiotomy or caesarean section wounds are examples. Frank discussion and good records are important again.

INSTRUMENTAL DELIVERY

The indications for forceps delivery are given in Chapter 8 and the criteria which must be observed before the forceps are applied. There is no place in modern obstetrics for a difficult forceps delivery and certainly there is no place for a junior doctor applying forceps without the closest supervision. Damage to the fetus may range from tentorial tears and intracerebral haemorrhage to superficial trauma to the face. Any of these may lead to litigation. A baby who is brain damaged during delivery carries huge awards against the obstetrician. Rotation and delivery with Kielland's forceps demands particular expertise and should only be undertaken by a fully trained and experienced operator.

BREECH DELIVERY

Many obstetricians feel that because of the potential risk to the fetus born as a breech and in the present climate of litigation, delivery should always be by caesarean section. But caesarean section carries its own intrinsic risk to mother and baby and with careful antenatal assessment of the mother's pelvic size, the size of the fetus and any other complicating features there would seem to be no reason why a vaginal delivery should not be allowed. The most careful watch must be kept over the progress of the labour by a team consisting of a senior midwife and assistant, an experienced obstetrician (usually of registrar or senior registrar status) and the junior doctor. The most experienced obstetrician should be in charge of the delivery, at which there should also be present an anaesthetist and a paediatrician.

One useful feature of modern labour ward practice is the protocol in which is written out the policy of a department largely for the benefit of new doctors and midwives. It should not supersede adequate record keeping but does at least provide some evidence of what is normal practice in that department.

CAESAREAN SECTION

There are many potential sources of litigation connected with this procedure. Failure to carry out the operation soon enough; carrying it out against the wishes of the patient; trauma (especially to the bladder) during the operation and postoperative complications, to name but a few. Full explanation of the reasons for caesarean section should be made to the woman so that her signed consent can be said to be informed. Careful surgery by an experienced operator will avoid complaint but complete documentation of the details of the surgery is also important.

ANAESTHETIC PROBLEMS IN OBSTETRICS

Inadequate pain relief from an epidural may be a source of complaint. Provided the procedure is explained to the woman, rash promises are not made and consent is obtained, then problems should be avoided.

Awareness during anaesthesia is well publicized and particularly refers to awareness during caesarean section. It is, of course, an anaesthetic problem but it is worth observing once more that an experienced anaesthetist should always be available for the labour ward and full details of anaesthetic agents used should be recorded.

FETAL HYPOXIA

At the present time fetal hypoxia and its relation to brain damage is giving rise to much discussion.

The huge awards that fetal hypoxia carries against the medical profession is a source of difference of opinion. Until such time as a clear causal link between fetal hypoxia and subsequent handicap, and factors other than prolonged fetal distress in labour or instrumental delivery for example, can be established, it is difficult for a court not to link these factors to the handicap on the basis of *res ipse loquitur*.

Congenital abnormality, intrauterine growth retardation, prematurity, maternal health and particularly smoking and alcohol habits are all possible causal background factors for subsequent handicap in the baby. But the final insult of labour and delivery may seem to be the precipitating factors and doctors, lawyers and the public tend to pin the entire blame on what is seen as mismanagement. A baby handicapped by cerebral palsy is indeed a tragedy for the infant and parents alike and compensation often seems right. Whether this should be made virtually on the automatic assumption of obstetric blame is another matter.

Fetal hypoxia may be avoided by detecting and paying special attention to high risk cases in the antenatal clinic, including the small-for-dates baby, women who have had a previous premature delivery and those with associated medical problems such as hypertension or diabetes.

The following steps are also helpful:

- judicious and justified induction of labour
- careful fetal monitoring during labour. Where any doubt on a fetal heart tracing exists, referral to a senior, more experienced doctor
- the wider use of fetal blood acid–base measurements
- skilled instrumental delivery
- earlier recourse to caesarean section when fetal distress arises
- good anaesthesia
- experienced paediatric attendance at delivery to provide adequate and immediate resuscitation.

Such measures will ensure a minimum risk of cerebral handicap to the baby and if they are carefully recorded and communicated adequately to the parents litigation may be avoided.

THE POSTNATAL PERIOD

Once a baby has been safely delivered and proves to be sound and healthy many a criticism of the conduct of labour and delivery will disappear. Usually discomforts and disappointments quickly recede, but not always, and it may be some time after the mother and baby have been discharged from hospital that murmurs of discontent reach the hospital authorities.

These frequently relate to what the parents consider to be unkind or inconsiderate handling by the staff, inadequate explanations and neglect of what the public feel are minimal standards of care. They almost always reflect a poor level of communication, a recurring word in all medico-legal considerations. This kind of complaint can usually be dealt with by a swift response, giving explanations and offering a face-to-face meeting between the complainant and the obstetrician involved. At such meetings, whatever may be the approach of the parents, the medical and nursing staff should always remain polite, understanding and sympathetic, not aggressively on the defensive. Careful records of such meetings should be kept.

But many complaints in the postnatal period may reach litigation level. Examples are the broken down or poorly healed episiotomy with resultant dyspareunia or even apareunia, understandably a source of discontent; a secondary postpartum haemorrhage or uterine infection necessitating a further admission to hospital; failure to detect problems in the new born baby which may become worse after discharge from hospital (jaundice for example). These complications are more likely to arise where early discharge has been inappropriately permitted.

Vigilance in the postnatal period must be scrupulous if such problems are to be avoided. Discussions with the mother, taking time to give proper explanations, careful note keeping and sharing of responsibility are important if medico-legal difficulties are to be avoided, however difficult they are to maintain in the midst of a busy programme.

HOME CONFINEMENT

It is necessary to consider this topic under litigation since requests are often made by mothers to be allowed to deliver at home particularly following the Government's Select Committee's Report of 1990. There seems no doubt that hospital is safer for delivery, as there is immediate access to medical staff, anaesthetics, blood transfusion and operating theatres. But it does not necessarily provide the ideal psychological environment and to some women this is of overriding concern.

But who bears legal responsibility if something goes wrong at a home confinement? Health Authorities have an obligation to ensure that midwifery services are available for home delivery. General practitioners can opt out of attendance during labour but in an emergency the doctor on call must attend. In most districts the midwives have direct access to a hospital emergency service (flying squad) which will respond immediately to a request for help. So on the face of it legal responsibility may devolve on the midwife on the spot, which in effect means the Health Authority; on any general practitioner called to attend the case; on the flying squad team and on the obstetric doctor who is usually of registrar status.

Most authorities have looked at the question of home confinements bearing in mind their possible medico-legal consequences. Domino (Domicillary In and Out) schemes for early discharge after delivery and the provision of general practitioner delivery suites attached to the hospital labour ward are offered as an alternative. In the latter a mother can be delivered by her midwife and attended by her own doctor and return home a few hours after delivery. The advantage is that all the facilities of the labour ward and specialist obstetric expertise are at hand. But the obstetric team of course depends on being notified of problems by the general practitioner or district midwife and it is upon them that initial responsibility rests.

As far as home confinements are concerned two points must be made. Adequate and proper advice must be given to the woman requesting a home confinement including some discussion about the limitations, for example limited pain relief and infant resuscitation facilities. In the case of the woman pursuing her demands against advice, clear notes to this effect must be made. In some parts of the country general practitioners may have direct access to consultant obstetricians to approve of an arrangement or to advise otherwise. In a few District Health Authorities a nominated consultant is available for reference during such a home labour if any problems arise. The responsibility then becomes the obstetrician's.

Where does the parents' responsibility lie when harm results to a baby when it can be shown that they were acting contrary to medical advice? The following statement from the Congenital Disabilities (Civil Liability) Act 1976 summarizes the position:

'If it is shown that the parent affected shared the responsibility for the child being born disabled, the damages are to be reduced to such an extent as the Court thinks just and equitable, having regard to the extent of the parents' responsibility'.

The obstetrician's duty is to the mother and the unborn child with the aim of improving care both in the antenatal period and during delivery. It would be a tragedy if increasing litigation deflected from the proper application of skills and experience, or worse still caused young doctors to turn away from entering the specialty. Viewed in its proper context the knowledge of potential litigation should raise the standard of the obstetrician's care of women. Meticulous recording, close observation, ready recourse to other opinions, maintenance of a high standard of postgraduate education, communication and counselling skills are the basis for good practice and certainly the basis for avoiding litigation.

FURTHER READING

Clements R. (1995) *Practice in Obstetrics and Gynaecology. A Medico-legal Handbook*. Churchill Livingstone, London.

Chamberlain G. (1992) *How to Avoid Medico Legal Problems in Obstetrics and Gynaecology*. Royal College of Obstetricians and Gynaecologists, London.

11

VITAL STATISTICS

Vital statistics are essential for the planning of health services, and are used by hospitals, regions and countries for this purpose. They allow obstetricians to compare their work with that of others, so as to show where progress might be made or where failures have occurred. Because childbirth is usually normal and any one person has limited experience of abnormal cases, pooling of results is necessary if false impressions are to be avoided.

THE COLLECTION OF OBSTETRICAL STATISTICS

The following account relates to England and Wales, but many other countries collect vital statistics to a similar degree.

In England and Wales a national census is taken every 10 years. All births, marriages and deaths are notified to the Registrars, and from these notifications statistics are computed and regularly published. Many maternity departments in hospitals compile their own statistics and prepare an annual report. Even when this information is not published it is of interest to those concerned and may help them to improve their practice. Audit procedures in UK hospitals will strengthen the checks on these data collections.

Birth rates

In the 1990s the average number of children per family in the UK is less than 2.1; this no longer replaces the population. The birth rate is calculated from the Registrars' returns. The *crude birth rate* is the number of live births per 1000 total population. Since this population includes men, children and women beyond childbearing age it is more informative to relate the number of live births to the number of women aged between 15 and 44. This gives the *general fertility rate*, which is expressed as the number of live births per 1000 women of these ages (*see* Figs 11.1 and 11.2).

There was a fall in the total annual number of births from 1885 to 1940, then there was an increase each year until 1964, after which the annual number fell. There has been a slight rise since 1976. There were 693 100 total births (live and still) in England and Wales in 1986, and the total fertility rate was 66/1000.

In less sophisticated areas of the world, where little contraception is practised, the crude birth rate is of the order of 50, against which the rate of 13.8 for England and Wales is small. Yet in the developing countries the total population may not rise much, for there are high stillbirth and infant mortality rates. When these fall because of better medical services and greater wealth, the community reaction is often a wider use of contraception, in the knowledge that then most children born alive will survive.

Obstetrical statistics

From the obstetrician's point of view the maternal mortality and perinatal mortality rates are of obvious importance, as they give some indication of the standard of practice and of the health of women. The number of deaths during infancy is an index of

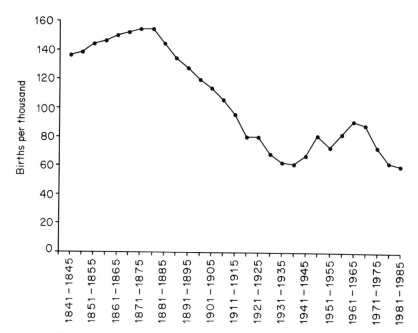

Fig. 11.1 The general fertility rate for England and Wales from 1841–1985 showing numbers of births per thousand women aged 15–44

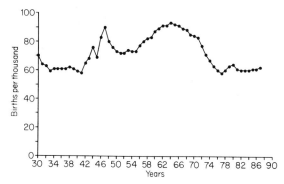

Fig. 11.2 The general fertility rate for England and Wales from 1930–87 showing the numbers of births per thousand women aged 15–44

the work of obstetricians, paediatricians and also of public health measures directed against infectious diseases. The death of an infant in the 1st week after delivery often has an obstetric association.

MATERNAL MORTALITY

In the nineteenth century the number of deaths of women from childbearing was appalling. Deaths were chiefly from sepsis but also from eclampsia,

disproportion and other complications of labour, and haemorrhage. Even in 1928, long after the introduction of aseptic surgery, for every 1000 births 4.28 women died; that is to say that any woman embarking on pregnancy had a 1 in 250 chance of dying. With such results it is not surprising that the first task of obstetricians was to make childbearing safer for the mother, and the fate of the fetus was, at that time, of lesser consideration. In recent years the maternal mortality has fallen to such low levels that it is no longer as useful as an index of obstetric failure or success, and now more attention is given to fetal and neonatal mortality. Nevertheless, every maternal death is still a major tragedy, and maternal mortality studies are of great importance for everyone who practises obstetrics (*see* Fig. 11.3).

In the 66 years since 1928, the maternal mortality has fallen from over 4 to 0.08 per 1000 total births. This has been due to several groups of factors:

- the control of infection
- blood transfusion
- a readier recourse to operative treatment which has been made safer by advances in anaesthesia and resuscitation

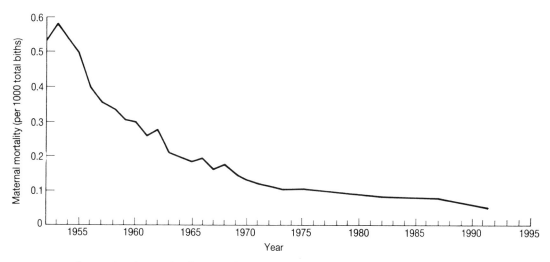

Fig. 11.3 Maternal mortality for England and Wales per 100 total births (excluding abortion) for the years 1952–92

• the improvement in the health and nutrition of women, who are therefore in better health when they become pregnant.

Nevertheless there is still room for improvement, as a closer look at the causes of death will show.

Statistical data about maternal mortality based on death certification are reported in triennial reports by the Registrar General in England and Wales and, from 1985, in the rest of the UK. Much additional information about maternal deaths is derived from these *Reports on Confidential Enquiries into Maternal Deaths*. When a woman who is or has recently been pregnant dies, the District Medical Officer or District Public Health Consultant sends a detailed form to be filled in by the doctor who was in charge of the woman. Any midwives or nurses who have knowledge of the case may also write reports. At no time is the identity of the person connected with the medical or nursing care of the woman disclosed, and because of this each person giving evidence may be completely frank and unbiased; nobody filling in the form need fear recriminations of any kind from anybody.

After the form is completed it is sent to a regional assessor, a senior obstetrician in the region, who gives his or her opinion about the cause of death. A senior pathologist assesses the autopsy findings and if the woman had an anaesthetic, the opinion of a regional assessor or anaesthetist will also be sought. In their reports the assessors try to judge whether the death was avoidable or unavoidable. These terms are not meant to imply criticism of those concerned with the case, but to ask whether the patient, her family, the doctor and the midwife, and anyone else concerned, had made the fullest possible use of all available services and help. If there had been shortcomings in the use of services and administration, or if the standard of professional care had been lower than should be expected, then the death would be classified as avoidable. This is not to say that it could have been avoided, but only that some factor was present which, if foreseen, might have made the outcome different. The results of these enquiries are published at 3-yearly intervals.

Table 11.1 summarizes some of the conclusions of the last Report.

Hypertensive diseases of pregnancy

The incidence of these conditions has been slowly falling during the last 30 years in Britain, but they are still one of the greatest single causes of maternal death. The slow improvement may have resulted from more efficient antenatal care, also from better living standards and better general health. One of the major aims of antenatal care from the earliest days has been the prevention of eclampsia by early recognition and treatment of pre-eclamptic hypertension and proteinuria.

Deaths from eclampsia and hypertension were associated with an avoidable factor in three-

Table 11.1   Causes of Direct maternal deaths per million estimated pregnancies, England and Wales 1973–90

Cause of Direct maternal death

Causes per million	1973–75	1976–78	1979–81	1982–84	1985–87	1988–90
Pulmonary embolism	12.8	18.5	9.0	10.0	9.1	8.0
Hypertensive disorders of pregnancy	13.2	12.5	14.2	10.0	9.4	8.6
Anaesthesia	10.5	11.6	8.7	7.2	1.9	1.0
Amniotic fluid embolism	5.4	4.7	7.1	5.6	3.4	3.5
Abortion	10.5	6.0	5.5	4.4	2.3	2.4
Ectopic pregnancy	7.4	9.0	7.9	4.0	4.1	5.2
Haemorrhage	8.1	10.3	5.5	3.6	3.8	7.3
Sepsis. excluding abortion	7.4	6.5	3.1	1.0	2.3	2.1
Ruptured uterus	4.3	6.0	1.6	1.2	1.9	0.7
Other direct causes	8.5	8.2	7.5	8.4	7.5	8.3
All deaths	88.0	93.4	70.0	55.0	45.6	47.0

With kind permission of HMSO, London.

quarters of the cases where assessment was possible. The most common group were women who refused antenatal care or who would not cooperate by accepting advice about admission to hospital. Doctors were sometimes at fault by delay in acting on signs of severe hypertension and instituting efficient treatment. These women must be under the care of a consultant obstetrician and not under the general practitioner. Those in charge at smaller units must be prepared to transfer those women affected by the condition to larger, regional centres for better care of the mother and the baby.

Pulmonary embolism

This is a sudden and dramatic cause of maternal death. In the reports from 1964 onwards a third of the deaths occurred during pregnancy, while two-thirds followed delivery. Of the postpartum deaths, one-third followed caesarean section, although only one-twentieth of all the women were delivered by this method, so that the risk of pulmonary embolism is much greater after section than after vaginal delivery. The risk of pulmonary embolism is higher in women aged over 34 years, in those who are overweight, and in those who have an operative delivery of any kind. Other important risk factors are a previous history of thromboembolism, and immobilization (which is of course more likely after operative deliveries, particularly in obese women). Most of the antenatal deaths occurred suddenly and unexpectedly,

and few were preceded by obvious signs of deep venous thrombosis. Therefore it is difficult at present to see how they could be prevented by improved diagnosis. Over the last 5 or so years, the relative safety of prophylaxis with subcutaneous heparin has become firmly established, and there can now be little doubt that more women in high risk categories should be given prophylactic anti-coagulant treatment.

The use of oestrogens to inhibit lactation is one cause of deep venous thrombosis and pulmonary embolism; their use for this purpose has largely been abandoned.

Abortion

In 1968 a major change in practice occurred when the Abortion Act of 1967 came into operation. Since then there has been a steady fall in the number of deaths following abortion, without any great rise in the number of deaths following therapeutic termination of pregnancy, despite over 165 000 terminations being notified each year among women resident in England and Wales.

In the 3 years covered by the latest *Report on Confidential Enquiries into Maternal Deaths (1988–90)* abortion accounted for 5 per cent of all deaths. There were 77 deaths reported in the 1970–72 report just after the Abortion Act came in, but only six deaths in the 1985–87 report. Of these only one was after legal abortion while five occurred after spontaneous abortion. There have been no deaths reported from illegal abortion since 1982; the UK is the only country in which accurate statistics are kept in which no such deaths have been reported.

Haemorrhage

Maternal deaths associated with both antepartum and postpartum haemorrhage have fallen steadily since 1950. Placental abruption remains one of the most serious complications of pregnancy. For these cases monitoring of the central venous pressure is essential, and large amounts of blood are required to treat hypovolaemia. Coagulopathy commonly occurs and should be treated vigorously with fresh frozen plasma and platelet transfusions as necessary. Placenta praevia used to cause a significant proportion of the deaths but in the period 1985–87 there was none from this cause. Increased diagnostic accuracy through the use of ultrasound scanning is probably a major factor in the improvement, although the diagnosis is still far from totally reliable.

The number of deaths from postpartum haemorrhage was sharply reduced in the 1950s by the prophylactic use of ergometrine in the third stage of labour and by increased availability of efficient blood transfusion systems developed during the Second World War. Mortality continued to fall in the 1980s. However, there has recently been a tendency for postpartum haemorrhage to become more common due to a fashion for women to opt for a more physiological management of the third stage, and refuse the use of oxytocics. All scientific data (including the results of a large prospective randomized trial carried out in Bristol in the 1980s) continue to support the use of routine oxytocics in the management of the third stage. Most symptoms from the use of oxytocics, such as headache and continuing uterine cramping, result from the use of ergometrine, and there is an increasing trend to use only oxytocin, which the evidence suggests is sufficient in most cases. Ergometrine should in any case be avoided in hypertensive women.

During the late 1980s and early 1990s there has been an increasing trend for a small proportion of women to revert to having their babies at home, because of the social and emotional advantages they perceive from this, and often their desire to avoid medical intervention. In some areas, the proportion of home births has risen from the national average of about 1 per cent to almost 4 per cent. Because the incidence of postpartum haemorrhage is often unpredictable (leaving aside women with an abruption, placenta praevia and multiple pregnancy, who are likely to be delivered in hospital anyway), this has emphasized the need to maintain appropriate emergency services. The old concept of the flying squad, in which an obstetrician, paediatrician, midwife and anaesthetist went out to the woman's home, is increasingly being abandoned, because it leaves the labour ward denuded of key staff and the need to collect them causes delay in transferring the mother to hospital. Instead, ambulances with trained paramedics are being used to provide resuscitation and transfer, and a national training scheme for such paramedics has been introduced.

Ectopic pregnancy

Deaths from this cause have been reduced, but not at the same rate as deaths from other causes. The number of deaths in each triennial report has fallen from 59 in 1952–54 to 10 in 1982–84. Half the deaths in this report were associated with delay or failure of diagnosis. To quote from the report, probably the most important contribution to reducing the risk of death from ectopic pregnancy is an awareness by medical attendants that in any woman of reproductive age, an ectopic pregnancy may be the cause of lower abdominal pain particularly when of sudden onset. All medical practitioners must be aware of the fallibility of urine pregnancy tests, particularly when carried out at home by the woman herself with commercial kits bought over the counter. A negative test using one of the kits does not rule out the possibility of an ectopic pregnancy. Even sensitive tests carried out in hospitals can be wrong if carried out by junior or inexperienced medical or nursing staff. Similar strictures apply to the interpretation of ultrasound scans. If there is any doubt about the diagnosis, early recourse to laparoscopy is still the best policy.

It is essential to operate immediately the diagnosis of ectopic pregnancy has been made. The operation should not await resuscitative measures; these should proceed while the operation is being performed, because once the bleeding vessels are secured improvement usually follows quickly.

Infection

Before 1936 puerperal sepsis was the largest single cause of maternal mortality. Since the introduction of sulphonamides and then antibiotics, and with better understanding of the mode of spread of bacteria, infection is a much less common cause of death. However, it is often still an associated factor in fatal cases. The 1985–87 report recorded three deaths due directly to sepsis, and three more where sepsis was a major factor. In four of these cases, the beta haemolytic streptococcus was the main pathogenic organism. There has been some suggestion that this organism is making a come back in terms of virulence; it is a particular problem in black women. Vigilance must be maintained if sepsis is not again to become an important cause of maternal death.

Amniotic fluid embolism

In the 3 years covered by the 1985–87 report there were nine deaths from this cause, and about half of these women had an associated coagulation disorder. The common symptoms were sudden collapse soon after rupture of the membranes, particularly when the uterine contractions were strong. The collapse was sometimes accompanied by a fit, and the woman was usually dyspnoeic and cyanosed, with frothy blood-stained sputum. Six of the nine women had received prostaglandin or Syntocinon. Confirmation of the diagnosis is possible if amniotic cells are recognized in the sputum. This condition should be kept in mind in cases of sudden collapse, for it is a difficult diagnosis to make. Treatment with oxygen and steroids can save some women, while the ensuing coagulation disorder needs correction.

Uterine rupture

In the last report there were only two deaths from this cause, although there were four other deaths from cervical laceration. With persisting haemorrhage and shock the cervix and uterine cavity should always be examined, if necessary under anaesthesia, and always with a good light and assistance. If there is any possibility of uterine rupture, laparotomy should not be delayed.

Anaesthesia

In the Report on Confidential Enquiries into Maternal Deaths many of the deaths associated with anaesthesia are classified under the headings of the conditions for which treatment was required. However, in the report for 1985–87 it is stated that six deaths were directly due to anaesthesia; this is a welcome reduction from the 18 in 1982–84. In all but one the problem was a misplaced endotracheal tube; difficulty with intubation in pregnant women remains a serious problem. It is possible that increasing experience with laryngeal airways may help in this regard. There were a further 16 deaths in which anaesthetic factors were contributory.

For general anaesthesia during labour a skilled and experienced anaesthetist is required. Women may be dehydrated after a long labour, and although they may not have been given food for some hours, the stomach may still contain food taken before that and gastric juice will still have been produced. Such women should be intubated in the head-up position, with controlled pressure over the cricoid. More frequent use of epidural block for operative deliveries will help to reduce mortality still further.

Caesarean section

In the report for 1985–87 there were 76 deaths associated with caesarean section. Unfortunately, due to reductions in government funding for statistical services, the total number of caesarean sections carried out is no longer recorded and therefore a fatality rate for the procedure cannot be calculated. However, in the previous triennium, when such figures were still available, there was a mortality rate of 0.24 per 1000 operations. This shows that the mortality following section is about four times greater than that of vaginal delivery but, of course, the conditions calling for treatment by caesarean section are themselves dangerous.

Further analysis shows that in the 1982–84 triennium the mortality rate for emergency caesarean section was 0.37 compared with 0.09 per 1000 for elective operations (a ratio of 4:1). The conditions under which emergency surgery is still performed carry greater risks and a shift from emergency to elective operations could reduce some of the maternal mortality from caesarean sections.

Deaths from associated and intercurrent conditions

There has been a fall in the number of deaths from cardiac disease in each triennial report. That for 1952–54 included 121 deaths, whereas the report for 1985–87 included only 23, 9 acquired and 10 from congenital heart disease. The latter figure is the highest since 1970–72 and there is likely to be a rising trend as more and more women come into the reproductive age group having benefited from the major advances in corrective surgery in infancy and childhood which have occurred over the last 30 years. It is important to note that six women who died became pregnant after having had strong medical advice that to do so would risk their life, thus illustrating how powerful is the procreative urge.

A wide variety of medical and surgical disorders and accidents caused the remaining deaths, but in most of them there was no avoidable obstetric factor. It is important to note that there were four deaths from acute fatty liver of pregnancy, and four from pneumonitis secondary to varicella zoster infection (*see* in Medical Disorders, Chapter 3).

STILLBIRTHS AND NEONATAL DEATHS: PERINATAL MORTALITY

Stillbirths and neonatal deaths are notifiable in Britain, and for statistical accuracy it is essential to use correct definitions.

Stillbirth

The term stillbirth referred to any child delivered after the 24th week of pregnancy that did not afterwards breathe or show any sign of life. In the UK from the first of October 1992, the gestation after which a child would be considered stillborn was changed legally, by Act of Parliament, to 24 completed weeks of gestation thus bringing the UK in line with other western countries. If the heart is beating after delivery, even if there is no sign of respiration, the death should not be recorded as a stillbirth but as a neonatal death. Unfortunately there is no international agreement on these definitions; the World Health Organization suggests a gestational age of greater than 24 weeks or a birth weight greater than 500 g. In many ways this

is more practical, as in many parts of the world accurate gestational age is not known for most women, while weighing the baby is a relatively easy and a more precise procedure.

The stillbirth rate is defined as the number of stillbirths per 1000 total births (live and still).

Neonatal death

The definition of the neonatal period is unfortunately confusing because obstetricians regard it as the 1st week of life whereas paediatricians use it to refer to the 1st month (four weeks). For clarity, it is sometimes useful to refer to the former as the early neonatal period. The early neonatal death rate (i.e. in the 1st week of life) and the stillbirth rate together make up the perinatal mortality rate, expressed with a denominator of 1000 total births, both live and still, whereas the neonatal death rate is defined as the number of infants dying in the 1st month per 1000 live births.

Infant mortality rate

This is defined as the number of infants dying in the first year of life per 1000 *live* births. The neonatal period is included and some three quarters of infant deaths occur in the first 4 weeks.

Perinatal mortality

The perinatal mortality rate is defined as the number of stillbirths (born dead after 24 weeks) together with the number of neonatal deaths in the 1st week of life per 1000 total births. This is a more useful index for obstetric purposes because deaths in the 1st week are often related to factors occurring during pregnancy or delivery, whereas deaths in the rest of the neonatal period are more often due to paediatric causes. It should be noted that because the definition of stillbirth changed in 1992 rates before and after that date are not directly comparable.

In the last 30 years the stillbirth rate has fallen from about 19 to about 4 per 1000 total births; the neonatal death rate from 20 to 4 per 1000 live births; the infant mortality rate from about 34 to about 10 per 1000 live births and the perinatal mortality rate from about 39 to about 8 per 1000 total births (*see* Fig. 11.4).

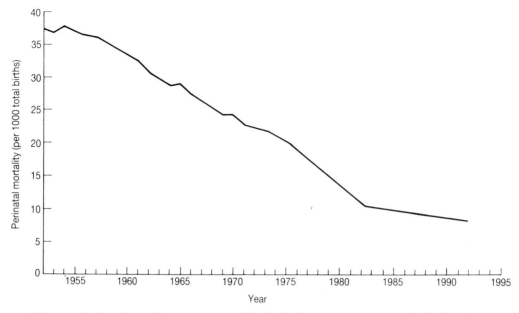

Fig. 11.4 Perinatal mortality per 1000 total births for the years 1952–7 (England and Wales)

CAUSES OF PERINATAL DEATH

Antepartum

In about 25 per cent of cases the fetus is macerated at birth, and a precise cause of death cannot be established. Congenital abnormality is present in a third of cases. However, the incidence of some congenital anomalies is declining rapidly, for example spina bifida and anencephaly. The reasons for this may be partly improvements in maternal nutrition and socio-economic conditions; ultrasound screening has also played a large part in relation to some anomalies, such as anencephaly. The incidence of congenital anomalies varies widely according to region, with, for example, a 10-fold variation in the incidence of neural tube defect. Also, recent studies have shown that detection of fetal anomaly ranges from 70 per cent in some areas, to as low as 20 per cent in others where there is poor provision of ultrasound facilities and personnel. It should be noted, of course, that antenatal diagnosis does not decrease the number of babies with congenital abnormality, it merely allows the selective termination of pregnancies where the baby is found to be malformed. This is of particular benefit to parents in those cases where the baby would not have died, but would have been grossly handicapped.

About half of all stillbirths are small for gestational age, with a birth weight less than the 10th centile for gestation. This does not mean that they are all growth retarded, as some will have been genetically programmed to be small. Equally, however, some babies weighing more than the 10th centile will be growth retarded. Overall, it seems likely that a high proportion, perhaps 50 per cent, of stillbirths are associated with failure of fetal growth. This may be associated with maternal disease, such as pre-eclampsia, congenital heart disease with restriction of cardiac output, or other systemic disease such as systemic lupus erythematosus, poor nutrition shown by maternal underweight, primary placental disease such as abruption of the placenta, fetal infection, or congenital abnormality.

Other occasional causes of stillbirth include diabetes, rhesus isoimmunization and cord entanglements.

Intrapartum death

Intrapartum stillbirths now occur in only about 1 labour per 1000 in the UK; 20 years ago the corresponding figure was 5. There is great controversy about the usefulness of continuous electronic fetal monitoring (EFM) in this improvement. The Dublin prospective randomized controlled trial of continuous EFM versus monitoring with the Pinard stethoscope suggested that both were equally effective at preventing intrapartum stillbirth, although use of EFM halved the number of babies suffering from neonatal convulsions. However, there can be no doubt that the knowledge we have gained from use of EFM, fetal blood sampling and pH analysis has improved our understanding of intrapartum events, which may have contributed to global improvements in management. In particular, very long labours (48 hours plus) are no longer tolerated either by women or staff, and this has removed one potential cause of mortality and morbidity. In addition, the rising caesarean section rate, although resulting in some maternal morbidity, has at least limited fetal impairment in complicated labour.

Some acute events in labour, such as placental abruption, cord prolapse and amniotic fluid embolism, remain unpredictable and are therefore likely to continue to be responsible for a number of intrapartum deaths still.

Neonatal death

About 80 per cent of neonatal deaths are now due to preterm labour. Despite dramatic improvements in neonatal intensive care, babies born before 26 weeks' gestation have still less than a 50 per cent chance of survival, and neonatal death rates remain high up to 28–30 weeks. The rate of preterm delivery has declined surprisingly little over the last 30 years, and the causes in many cases remain obscure, so immaturity seems likely to remain a major problem.

The usual cause of death in immature babies is the respiratory distress syndrome (RDS or hyaline membrane disease); some inroads into mortality from this cause have been made by the widespread use of maternally administered steroids in threatened and actual preterm labour. The use of steroids can reduce mortality from RDS by 50 per cent at any given gestation by encouraging the release of surfactant into the future airways of the fetal lung. Other important causes of death include infection, intracranial haemorrhage, and congenital anomaly.

The role of hypoxia or asphyxia

In many accounts of perinatal mortality, birth

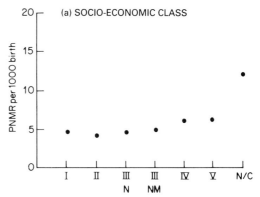

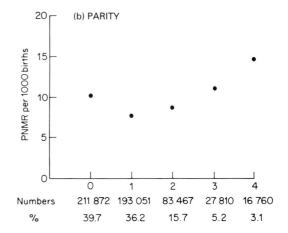

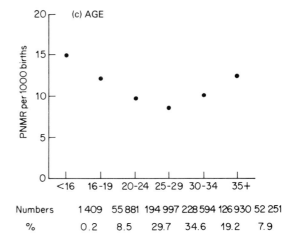

Fig. 11.5 Effects of age, parity and social; class on perinatal mortality (England and Wales, 1985)

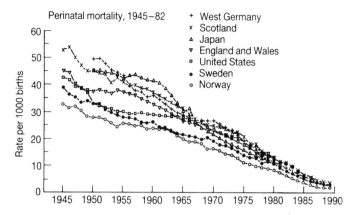

Fig. 11.6 The comparison of perinatal mortality rates in certain countries per 1000 total births since 1945

asphyxia is given as a major cause. Of course, almost all babies who die are by definition asphyxiated, if by that is meant having a low pO_2, a high pCO_2 and a low pH. The only exceptions are those babies that die from sudden cardiac asystole or intracranial haemorrhage not secondary to hypoxia. However, such a description does not identify the primary cause, whether this is poor placental function or functional immaturity. Over the last 10 years it has become apparent that traditional measures such as fetal heart rate, acid–base balance and the Apgar score, are not closely related and can be differentially affected by various conditions. For example, a baby with meconium aspiration in labour may be born with normal acid–base status, but a low Apgar score, and go on to become hypoxic because of difficulty with ventilating its lungs. In addition, it has become apparent that most babies even with severe acidosis at birth develop normally provided they are resuscitated properly, and intrapartum events are not a major cause of long-term mental handicap, as was once thought. Thus simplistic concepts such as fetal distress and birth asphyxia are better avoided until we have a better understanding of their pathophysiology.

Factors affecting perinatal mortality

There are very few risk factors which predict perinatal mortality with much sensitivity and specificity; in most cases the maximum relative risk between groups with the risk factor and groups without is 2:1. Those risk factors which are very specific are usually also very rare, for example, Eisenmenger's syndrome (Fig. 11.5).

Many of the important risk factors we know about relate to the social condition. There is a two-fold difference in perinatal mortality rate (PMR) between women in social class one (professionals at low risk) and social class five which consists of unskilled manual workers, who are at higher risk. This differential has remained unchanged for many years despite overall improvements in the PMR. The lowest PMR is seen in the 25–35 age group; in women aged <16 years or >40 years it is increased by about 60 per cent. Women who are married have a lower PMR than single women, even when the latter are in a stable relationship. If the woman is single and unsupported, the PMR is 2.5 times greater. High parity has in the past been linked with increased PMRs, but recent studies suggest that this is because large families tend to be of lower socio-economic status. PMR in the first pregnancy is also slightly higher than the second, probably because of the longer and more difficult labours associated with the untried pelvis.

Smoking cigarettes is also an important risk factor. A woman smoking 20 or more cigarettes a day is 30 per cent more likely to have a preterm or growth retarded baby than her identical twin who does not smoke.

Racial origin can also be an important factor. Women of Pakistani origin have double the risk of having a baby with a congenital abnormality than the rest of the population. This may be in part associated with the high incidence of arranged marriages which results in about 40 per cent of marriages being consanguineous (e.g. between cousins) with the resultant unmasking of recessive genes for abnormality (Fig 11.6).

Reductions in perinatal mortality rates have probably been very substantially effected by improved socio-economic conditions, and future changes are likely to continue to reflect social conditions. Action to reduce cigarette smoking in pregnancy is likely to be as important as further improvements in obstetric technology.

FURTHER READING

Annual Report of Confidential Enquiry with Still births and Deaths in Infancy for 1993 (1995). HMSO, London.

Report on Confidential Enquiry into Maternal Deaths in the United Kingdom for 1988–90 (1994). HMSO, London

INDEX